T0238864

Modern Fluid Dynamics

FLUID MECHANICS AND ITS APPLICATIONS
Volume 87

Series Editor: R. MOREAU
MADYLAM
Ecole Nationale Supérieure d'Hydraulique de Grenoble
Boîte Postale 95
38402 Saint Martin d'Hères Cedex, France

Aims and Scope of the Series

The purpose of this series is to focus on subjects in which fluid mechanics plays a fundamental role.

As well as the more traditional applications of aeronautics, hydraulics, heat and mass transfer etc., books will be published dealing with topics which are currently in a state of rapid development, such as turbulence, suspensions and multiphase fluids, super and hypersonic flows and numerical modeling techniques.

It is a widely held view that it is the interdisciplinary subjects that will receive intense scientific attention, bringing them to the forefront of technological advancement. Fluids have the ability to transport matter and its properties as well as to transmit force, therefore fluid mechanics is a subject that is particularly open to cross fertilization with other sciences and disciplines of engineering. The subject of fluid mechanics will be highly relevant in domains such as chemical, metallurgical, biological and ecological engineering. This series is particularly open to such new multidisciplinary domains.

The median level of presentation is the first year graduate student. Some texts are monographs defining the current state of a field; others are accessible to final year undergraduates; but essentially the emphasis is on readability and clarity.

For other titles published in this series, go to
www.springer.com/series/5980

Clement Kleinstreuer

Modern Fluid Dynamics

Basic Theory and Selected Applications in Macro- and Micro-Fluidics

Clement Kleinstreuer
Department of Mechanical and
Aerospace Engineering
North Carolina State University
Raleigh, NC 27695-7910
USA
ck@eos.ncsu.edu

ISBN 978-94-007-3130-1 ISBN 978-1-4020-8670-0 (eBook)
DOI 10.1007/978-1-4020-8670-0
Springer Dordrecht Heidelberg London New York

Softcover reprint of the hardcover 1st edition 2009

Printed on acid-free paper

Springer is part of Springer Science+Business Media (www.springer.com)

To my family,
Christin, Nicole, and Joshua

Contents

Preface

This textbook covers essentials of traditional and modern fluid dynamics, i.e., the fundamentals of and basic applications in fluid mechanics and convection heat transfer with brief excursions into fluid particle dynamics and solid mechanics. Specifically, it is suggested that the book can be used to enhance the knowledge base and skill level of engineering and physics students in *macro-scale* fluid mechanics (see Chaps. 1–5 and 10), followed by an introductory excursion into *micro-scale* fluid dynamics (see Chaps. 6 to 9). These ten chapters are rather self-contained, i.e., most of the material of Chaps. 1–10 (or selectively just certain chapters) could be taught in one course, based on the students' background. Typically, serious seniors and first-year graduate students form a receptive audience (see sample syllabus). Such as target group of students would have had prerequisites in thermodynamics, fluid mechanics and solid mechanics, where Part A would be a welcomed refresher. While introductory fluid mechanics books present the material in progressive order, i.e., employing an *inductive* approach from the simple to the more difficult, the present text adopts more of a *deductive* approach. Indeed, understanding the derivation of the basic equations and then formulating the system-specific equations with suitable boundary conditions are two key steps for proper problem solutions.

The book reviews in more depth the essentials of fluid mechanics and stresses the fundamentals via detailed derivations, illustrative examples and applications covering traditional and modern topics. Similar to learning a language, frequent repetition of the essentials is employed as a pedagogical tool. *Understanding of the fundamentals and independent application skills are the main learning objectives.* For students to gain confidence and independence, an instructor may want to be less of a "sage on the stage" but more of a "guide on the side". Specifically, "white-board performances", tutorial presentations of specific topics in Chaps. 4–10 and associated journal articles by students are highly recommended.

The need for the proposed text evolved primarily out of industrial demands and post-graduate expectations. Clearly, industry and government recognized that undergraduate fluid mechanics education had to change measurably due to the availability of powerful software which runs on PCs and because of the shift towards more complicated and interdisciplinary tasks, tomorrow's engineers are facing (see NAS "The Engineers of 2020" at http://national-academics.org). Also, an increasing number of engineering firms recruit only MS and Ph.D. holders having given up on BS engineers being able to follow technical directions, let alone to build mathematical models and consequently analyze and improve/design devices related to fluid dynamics, i.e., here: fluid flow, heat transfer, and fluid–particle/fluid–structure interactions. In the academic environment, a fine knowledge base and solid skill levels in modern fluid dynamics are important for any success in emerging departmental programs and for new thesis/dissertation requirements responding to future educational needs. Such application areas include microfluidics, mixture flows, fluid–structure interactions, biofluid dynamics, thermal flows, and fluid-particle flows. Building on courses in thermodynamics, fluid mechanics and solid mechanics as prerequisites as well as on a junior-level math background, a differential approach is most insightful to teach the fundamentals in fluid mechanics, to explain traditional and modern applications on an intermediate level, and to provide sufficient physical insight to understand results, providing a basis for extended homework assignments, challenging course projects, and virtual design tasks.

Pedagogical elements include a consistent 50/50 physics-mathematics approach when introducing new material, illustrating concepts, showing flow visualizations, and solving problems. The problem solution format follows strictly: System Sketch, Assumptions, and Concept/Approach – *before* starting the solution phase which consists of symbolic math model development (App. A), numerical solution, graphs, and comments on "physical insight". After some illustrative examples, most solved text examples have the same level of difficulty as suggested assignments and/or exam problems. The ultimate goals are that the more serious student can solve basic fluid dynamics problems *independently*, can provide *physical insight*, and can suggest, via a course project, system *design improvements*.

The proposed textbook is divided into three parts, i.e., a review of essentials of fluid mechanics and convection heat transfer (Part A) as well as traditional (Part B) and modern fluid dynamics applications (Part C). In Part A, the same key topics are discussed as in the voluminous leading texts (i.e., White, Fox et al., Munson et al., Streeter et al., Crowe et al., Cengle & Cimbala, etc.); but, stripped of superfluous material and presented in a concise streamlined form with a different pedagogical approach. In a nutshell, quality of education stressing the fundamentals is more important than providing high quantities of material trying to address everything.

Chapter 1 starts off with brief comments on "fluid mechanics" in light of classical vs. modern physics and proceeds with a discussion of the basic concepts. For example, the amazing thermal properties of "nanofluids"; i.e., very dilute nanoparticle suspensions in liquids, are discussed in Sect. 1.4 in conjunction with the properties of more traditional fluids. Derivations of the conservation laws are so important that three approaches are featured, i.e., integral, transformation to differential, and representative-elementary-volume (Chap. 2). On the other hand, tedious derivations are relegated to App. C in order to maintain text fluidity. Each section of Chap. 2 contains illustrative examples to strengthen the student's understanding and problem-solving skills. Appendix A provides a brief summary of analytical methods as well as an overview of basic approximation techniques. Chapter 3 continues to present typical 1st-year case studies in fluid mechanics; however, some 2nd-level fluids material appears already in terms of exact/approximate solutions to the Navier–Stokes equations as well as solutions to scalar transport equations. The concept of entropy generation in internal thermal flow systems for waste minimization is discussed as well.

Part B is a basic discourse focusing especially on practical pipe flows as well as boundary-layer flows. Specifically, applications to the bifurcation and slit flows as well as laminar or turbulent pipe flow, lubrication and compartmental system analysis are presented in Chap. 4, while Chap. 5 deals with boundary-layer and thin-film flows, including coating as well as drag computations.

Part C introduces some modern fluid dynamics applications for which the fundamentals presented in the previous chapters plus App. A form necessary prerequisites. Specifically, Chap. 6 discusses simple two-phase flow cases, stressing power-law fluids and homogeneous mixture flows, previously the domain of only chemical engineers. Chapter 7 is very important. It deals with fluid flow in microsystems, forming an integral part of nanotechnology, which is rapidly penetrating many branches of industry, academia, and human health. After an overview of microfluidic systems given in the Introduction, Sect. 7.2 reviews basic modeling equations and necessary submodels. Then, in Sects. 7.3 to 7.5 key applications of microfluidics are analyzed, i.e., electrokinetic flows in microchannels, nanofluid flow in microchannels, and convective heat transfer with entropy generation in microchannels. Chapter 8 deals with fluid–structure interaction (FSI) applications for which a brief solid-mechanics review may be useful (Sect. 8.2). Clearly, fluid flows interacting with structural elements occur frequently in nature as well as in industrial and medical applications. The two-way coupling is a true multiphysics phenomenon, ultimately requiring fully coupled FSI solvers. Thus, young engineers should have had an exposure to the fundamentals of FSI before using such multiphysics software for R&D work. Chapter 9 deals with biofluid dynamics, i.e., stressing its unique transport processes and focusing on the three major applications of blood flow in arteries, air-particle flow in lung airways, and tissue heat transfer. An overview of CFD tools and solved examples with flow visualizations are given in Chap. 10, stressing computer simulations of internal and external flow examples.

As all books, this text relied on numerous sources as well as contributions provided by the author's colleagues, research associates, former graduate students and the new MAE589K-course participants at NC State. Special thanks go to Mrs. Joyce Sorensen and Mrs. Joanne Self for expertly typing the first draft of the manuscript. Seiji Nair generated the system sketches and figures, while Christopher Basciano provided the computer simulations of Sects. 10.3 to 10.5. Dr. Jie Li then helped checking the content of all chapters after he generated result graphs, obtained the cited references, generated the index, and formatted the text. The critical comments and helpful suggestions provided by the expert reviewers Alex Alexeev (Georgia Tech, GA), Gad Hetsroni (Technion, Israel), and Alexander Mitsos (MIT, MA) are gratefully acknowledged as well. Many thanks for their support go also to the editorial staff at Springer Verlag, especially the Publishing Editor Nathalie Jacobs, to the professionals in the ME Department at Stanford University and in the Engineering Library.

A Solutions Manual, authored by Dr. Jie Li, is available for instructors adopting the textbook. For technical correspondence, please contact the author via e-mail ck@eos.ncsu.edu or fax 919.515.7968.

Raleigh, NC, 2009 Clement Kleinstreuer

NC State University, MAE Dept. C. Kleinstreuer
Spring 2009 BR4160; by appointment
Library Reserve for MAE589K Website ck@eos.ncsu.edu (any time)

MAE 589K "Modern Fluid Dynamics"

(Tu & Th 13:30-14:45 in BR 3218)

Prerequisites: MAE 301, 308, 310, 314 (or equivalent); also: math and computer skills, including use of software (e.g., Matlab or Mathcad or MAPLE, and desirable: COMSOL, etc.)

Text: C. Kleinstreuer (2009) "Modern Fluid Dynamics" Springer Verlag, Dordrecht, The Netherlands

Objectives: To strengthen the background in fluid dynamics (implying fluid mechanics plus heat transfer) and to provide an introduction to modern academic/industrial fluid dynamics topics. Report writing and in-class presentations are key preparations for GR School and the job market.

Wks	Topics	Assignments
4	1. Review of Fluid Dynamics Essentials 1.1 Definitions and Concepts 1.2 Conservation Laws 1.3 Basic Fluid Dynamics Applications	• Review Chaps. 1–4 • Solve Book Examples and Problems *independently* • HW Sets #1 and #2 • White Board presentations
7	2. Modern Fluid Dynamics Topics 2.1 Film Drawing and Surface Coating 2.2 Dilute Fluid-Particle Suspensions 2.3 Microfluidics 2.4 Fluid–Structure Interactions 2.5 Biofluid Mechanics	• Study Chaps. 5–10 • Solve selected Book Examples and Problems • White Board presentations • HW Set #3 • Journal Article presentations
3	3. Modern Fluid Dynamics Projects 3.1 Math Modeling and Computer Simulation 3.2 Nanofluid Flow in Microchannels 3.3 Microfluidics and Medical Devices	• Revisit Chaps. 7–10 • Course Project outlines • Course Project presentations

Grading Policy: Three HW Sets plus two Tests: 70%; Presentations and Course Project: 30%

Part A

Fluid Dynamics Essentials

Part A: Fluid Dynamics Essentials

Chapter 1

Review of Basic Engineering Concepts

"Fluid dynamics" implies fluid flow and associated forces described by vector equations, while convective heat transfer and species mass transfer are described by scalar transport equations. Specifically, this chapter reiterates some basic definitions and continuum mechanics concepts with an emphasis on how to describe standard fluid flow phenomena. Readers are encouraged to occasionally jump ahead to specific sections of Chaps. 2 and 3. After refreshing his/her knowledge base, the student should solve the assigned Homework Problems *independently* (see Sect. 1.5) in conjunction with Appendix A (see Table 1.1 for acquiring good study habits).

It should be noted that the material of Part A is an extension of the introductory chapters of the author's "Biofluid Dynamics" text (CRC Press, Taylor & Francis Group, 2006; with permission).

1.1 Approaches, Definitions and Concepts

A sound understanding of the physics of fluid flow with mass and heat transfer, i.e., transport phenomena, as well as statics/dynamics, stress–strain theory and a mastery of basic solution techniques are important prerequisites for studying, applying and improving engineering systems. As always, the objective is to learn to develop mathematical models; here, establish approximate representations of actual transport phenomena in terms of differential or integral equations. *The* (analytical or numerical) *solutions to the describing*

C. Kleinstreuer, *Modern Fluid Dynamics: Basic Theory and Selected Applications in Macro- and Micro-Fluidics*, Fluid Mechanics and Its Applications 87,
DOI 10.1007/978-1-4020-8670-0_1, © Springer Science+Business Media B.V. 2010

equations should produce testable predictions and allow for the analysis of system variations, leading to a deeper understanding and possibly to new or improved engineering procedures or devices. Fortunately, most systems are governed by continuum mechanics laws. Notable exceptions are certain micro- and nano-scale processes, which require modifications of the classical boundary conditions (see Sect. 7.4) or even molecular models solved via statistical mechanics or molecular dynamics simulations.

Clearly, transport phenomena, i.e., mass, momentum and heat transfer, form a subset of *mechanics* which is part of *classical* (or Newtonian) *physics* (see Fig. 1.1). Physics is the mother of all hard-core sciences, engineering and technology. The hope is that one day advancements towards a "universal theory" will unify classical with modern physics, i.e., resulting in a fundamental equation from which all visible/detectable phenomena can be derived and described.

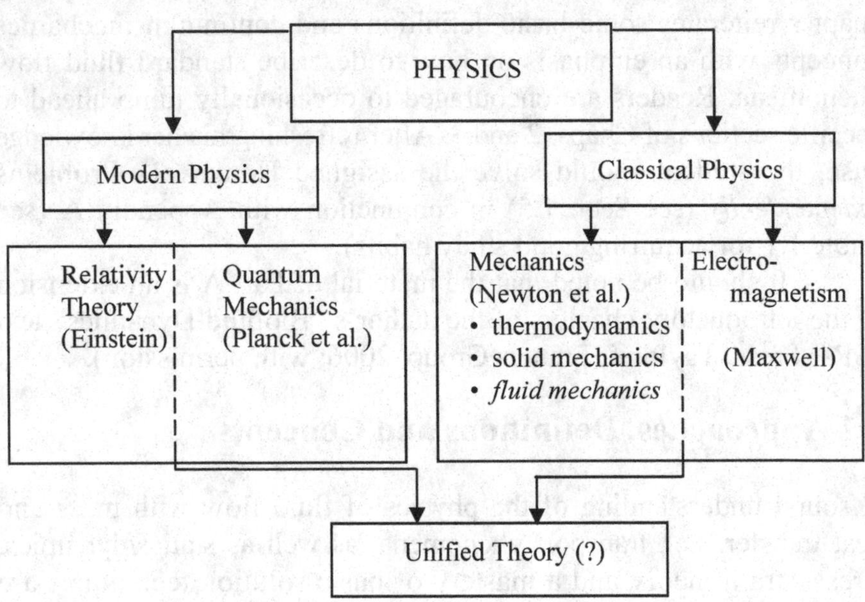

Fig. 1.1 Subsets of Physics and the quest for a Unifying Theory In any case, staying with Newtonian physics, the continuum mechanics assumption, basic definitions, equation derivation methods and problem solving goals are briefly reviewed next – in reverse order.

Approaches to Problem Solving Traditionally, the answer to a given problem is obtained by copying from available sources suitable equations, needed correlations (or submodels), and boundary conditions with their appropriate solution procedures. This is called "matching" and may result in a good first-step learning experience.

Table 1.1 Suggestions for students interested in understanding fluid mechanics and hence obtaining a good grade

1. Review topics:

Eng. Sciences **(Prerequisites)**	***Math Background*** **(see App. A)**
• Problem Solution FORMAT: System Sketch, Assumptions, Approach/Concepts; Solution, Properties, Results; Graphing Analysis, & Comments	• Algebra, Vector Analysis & Taylor Series Expansion • Calculus & Functional including Graphing
• Differential Force, Energy & Mass Balances (i.e., free-body diagram, control volume analysis, etc.)	• Surface & Volume Integrals • Differential Equations subject to Boundary Conditions
• Symbolic Math Analyses, where # of Unknowns $\hat{=}$ # of Equations	

2. Preparation

- Study Book Chapters, Lecture Notes, and Problem Assignments
- Learn from solved Book Examples, Lecture Demos, and Review Problem Solutions (work independently!)
- Practice graphing of results and drawing of velocity or temperature profiles and streamlines
- Ask questions (in-class, after class, office, email)
- Perform "Special Assignments" in-class, such as White-board Performance, lead in small-group work, etc.
- Solve Old Test Problems with your group
- Solve test-caliber questions & problems: well-paced and INDEPENDENTLY

3. Participation

- Enrich your knowledge base and sharpen your communication skills via Presentations
- Understand some Fluid Mechanics Topics in more depth from exploring Flow Visualizations as well as doing Computer Project Work, and Report Writing.

However, it should be augmented later on by more *independent work*, e.g., deriving governing equations, obtaining data sets, plotting and visualizing results, improving basic submodels, finding new, interdisciplinary applications, exploring new concepts, interpreting observations in a more generalized form, or even pushing the envelope of existing solution techniques or theories. In any case, the triple pedagogical goals of advanced *knowledge, skills, and design* can be achieved only via independent practice, hard work, and creative thinking. To reach these lofty goals, a deductive or "top-down" approach is adopted, i.e., from-the-fundamental-to-the-specific, where the general transport phenomena are recognized and mathematically described, and then special cases are derived and solved. For the reader's convenience and pedagogical reasons, specific (important) topics/definitions are several times repeated throughout the text.

While a good grade is a primary objective, a thorough understanding of the subject matter and mastery in solving engineering problems should be the main focus. Once that is achieved, a good grade comes as a natural reward (see Table 1.1).

Derivation Approaches There are basically four ways of obtaining specific transport equations reflecting the conservation laws. The points of departure for each of the four methods are either given (e.g., Boltzmann equation or Newton's second law) or derived based on differential mass, momentum and energy balances for a representative elemental volume (REV).

(i) *Molecular Dynamics Approach*: Fluid properties and transport equations can be obtained from kinetic theory and the Boltzmann equation, respectively, employing statistical means. Alternatively, $\sum \vec{\mathbf{F}} = m\vec{\mathbf{a}}$ is solved for each molecule using direct numerical integration (see Sect. 1.3).

(ii) *Integral Approach*: Starting with the Reynolds Transport Theorem (RTT) for a fixed open control volume (Euler), specific transport equations in integral form can be obtained (see Sect. 2.2).

(iii) *Differential Approach*: Starting with 1-D balances over an REV and then expanding them to 3-D, the mass, momentum and energy transfer equations in differential form can be formulated. Alternatively, the RTT is transformed via the divergence theorem, where in the limit the field equations in differential form are obtained (see Sects. 2.3–2.5).

(iv) *Phenomenological Approach*: Starting with balance equations for an open system, i.e., a control volume, transport phenomena in complex flows are derived largely based on empirical correlations and dimensional analysis considerations. A very practical example is the description of transport phenomena with compartment models (see Sect. 4.4). These "compartments" are either well-mixed, i.e., transient lumped-parameter models without any spatial resolution, or they are transient with a one-dimensional resolution in the axial direction.

Definitions Elemental to transport phenomena is the description of fluid flow, i.e., the equation of motion, which is also called the momentum transfer equation. It is an application of Newton's second law, $\sum \vec{\mathbf{F}}_{\text{ext.}} = m\,\vec{\mathbf{a}}$, which Newton postulated for the motion of a particle. For most engineering applications the equation of motion is nonlinear but independent of the mass and heat transfer equations, i.e., fluid properties are not measurably affected by changes in solute concentration and temperature. Hence, the major emphasis in Chap. 1 is on the description, solution and understanding of the physics of fluid flow. Here is a review of a few definitions:

- A *fluid* is an assemblage of gas or liquid molecules which deforms continuously, i.e., it *flows* under the application of a shear stress. Note, solids do not behave like that; but, what about borderline cases, i.e., the behavior of materials such as jelly, grain, sand, etc.?
- Key fluid properties are density ρ, dynamic viscosity μ, species diffusivity $\mathcal{D}$, heat capacities c_p and c_v, and thermal conductivity k. In general, all six are temperature and species concentration dependent. Most important is the viscosity (see

also kinematic viscosity $\nu \equiv \mu/\rho$) representing frictional (or drag) effects. Certain fluids, such as polymeric liquids, blood, food stuff, etc., are also shear-rate dependent and hence called *non-Newtonian fluids* (see Sect. 6.3).

- *Flows* can be categorized into:

Internal flows	and	External flows
- Oil, air, water or steam in pipes and inside devices		- Air past vehicles, buildings and planes
- Blood in arteries/veins or air in lungs		- Water past pillars, submarines, etc.
- Water in rivers or canals		- Polymer coating on solid surfaces

- *Driving* forces for fluid-flow include gravity, pressure differentials or gradients, temperature gradients, surface tension, electromagnetic forces, etc.
- Any fluid-flow is described by its *velocity* and *pressure* fields. The velocity vector of a fluid element can be written in terms of its three scalar components:

$$\vec{\mathbf{v}} = u\,\hat{i} + v\,\hat{j} + w\,\hat{k} \qquad \langle\text{rectangular coordinates}\rangle \qquad (1.1a)$$

or

$$\vec{\mathbf{v}} = v_r\,\hat{e}_r + v_\theta\,\hat{e}_\theta + v_z\,\hat{e}_z \qquad \langle\text{cylindrical coordinates}\rangle \qquad (1.1b)$$

Its total time derivative is the fluid element acceleration (see App. A):

$$\frac{d\vec{\mathbf{v}}}{dt} \cong \frac{D\vec{\mathbf{v}}}{Dt} = \vec{\mathbf{a}}_{\text{total}} = \vec{\mathbf{a}}_{\text{local}} + \vec{\mathbf{a}}_{\text{convective}} \qquad (1.2)$$

where Eq. (1.2) is also known as Stokes, material or substantial time derivative.

- *Streamlines* for the visualization of flow fields are lines to which the local velocity vectors are tangential. For example, for steady 2-D flow:

$$\frac{dy}{dx} = \frac{v}{u} \tag{1.3}$$

where the 2-D velocity components $\vec{v} = (u, v, 0)$ have to be given to obtain, after integration, the streamline equation y(x).

- Forces acting on a fluid element can be split into *normal* and *tangential forces* leading to pressure and normal/shear stresses. Clearly, on any surface element:

$$p \text{ or } \tau_{normal} = \frac{F_{normal}}{A_{surface}} \tag{1.4}$$

while

$$\tau_{shear} = \frac{F_{tangential}}{A_{surface}} \tag{1.5}$$

As Stokes postulated, the stress can be viewed as a linear derivative, i.e., $\vec{\vec{\tau}} \sim \nabla \vec{v}$ (see App. A), where relative motion of viscous fluid elements (or layers) generate a shear stress, τ_{shear}. In contrast, the total pressure sums up the mechanical (or thermodynamic) pressure, which is experienced when moving with the fluid (and therefore labeled "static" pressure and measured with a piezometer). The dynamic pressure is due to the fluid motion (i.e., $\rho\, v^2/2$), and the hydrostatic pressure is due to gravity (i.e., ρgz):

$$p_{total} = p_{static} + p_{dynamic} + p_{hydro-static}$$

$$= p_{static} + \frac{\rho}{2} v^2 + \rho gz = \not{c} \tag{1.6a, b}$$

where

$$p_{static} + p_{dynamic} = p_{stagnation} \tag{1.7}$$

Recall for a stagnant fluid body (i.e., a reservoir), where h is the depth coordinate:

$$p_{hydro-static} = p_0 + \rho gh \tag{1.8}$$

Clearly, the hydrostatic pressure due to the fluid weight appears in the momentum equation as a body force per unit volume, i.e., $\rho\vec{g}$ (see Example 1.1).

- *Dimensionless groups*, i.e., ratios of forces, fluxes, process or system parameters, indicate the importance of specific transport phenomena. For example, the Reynolds number is defined as (see Example 1.1):

$$\mathrm{Re}_L \equiv \frac{F_{\text{inertia}}}{F_{\text{viscous}}} := vL/\nu \tag{1.9}$$

where v is an average system velocity, L is a representative system "length" scale (e.g., the tube diameter D), and $\nu \equiv \mu/\rho$ is the kinematic viscosity of the fluid. Other dimensionless groups with applications in engineering include the Womersley number and Strouhal number (both dealing with oscillatory/transient flows), the Euler number (pressure difference), the Weber number (surface tension), the Stokes number (particle dynamics), Schmidt number (diffusive mass transfer), Sherwood number (convective mass transfer) and the Nusselt number, the ratio of heat conduction to heat convection. The most common source, or derivation, of these numbers is the non-dimensionalization of partial differential equations describing the transport phenomena at hand as well as scale analysis (see Example 1.1).

Example 1.1: Generation of Dimensionless Groups

(A) Scale Analysis

As outlined in Sect. 2.4, the Navier–Stokes equation (see Eq. (2.22)) describes fluid element acceleration due to several forces per unit mass, i.e.,

$$\vec{\mathbf{a}}_{\text{total}} \equiv \underbrace{\frac{\partial \vec{\mathbf{v}}}{\partial t}}_{\substack{\textit{transient}\\ \textit{term}}} + \underbrace{(\vec{\mathbf{v}} \cdot \nabla)\vec{\mathbf{v}}}_{\substack{\textit{inertia}\\ \textit{term}}} = \underbrace{-\frac{1}{\rho}\nabla p}_{\substack{\text{pressure}\\ \text{force}}} + \underbrace{\nu\nabla^2\vec{\mathbf{v}}}_{\substack{\text{viscous}\\ \text{force}}} + \underbrace{\vec{\mathbf{g}}}_{\text{gravity}}$$

Now, by definition:

$$\mathrm{Re} = \frac{\text{inertial force}}{\text{viscous force}} := \frac{(\vec{\mathbf{v}} \cdot \nabla)\vec{\mathbf{v}}}{\nu \nabla^2 \vec{\mathbf{v}}}$$

Employing the scales $\vec{\mathbf{v}} \sim \mathrm{v}$ and $\nabla = \left(\frac{\partial}{\partial x}, \frac{\partial}{\partial y}, \frac{\partial}{\partial z}\right) \sim \frac{1}{L}$ where v may be an average velocity and L a system-characteristic dimension, we obtain:

$$\mathrm{Re} = \frac{\left(\mathrm{v} \cdot \frac{1}{L}\right)\mathrm{v}}{\nu L^{-2} \mathrm{v}} = \frac{\mathrm{v}L}{\nu}$$

Similarly, taking

$$\frac{\text{local acceleration}}{\text{convective acceleration}} \equiv \frac{\text{transient term}}{\text{inertia term}} = \frac{\partial \vec{\mathbf{v}} / \partial t}{(\vec{\mathbf{v}} \cdot \nabla)\vec{\mathbf{v}}}$$

we can write with system time scale T (e.g., cardiac cycle: $\mathrm{T} = 1\mathrm{s}$)

$$\frac{\mathrm{v}/\mathrm{T}}{\mathrm{vL}^{-1}\mathrm{v}} = \frac{\mathrm{L}}{\mathrm{vT}} = \mathrm{Str}$$

which is the *Strouhal number*. For example, when T >> 1, $\mathrm{Str} \to 0$ and hence the process, or transport phenomenon, is quasi-steady.

(B) Non-dimensionalization of Governing Equations

Taking the transient boundary-layer equations (see Sect. 2.4, Eq. (2.22)) as an example,

$$\rho\left(\frac{\partial u}{\partial t}+u\frac{\partial u}{\partial x}+v\frac{\partial u}{\partial y}\right)=-\frac{\partial p}{\partial x}+\mu\frac{\partial^2 u}{\partial y^2}$$

we nondimensionalize each variable with suitable, constant reference quantities. Specifically, approach velocity U_0, plate length ℓ, system time T, and atmospheric pressure p_0 are such quantities. Then,

$$\hat{u}=u/U_0,\ \hat{v}=v/U_0;\ \hat{x}=x/\ell,\ \hat{y}=y/\ell;\ \hat{p}=p/p_0 \text{ and } \hat{t}=t/T.$$

Note: In Sect. 5.2 $\hat{y}$ is defined as $\hat{y}=y/\delta(x)$, where $\delta(x)$ is the varying boundary-layer thickness.

Inserting all variables, i.e., $u=\hat{u}U_0, t=\hat{t}T$, etc., into the governing equation yields

$$\frac{\rho U_0}{T}\frac{\partial\hat{u}}{\partial\hat{t}}+\left[\frac{\rho U_0^2}{\ell}\right]\left(\hat{u}\frac{\partial\hat{u}}{\partial\hat{x}}+\hat{v}\frac{\partial\hat{u}}{\partial\hat{y}}\right)=\left[\frac{p_0}{\ell}\right]\frac{\partial\hat{p}}{\partial\hat{x}}+\left[\frac{\mu U_0}{\ell^2}\right]\frac{\partial^2\hat{u}}{\partial y^2}$$

Dividing the entire equation by, say, $\left[\frac{\rho U_0^2}{\ell}\right]$ generates:

$$\underbrace{\left[\frac{\ell}{Tu_0}\right]}_{\text{Strouhal \#}}\frac{\partial\hat{u}}{\partial\hat{t}}+\hat{u}\frac{\partial\hat{u}}{\partial\hat{x}}+\hat{v}\frac{\partial\hat{u}}{\partial\hat{y}}=-\underbrace{\left[\frac{p_0}{\rho U_0^2}\right]}_{\text{Euler \#}}\frac{\partial\hat{p}}{\partial\hat{x}}+\underbrace{\left[\frac{\mu}{\rho U_0\ell}\right]}_{\substack{\text{inverse}\\ \text{Reynolds \#}}}\frac{\partial^2\hat{u}}{\partial\hat{y}^2}$$

Comments:

In a way three goals have been achieved:

- The governing equation is now dimensionless.
- The variables vary only between 0 and 1.
- The overall fluid flow behavior can be assessed by the magnitude of three groups, i.e., Str, Eu and Re numbers.

1.2 The Continuum Mechanics Assumption

Fundamental to the description of all transport phenomena are the conservation laws, concerning mass, momentum and energy, as well as their applications to continua. For example, Newton's second law of motion holds for both molecular dynamics, i.e., interacting molecules, and continua, like air, water, plasma, and oils. Thus, *solid structures* and *fluid flow fields* are assumed to be continua as long as the local material properties can be defined as averages computed over material elements/volumes sufficiently large when compared to microscopic length scales of the solid or fluid, but small relative to the macroscopic structure. Variations in solid-structure or fluid-flow quantities can be obtained via differential equations. The continuum mechanics method is an effective tool to physically explain and mathematically describe various transport phenomena without detailed knowledge of their internal nano/micro structures. Specifically, fluids are treated as continuous media characterized by certain field quantities associated with the internal structure, such as density, temperature and velocity. In summary, continuum mechanics deals with three aspects:

- *Kinetics*, i.e., fluid element motion regardless of the cause
- *Dynamics*, i.e., the origin and impact of forces and fluxes generating fluid motion and waste heat, e.g., the stress tensor, heat flux vector, and entropy
- *Balance Principles*, i.e., the mass, momentum and energy conservation laws

Also, all flow properties are in *local thermodynamic equilibrium*, implying that the macroscopic quantities of the flow field can adjust swiftly to their surroundings. This local adjustment to varying conditions is rapidly achieved if the fluid has very small characteristic length and time scales of molecular collisions, when compared to the macroscopic flow variations.

However, as the channel (or tube) size, typically indicated by the hydraulic diameter D_h, is reduced to the *micro-scale*, the surface-area-to-volume ratio becomes larger because $A/V \sim D_h^{-1}$. Thus, wall surface effects may become important; for example, wall roughness and surface forces as well as discontinuities in fluid (mainly gas) velocity and temperature relative to the wall. When flow micro-conduits are short as in micro-scale cooling devices and MEMS, nonlinear entrance effects dominate, while for long microconduits viscous heating (for liquids) or compressibility (for gases) may become a factor (see Chap. 7). In such cases, the validity of the continuum mechanics assumption may have to be re-examined.

1.3 Fluid Flow Description

Any flow field can be described at either the microscopic or the macroscopic level. The *microscopic* or molecular models consider the position, velocity, and state of every molecule of a single fluid or multiple 'fluids' at all times. Averaging discrete-particle information (i.e., position, velocity, and state) over a local fluid volume yields macroscopic quantities, e.g., the velocity field $\vec{\mathbf{v}}(\vec{\mathbf{x}}, t)$, at any location in the flow. The advantages of the molecular approach include general applicability, i.e., no need for submodels (e.g., for the stress tensor, heat flux, turbulence, wall conditions, etc.), and an absence of numerical instabilities (e.g., due to steep flow field gradients). However, considering myriads of molecules, atoms, and nanoparticles requires enormous computer resources, and hence only simple channel or stratified flows with a finite number of interacting molecules (assumed to be solid spheres) can be presently analyzed. For example, in a 1-mm cube there are about 34 billion water molecules (about a million air molecules at STP), which make molecular dynamics simulation prohibitive, but on the other hand, intuitively validates the continuum assumption (see Sect. 1.2).

Here, the overall goal is to find and analyze the interactions between *fluid forces*, e.g., pressure, gravity/buoyancy, drag/friction, inertia, etc., and *fluid motion*, i.e., the velocity vector field and pressure distribution from which everything else can be directly obtained or derived (see Fig. 1.2a, b). In turn, scalar transport equations, i.e., convection mass and heat transfer, can be solved based on the velocity field to obtain critical magnitudes and gradients (or fluxes) of species concentrations and temperatures.

In summary, *unbalanced surface/body forces and gradients cause motion in form of fluid translation, rotation, and/or deformation, while temperature or concentration gradients cause mainly heat or species-mass transfer.* Note that flow visualization CDs plus web-based university sources provide fascinating videos of complex fluid flow, temperature and species concentration fields.

(a) Cause-and-effect dynamics:

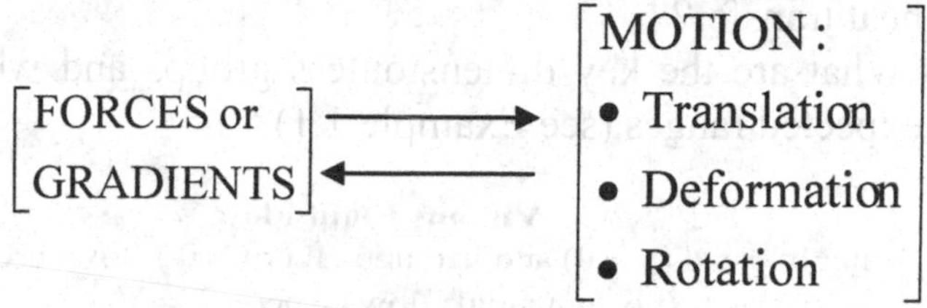

(b) Kinematics of a 2-D fluid element (Lagrangian frame):

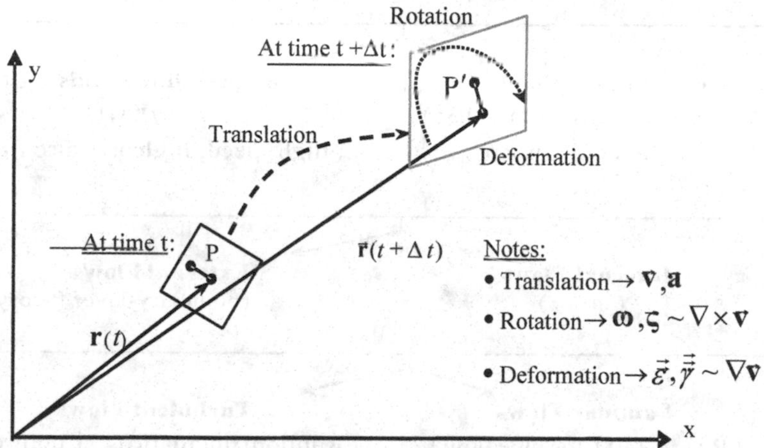

Fig. 1.2 Dynamics and kinematics of fluid flow: (a) force-motion interactions; and (b) 2-D fluid kinematics

Exact *flow problem identification*, especially in industrial settings, is one of the more important and sometimes the most difficult first task. After obtaining some basic information and reliable data, it helps to think and speculate about the physics of the fluid flow, asking:

(i) What category does the given flow system fall into, and how does it respond to normal as well as extreme changes in operating conditions? Figure 1.3 may be useful for categorization of real fluids and types of flows.
(ii) What variables and system parameters play an important role in the observed transport phenomena, i.e., linear or angular momentum transfer, fluid-mass or species-mass transfer, and heat transfer?
(iii) What are the key dimensionless groups and what are their expected ranges (see Example 1.1)?

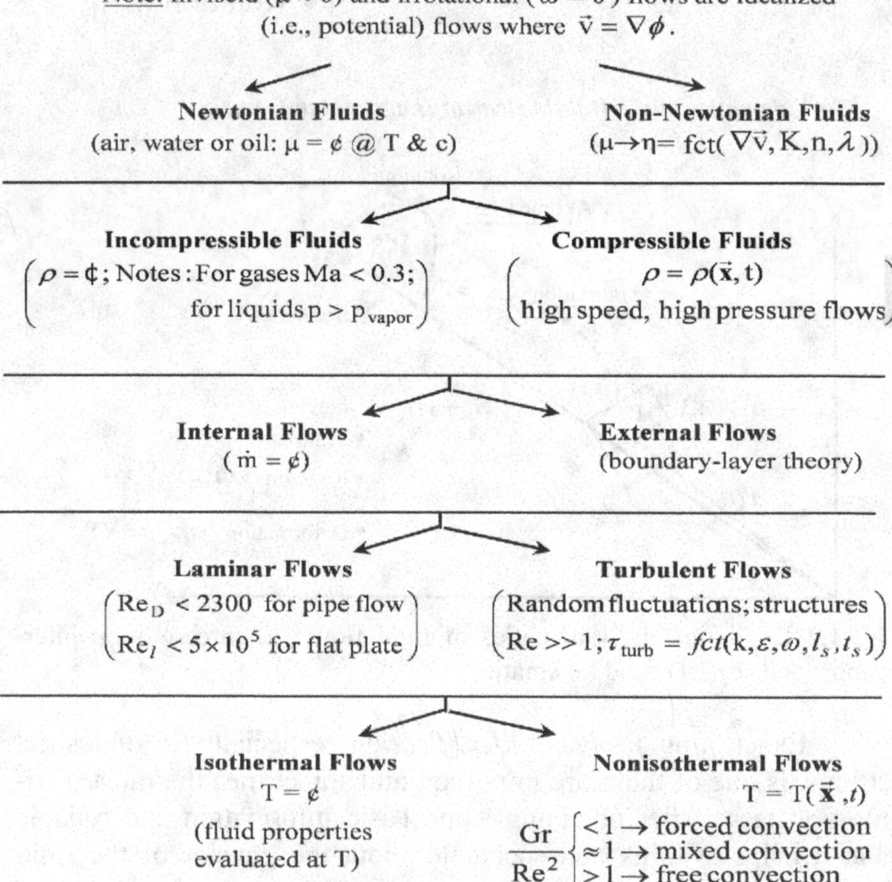

Fig. 1.3 Special cases of viscous fluid flows

Answers to these questions assist in grouping the flow problem at hand. For example, with the exception of "superfluids", all others are viscous, some more (e.g., syrup) and some less (e.g., rarefied gases). However, with the advent of *Prandtl's boundary-layer concept* (Sect. 2.4) the flow field, say, around an airfoil has been traditionally divided into a very thin (growing) viscous layer and beyond that an unperturbed *inviscid region*. This paradigm helped to better understand actual fluid mechanics phenomena and to simplify velocity and pressure as well as drag and lift calculations. Specifically, at sufficiently high approach velocities a fluid layer adjacent to a submerged body experiences steep gradients due to the "no-slip" condition and hence constitutes a *viscous flow region*, while outside the boundary layer frictional effects are negligible (see Prandtl equations vs. Euler equation in Sect. 2.4). Clearly, with the prevalence of powerful CFD software and engineering workstations, such a fluid flow classification is becoming more and more superfluous (see discussion in Sect. 5.2).

While in addition to air and water almost all oils are *Newtonian*, some synthetic motor-oils are shear-rate dependent and that holds as well for a variety of new (fluidic) products. This implies that modern engineers have to cope with the analysis and computer modeling of *non-Newtonian fluids* (see Sect. 6.3). For example, Latex paint is shear-thinning, i.e., when painting a vertical door rapid brush strokes induce high shear rates ($\dot{\gamma} \sim$ dw/dz) and the paint viscosity/resistance is very low. When brushing stops, locally thicker paint layers (due to gravity) try to descent slowly; however, at low shear rates the paint viscosity is very high and hence "tear-drop" formation is avoided and a near-perfect coating can dry on the door.

All natural phenomena change with time and hence are *unsteady (i.e., transient)* while in industry it is mostly desirable that processes are steady, except during production line start-up, failure, or shut-down. For example, turbines, compressors and heat exchangers operate continuously for long periods of time and hence are labeled "steady-flow devices"; in contrast, pacemakers, control systems and drink-dispensers work in a time-dependent fashion. In some cases, like a heart valve, devices change their orientation periodically and the associated flows oscillate about a mean value. In contrast, it should be noted that the term *uniform* implies "no change with system *location*", as in uniform (i.e., constant over a cross-section) velocity or uniform particle distribution, which all could still vary with time.

Mathematical flow field descriptions become complicated when *laminar flow* turns unstable due to high speed and/or geometric irregularities ranging from surface roughness to complex conduits. The deterministic laminar flow turns *transitional* on its way to become *fully turbulent*, i.e., chaotic, transient 3-D with random velocity fluctuations, which help in mixing but also induce high apparent stresses. As an example of "flow transition", picture a group (on bikes or skis) going faster and faster down a mountain while the terrain gets rougher. The initially quite ordered group of riders/skiers may change swiftly into an unbalanced, chaotic group. So far no *universal model for turbulence*, let alone for the transitional regime from laminar to turbulent, has been found. Thus, major efforts focus on direct numerical simulation (DNS) of turbulent flows which are characterized by relatively high Reynolds numbers and chaotic, transient 3-D flow pattern (see Sects. 4.2 and 5.2).

Basic Flow Assumptions and Their Mathematical Statements
Once a given fluid dynamics problem has been categorized (Fig. 1.3), some justifiable assumptions have to be considered in order to simplify the general transfer equations, as exemplified here:

Flow assumption: Consequence:

- Time-dependence → $\frac{\partial}{\partial t} = 0$ i.e., steady-state;
  $\vec{\mathbf{v}} = \vec{\mathbf{v}}(t)$ i.e., transient flow
- Dimensionality → Required number of space coordinates $\vec{\mathbf{x}} = (x, y, z)$
- Directionality → Required number of velocity components $\vec{\mathbf{v}} = (u, v, w)$
- Unidirectional flow → Special case when all but one velocity component are zero
- Development phase → $\frac{\partial v}{\partial s} = 0$ i.e., fully developed flow, where s is the axial coordinate

- Symmetry $\rightarrow \begin{cases} \dfrac{\partial}{\partial n} = 0 : \text{midplane (n is the normal coordinate)} \\ \dfrac{\partial}{\partial \theta} = 0 : \text{axisymmetry} \end{cases}$

Closed vs. Open Systems Information on a given flow problem, in terms of the viscous flow grouping (see Fig. 1.3) and in conjunction with a set of proper assumptions, allows for the selection of a suitable solution technique (see App. A). That decision, however, requires first a brief review of possible flow field descriptions in terms of the *Lagrangian vs. Eulerian framework* in continuum mechanics.

Within the continuum mechanics framework, two basic flow field descriptions are of interest, i.e., the Lagrangian viewpoint and the Eulerian (or control-volume) approach (see Fig. 1.4, where C.∀. ≙ control volume and C.S. ≙ control surface).

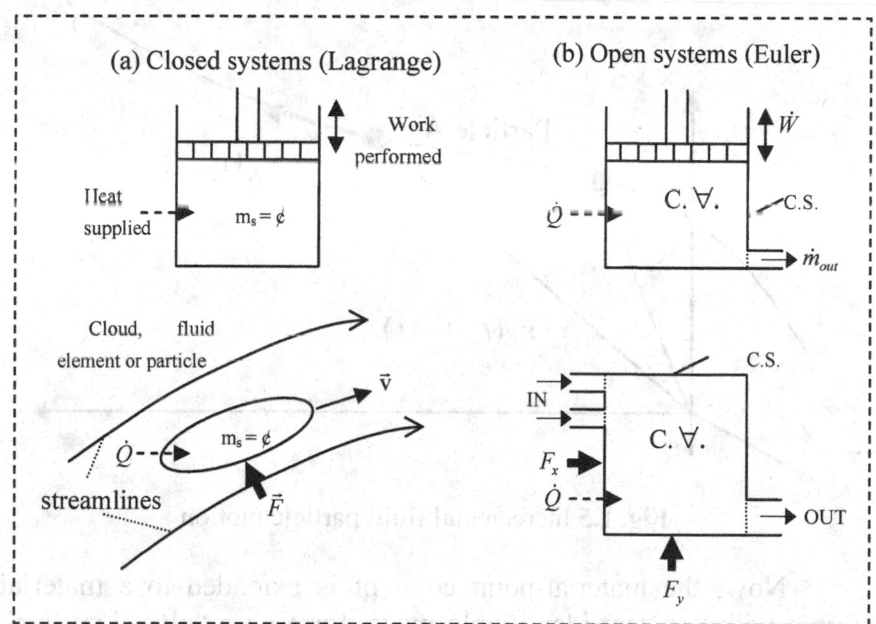

Fig. 1.4 Closed vs. Open Systems

For the *Lagrangian description* consider particle A moving on a path line with respect to a fixed Cartesian coordinate system.

Initially, the position of A is at $\vec{\mathbf{r}}_{\mathbf{o}} = \vec{\mathbf{r}}_{\mathbf{o}}(\vec{\mathbf{x}}_o, t_o)$ and a moment later at $\vec{\mathbf{r}}_{\mathbf{A}} = \vec{\mathbf{r}}_{\mathbf{A}}(\vec{\mathbf{r}}_{\mathbf{o}}, t_o + \Delta t)$ as depicted in Fig. 1.5, where $\vec{\mathbf{r}}_{\mathbf{A}} = \vec{\mathbf{r}}_{\mathbf{o}} + \Delta\vec{\mathbf{r}}$. Considering all distinct points and following their motion for $t > t_o$, solid particle (or fluid element) motion can be described with the position vector

$$\vec{\mathbf{r}} = \vec{\mathbf{r}}(\vec{\mathbf{r}}_{\mathbf{o}}, t) \tag{1.10}$$

where in the limit we obtain the fluid velocity and acceleration, i.e.,

$$\frac{d\vec{\mathbf{r}}}{dt} = \vec{\mathbf{v}} \tag{1.11}$$

and

$$\frac{d^2\vec{\mathbf{r}}}{dt^2} = \frac{d\vec{\mathbf{v}}}{dt} = \vec{\mathbf{a}} \tag{1.12}$$

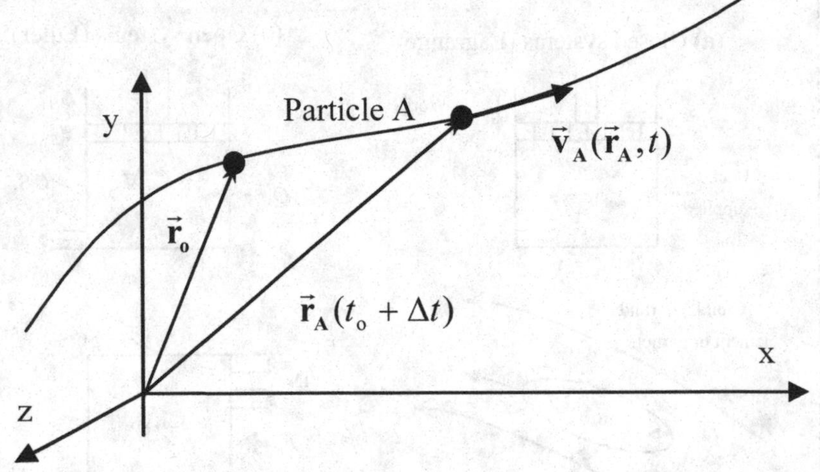

Fig. 1.5 Incremental fluid particle motion

Now, the material-point concept is extended to a material volume with constant identifiable mass, forming a "closed system" that moves and deforms with the flow but no mass crosses the material volume surface because it is closed (see Fig. 1.4a). Again, the system is tracked through space and as time expires, it is of interest to know what the changes in system mass, momentum, and energy are. This can be expressed in terms of the system's extensive property N_S which is either mass *m*, momentum $m\vec{v}$, or total energy E. Thus, the key question is: "How can we express the fate of N_S", or in mathematical shorthand, what is "DN_S/Dt"? Clearly, the material time (or Stokes) derivative $D/Dt \equiv \partial/\partial t + \vec{v}\cdot\nabla$ follows the closed system and records the total time-rate-of-change of whatever is being tracked (see Sect. 2.2).

Now, a brief illustration of the various *time derivatives*, i.e., $\partial/\partial t$ (local), d/dt (total of a material point or solid particle), and D/Dt (total of a fluid element) is in order. Their differences can be illustrated using acceleration (see also App. A):

- $a_{x,\text{local}} = \frac{\partial u}{\partial t}$, where *u* is the fluid element velocity in the x-direction.
- $\vec{a}_{\text{particle}} = \frac{d\vec{v}}{dt}$ is employed in solid particle dynamics.

  whereas
- $\vec{a}_{\substack{\text{fluid}\\\text{element}}} = \frac{D\vec{v}}{Dt} = \underbrace{\frac{\partial \vec{v}}{\partial t}}_{\vec{a}_{\text{local}}} + \underbrace{(\vec{v}\cdot\nabla)\vec{v}}_{\vec{a}_{\text{convective}}}$ is the total fluid element acceleration.

The derivation of $D\vec{v}/Dt$ is given in the next example.

Example 1.2: Derive the material (or Stokes) derivative, $\frac{D}{Dt}$ operating on the velocity vector, describing the "total time-rate-of-change" of a fluid flow field.

Hint: For illustration purposes, use an arbitrary velocity field, $\vec{v} = \vec{v}(x, y, z; t)$, and form its total differential.

Recall: The total differential of any continuous and differentiable function, such as $\vec{v} = \vec{v}(x, y, z; t)$, can be expressed in terms of its infinitesimal contributions in terms of changes of the independent variables.

$$d\vec{v} = \frac{\partial \vec{v}}{\partial x}dx + \frac{\partial \vec{v}}{\partial y}dy + \frac{\partial \vec{v}}{\partial z}dz + \frac{\partial \vec{v}}{\partial t}dt$$

Solution:

- Dividing through by dt and recognizing that dx/dt = u, dy/dt = v and dz/dt = w are the local velocity components, we have:

$$\frac{d\vec{v}}{dt} = \frac{\partial \vec{v}}{\partial x}u + \frac{\partial \vec{v}}{\partial y}v + \frac{\partial \vec{v}}{\partial z}w + \frac{\partial \vec{v}}{\partial t}$$

- Substituting the "particle dynamics" differential with the "fluid element" differential yields:

$$\frac{d\vec{v}}{dt} \hat{=} \frac{D\vec{v}}{Dt} = \frac{\partial \vec{v}}{\partial t} + u\frac{\partial \vec{v}}{\partial x} + v\frac{\partial \vec{v}}{\partial y} + w\frac{\partial \vec{v}}{\partial z} \equiv \frac{\partial \vec{v}}{\partial t} + (\vec{v} \cdot \nabla)\vec{v} = \vec{a}_{local} + \vec{a}_{conv.}$$

In the *Eulerian frame,* an "open system" is considered where mass, momentum and energy may readily cross boundaries, i.e., being convected across the control volume surface and local fluid flow changes may occur within the control volume over time (see Fig. 1.4b). The fixed or moving control volume may be a large system/device with inlet and outlet ports, it may be small finite volumes generated by a computational mesh, or it may be in the limit a "point" in the flow field. In general, the Eulerian, observer fixed to an inertial reference frame records temporal and spatial changes of the flow field at all "points" or in case of a control volume, transient mass, momentum and/or energy changes inside and fluxes across its control surfaces.

In contrast, the Lagrangian observer stays with each fluid element or material volume and records its basic changes while moving through space. Section 2.2 employs both viewpoints to describe mass, momentum, and heat transfer in integral form, known as the Reynolds Transport Theorem (RTT). *Thus, the RTT simply links the conservation laws from the Lagrangian to the Eulerian frame.* In turn, a surface-to-volume integral transformation then yields the conservation laws in differential form in the Eulerian framework, also known as the *control-volume approach.*

Example 1.3: Lagrangian vs. Eulerian Flow Description of River Flow

In the Eulerian fixed coordinate frame, river flow is approximated as steady 1-D, i.e.,

$$v(x) = v_0 + \Delta v\left(1 - e^{-ax}\right)$$

which implies that at x=0, say, the water surface moves at v_0 and then accelerates downstream to $v(x\to\infty) = v_0 + \Delta v$. Derive an expression for $v = v(v_0, t)$ in the Lagrangian frame.

Recall: $\vec{v} = d\vec{r}/dt$ and in our 1-D case

$$\frac{dx}{dt} = v(x) = v_0 + \Delta v\left(1 - e^{-ax}\right)$$

Solution: Separation of variables and integration yield:

$$\int_0^x \frac{dx}{(v_0 + \Delta v) - \Delta v e^{-ax}} = \int_0^t dt$$

so that

$$x + \frac{1}{a}\ln\left[1 + \frac{\Delta v}{v_0}\left(1 - e^{-ax}\right)\right] = (v_0 + \Delta v)t$$

Now, replacing the two x-terms with expressions from the v(x)-equation, i.e., $x = -\frac{1}{a}\ln\left(1-\frac{v-v_0}{\Delta v}\right)$ and $e^{-ax} = 1-\frac{v-v_0}{\Delta v}$, we can express the Lagrangian velocity as:

$$v(t) = \frac{v_0(v_0 + \Delta v)}{v_0 + \Delta v \exp[-a(v_0 + \Delta v)t]}$$

Graphs:

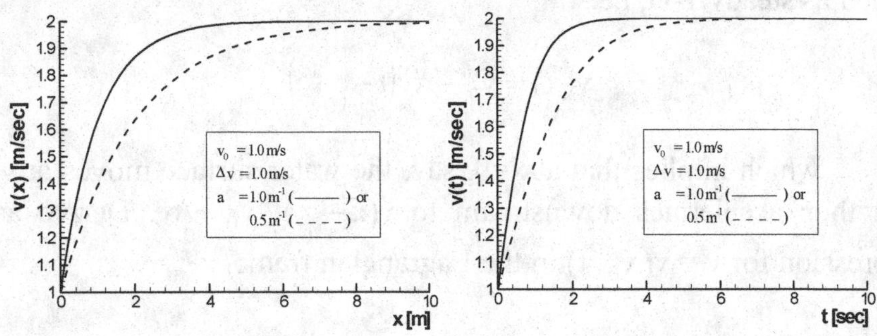

Comments:

Although the graphs look quite similar because of the rather simple v(x)-function considered, subtle differences are transparent when comparing the velocity gradients (i.e., dv/dx and dv/dt) rather than just the magnitudes v(x) and v(t). Clearly, the mathematical river flow description is much more intuitive in the Eulerian frame-of-reference.

1.4 Thermodynamic Properties and Constitutive Equations

Thermodynamic properties, such as mass and volume (extensive properties) or velocity, pressure and temperature (intensive properties), characterize a given system. In addition, there are *transport properties*, such as viscosity, diffusivity and thermal conductivity, which are all temperature-dependent and may greatly influence, or even largely determine, a fluid flow field. Any extensive, i.e., mass-dependent, property divided by unit mass is called a *specific property,* such as the specific volume v =V/m (where its inverse is the fluid density) or the specific energy e=E/m (see Sect. 2.2). An *equation of state* is a correlation of independent intensive properties, where for a simple compressible system just two describe the state of such a system. A famous example is the ideal-gas relation, pV=mRT, where m= ρ V and R is the gas constant.

Constitutive Equations When considering the conservation laws derived in Chap. 2 for fluid flow and heat transfer, it is apparent that additional relationships must be found in order to solve for the field variable $\vec{\mathbf{v}}$, p and T as well as $\vec{\mathbf{q}}$ and $\vec{\vec{\tau}}$. Thus, this is necessary for reasons of: (i) mathematical closure, i.e., a number of unknowns require the same number of equations, and (ii) physical evidence, i.e., additional material properties other than the density ρ are important in the description of system/material/fluid behavior. These additional relations, or *constitutive equations,* are fluxes which relate via "material properties" to gradients of the principle unknowns. Specifically, for basic *linear* proportionalities we recall:

- Hooke's law, i.e., the stress-strain relation (see Sect. 8.2):

$$\sigma_{ij} = D_{ijkl}\,\varepsilon_{kl} \tag{1.13}$$

  where D_{ijkl} is the Lagrangian elasticity tensor;
- Fourier's law, i.e., the heat conduction flux (see Sect. 2.5):

$$\vec{\mathbf{q}} = -k\nabla T \tag{1.14}$$

  where k is the thermal conductivity;
- Binary diffusion flux (see Sect. 2.5):

$$\vec{\mathbf{j}}_c = -\mathcal{D}_{AB}\nabla c \tag{1.15}$$

  where $\mathcal{D}_{AB}$ is the species-mass diffusion coefficient;

- Stokes' postulate, i.e., the fluid shear stress tensor

$$\vec{\vec{\tau}} = \mu(\nabla\vec{v} + \nabla\vec{v}^{T}) \tag{1.16}$$

where μ is the dynamic viscosity. Equations (1.14)–(1.16) are illustrated next.

Viscosity and the Basic Shear Stress Component To move fluid elements relative to each other, a shear force $F_{tangential} = \tau_{shear} A_{interface}$ (see Eq. (1.5)) is necessary. The shear stress is proportional to $\nabla\vec{v}$ (see Fig. 1.6) and the dynamic viscosity which is just temperature-dependent for Newtonian fluids (e.g., air, water and oil) or shear-rate dependent for polymeric liquids, paints, blood (at low shear rates), food stuff, etc. Specifically, for incompressible fluid flow Stokes postulated:

$$\vec{\vec{\tau}} = \mu(\nabla\vec{v} + \nabla\vec{v}^{T}) \tag{1.17a}$$

where for simple shear flow:

$$\tau_{yx} = \mu\, du/dy \tag{1.17b}$$

Physical insight to Eq. (1.17b) is given with Fig. 1.6.

(a) Tangential force $F_{pull} = -F_{drag}$**:**

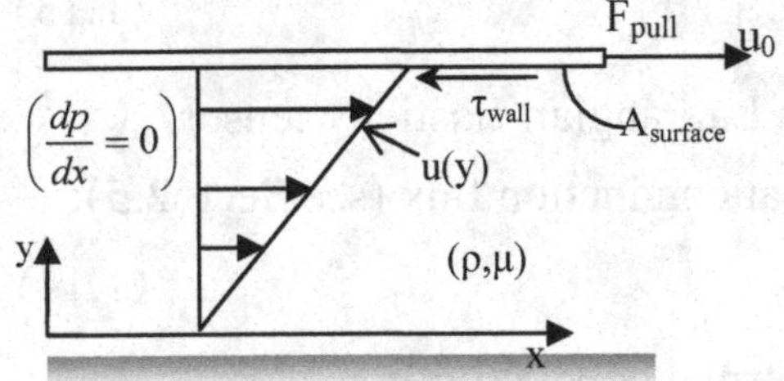

We observe:

$$\tau_{wall} = \frac{F_{drag}}{A_{surface}}$$

or anywhere in the fluid:

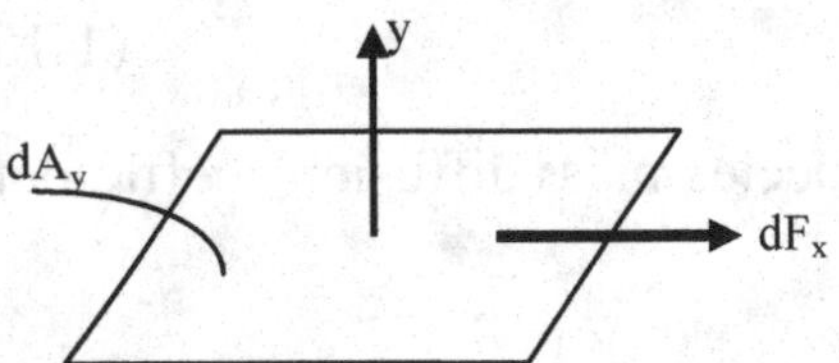

$$\tau_{yx} = \frac{dF_x}{dA_y}$$

(b) Resistance to fluid element deformation:

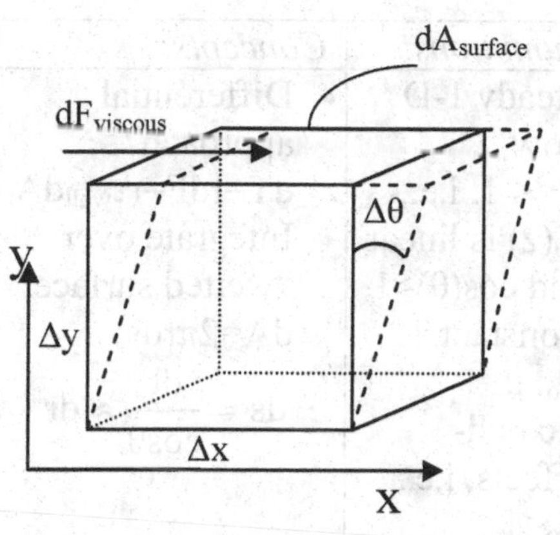

- Physics:

$$\tau_{ij} = \lim_{\delta A_j \to 0} \frac{\delta F_i}{\delta A_j} \hat{=} \frac{\text{surface force}}{\text{unit area}}$$

$$\tau = \frac{dF_v}{dA_s} \sim \frac{\Delta\theta}{\Delta t}$$

- Geometry:

$$\tan \Delta\theta \approx \Delta\theta = \frac{\Delta s}{\Delta y} = \frac{\Delta u \cdot \Delta t}{\Delta y}$$

Combining both:

$$\tau \sim \frac{\Delta u}{\Delta y}$$

where in the limit with the proportionality factor, μ, for unidirectional flow:

$$\tau_{yx} = \mu \frac{du}{dy}$$

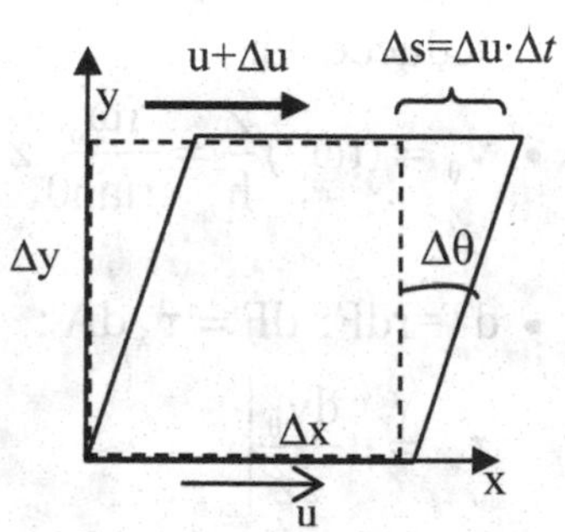

Fig. 1.6 Illustration of the shear stress derivation for simple Couette-like flow

Comment: Because u(y) is linear, $\tau_{yx} = \tau_{wall} = \not{c}$.

Example 1.4: Determine the viscosity of a fluid in a basic cone-plate viscometer with given cone angle and radius, applied torque and resulting constant angular velocity. Plot the result as $T = T(R, \mu)$.

Sketch:	*Assumptions:*	*Concepts:*
Torque T, ω_o R Cone μ z θ r Plate	• Steady 1-D flow • $\theta << 1$, i.e., $v_\theta(z)$ is linear and $\cos(\theta) \approx 1$ • Constant T, ω_o, μ • No end-effects, i.e., $\omega_o < 1$	• Differential approach $dT = rdF = r\tau_{wall}dA$ • Integrate over "wetted surface" $dA = 2\pi rds$; $ds = \dfrac{dr}{\cos\theta} \approx dr$

Solution:

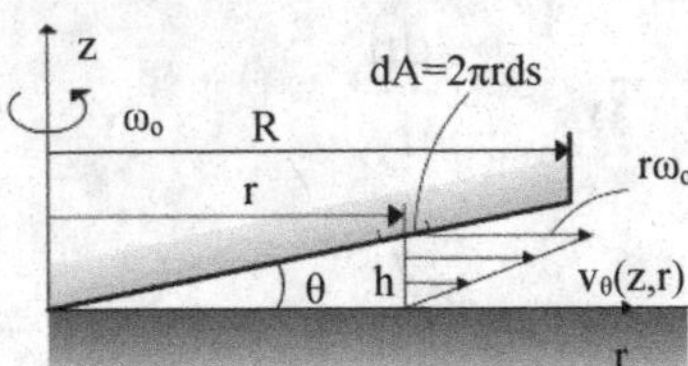

- From the graph, the linear circumferential velocity v_θ can be deduced as:
- $v_\theta = (r\omega_o)\dfrac{z}{h} = \dfrac{r\omega_o}{r\tan\theta}z$
- $dT = rdF$; $dF = \tau_w dA$;

$$\tau_w = \mu \left.\frac{dv_\theta}{dz}\right|_{z=h}$$

- $\therefore dT = 2\pi\mu \dfrac{\omega_o}{\tan\theta} r^2 dr$
- or
- $T = K\displaystyle\int_0^R r^2 dr = \dfrac{K}{3}R^3$

- Finally, with $K \equiv 2\pi\mu\omega_0 / \tan\theta$, the test-fluid viscosity is:
- $\mu = \dfrac{\frac{3}{2} T \tan\theta}{\pi\omega_0 R^3}$

Graph:

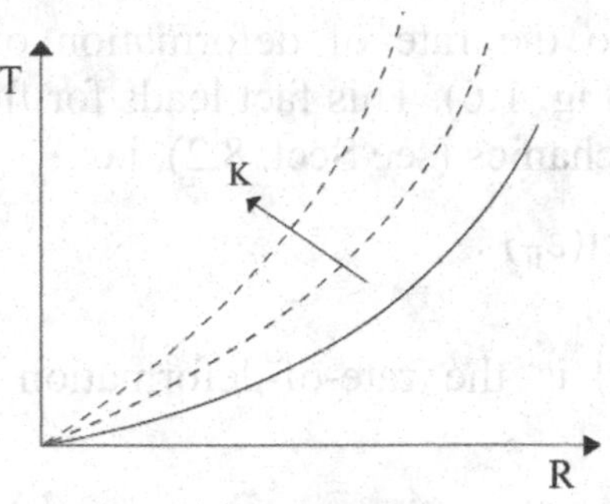

Comments:

- Because of the stated assumptions, $v_\theta = v_\theta(z)$ only and $T = T(R, \mu, \omega_o, \theta)$.
- As expected, the device size, in terms of R, has the strongest influence on T.

Stress Tensors and Stress Vectors In the equation of motion, as based on a macroscopic linear momentum balance (see Sect. 2.4), the Cauchy or *total stress tensor* $\vec{\vec{T}}$ is an unknown. That constitutes a closure problem, i.e., $\vec{\vec{T}}$ has to be related to the principal variable $\vec{\mathbf{v}}$ or its derivatives. In expanded form:

$$\vec{\vec{T}} = -p\vec{\vec{I}} + \vec{\vec{\tau}} \tag{1.18a}$$

where p is the *thermodynamic pressure* (or the hydro-static pressure for $\Delta\vec{\mathbf{v}} = 0$), $\vec{\vec{I}}$ is the necessary unit tensor for homogeneity, and $\vec{\vec{\tau}}$ is the stress tensor. For any coordinate system, the stress vector $\vec{\tau}$ relates to the *symmetric* second-order tensor $\vec{\vec{T}}$ as:

$$\vec{\boldsymbol{\tau}} = \hat{\mathbf{n}} \cdot \vec{\vec{T}} = \vec{\vec{T}} \cdot \hat{\mathbf{n}} \tag{1.19}$$

where $\hat{\mathbf{n}}$ is the normal (unit) vector. Without tensor symmetry, i.e., $T_{ij} \neq T_{ji}$, an infinitesimal fluid element ($\Delta\forall \to 0$) would spin as $|\overline{\omega} \to \infty|$. It is more insightful to write Eq. (1.18a) in tensor notation so that the total stress tensor reads:

$$T_{ij} = -p\delta_{ij} + \tau_{ij} \tag{1.18b}$$

where $-p\delta_{ij}$ is interpreted as the isotropic part (e.g., fluid statics and inviscid flow) and τ_{ij} is the deviatoric part for which a constitutive equation has to be found. Physically, $\tau_{ij} \hat{=} \tau_{(i=surface-normal)(j=stress-direction)}$ represents a force field per unit area as a result of the resistance to the rate of deformation of fluid elements, i.e., internal friction (Fig. 1.6). This fact leads for fluids to the postulate related to solid mechanics (see Sect. 8.2), i.e.,

$$\tau_{ij} = \text{fct}(\varepsilon_{ij}) \tag{1.20}$$

where $\varepsilon_{ij} = \frac{1}{2}\left(v_{i,j} + v_{j,i}\right)$ is the rate-of-deformation tensor (see Sect. 8.2).

The relation between τ_{ij} and ε_{ij} (plus vorticity tensor ζ_{ij}) can be more formally derived, starting with a fluid element displacement from point P (with $\vec{\mathbf{v}}$ at t) to point P′ ($\vec{\mathbf{v}} + d\vec{\mathbf{v}}$ at $t + dt$) a distance ds apart. Expanding the total derivative in Cartesian coordinates:

$$d\vec{\mathbf{v}} = \begin{bmatrix} \frac{\partial u}{\partial x} & \frac{\partial u}{\partial y} & \frac{\partial u}{\partial z} \\ \frac{\partial v}{\partial x} & \frac{\partial v}{\partial y} & \frac{\partial v}{\partial z} \\ \frac{\partial w}{\partial x} & \frac{\partial w}{\partial y} & \frac{\partial w}{\partial z} \end{bmatrix} d\vec{\mathbf{s}}_i = \nabla\vec{\mathbf{v}}d\vec{\mathbf{s}} \tag{1.21}$$

The spatial changes, or deformations, the fluid element is experiencing can be expressed as the "rate-of-deformation" tensor, i.e.,

$$\frac{d\vec{v}}{d\vec{s}} = \nabla\vec{\mathbf{v}} := \frac{\partial v_i}{\partial x_j} \tag{1.22a}$$

Equation (1.22) can be decomposed into the strain-rate tensor ε_{ij} <symmetrical part> and the vorticity (or rotation tensor) ζ_{ij}:

$$\frac{\partial v_i}{\partial x_j} = \varepsilon_{ij} + \zeta_{ij} \tag{1.22b}$$

It can be readily shown that:

- $\zeta_{yx} = -\zeta_{xy} = \omega_z, \zeta_{xz} = -\zeta_{zx} = \omega_y,$ and $\zeta_{zy} = -\zeta_{yz} = \omega_x$, where $\omega_z = \frac{1}{2}\left(\frac{\partial v}{\partial x} - \frac{\partial u}{\partial y}\right)$, etc.; thus, the tensor ζ_{ij} collapses to:
  $2\vec{\omega} = \nabla \times \vec{v} = \zeta$, the *vorticity vector*
- $\varepsilon_{ii} := \dot{\gamma}_{ii} \equiv \partial v_i / \partial x_i$ indicates volume change (dilation)
- $\varepsilon_{ij} := \frac{1}{2}\dot{\gamma}_{ij}, i \neq j$, represents element distortion
- The shear-rate tensor $\dot{\gamma}_{ij} \equiv \frac{\partial v_i}{\partial x_j} + \frac{\partial v_j}{\partial x_i}$ and hence the $\frac{1}{2}$ in $\varepsilon_{ij} = \frac{1}{2}\dot{\gamma}_{ij}$ is mathematically necessary in order to match Eq. (1.22b).

As mentioned, Stokes suggested that $\vec{\vec{\tau}}$ is a *linear* function of $\vec{\vec{\varepsilon}}$, which is not the case for non-Newtonian fluids, rarefied gases, and some fluid flows in microscale devices, e.g., bioMEMS. Specifically, for Newtonian fluids, i.e., air, water and most oils:

$$\vec{\vec{\tau}} = \lambda(\nabla \cdot \vec{v})\vec{\vec{I}} + 2\mu\vec{\vec{\varepsilon}} \tag{1.23}$$

where the viscosity coefficients λ and μ depend only on the thermodynamic state of the fluid. For incompressible flow $\nabla \cdot \vec{v} = 0$ (see Sect. 2.3) and the total stress tensor reduces to:

$$T_{ij} = -p\delta_{ij} + 2\mu\varepsilon_{ij} \tag{1.24}$$

where

$$2\mu\varepsilon_{ij} \equiv \tau_{ij} = \mu\left(\frac{\partial v_i}{\partial x_j} + \frac{\partial v_j}{\partial x_i}\right) := \mu\dot{\gamma}_{ij} \quad (1.25)$$

Here, $\dot{\gamma}_{ij} \equiv 2\varepsilon_{ij}$ is called the shear-rate tensor (see App. A, for all stress and shear-rate components in rectangular and cylindrical coordinates).

Of great importance is the wall shear stress vector ($\vec{\boldsymbol{\tau}}_{wall} \equiv$ WSS) as a result of frictional (or viscous) effects and the *no-slip condition* for macro-scale systems, i.e., at any solid surface:

$$\vec{\mathbf{v}}_{fluid} = \vec{\mathbf{v}}_{wall} \quad (1.26)$$

Typically, $\vec{\mathbf{v}}_{wall} = 0$, i.e., the wall is stationary and impermeable. The experimentally verified no-slip condition for macro-flow (see Sect. 7.3 for deviations) generates velocity gradients normal to the wall at all axial flow speeds. As illustrated in Fig. 1.4,

$$\tau_{wall} \equiv \text{WSS} \sim \begin{cases} \partial u / \partial n & \text{(due to the no-slip condition)} \\ \mu & \text{(due to viscous fluid effects)} \\ F_{tang} / A & \text{(tangential force per unit surface area)} \end{cases}$$

Very high or low WSS-values have been related to device malfunctions and arterial diseases (Kleinstreuer 2006). Integration of $\vec{\tau}_{wall}$ over the entire surface of a submerged body or inside a conduit yields the frictional drag:

$$F_{viscous} = \int_A \tau_{wall}\, dA \quad (1.27)$$

Viscous drag (frictional effect) plus form drag (pressure effect) make up the total drag:

$$F_{drag} = \int_A (\tau_w + p)\, dA \quad (1.28)$$

which, for most cases, would require elaborate CFD (computational fluid dynamics) analysis to evaluate the WSS and pressure distributions on the submerged body surface and then integrate (see Chap. 10).

Flux Vectors "Flux" implies the transport of a quantity "per unit area and time". For example, in heat transfer, energy moves, i.e., heat conducts, by molecular interaction from a point of high temperature to one at lower temperature in a fluid or solid. A simple steady 1-D experiment of a homogeneous plate (of area A and thickness L) with one side at T_1 and the other one at T_2 (where $T_1 > T_2$) would show that the heat flow rate $Q \sim A (T_1 - T_2)/L$. Thus, with Q/A being the conductive heat flux in the x-direction, q_x [kJ/(s m^2)], and recognizing that in the limit $\Delta T/L$ is the temperature gradient in the x-direction, we have $q_x \sim dT/dx$ or:

$$q_x = -k dT/dx \tag{1.29}$$

where k is the thermal conductivity which may vary with temperature (see App. B) and direction, i.e., $k_x \neq k_y \neq k_z$. In general, $k_{metal} > k_{liquid} > k_{gas}$ largely due to differences in intermolecular spacing for the three states. In 3-D, the *law of heat conduction* after Fourier reads:

$$\mathbf{q} = -k \nabla T \tag{1.30}$$

Clearly, the negative sign assures that in light of the negative temperature gradient, the heat flows in the positive direction, i.e., of decreasing temperature (see 2nd law of thermodynamics).

It should be recalled that $k/(\rho c_p)$ is the *thermal diffusivity* α, which has the same dimensions [L^2/T] as the kinematic viscosity (or momentum diffusivity) and the (binary) mass diffusivity D_{AB}. The ratio of $\frac{\nu}{\alpha}$ = Pr, is the *Prandtl number*. So it is not surprising that for ordinary diffusion, i.e., one species A (a dye) in "a solvent" B (say, water), a somewhat similar transfer process takes place, driven by the concentration gradient of species c:

$$\mathbf{j}_c = -\mathcal{D}_{AB} \nabla c \tag{1.31a}$$

where $\mathbf{j}_c$ [kg/(s m^2)] is the mass flux vector of species c and $\mathcal{D}_{AB}$ is the binary diffusion coefficient, i.e., for dilute suspensions of nanoparticles (see Sects. 6.4 and 7.5) $\mathcal{D}_{AB} = \mathcal{D}$ can be evaluated from the Stokes–Einstein equation as:

$$\mathcal{D} = \frac{\kappa_B T}{3\pi\mu d_p} \tag{1.31b}$$

where κ_B is the Boltzmann constant, T is the temperature, and d_p is the particle diameter.

Example 1.5: Consider heat transfer through a tube wall (R_1, T_1 and R_2, T_2) with length L and of conductivity k, where $T_1 > T_2$. Derive the wall temperature profile T(r) based on a heat flux balance and find an expression for the heat transfer rate $\dot{Q}$.

Sketch	*Assumptions*	*Concepts*
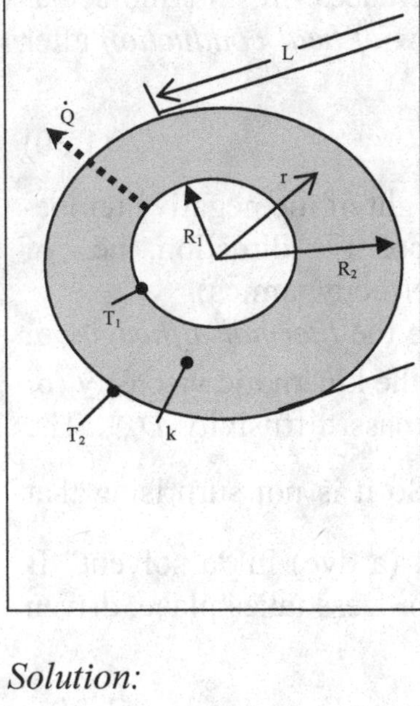 L, $\dot{Q}$, r, R_1, R_2, T_1, T_2, k	• Steady 1-D (radial) heat conduction • Very long tube, • i.e. T=T(r) only • Constant properties • No internal heat generation	• Steady 1-D heat balance (as shown) • OR • Reduced heat transfer equation (Sect. 2.5) • Fourier's Law $q = -k\frac{dT}{dr}$ • Heat flow rate: $\dot{Q} = qA$

Solution:

- In cylindrical coordinates: $q_r = -k\frac{dT}{dr}$
- Heat flow balance for a thin shell (i.e., the wall) $R_1 \le r \le R_2$ is: $\left.qA\right|_r - \left.(qA)\right|_{r+dr} = 0$
- Expanded (see truncated Taylor series in App. A):

$$-\frac{1}{r}\frac{d}{dr}(rq_r) = 0 \text{ or } \frac{1}{r}\frac{d}{dr}(kr\frac{dT}{dr}) = 0$$

Thus, solve

$$\frac{d}{dr}(r\frac{dT}{dr}) = 0 \text{ subject to } T = T_i \text{ at } r = R_i\text{; } i\text{=1,2}$$

The general solution is $T(r) = C_1 \ln r + C_2$, so that with the B.C.'s invoked,

$$T(r) = \frac{T_1 - T_2}{\ln(\frac{R_1}{R_2})}\ln(\frac{r}{R_2}) + T_2$$

Recalling that $\dot{Q} = qA = -kA\frac{dT}{dr} = \mathbb{C}$, with $A_i = 2R_i\pi L$ we have

$$\dot{Q} = \frac{2\pi kL}{\ln(\frac{R_2}{R_1})}(T_1 - T_2)$$

Graph:

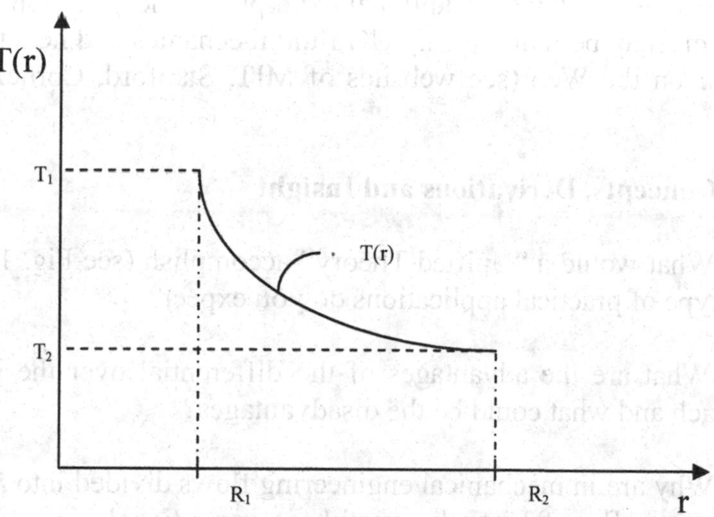

Comments:

The solution represents heat loss through a pipe wall of thickness $R_2 - R_1$ without internal/external convection.

1.5 Homework Assignments

Solutions to homework problems done individually or in, say, three-person groups should help to further illustrate fluid dynamics concepts, and in conjunction with App. A, sharpen the readers' math skills. Note, there is no substantial correlation between good HSA results and fine test performances, just *vice versa*. Table 1.1 summarizes three suggestions for students to achieve a good grade in fluid dynamics – for that matter in any engineering subject. The key word is "independence", i.e., equipped with an equation sheet (see App. A), the student should be able: (i) to satisfactory answer all concept questions and (ii) to solve correctly all basic fluid dynamics problems.

The "Insight" questions emerged directly out of the Chap. 1 text, while the "Problems" were taken from lecture notes in modified form when using White (2006), Cimbala & Cengel (2008), and Incropera et al. (2007). Additional examples, concept questions and problems may be found in any UG fluid mechanics and heat transfer text, or on the Web (see websites of MIT, Stanford, Cornell, UM, etc.).

1.5.1 Concepts, Derivations and Insight

1.5.1 What would a "Unified Theory" accomplish (see Fig. 1.1) and what type of practical applications do you expect?

1.5.2 What are the advantages of the differential over the integral approach and what could be the disadvantages?

1.5.3 Why are in mechanical engineering flows divided into *internal* and *external* flows? List other useful categorizations!

1.5.4 Streamlines: (a) Derive Eq. (1.3) and provide an example; (b) draw streamlines in a channel partially occluded by a block; (c) draw streamlines behind a sphere in uniform flow at low and high Reynolds numbers.

1.5.5 Considering Eq. (1.4), how does τ_{normal} differ physically from *p*; give an illustrative example.

1.5.6 Derive Eq. (1.8) for a prismatic fluid element and show that in any static fluid container p = p(h) only.

1.5.7 What is the usefulness of dimensionless groups? Provide three applications.

1.5.8 Describe the math conditions for: (a) the continuum hypothesis; and (b) thermodynamic equilibrium.

1.5.9 In Sect. 1.3 it is claimed that "flow problem identification" (FPI) is an important and challenging task. Provide FPIs for: (a) loud noise in an industrial pipe network; (b) scatter-marks on a machined part (e.g., a cylinder shaped with a lathe); (c) a person's left arm turning cold and losing any feeling.

1.5.10 Where and why are inviscid flow calculations still being carried out?

1.5.11 Provide two examples where, even by engineers, the Lagrangian modeling approach is preferred over the Eulerian approach.

1.5.12 Derive the material derivative D/Dt from a geometric viewpoint, and explain with an illustration $\vec{a}_{local}$ vs. $\vec{a}_{conv.}$.

1.5.13 Why was "enthalpy" dh = du + d(pv) introduced in thermodynamics and why is h = h(T) only for an ideal gas?

1.5.14 Equation (1.17) is sometimes written with a "minus" sign; on what physical grounds?

1.5.15 In Eqs. (1.18a, b), why exactly is the unit tensor necessary and what happens when $\Delta\vec{v} = 0$?

1.5.16 Consider steady laminar unidirectional flow in a pipe of radius R and length L with maximum centerline velocity u_{max}, i.e.,

$$u(r) = u_{max}[1 - (\frac{r}{R})^n]$$

(a) What is u_{max} dependent upon?
(b) Draw velocity profiles for n = 0.5, 1.0 and 2.0 and comment!
(c) Develop an equation for τ_{wall} and compute the drag force exerted by the fluid onto the pipe wall. Why is F_D independent of R?

1.5.17 Categorize the flow described by

$$\vec{v} = (u_0 + bx)i - byj$$

in terms of time-dependence, compressibility, dimensionality, and fluid-element spin.

1.5.18 Explain the rational for Eq. (1.22b), where mathematics merges into physics: (a) prove that $2\vec{\omega} = \nabla \times \vec{v} \equiv \vec{\zeta}$; and (b) compare two circular flows, i.e., $v_\theta = \omega r$ and $v_\theta = \frac{C}{r}$ $(r \neq 0)$, compute the vorticity fields and sketch them.

1.5.19 Having the strain-rate tensor already, why was the shear-rate tensor introduced, and how would you (alternatively to the book) derive the total stress tensor $T_{ij} = -p\delta_{ij} + \mu\dot{\gamma}_{ij}$ for incompressible flow of Newtonian fluids.

1.5.20 Heat flux and mass flux are standard flux vector examples. What makes them "vectors" and what is the (associated) momentum flux?

1.5.2 Problems

1.5.21 A car (with a door 1.2 m × 1 m wide) plunges into a lake, i.e., 8 m deep to the top of the door: (a) Find the hydrostatic force on the door and the point-force location; (b) can a strong driver who generates $1\text{kN}\cdot\text{m}$ torque (or moment) open the door under water?

1.5.22 An inverted cone (D =12 cm, d = 4 cm, L =12 cm) rotates at $\omega_0 = 200$ rad/s in a tight housing with all around clearance of h =1.2 mm filled with oil, where $\mu_1(20°\text{C}) = 0.1\text{N}\cdot\text{s}/\text{m}^2$ and $\mu_2(80°\text{C}) = 0.0078$ Pa s. Assuming linear velocity profiles, find the total power requirement $\text{P}_{\text{total}} = \text{P}_{\text{top}} + \text{P}_{\text{bottom}} + \text{P}_{\text{side}}$ (where $\text{dP} = \omega\text{dT}$ and $\text{dT} = \text{rdF}$) for the two viscosities and comment.

1.5.23 Consider the velocity field described by $\vec{v} = (0.5 + 0.8x)\hat{i} + (1.5 - 0.8y)\hat{j}$, where $\hat{i}$ and $\hat{j}$ are unit vectors in the x and y-direction: (a) Classify the velocity flow field; (b) find the coordinates of the stagnation point; (c) calculate the material accelerations at point x = 2 m and y = 3 m; (d) draw some streamlines and fluid acceleration vectors in the domain, say, $-3 \le x \le 3$ m and $-1 \le y \le 6$ m.

1.5.24 Consider simple shear flow, such as the Couette profile $u(y) = u_0 y/h$, where h is the parallel-plate spacing. Calculate the vorticity component in the z-direction, i.e., ζ_z, and determine the direction of rotating fluid particles, if any.

1.5.25 Compute the temperature in a very thin silicon chip (which receives $q_{chip} = 10^4 W/m^2$ and where the allowable $T_{max} = 85°\text{C}$) for the following system: The chip sits via an epoxy joint (thermal resistance $R_{Ej} \approx 10^{-4} m^2 \cdot K/W$) on an aluminum block (8 mm high, $k = 240 \text{W}/\text{m}\cdot\text{K}$) and both sides (i.e., top of chip + bottom of block) are exposed to moving air ($T_\infty = 25°C$, $h \approx 100$ W/m^2 K).

Note: This problem is adaped from Incropera et al. (2008)

Recall: The 1-D heat flux is $q_x = -kdT/dx$ and hence the heat flow rate for linear conduction $Q = qA = \frac{kA}{L}(T_1 - T_2)$, where L is the wall thickness, A is the surface area and $T_{1,2}$ are the surface temperatures. Now, with the thermal resistance $R_{th} = \frac{L}{kA}$, $Q = \frac{\Delta T}{R_{th}}$ and for several resistances in series:

$$Q = \frac{\Delta T}{\sum R_{th}}$$

where $\Delta T = T_{\infty,1} - T_{\infty,n}$ and $\sum R_{th} = \frac{1}{Ah_1} + \sum_1^n (\frac{L}{kA})_i + \frac{1}{Ah_n}$.

Chapter 2

Fundamental Equations and Solutions

2.1 Introduction

Every other day one may observc puzzling fluid mechanics phenomena. Such counter-intuitive examples include:

(a) Keeping the tailgate of a pick-up truck *up* reduces aerodynamic drag (why?) and hence saves gasoline; although, most drivers intentionally keep it *down* and even install "airflow" nets to retain cargo when accelerating.
(b) Under otherwise identical conditions, it is easy to *blow* out a candle but nearly impossible to *suck* it out. Why?
(c) Very high (horizontal) winds can *lift* pitched roofs *off* houses. How?
(d) When bringing a spoon near a jet, e.g., faucet stream, it gets *sucked into* the stream. Try it out and explain!
(e) Chunks of metal are *torn out* from ship propellers at high speeds after a long period of time in operation. Why?
(f) The long hair of a girl driving a convertible is being pushed *into her face* rather than *swept back*. How come?
(g) A snowstorm leaves a cavity *in front* of a pole or tree and deposits snow *behind* the "vertical cylinder." Impossible?
(h) Three-dimensional effects in river bends create unusual (axial) velocity profiles right after the bend and subsequently, lateral material transport results in shifting riverbeds. Explain!

C. Kleinstreuer, *Modern Fluid Dynamics: Basic Theory and Selected Applications in Macro- and Micro-Fluidics*, Fluid Mechanics and Its Applications 87,
DOI 10.1007/978-1-4020-8670-0_2, © Springer Science+Business Media B.V. 2010

(i) Certain non-Newtonian fluids when stirred in an open container *climb up* the rotating rod, rather than forming a depressed, parabolic free surface. Weird!
(j) Airplanes flying through microbursts (or high up in the blue sky) may crash. What is happening during these two very different weather types?
(k) Racecar (and motorcycle) tires are hardly threaded but passenger cars are. Why?
(l) Consider a tsunami (Japanese for "great harbor wave") hitting either a very shallow shore or a deep sea near the shoreline. Describe cause-and-effect for these two scenarios.
(m) Wildfires spread rapidly because of their own local weather pattern they create. Describe the underlying convection system, and how "back-fires" work.
(n) A very small amount of carbon nanotubes added to a liquid increases measurably the apparent (or effective) thermal conductivity, k, of the dilute mixture (called a nanofluid) when compared to k [W/(m K)] of the pure base fluid. Why?
(o) Gas flow in *microchannels* may exhibit significantly higher flow rates than predicted by conventional theory. What's happening?

What are the underlying physical explanations and mathematical descriptions of these and much more ordinary phenomena of fluid flow and fluid-particle dynamics? Some of these questions (a)–(o) can be quickly answered by visualizing the unique fluid flow pattern via streamline drawings, assuming steady laminar flow, and applying basic definitions or Bernoulli's equation. Others require some background reading and sharp thinking. In any case, the answers rely on an equal dose of physics, i.e., insight, and applied mathematics, i.e., modeling.

The *objective* of the next sections plus Chap. 3 is to provide physical insight, mathematical modeling tools and application skills to solve basic fluid-flow problems. This is accomplished, first in form of derivations of the mass, momentum and energy conservation laws and then via special case studies, employing simplified forms of the conservation equations.

Chapter Overview *Derivations* of the fluid dynamics equations (Sects. 2.3–2.5) are very important because they provide a deeper understanding of the physics, mathematically represented by each term in the final equations, and a sense of the underlying assumptions, i.e., the limitations of a particular mathematical model. Of course, derivations are regarded by most as boring and mathematically quite taxing; however, for those, it's time to become a convert for the two beneficial reasons stated. One should not forget the power of *dimensional analysis* (DA) which requires only simple algebra when nondimensionalizing governing equations and hence generating dimensionless groups. Alternatively, *scale analysis* (SA) is a nifty way of deriving dimensionless groups as demonstrated with Example 1.1 in Chap. 1 and Example 2.11. Both DA and SA are standard laboratory/computational tools for estimating dominant transport phenomena, graphing results, to evaluate engineering systems, and to test kinematic/dynamic similarities between a physical model and the actual prototype.

Outside the cutting-edge research environment, fluid mechanics problems are solved as special cases, i.e., the conservation equations are greatly reduced based on justifiable assumption on a case-by-case basis (see Sect. 1.3). As part of Sect. 2.6, the simplest case is *fluid statics* where the fluid mass forms a "whole body," either stationary or moving without any *relative* velocities. The popular (because very simple) Bernoulli equation, for frictionless fluid flow along a representative streamline, balances kinetic energy $(\sim \rho \bar{v}^2)$, flow work $(\sim \Delta p)$, and potential energy $(\sim \rho g z)$ and hence in some cases provides useful pressure-velocity-elevation correlations. In order to make Sects. 2.3 to 2.5 amiable, the featured problems have analytic solutions because they are basically one-dimensional. Thus, without being mathematically challenging, the solved example problems are insightful demonstrations of the conservation principles with direct applications in engineering mechanics.

In summary, Chaps. 2 and 3 problem solutions as well as the material of Parts B and C should broaden the student's knowledge base and provide a higher skill level, already necessary at the undergraduate-level to cope with today's engineering problems encountered in industry or graduate school.

Approach to Problem Solving As discussed in Chap. 1, for setting up and solving fluid mechanics problems, we follow the three-step approach (see Fig. 2.1):

(i) Classification of the fluid-flow system
(ii) Mathematical description of the system
(iii) Solution of the modeling equations and result graphing plus comments

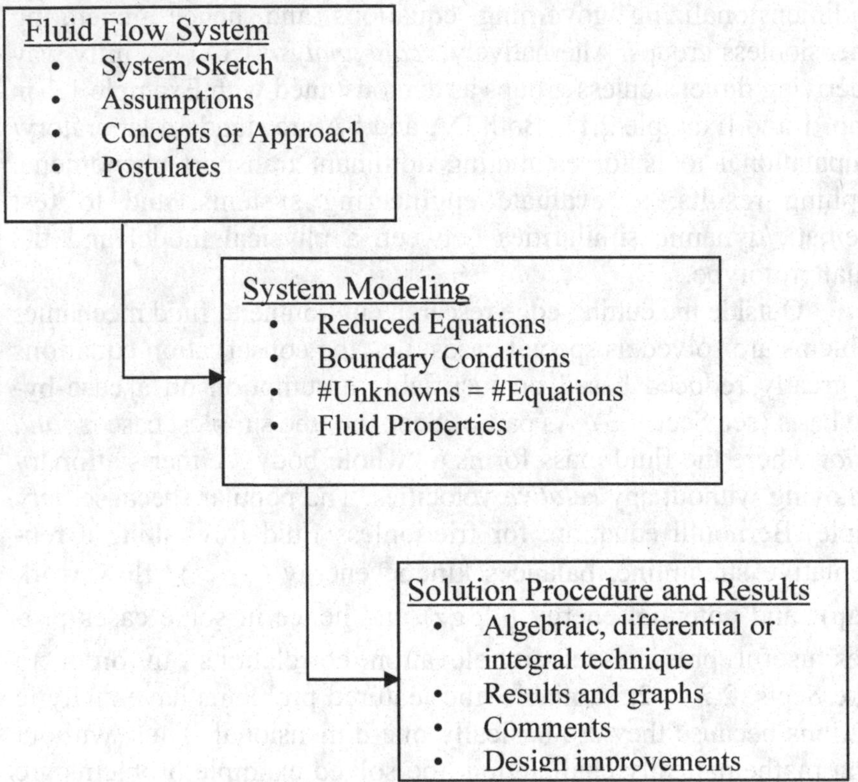

Fig. 2.1 Sequential steps in problem solving

While Chap. 1 dealt with basic concepts of *fluid-flow systems*, this section provides the fluid dynamics equations for the second step, *system modeling*, and brief applications of standard *solution techniques* (see App. A).

The conservation equations for mass, momentum and energy transfer are first repeated in *integral* form in terms of the Reynolds

Transport Theorem (RTT), linking the Lagrangian closed system with the Eulerian control volume (i.e., open system). Then, via a straight integral transformation, using the Divergence Theorem, the fundamental transport equations in *differential* form are obtained. In order to provide additional physical insight, a micro-scale derivation approach is illustrated, i.e., balancing mass, momentum and energy for a representative elemental volume $\Delta\forall$ (open system).

Fundamental Assumptions While every solution approach requires a list of system-specific assumptions, the fundamental ones of classical physics apply to all the basic equations derived (see also Chap. 1).

A) Classical vs. Modern Physics

- The gravity vector $\vec{g} = \vec{g}(\vec{x})$ only, i.e., no space-time curvature effects
- $v_{fluid} << c_{light}$, i.e., no time dilation
- No mass↔energy conversions
- No uncertainty in measuring particle (or fluid element) position and velocity exactly at the same time
- No probabilistic (or random) effects, even in "turbulence" analysis

(B) Definitions and Approaches

- A fluid is an aggregation of moving (gas or liquid) molecules, which cannot resist any shear stress, i.e., it flows.
- The fluid is treated as a continuum; for example, the density $\rho = \lim_{\Delta\forall \to \Delta\forall_{min}} \left(\frac{\Delta m}{\Delta\forall}\right)$, where $\Delta\forall_{min} \leq 10^{-9}\ \text{mm}^3$.
- Fluid flow analyses are done via:
  - Fluid element, i.e., identifiable mass, tracking of a "closed system" (Lagrange);

  or most frequently via:

  - Fluid mass, momentum and energy transfer through an "open system" (Euler), i.e., a control volume approach

Derivation Approaches As mentioned in Chap. 1, there are basically four ways of obtaining specific transport equations reflecting the conservation laws. The points of departure for each of the four methods are either given (e.g., the Boltzmann equation or Newton's second law) or derived based on differential balances for a representative elemental volume (REV).

(i) *Molecular Dynamics Approach*: Fluid properties and transport equations can be obtained from kinetic theory and the Boltzmann equation, respectively, employing statistical means. Alternatively, $\sum \vec{\mathbf{F}} = m\vec{\mathbf{a}}$ is solved for each molecule using direct numerical integration.

(ii) *Integral Approach*: Starting with the Reynolds Transport Theorem (RTT) for a fixed open control volume (Euler), specific transport equations in integral form can be obtained (see Sect. 2.2).

(iii) *Differential Approach*: Starting with 1-D balances over an REV and then expanding them to 3-D, the mass, momentum and energy transfer equations in differential form can be formulated. Alternatively, the RTT is transformed via the divergence theorem, where in the limit the field equations in differential form are obtained (see Sects. 2.3–2.5).

(iv) *Phenomenological Approach:* Starting with balance equations for an open system, transport phenomena in complex transitional, turbulent or multiphase flows are derived largely based on empirical correlations and dimensional analysis considerations. A very practical example is the description of transport phenomena with fluid-compartment models. These "compartments" are either well-mixed, i.e., *transient lumped-parameter models* without any spatial resolution, or they are transient with a one-dimensional resolution in the axial direction (see Sect. 4.4).

Especially for the (preferred) differential approach (iii) the system-specific fluid-flow assumptions have to be carefully stated and justified. As mentioned in Sect. 1.3, we have to determine/assume most all the type of flow require, time-dependence and dimensionality of the given system or problem.

2.2 The Reynolds Transport Theorem

Consider ***B*** to be an arbitrary extensive quantity of a *closed system* (or a moving material volume). Such a system may include an ideal piston-cylinder device with enclosed (constant) gas mass, a rigid tank without any fluid leaks, or an identifiable pollutant cloud – all subject to forces and energy transfer. In any case, ***B*** could be the system's mass, momentum, or energy.

Task 1 is to express in the Lagrangian frame the fate of ***B*** in terms of the material derivative, D***B***/Dt, i.e., the total time-rate-of change of $\boldsymbol{B}_{\text{system}}$ (see Sect. 1.3 reviewing the two system approaches and App. A discussing the operator D/Dt). Clearly, for:

- $$\boldsymbol{B} \equiv m_{\text{system}} = \text{¢} \rightarrow \frac{Dm}{Dt} = 0 \qquad (2.1a)$$
  (conservation of mass)
- $$\boldsymbol{B} \equiv (\text{mv})_{\text{system}} \rightarrow \text{m}\frac{\text{D}\vec{\text{v}}}{\text{Dt}} = \text{m}\vec{\text{a}}_{\text{total}} = \Sigma\vec{\text{F}}_{\text{surface}} + \Sigma\vec{\text{F}}_{\text{body}} \qquad (2.1b)$$
  (conservation of momentum or Newton's second law)
- $$\boldsymbol{B} = \text{E}_{\text{system}} \rightarrow \frac{\text{DE}}{\text{Dt}} - \dot{\text{Q}} - \dot{\text{W}} \qquad (2.1c)$$
  (conservation of energy or first law of thermodynamics)

In *Task 2* the conservation laws, in terms of D***B***/Dt, are related to an *open system*, i.e., in the Eulerian frame. Here, for a fixed control volume (C.∀.) with material streams flowing across the control surface (C.S.), and possibly accumulating inside *C.∀.*, we observe with specific quantity $\beta \equiv \mathbf{B}/\text{m}$:

$$\left\{\begin{matrix}\textit{Total}\text{ time - rate - of}\\ \text{change of system}\\ \text{property }\boldsymbol{B}\end{matrix}\right\} = \left\{\begin{matrix}\textit{Local}\text{ time - rate - of}\\ \text{change of }\boldsymbol{B}/\forall \equiv \rho\beta\\ \text{within the C.}\forall.\end{matrix}\right\} + \left\{\begin{matrix}\textit{Net efflux}\text{ of }(\rho\beta),\\ \text{i.e., net material}\\ \text{convection across}\\ \text{control surface C.S.}\end{matrix}\right\}$$

or in mathematical shorthand:

$$\left.\frac{D\boldsymbol{B}}{Dt}\right|_{\text{closed system}} = \frac{\partial}{\partial t}\iiint_{C.\forall.} (\rho\beta)dV + \iint_{C.S.}(\rho\beta)\vec{\mathbf{v}}\cdot d\vec{\mathbf{A}} \qquad (2.2a)$$

Equation (2.2a), which is formally derived in any UG fluids text, is the Reynolds Transport Theorem (RTT) for a control volume at rest. Clearly, the specific quantity β can be expressed as:

$$\beta \mathrel{\hat{=}} \frac{\boldsymbol{B}}{m} := \begin{cases} 1 & \text{(mass per unit mass)} \\ \vec{\mathbf{v}} & \text{(momentum per unit mass)} \\ e & \text{(energy per unit mass)} \end{cases} \qquad (2.2b\text{–}d)$$

Extended Cases For a *moving* control volume the fluid velocity $\vec{\mathbf{v}}$ is replaced by $\vec{\mathbf{v}}_{relative} = \vec{\mathbf{v}}_{fluid} - \vec{\mathbf{v}}_{C.\forall.}$ (see Example 2.5). The operator $\partial/\partial t$, acting on the first term on the R.H.S., has to be replaced by d/dt when the control volume is *deformable*, i.e., the C.S. moves with time (see Example 2.3). For a *noninertial* coordinate system (see Fig. 2.2a), for example when tracking an accelerating rocket, $\sum \vec{\mathbf{F}}_{external}$ of Eq. (2.1b) is expressed as:

$$m\vec{\mathbf{a}}_{abs} = m\left(\frac{d\vec{\mathbf{v}}}{dt} + \vec{\mathbf{a}}_{rel}\right) \qquad (2.2e)$$

where $m\vec{\mathbf{a}}_{rel}$ accounts for noninertial effects (e.g., arbitrary C.∀. acceleration as in Example 2.7):

$$\vec{\mathbf{a}}_{rel} = \frac{d^2\vec{\mathbf{R}}}{dt^2} \qquad (2.2f)$$

In case of C.∀. rotation,

$$\vec{\mathbf{a}}_{rel} = \frac{d\vec{\boldsymbol{\omega}}}{dt} \times \vec{\mathbf{r}} \qquad (2.2g)$$

Specifically, for a *rotating* material volume, the fluid angular momentum per unit volume $\beta \equiv \rho(\vec{\mathbf{r}} \times \vec{\mathbf{v}})$ has to be considered. The law of *conservation of angular momentum* states that the rate-of-change of angular momentum of a material volume is equal to the resultant moment on the volume (see any UG fluids text for details).

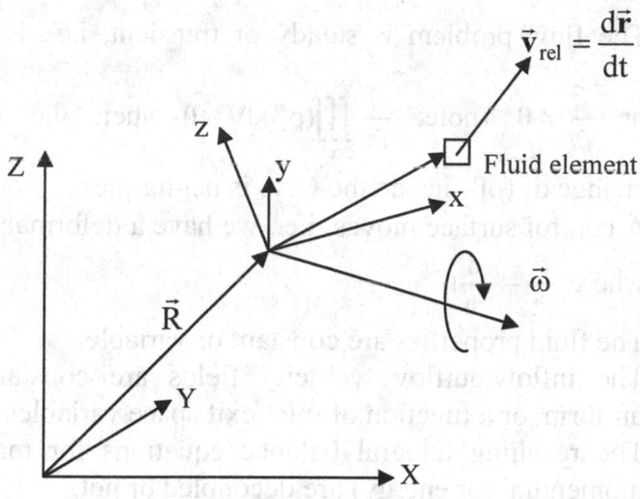

Fig. 2.2a Inertial coordinate system and moving/rotating frame

Setting-up the Reynolds Transport Theorem There are a few sequential steps necessary for tailoring the general RTT toward a specific flow system description and solving the resulting integral equations:

(i) Identify the extensive quantity $\boldsymbol{B}_{\text{systems}}$, e.g., the mass of the identifiable material, linear (or angular) momentum, or total energy. As a result, the specific property of the closed system $\beta = \boldsymbol{B} / m_{\text{system}}$, is known (see Eqs. (2.1 and 2.2)).

(ii) Determine $\left. D\boldsymbol{B}/Dt \right|_{\text{system}}$ for each conservation case, i.e., mass, momentum, or energy (see Eq. (2.1)).

(iii) Select a "smart" control volume and determine *if*:

- the control volume is fixed, or moving at $\vec{v}_{C.\forall.} = ¢$, or accelerating at $a_{rel} = ¢$, or accelerating and rotating (see "Extended cases");

- The flow problem is steady or transient, i.e., is $\frac{\partial}{\partial t} = 0$ or $\frac{\partial}{\partial t} \neq 0$ (note, $\frac{\partial}{\partial t} \iiint_{C.\forall.} (\rho\beta) d\forall \cong 0$ when the rate-of-change of $(\rho\beta)$ inside the C.∀. is negligible)
- A control surface moves, i.e., we have a deformable C.∀. where $\frac{\partial}{\partial t} \rightarrow \frac{d}{dt}$
- The fluid properties are constant or variable
- The inflow/outflow velocity fields are constant, i.e., uniform, or a function of inlet/exit space variables
- The resulting integral balance equations for mass and momentum (or energy) are decoupled or not.

(iv) Set up the momentum, i.e., force balance, equation for each coordinate direction.

(v) Solve the volume and/or surface integrals (use integration tables, if necessary)

(vi) Follow the inflow/outflow sign convention (see Fig. 2.2b), i.e., IN $\hat{=}$ "-" and OUT $\hat{=}$ "+"

(vii) Check the results for correctness, i.e., apply common sense

Inflow: $\vec{\mathbf{v}} \cdot d\vec{\mathbf{A}} = -v_n dA$

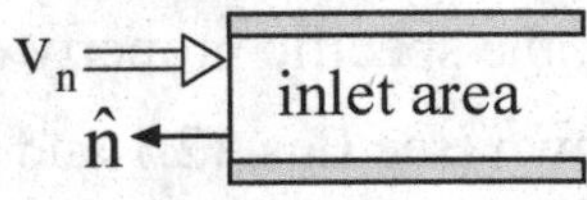

Outflow: $\vec{\mathbf{v}} \cdot d\vec{\mathbf{A}} = v_n dA$

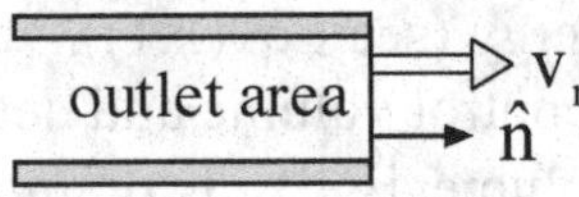

Fig. 2.2b Sign convention for the "net efflux" RTT term (Recall: $\cos 180° = -1$; and $\cos 0° = 1$)

2.3 Fluid-Mass Conservation

Conservation of mass is very intuitive and standard in daily life observations where a given mass of a fluid may change its thermodynamic state, i.e., liquid or gaseous, but it can neither be destroyed nor created. This, as the other two conservation laws, can be expressed in *integral form* for a control volume of any size and shape or derived in *differential form.*

2.3.1 Mass Conservation in Integral Form

In order to track within the Lagrangian frame an identifiable constant mass of fluid, we set (see Eq. (2.2b))

$$\boldsymbol{B}_{\text{system}} \equiv \text{m} \text{ and hence } \beta \equiv 1$$

The conversation principle requires that with $m = ¢$, $Dm/Dt = 0$ and hence Eq. (2.2a) reads:

$$0 = \frac{\partial}{\partial t} \iiint_{C.\forall.} \rho d\forall + \iint_{C.S.} \rho \vec{v} \cdot d\vec{A} \tag{2.3}$$

Thus, we just completed Steps (i) and (ii) of the "**setting-up-the-RTT**" procedure. Aspects of Step (iii) are best illustrated with a couple of examples.

Example 2.1: Volumetric Flow Rate

Consider a liquid-filled tank (depth H) with a horizontal slot outlet (height 2h and width w) where the locally varying outlet velocity can be expressed as (see Sketch):

$$u \approx \sqrt{2g\,(H - z)}$$

A constant fluid mass flow rate, $\dot{m}_{in}$, is added to maintain the liquid depth H. The z-coordinate indicates the location of the center of the outlet. Find Q_{outlet} as a function of H and h.

Sketch	*Assumptions*	*Concepts*
$\dot{m}_{in}$, $\vec{g}$, $H = ¢$, z, 2h, u(z)	• Steady (why?) incompressible flow • Outflow velocity u(z) $u(z) = \sqrt{2g(H-z)}$ base on Toricelli's law	• Mass RTT • Fixed, non-deforming C.∀. • Toricelli's law

Solution:

The u(z)-equation will be derived in Sect. 2.3.2. Here, we apply a reduced version of Eq. (2.3). Specifically, with $\frac{\partial}{\partial t}\iiint \rho d\forall = 0$ (steady-state because no system parameter changes with time), we have with ρ =¢ (incompressible fluid):

$$0 = \iint_{C.S.} \vec{v} \cdot d\vec{A} \qquad \text{(E.2.1.1)}$$

Graph:

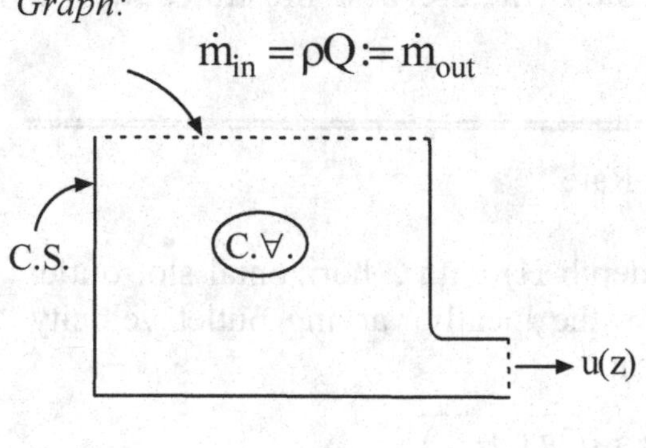

Fluid mass crosses the control surface at two locations (see graph). Recalling that "inflow" is negative and "outflow" positive (see Fig. 2.2b), Eq. (E.2.1.1) reads with $\frac{\dot{m}_{in}}{\rho} = Q$,

$$Q + \iint_{A_{slot}} \vec{v} \cdot d\vec{A} = 0$$

or

$$Q = \int_A v_n dA \qquad \text{(E.2.1.2)}$$

Here, $Q \equiv Q_{outlet}$, $v_n = u(z) = \sqrt{2g(H-z)}$ and dA = wdz so that

$$Q_{outlet} = w\sqrt{2g}\int_{-h}^{h}\sqrt{H-z}\,dz$$

which yields

$$Q_{outlet} = K[(H+h)^{3/2} - (H-h)^{3/2}] \qquad (E.2.1.3)$$

where $K \equiv \frac{2}{3}w\sqrt{2g}$.

For h << H, as it is often the case, Eq. (E.2.1.3) can be simplified to (see HW Assignment):

$$Q_{outlet} \approx 2\,wh\sqrt{2gH} \qquad (E.2.1.4)$$

Comment: Equation (E.2.1.4) is known as *Toricelli's law.*

What can be deduced from Example 2.1 is that for *incompressible* fluid flow through a conduit,

$$\iint_{C.S.} \vec{v}\cdot d\vec{A} = 0 \qquad (2.4)$$

or $-Q_{in} + \int_{A_{outlet}} \vec{v}\cdot d\vec{A} = 0$, i.e., in general:

$$Q = \int_A \vec{v}\cdot d\vec{A} = \int_A v_n dA = \bar{v}A \qquad (2.5)$$

where $\bar{v} = Q/A$ is the area-averaged velocity. We also recall that the mass flow rate at any point in a conduit is:

$$\dot{m} = \rho\bar{v}A = \rho Q = \text{constant} \qquad (2.6)$$

which holds for any fluid and is a key "internal flow" condition, reflecting conservation of mass. Clearly, for a C.∀. with multiple inlets and outlets:

$$\Sigma Q_{in} = \Sigma Q_{out} \qquad (2.7)$$

as illustrated in Example 2.2.

Example 2.2: Multi-port Flow Junction

Consider a feed pipe (A, v_1) bifurcating into two outlet pipes $(A_2, v_2 \text{ and } A_3, v_3)$ where a small hole (A_4) has been detected in the junction area. Develop an equation for the leak Q_4.

Sketch	*Assumptions*	*Approach*
① ④(leak) C.∀. ③ ②	• Steady incompressible flow • Fixed, non-deformable C.∀. • Constant velocities	• Reduced mass RTT, i.e., direct use of Eq. (2.7) • Sign convention (Fig. 2.2b)

Solution: The fact that the inlet/outlet velocities are all constant, simplifies Eqs. (2.4) to (2.7), i.e.,

$$-Q_1 - Q_2 + Q_3 + Q_4 = 0 \text{ where } Q_i = v_i A_i$$

Thus,

$$Q_{leak} \equiv Q_4 = v_1 A_1 + v_2 A_2 - v_3 A_3$$

The next example considers a *deforming* control volume inside a tank due to a single outflow in terms of a variable velocity. Thus, Eq. (2.3) has to be rewritten as:

$$0 = \frac{d}{dt} \iiint_{C.\forall.} \rho d\forall + \iint_{C.S.} \rho \vec{\mathbf{v}} \cdot d\vec{\mathbf{A}} \tag{2.8a}$$

For incompressible fluid flow, we have:

$$\left.\frac{d\forall}{dt}\right|_{C.\forall.} = -\iint_{C.S.} \vec{\mathbf{v}} \cdot d\vec{\mathbf{A}} \tag{2.8b}$$

Example 2.3: Draining of a Tank: A "deformable C.∀." because the fluid level decreases and hence we have a shrinking C.∀. with moving C.S

Consider a relatively small tank of diameter D and initially filled to height h_0. The fluid drains through a pipe of radius (r_0) according to:

$$u(r) = 2\bar{u}\,[1-(\frac{r}{r_0})^2] \qquad \text{(see Example 2.10)}$$

where $\bar{u} = \sqrt{2gh}$ (see Example 2.1), r_0 is the outlet pipe radius, and r is its variable radius, $0 \le r \le r_0$. The fluid depth was h_0 at time t = 0. Find h(t) for a limited observation time Δt.

Sketch	*Assumptions*	*Control Volume*
D, h_0, h(t), $v_{surface} = -\frac{dh}{dt}$, r_0, r, u(r)	• Transient incompressible flow • Stationary C.∀. (tank) but deforming liquid volume	v_s, C.S., C.∀., u(r)

Solution: Equation (2.8b) can be expanded for this problem with $\forall_{C.\forall.} = \frac{D^2\pi}{4}h$ to

$$(\frac{D^2\pi}{4})\frac{dh}{dt} = -\int_0^{r_0} u(r)dA \qquad \text{(E.2.3.1)}$$

where dA = 2πrdr *is the variable ring element* as part of the cross-sectional area of the outlet pipe. Thus,

$$\frac{dh}{dt} = -\frac{16\sqrt{2gh}}{D^2}\int_0^{r_0}[1-(\frac{r}{r_0})^2]\,rdr \qquad \text{(E.2.3.2a)}$$

or

$$\frac{dh}{dt} = -4\sqrt{2g}(\frac{r_0}{D})^2\sqrt{h} \qquad \text{(E.2.3.2b)}$$

subject to h (t = 0) = h_0. Separation of variables and integration yields:

$$\sqrt{h_0} - \sqrt{h} = \sqrt{8g}(\frac{r_0}{D})^2 t$$

or

$$\frac{h(t)}{h_0} = \left[1 - \sqrt{\frac{8g}{h_0}}(\frac{r_0}{D})^2 t\right]^2 \quad \text{(E.2.3.3)}$$

Graph:

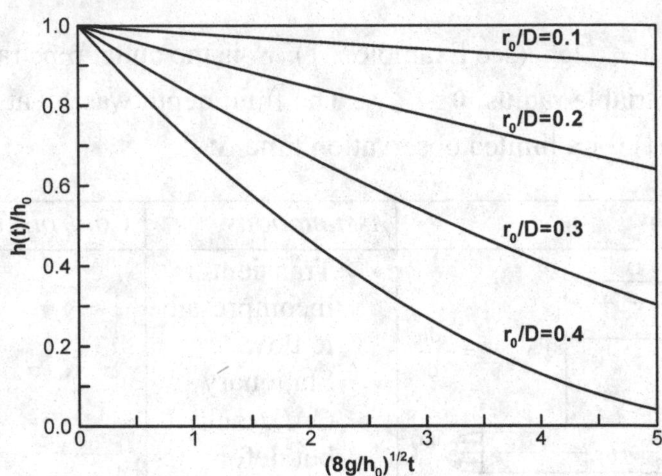

Comments:

- The standard assumption that h = const. <reservoir> is only approximately true when $r_0 / D < 0.1$
- A variable speed dh/dt, i.e., accelerated tank-draining, occurs when $r_0 / D > 0.2$

2.3.2 Mass Conservation in Differential Form

In general, the RTT is great for computing global (or integral) quantities, such as flow rates and mass fluxes or forces and energies without knowledge of the detailed fluid flow field *inside* the open system (i.e., the control volume). However, if it is necessary to find point-by-point density variations as well as velocity and pressure distributions in order to analyze fluid flow patterns, the conservation laws in *differential* rather than integral form have to be solved.

The conservation equations can be readily derived from the RTT, e.g., Eq. (2.2), by considering an infinitesimally small control volume $d\forall$ and then expressing each term in the form of a volume integral. For example, Eq. (2.3) contains a surface integral which has to be transformed into a volume intcgral, employing Gauss' divergence theorem (App. A):

$$\iint_S \vec{v} \cdot d\vec{S} \equiv \iiint_\forall \nabla \cdot \vec{v} d\forall \tag{2.9}$$

where $\vec{v}$ is a vector in surface S, ∇ is the del-operator (see App. A), and $(\nabla \cdot \vec{v})$ is the divergence of the vector field.

The Continuity Equation Using Eq. (2.9) to express the surface integral in Eq. (2.3) as a volume integral, Eq. (2.3) can be written as:

$$\frac{\partial}{\partial t}\iiint \rho d\forall + \iiint \nabla \cdot (\rho\vec{v})\, d\forall = 0$$

or, following Leibniz's rule (App. A) and switching the operation for the first term, we have

$$\iiint [\frac{\partial \rho}{\partial t} + \nabla \cdot (\rho\vec{v})] d\forall = 0$$

Clearly, either $d\forall = 0$ (not physical) or

$$\frac{\partial \rho}{\partial t} + \nabla \cdot (\rho\vec{v}) = 0 \tag{2.10}$$

Equation (2.10) is known as the *continuity equation*, stating fluid-mass conservation on a differential basis. Note the special cases:

- For steady flow $(\frac{\partial}{\partial t} \equiv 0)$: $\nabla \cdot (\rho\vec{v}) = 0$ (2.11)
- For incompressible fluids $(\rho = ¢)$: $\nabla \cdot \vec{v} = 0$ (2.12)

It should be noted that the widely applicable Eq. (2.12) holds for transient flow as well.

2.3.3 Continuity Derived from a Mass Balance

In order to gain more physical insight, Eq. (2.10) is now derived based on a 3-D mass balance. A fluid mass balance over an open system, say, a cube of volume $\Delta\forall = \Delta x \Delta y \Delta z$, yields:

$$\sum \dot{m}_{in} - \sum \dot{m}_{out} = \left.\frac{\partial m}{\partial t}\right|_{\Delta\forall} \tag{2.13}$$

from a global perspective. However, in 1-D on a differential basis (see Fig. 2.3)

$$\left[(\rho u)\big|_{x} - (\rho u)\big|_{x+\Delta x}\right]\Delta y \Delta z = \frac{\partial \rho}{\partial t}\Delta\forall$$

Now, with Taylor's truncated series expansion (see App. A)

$$f\big|_{x+\Delta x} = f\big|_{x} + \frac{\partial f}{\partial x}\Delta x + \cdots\cdots$$

we obtain

$$-\frac{\partial(\rho u)}{\partial x}\Delta x \Delta y \Delta z = \frac{\partial \rho}{\partial t}\Delta x \Delta y \Delta z$$

Adding the other two net fluxes (ρv) and (ρw) in the y- and z-direction, respectively, and dividing by the arbitrary volume, $\Delta\forall$, yields:

$$-\frac{\partial(\rho u)}{\partial x} - \frac{\partial(\rho v)}{\partial y} - \frac{\partial(\rho w)}{\partial z} = \frac{\partial \rho}{\partial t}$$

or (see Eq. (2.10)):

$$\frac{\partial \rho}{\partial t} + \nabla \cdot (\rho \vec{v}) = 0$$

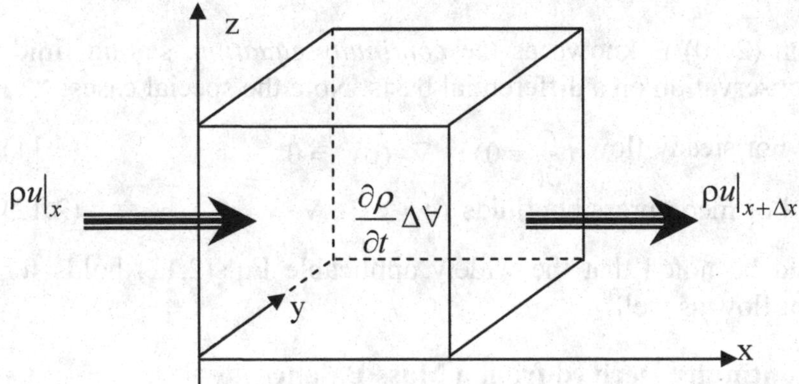

Fig. 2.3 One-dimensional fluid mass balance for a 3-D control volume

Example 2.4: Use of the Continuity Equation (Two Problems: A and B)

(A) For steady laminar fully-developed pipe flow of an incompressible fluid (see Sect. 2.4), the axial flow is:

$$v_z(r) = v_{max}[1-(\frac{r}{r_0})^2] \tag{E.2.4.1}$$

Show that the radial (or normal) velocity $v_r = 0$.

Sketch	*Assumptions*
$r = r_0$; r; $v_z(r)$; $r = 0$; z; v_{max} Note: $v_{max} = 2v_{average}$	• Steady implies $\frac{\partial}{\partial t} \equiv 0$ • Incompressible fluid: $\rho = ¢$ • Axisymmetric pipe: $\frac{\partial}{\partial \Theta} \equiv 0$ • Fully developed flow: $\frac{\partial}{\partial z} \equiv 0$

Solution: Based on the assumptions, Eq. (2.10) is appropriate and reads in cylindrical coordinates (App. A):

$$\frac{1}{r}\frac{\partial(rv_r)}{\partial r} + \frac{1}{r}\frac{\partial v_\Theta}{\partial \Theta} + \frac{\partial v_z}{\partial z} = 0 \tag{E.2.4.2}$$

Clearly, the given velocity profile $v_z(r)$ (see Eq. (E.2.4.1)) is not a function of z, i.e., $\partial v_z / \partial z = 0$ which implies with $\partial / \partial \Theta = 0$ (axisymmetry) that Eq. (E.2.4.2) reduces to:

$$\frac{\partial(rv_r)}{\partial r} = 0$$

Partial integration yields:

$$v_r = f(z)/r$$

where $0 \le r \le r_0$, i.e., r could be zero. That fact and the boundary condition $v_r(r = r_0) = 0$, i.e., no fluid penetrates the pipe wall, forces the physical solution $v_r \equiv 0$. Indeed, if $v_r \neq 0$, such a radial velocity component would alter the axial velocity profile to $v_z = v_z(r,z)$ which implies *developing* flow; that happens, for example, in the pipe's entrance region or due to a porous pipe wall through which fluid can escape or is being injected.

(B) Consider 2-D steady laminar symmetric flow in a smooth converging channel where axial velocity values were measured at five points (see Sketch). Estimate the fluid element acceleration a_x at point C as well as the normal velocity v at point B'. All distances are 2 cm and the centerline velocities are 5 m/s at A; 7 m/s at B; 10 m/s at C; and 12m/s at D.

Sketch	*Concept*
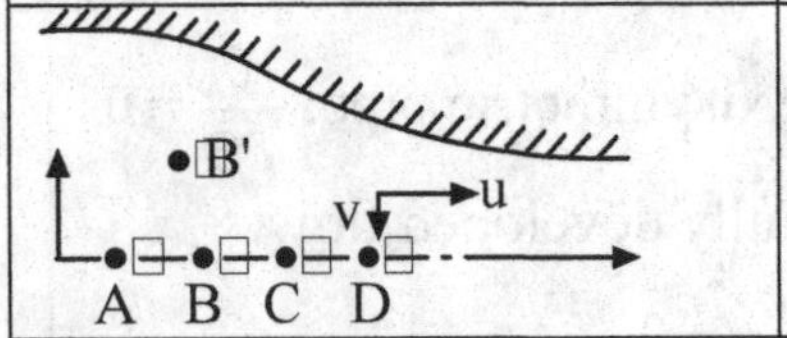 B' v u A B C D	• Approximate via finite differencing the reduced acceleration and continuity equations.

Solution: From Sect. 1.3 (Eq. 1.12) the axial acceleration can be written as:

$$a_x = \frac{\partial u}{\partial t} + u\frac{\partial u}{\partial x} + v\frac{\partial u}{\partial y} + w\frac{\partial u}{\partial z} \tag{E.2.4.3}$$

Based on the stated assumptions $(\frac{\partial}{\partial t} = 0, \frac{\partial}{\partial y} = 0, w = 0)$:

$$a_x = u\frac{\partial u}{\partial x} \approx u\frac{\Delta u}{\Delta x} = u_c \frac{u_D - u_B}{x_D - x_B} = 10\frac{12-7}{0.04}$$

$$\therefore \qquad \boxed{a_x\big|_C \approx 1250\ m/s^2}$$

In order to find $v\big|_B$, we employ the 2-D continuity equation in rectangular coordinates:

$$\frac{\partial u}{\partial x} + \frac{\partial v}{\partial y} = 0 \tag{E.2.4.4}$$

which can be approximated as:

$$\frac{\Delta v}{\Delta y} = -\frac{\Delta u}{\Delta x}$$

or

$$\Delta v = -\frac{10-5}{0.04} 0.02 = -2.5 \text{ m/s}$$

Recall that $\Delta v = v|_{B'} - v|_{B}$ where $v_B \equiv 0$ <symmetry> so that

$$\boxed{v|_{B'} = -2.5 \text{ m/s}}$$

Comment: This is a very simple example of finite differencing where derivatives are approximated by finite differences of all variables. Discretization of the governing equations describing the conservation laws is the underlying principle of CFD (computational fluid dynamics) software (see Sect. 10.2).

2.4 Momentum Conservation

Focusing on linear momentum transfer, in contrast to angular momentum transfer, the momentum conservation law is again derived two ways, i.e., via integral and differential approaches. Not surprisingly, Newton's Second Law of Motion is the foundation of the final momentum equation, or "equation of motion".

2.4.1 Momentum Conservation in Integral Form

Forces acting on an identifiable *fluid* element accelerate it, as it is well known from Newton's second law of motion $(m\vec{a})_{fluid} = \Sigma\vec{F}_{external}$. Specifically, we set (see Eq. (2.1b)):

$$B_{system} \equiv m\vec{v} \text{ and hence } \beta \equiv \vec{v}.$$

As previously indicated, with m = ¢, $m\frac{D\vec{v}}{Dt} = \Sigma\vec{F}_{ext}$ and hence Eq. (2.2a) reads:

$$m\frac{D\vec{\mathbf{v}}}{Dt} = m\vec{\mathbf{a}}_{total} = \Sigma\vec{\mathbf{F}}_{body} + \Sigma\vec{\mathbf{F}}_{surface} = \frac{\partial}{\partial t}\int_{C.\forall.} \rho\vec{\mathbf{v}}d\forall + \int_{C.S.} \vec{\mathbf{v}}\rho\vec{\mathbf{v}}\cdot d\vec{\mathbf{A}} \quad (2.14a)$$

As discussed, for control volumes *accelerating* without rotation relative to inertial coordinates (see Fig. 2.2a), an additional force $\vec{\mathbf{F}}_{inertia} \sim \vec{\mathbf{a}}_{relative}$ appears, where $\vec{\mathbf{a}}_{rel}$ is the C.∀. acceleration relative to the fixed frame of reference X-Y-Z (See Fig. 2.2a). Thus, with

$$\vec{\mathbf{v}}_{XYZ} = \vec{\mathbf{v}}_{xyz} + \vec{\mathbf{v}}_{relative} \text{ or } \vec{\mathbf{a}}_{XYZ} = \vec{\mathbf{a}}_{absolute} = \frac{d\vec{\mathbf{v}}}{dt} + \vec{\mathbf{a}}_{rel}$$

we can express the sum of all forces as:

$$\Sigma\vec{\mathbf{F}}_{total} = m\vec{\mathbf{a}}_{abs} = m\left(\frac{D\vec{\mathbf{v}}}{Dt} + \vec{\mathbf{a}}_{rel}\right),$$

i.e., for an accelerating and deforming control volume:

$$m\frac{D\vec{\mathbf{v}}}{Dt} = \Sigma\vec{\mathbf{F}}_B + \Sigma\vec{\mathbf{F}}_S - \int_{C.\forall.} \vec{\mathbf{a}}_{rel}\, dm = \frac{d}{dt}\int_{C.\forall.} \rho\vec{\mathbf{v}}d\forall + \int_{C.S.} \vec{\mathbf{v}}\rho\,\vec{\mathbf{v}}_{rel} \cdot dA \quad (2.14b)$$

Example 2.5: Force on a Disk Moving into an Axisymmetric Jet

A steady uniform round jet impinges upon an approaching conical disk as shown. Find the force exerted on the disk, where v_{jet}, A_{jet}, v_{disk}, diameter D, angle θ, and fluid layer thickness t are given.

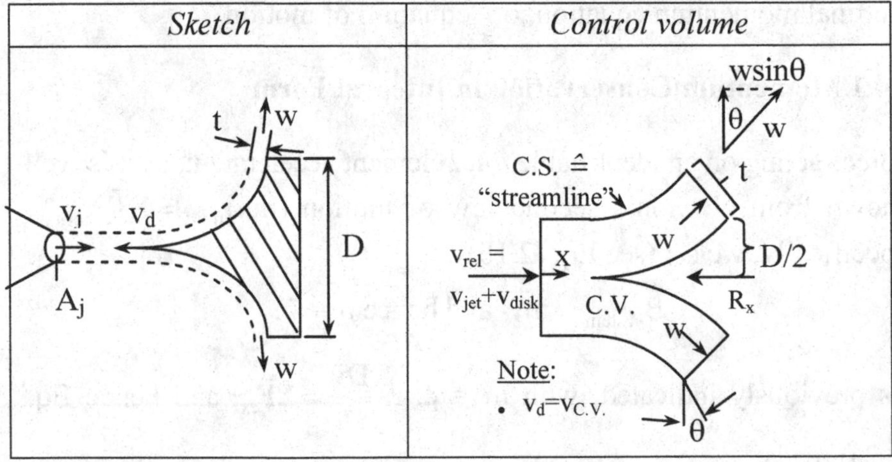

Assumptions: as stated; constant averaged velocities and properties

Approach: RTT (mass balance and 1-D force balance)

Solution:

(a) Mass Conservation

$$0 = \int_{C.S.} \rho \vec{v} \cdot d\vec{A} = -\int_{A_{jet}} \rho v_{rel} dA + \int_{A_{exit}} \rho w dA \quad \text{(E.2.5.1)}$$

where $v_{rel} = v_{jet} - (-v_{disk}) = v_j + v_d$ and $A_{exit} \approx \pi D\, t$.

$\therefore\ -\rho A_j (v_j + v_d) + \rho\, (\pi D\, t) w = 0$

Hence,

$$w = \frac{(v_j + v_d) A_j}{\pi D\, t} \quad \text{(E.2.5.2)}$$

(b) Momentum Conservation

$$B = (m\vec{v})_s;\ \beta = \vec{v};\ {}^{DB}\!/_{Dt} := \vec{F}_{surface} = -R_x.$$

$$\Sigma \vec{F}_{surf} = \int_{C.S.} \vec{v} \rho \vec{v}_{rel} \cdot d\vec{A} \xrightarrow{\text{x - component}} -R_x$$

$$= \int_{C.S.} u \rho v_{rel} dA \quad \text{(E.2.5.3a, b)}$$

Thus,

$$-R_x = -v_{rel} \rho v_{rel} A_{jet} + w \sin\theta\ \rho w A_{exit}$$

where with Eq. (E.2.5.2):

$$R_x = \rho A_j (v_j + v_d)^2 \left[1 - \frac{A_j \sin\theta}{\pi t D}\right] \quad \text{(E.2.5.4)}$$

Comment: The resultant fluid-structure force (E.2.5.4) can be rewritten as:

$$R_x = (\dot{m} v_x)_{out} - (\dot{m} v_x)_{in} \quad \text{(E.2.5.5)}$$

i.e., the result of a change in fluid flow momentum. If the disk would move away with $v_d = v_j$, i.e., escaping the jet, $v_{rel} = 0$ and hence $R_x = 0$.

Expanding on Eq. (E.2.5.5), we can generalize that the net momentum flux due to applied forces is:

$$\sum \vec{F} = \dot{m}(\alpha_2 \vec{v}_2 - \alpha_1 \vec{v}_1) \tag{2.15}$$

where the correction factor α accounts for the variation of v^2 across the inlet ① or outlet ② duct section. Specifically,

$$\rho \int_A v^2 dA = \alpha \dot{m} v_{average} = \alpha \rho A v_{av}^2$$

so that

$$\alpha \equiv \frac{1}{A} \int \left(\frac{v}{v_{av}} \right)^2 dA \tag{2.16}$$

Example 2.6: Force on a Submerged Body

Find the drag on a submerged elliptic rod of characteristic thickness h, based on a measured velocity profile downstream from the body, say,

$$u(y) = \frac{u_\infty}{4}\left[1 + \frac{y}{h}\right]; \quad 0 \le y \le 3h$$

Sketch	*Approach*
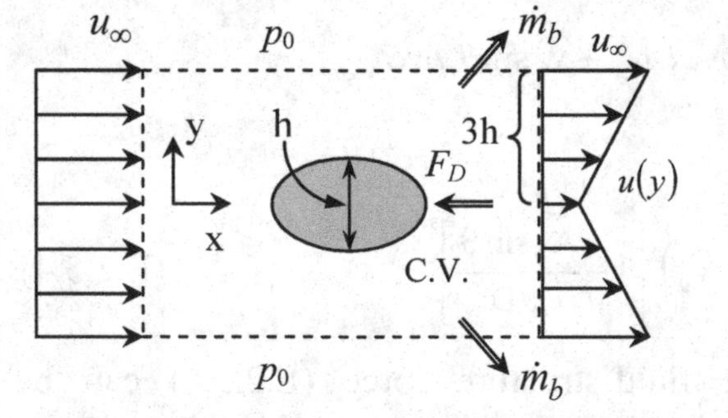 u_∞, p_0, $\dot{m}_b$, u_∞, y, h, 3h, F_D, $u(y)$, x, C.V., p_0, $\dot{m}_b$	• Integral mass and momentum balances (i.e., RTT)

Assumptions: Steady state, 2-D, constant properties, fixed C.∀., infinite extent in the z-direction.

Solution:

(a) Mass Conservation: $\dot{m}_{in} = \dot{m}_{boundary} + \dot{m}_{out}$ (E.2.6.1)

$$0 = \int_{C.S.} \rho\vec{v}\cdot dA = \rho u_\infty 2(3h) + \dot{m}_{boundary} + 2\rho\frac{u_\infty}{4}\int_0^{3h}\left(1+\frac{y}{h}\right)dy \quad \text{(E.2.6.2a)}$$

$$\therefore \quad \dot{m}_{boundary} = 6\rho h u_\infty - \frac{15}{4}\rho h u_\infty = \frac{9}{4}\rho h u_\infty \quad \text{(E.2.6.2b)}$$

(b) Momentum Conservation (x-momentum):

$$-F_D = \int_{C.S.} v_x \rho\vec{v}\cdot d\vec{A} := (\dot{m}v_x\big|_{exit} + \dot{m}v_x\big|_b) - \dot{m}v_x\big|_{inlet} \quad \text{(E.2.6.3a)}$$

$$-F_D = -u_\infty \rho u_\infty(6h) + u_\infty(\frac{9}{4}\rho h u_\infty) + 2\rho\int_0^{3h}[u(y)]^2 dy \quad \text{(E.2.6.3b)}$$

$$F_D = \frac{9}{8}\rho h u_\infty^2 \quad \text{(E.2.6.3c)}$$

Comment: The fluid flow field inside the C.∀., especially behind the submerged body, is very complex. The RTT treats it as a "black box" and elegantly obtains $F_D = F_{fixation}$ via "velocity defect" measurements, indicated with the given velocity profile u(y). Note that the given system is 2-D with $z \to \infty$; but, after integration the result, $F_x = -F_D$, is obtained in 1-D.

Example 2.7: Rocket with Air Drag

Set up the equation of motion for a vertically accelerating rocket using: (a) Newton's law and (b) the momentum RTT. Clearly, the weight of the missile changes as $W(t) = m_M g = W_0 - g\dot{m}_f t$, where $\dot{m}_f$ is the fuel mass flow rate (i.e., fuel consumption) exiting at v_e relative to the missile. Assume that $\dot{m}_f v_e = F_{thrust}$ = constant. The air drag is $F_D = \rho C_D D^2 v^2$ where C_D is the drag coefficient, D is the missile's mean diameter and v(t) is the missile velocity. Then, (c) neglecting air drag, find the rocket's initial acceleration and velocity after 3 s for $\dot{m}_f = 5\,\text{kg/s}$, and $v_e = 3{,}500\ \text{m/s}$.

Sketch	*Assumptions*	*Concepts*
v, a; $\vec{F}_D$; $\vec{g}$; D; W; z; Z, Y, X; $v_e, \dot{m}_f$	• Initial phase acceleration is constant • Constant uniform v_e and constant $\dot{m}_f$ • Negligible momentum change inside missile, i.e., $\frac{\partial}{\partial t} \approx 0$	• Newton's second law for a "particle" • Use of Eq. (2.15)

Solution:

(a) Particle Dynamics

$$m\frac{dv}{dt} = \Sigma\vec{F}_{ext} = -W(t) + F_{Thrust} - F_{Drag}(t) \qquad \text{(E.2.7.1)}$$

where

$$W(t) = \underbrace{(m_0 - \dot{m}_f t)}_{m_M(t)}g,\;\; F_T = \dot{m}_f v_e \text{ and } F_D = \rho C_D D^2 v^2$$

Equation (E.2.7.1) is a nonlinear first-order ODE to be solved for v(t) with a RUNGE-KUTTA routine.

(b) Momentum Conservation (x-momentum of Eq. (2.14b)):

$$F_{B,z} + F_{S,z} - \int_{C.\forall.} a_z\, dm = \frac{\partial}{\partial t}\int_{C.\forall.} v_z \rho d\forall + \int_{C.S.} v_z \rho \vec{v}_{rel} \cdot d\vec{A} \qquad \text{(E.2.7.2)}$$

With the given information and based on the listed assumptions, Eq. (E.2.7.2) can be reduced to:

$$-W F_D - am_M = -v_e \int_{C.S.} (\rho v_e dA) = -v_e \dot{m}_f \quad \text{(E.2.7.3)}$$

Neglecting F_D and inserting W and m_M yields the initial-phase rocket acceleration

$$a = \frac{\dot{m}_f v_e}{m_0 - \dot{m}_f t} - g \quad \text{(E.2.7.4)}$$

(c) Numerical Results

$$\text{At } t = 0\text{: } a_0 = \frac{\dot{m}_f v_e}{m_0} - g := 34 \text{ m/s}^2$$

In order to find the rocket velocity, we recall that

$$a = \frac{dv}{dt} = \frac{\dot{m}_f v_e}{m_0 - \dot{m}_f t} - g$$

so that after integration

$$v = -v_e \ln\left(\frac{m_0 - \dot{m}_f t}{m_0}\right) - gt \quad \text{(E.2.7.5)}$$

and at t = 3 s:

$$v|_{3s} = 104.34 \text{ m/s}$$

2.4.2 Momentum Conservation in Differential Form

In order to obtain the equation of motion describing any *point* in a fluid flow field, all terms in the RTT have to be again converted to volume integrals, employing Gauss' divergence theorem (see Eq. (2.9)).

The Equation of Motion First, body forces in Eq. (2.14) are logically expressed in terms of volume integrals, i.e.,

$$\vec{F}_B = \iiint_{C.\forall.} \rho \vec{f}_B d\forall$$

and surface forces in terms of surface integrals, i.e.,

$$\vec{F}_S = \iint_{C..S.} \vec{\vec{T}} \cdot d\vec{A}$$

where $\vec{\mathbf{f}}_B$ is a body force per unit mass and $\vec{\vec{T}}$ is the total stress tensor (see Sect. 1.4 and App. A). Now, for a stationary control volume the linear momentum equation in integral form reads:

$$\int_{C.\forall.} \rho\vec{\mathbf{f}}_B \, d\forall + \int_{C.S.} \vec{\vec{T}} \cdot d\vec{\mathbf{A}} = \frac{\partial}{\partial t} \int_{C.\forall.} \rho\vec{\mathbf{v}} \, d\forall + \int_{C.S.} \vec{\mathbf{v}}\rho\vec{\mathbf{v}} \cdot d\vec{\mathbf{A}} \qquad (2.17)$$

Recall: This is a (3-component) vector equation, in principal for the velocity field $\vec{v}$. It contains $\vec{\vec{T}} \equiv -p\vec{\vec{I}} + \vec{\vec{\tau}}$, i.e., the (9-component) total stress tensor (see Sect. 1.4 and App. A), as an additional unknown because in most cases $\vec{f}_B$ is simply weight ($\vec{\mathbf{g}}$) per unit mass. Thus, in order to solve this closure problem, we have to know the *thermodynamic pressure* p and an expression for the *stress tensor* $\vec{\vec{\tau}}$. Recall that $\vec{\vec{I}}$ is the unit tensor, i.e., only ones on the diagonal and zeros everywhere else in the 3 × 3 matrix, elevating the product $p\vec{\vec{I}}$ to a "pseudo-tensor" because p is just a scalar.

Now, converting all surface integrals into volume integrals yields

$$\iiint_{C.\forall.} \left[\frac{\partial(\rho\vec{\mathbf{v}})}{\partial t} + \nabla \cdot (\rho\vec{\mathbf{v}}\vec{\mathbf{v}}) + \nabla \cdot (p\vec{\vec{I}} - \vec{\vec{\tau}}) - \rho\vec{\mathbf{g}} \right] d\forall = 0$$

or

$$\frac{\partial(\rho\vec{\mathbf{v}})}{\partial t} + \nabla \cdot (\rho\vec{\mathbf{v}}\vec{\mathbf{v}}) = -\nabla p + \nabla \cdot \vec{\vec{\tau}} + \rho\vec{\mathbf{g}} \qquad (2.18)$$

which is the Cauchy's equation of motion (or linear momentum equation) for any fluid and with gravity as the body force. In order to reduce it in complexity and provide some physical meaning, let's consider *constant fluid properties* and express the unknown stress tensor, employing Stokes' hypothesis, in terms of the principal variable. In vector notation:

$$\vec{\vec{\tau}} = \mu(\nabla\vec{\mathbf{v}} + \nabla\vec{\mathbf{v}}^{tr}) = \mu\vec{\vec{\dot{\gamma}}} \qquad (2.19a)$$

and in index (or tensor notation):

$$\tau_{ij} = \mu(\frac{\partial v_i}{\partial x_j} + \frac{\partial v_j}{\partial x_i}) = \mu\dot{\gamma}_{ij} \quad (2.19b)$$

where $\dot{\gamma}$ is the shear rate.

Recall: Physically τ_{ij} represents a force field per unit area (see Sect. 1.4 and Example 2.8) as a result of the resistance to the rate of deformation of fluid elements, i.e., internal friction. This insight leads for Newtonian fluids, such as air, water, typical oils, etc., to the postulate:

$$\tau_{ij} = \mathrm{fct}(\varepsilon_{ij})$$

where $\varepsilon_{ij} = \frac{1}{2}(v_{i,j} + v_{j,i})$ is the rate-of-deformation tensor, discussed in Sects. 1.4 and 8.2. Now, Stokes suggested that $\vec{\vec{\tau}}$ is a *linear function* of $\vec{\vec{\varepsilon}}$ which is not the case for non-Newtonian fluids (Sect. 6.3), rarefied gases, and some fluid flows in micro-scale devices, e.g., MEMS. Specifically, for Newtonian fluids:

$$\vec{\vec{\tau}} = \lambda(\nabla \cdot \vec{v})\vec{\vec{I}} + 2\mu\vec{\vec{\varepsilon}} \quad (2.19c)$$

where the viscosity coefficients λ and μ depend only on the thermodynamic state of the fluid. For incompressible flow $\nabla \cdot \vec{v} = 0$ (see Eq. (2.12)) and the total stress tensor reduces to

$$\tau_{ij} = 2\mu\varepsilon_{ij} \quad (2.19d)$$

where

$$\varepsilon_{ij} = \frac{1}{2}\left(\frac{\partial v_i}{\partial x_j} + \frac{\partial v_j}{\partial x_i}\right) = \frac{1}{2}\dot{\gamma}_{ij}$$

so that

$$\tau_{ij} = \mu(\nabla\vec{v} + \nabla\vec{v}^{|tr}) = \mu\dot{\gamma}_{ij} \quad (2.19e, f)$$

Example 2.8: Shear Stress in Simple COUETTE Flow

Consider Couette-flow, i.e., a viscous fluid between two parallel plates a small gap h apart, where the upper plate moves with a constant velocity u_o, in general due to a tangential force, F_{pull}.

Note: The experimentally observed boundary condition for a conventional fluid at any solid surface demands that:

$$\vec{v}_{\text{fluid}} = \vec{v}_{\text{wall}} \tag{E.2.8.1}$$

where in rectangular coordinate $\vec{v} = (u, v, w)$ or $\vec{v} = u\hat{i} + v\hat{j} + w\hat{k}$. Applying Eq. (E.2.8.1) for the present case (see System Sketch), we have for a stationary solid wall $(y = 0)$:

$u_{\text{fluid}} = 0$ < "no - slip" condition > (E.2.8.2a)

For the moving wall $(y = h)$:

$u_{\text{fluid}} = u_0$ <"no-slip" condition> (E.2.8.2b)

and the normal velocity component at both walls is:

$v_{\text{fluid}} = 0$ <"no-penetration" condition> (E.2.8.2c)

Sketch	*Assumptions*	*Approach*
$u_o = ¢$; h; y; ρ, μ; x	• Steady laminar fully-developed (unidirectional) flow • Constant fluid properties	• Reduced N-S equations based on assumptions and boundary conditions

Solution: Translating the problem statement plus assumptions into mathematical shorthand, we have:

- Movement of the upper plate (u_0 = constant) keeps the viscous fluid between the plates in motion via frictional effects propagating normal to the plate; hence, the usual "driving force" $\frac{\partial p}{\partial x} \equiv 0$.
- *Steady flow*⇒all time derivatives are zero, i.e., $\frac{\partial}{\partial t} \equiv 0$.

- Laminar *unidirectional flow*⇒only one velocity component dependent on one dimension (1-D), is non-zero, i.e., $\vec{v} = (u,0,0)$, where u = u(y) only. This implies *parallel* or *fully-developed flow* where $\frac{\partial}{\partial x} \equiv 0$ and hence $v = 0$.

In summary, we can postulate that

- u = u(y), v = w = 0;
- $\frac{\partial u}{\partial x} = 0; \ \frac{\partial p}{\partial x} = 0; \ g_x = 0$

Checking Eqs. (2.23a–c) with these postulates, we realize the following:

- Continuity equation confirms: $0 + \frac{\partial v}{\partial y} = 0 \succ \underline{v = 0}$; (E.2.8.1)

or better, $v = 0 \Rightarrow \frac{\partial u}{\partial x} = 0$; i.e., fully-developed flow.

- x-momentum yields: $0 + 0 = 0 + \nu(0 + \frac{\partial^2 u}{\partial y^2}) + 0$ (E.2.8.2)

and

- y-momentum collapses to: $\frac{\partial p}{\partial y} = \rho g_y$ <fluid statics> (E.2.8.3)

Thus, Eq. (E.2.8.2) can be written as

$$\frac{d^2 u}{dy^2} = 0 \tag{E.2.8.4a}$$

subject to the "no-slip" conditions

$$u(y = 0) = 0 \ \text{ and } \ u(y = h) = u_0 \tag{E.2.8.4b,c}$$

Double integration of (E.2.8.4a) and inserting the B.C.s (E.2.8.4b,c) yields:

$$u(y) = u_0 \frac{y}{h} \tag{E.2.8.5}$$

Of the stress tensor (Eq. (2.19b)), $\tau_{ij} = \mu(\frac{\partial u_i}{\partial x_j} + \frac{\partial u_j}{\partial x_i})$, only τ_{yx} is non-zero, i.e.,

$$\tau_{xy} = \mu\left(\frac{\partial u}{\partial y} + \frac{\partial v}{\partial x}\right) \tag{E.2.8.6}$$

With $v \equiv 0$ and Eq. (E.2.8.5)

$$\tau_{xy} = \mu u_0 / h = \not c$$

Of the vorticity tensor $\omega_{ij} = \frac{\partial u_i}{\partial x_j} - \frac{\partial u_j}{\partial x_i}$ (App. A), only ω_{yx} is non-zero, i.e.,

$$\omega_{yx} = \frac{\partial u}{\partial y} - \frac{\partial v}{\partial x} := u_0 / h = \not c \tag{E.2.8.7}$$

which implies that the fluid elements between the plates rotate with constant angular velocity ω_{yx}, while translating with u(y).

Note: The wall shear stress at the upper (moving) plate is also constant, i.e.,

$$\tau_w = \mu \frac{\partial u}{\partial y}\bigg|_{y=h} = \mu u_0 / h$$

so that $F_{drag} = -\int \tau_w dA_{plate} = F_{pull} = \frac{\mu u_0}{h} A_{surface} = \text{constant.}$

Profiles:

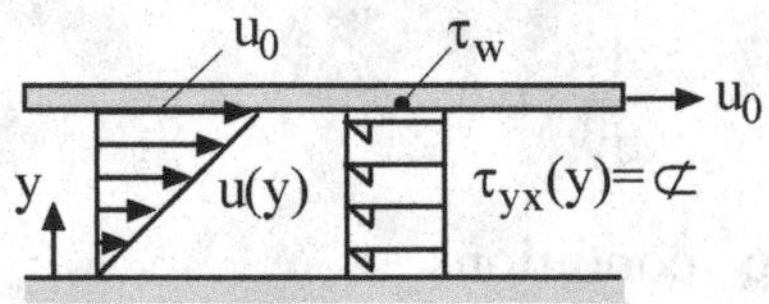

Comments: In the absence of a pressure gradient, only viscous effects set the fluid layer into (linear) motion. The necessary "pulling force" is inversely proportional to the gap height, i.e., the thinner the fluid layer the larger is the shear stress and hence F_{pull}.

Force Balance Derivation A more physical approach for deriving the (linear) momentum equation starts with a force balance for a representative elementary volume (REV). Employing rectangular coordinates and an incompressible fluid, external surface and body forces accelerate an REV of mass *m*, so that we can write Newton's second law of motion per unit volume as (cf. Fig. 2.4):

$$\rho \frac{D\vec{v}}{Dt} = \Sigma \vec{\mathbf{f}}_{\text{surface}} + \Sigma \vec{\mathbf{f}}_{\text{body}} \tag{2.20a}$$

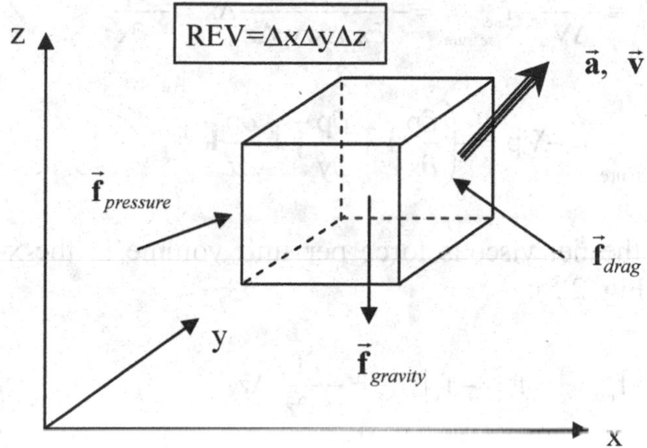

Fig. 2.4 Closed system, i.e., accelerating material volume (REV)

Using the definition of the material (or Stokes) derivative (App. A), from the Eulerian point of view (see Sect. 1.3), the REV is a control volume for which we record local and convective momentum changes due to *net* pressure, viscous, and gravitational forces, viz. Eq. (2.20b):

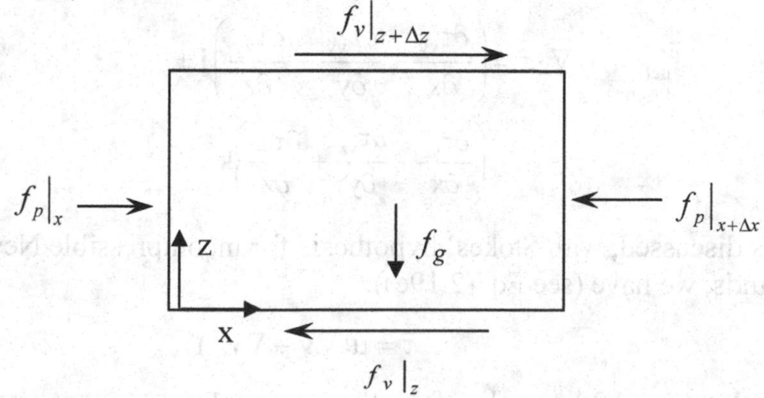

Fig. 2.5 Control volume for 1-D force balances

$$\rho\left[\frac{\partial \vec{\mathbf{v}}}{\partial t}+(\vec{\mathbf{v}}\cdot\nabla)\vec{\mathbf{v}}\right]=\vec{\mathbf{f}}_{\underset{\text{pressure}}{\text{net}}}+\vec{\mathbf{f}}_{\underset{\text{viscous}}{\text{net}}}+\vec{\mathbf{f}}_{\underset{\text{buoyancy}}{\text{net}}} \quad (2.20b)$$

- $f_{\underset{\text{pressure}}{\text{net}}}=f_p\big|_x-f_p\big|_{x+\Delta x}=-\frac{\partial f_p}{\partial x}\Delta x$

and with $f_p \equiv \frac{p\Delta A}{\Delta V}$, $f_{\underset{\text{pressure}}{\text{net}}}=-\frac{\partial p}{\partial x}\frac{\Delta y\Delta z}{\Delta x\Delta y\Delta z}\Delta x=-\frac{\partial p}{\partial x}$,
or in 3-D

- $\vec{\mathbf{f}}_{\underset{\text{pressure}}{\text{net}}}=-\nabla p=-\left(\frac{\partial p}{\partial x}\vec{\mathbf{i}}+\frac{\partial p}{\partial y}\vec{\mathbf{j}}+\frac{\partial p}{\partial z}\vec{\mathbf{k}}\right)$

Similarly, the net viscous force per unit volume in the x-direction reads (see Fig. 2.5):

$$f_{\underset{\text{viscous}}{\text{net}}}=f_v\big|_z-f_v\big|_{z+\Delta z}=-\frac{\partial f_v}{\partial z}\Delta z$$

and with $f_v \equiv \frac{\tau\Delta A}{\Delta V}$, $f_{\underset{\text{viscous}}{\text{net}}}=-\frac{\partial \tau}{\partial z}\frac{\Delta x\Delta z}{\Delta x\Delta y\Delta z}\Delta z=-\frac{\partial \tau}{\partial z}$,

In 3-D, the net frictional force can be expressed as:

$$\vec{\mathbf{f}}_{\underset{\text{viscous}}{\text{net}}}=\nabla\cdot\vec{\vec{\tau}}=\begin{matrix}\left(\frac{\partial \tau_{xx}}{\partial x}+\frac{\partial \tau_{yx}}{\partial y}+\frac{\partial \tau_{zx}}{\partial z}\right)\vec{\mathbf{i}}+\\ \left(\frac{\partial \tau_{xy}}{\partial x}+\frac{\partial \tau_{yy}}{\partial y}+\frac{\partial \tau_{zy}}{\partial z}\right)\vec{\mathbf{j}}+\\ \left(\frac{\partial \tau_{xz}}{\partial x}+\frac{\partial \tau_{yz}}{\partial y}+\frac{\partial \tau_{zz}}{\partial z}\right)\vec{\mathbf{k}}\end{matrix}$$

As discussed, with Stokes' hypothesis for incompressible Newtonian fluids, we have (see Eq. (2.19e)):

$$\vec{\vec{\tau}}=\mu(\nabla\vec{\mathbf{v}}+\nabla\vec{\mathbf{v}}^{tr})$$

so that Eq. (2.20a) reads (cf. N–S equation (Eq. (2.22))):

$$\rho\left[\frac{\partial \vec{v}}{\partial t}+(\vec{v}\cdot\nabla)\vec{v}\right]=-\nabla p+\mu\nabla^2\vec{v}+\rho\vec{g} \tag{2.21}$$

2.4.3 Special Cases of the Equation of Motion

Returning to Eq. (2.18), which generally is known as Cauchy's *equation of motion,* we now introduce simplifications of increasing magnitude. Fluid properties, i.e., density ρ and dynamic viscosity μ are typically constant; but, in general, ρ and μ are functions of temperature T, pressure p, and species concentration c. Thus, the underlying assumptions for $\rho = ¢$ and $\mu = ¢$ are that:

- Only relatively small temperature variations occur.
- The Mach number $M \equiv \frac{v_{fluid}}{a_{sound}} < 0.3$, which may be only violated by gases.
- Pressure drops in gas flow are relatively small and cavitation in liquid flow is avoided.
- Concentration variations of components in mixture flows are small.

(i) The Navier–Stokes Equations

Dividing Eq. (2.18) through by the constant density ρ and recalling that $\mu/\rho \equiv \nu$, the kinematic viscosity, we have with Stokes' hypothesis (Eq. (2.19)):

$$\underbrace{\frac{\partial \vec{v}}{\partial t}+(\vec{v}\cdot\nabla)\vec{v}}_{\frac{D\vec{v}}{Dt}=\vec{a}_{total}}=\underbrace{-\frac{1}{\rho}\nabla p}_{\vec{f}_{net\ pressure}}+\underbrace{\nu\nabla^2\vec{v}}_{\vec{f}_{net\ viscous}}+\underbrace{\vec{g}}_{\vec{f}_{body}} \tag{2.22}$$

Clearly, Eq. (2.22) is Newton's particle dynamics equation applied to fluid elements. Together with the continuity equation, $\nabla\cdot\vec{v}=0$, they are called the *Navier–Stokes* (N–S) *equations*.

For example, for steady two-dimensional (2-D) flows, the N–S equations read in rectangular coordinates (see App. A):

$$\text{(Continuity)} \qquad \frac{\partial u}{\partial x}+\frac{\partial v}{\partial y}=0$$

$$\text{(x-momentum)}\ u\frac{\partial u}{\partial x}+v\frac{\partial u}{\partial y}=-\frac{1}{\rho}\frac{\partial p}{\partial x}+\nu\left(\frac{\partial^2 u}{\partial x^2}+\frac{\partial^2 u}{\partial y^2}\right)+g_x \qquad (2.23a\text{–}c)$$

$$\text{(y-momentum)}\ \ u\frac{\partial v}{\partial x}+v\frac{\partial v}{\partial y}=-\frac{1}{\rho}\frac{\partial p}{\partial y}+\nu\left(\frac{\partial^2 v}{\partial x^2}+\frac{\partial^2 v}{\partial y^2}\right)+g_y$$

On a professional level this set of four partial differential equations (PDEs), subject to appropriate boundary conditions, is now being routinely solved for the four unknowns, u, v, w, and p, using numerical software packages on desktop workstations, HPC clusters and supercomputers (see Chap. 10). In a classroom environment, only reduced forms of Eqs. (2.23a–c) can be solved, as illustrated in Chaps. 3 to 9.

(ii) Prandtl's Boundary-Layer Equations

As indicated in Example 2.8, the fluid velocity is zero at a stationary wall. Now, considering relatively high-speed fluid flow past a (horizontal) solid surface, the quasi-uniform high velocity suddenly has to reduce, within a narrow region, to zero at the stationary wall. This region of high velocity gradients is called a "thin shear layer," or more generally a *boundary layer*. For example, Fig. 2.9 depicts such a (laminar) boundary layer of thickness δ(x), formed along a horizontal stationary flat plate (e.g., a giant razor-blade) which is approached by a uniform fluid stream of velocity u_∞ with $10^3 < Re < 10^5$. It can be readily demonstrated that the $\nu\frac{\partial^2 u}{\partial x^2}$-term (axial momentum diffusion) of Eq. (2.23b) is negligible (see Sect. 5.2) and that the y-momentum equation collapses to $-\frac{1}{\rho}\frac{\partial p}{\partial y}=0$; i.e., p = p(x) only. As a result, Eqs. (2.23a–c) reduce to:

$$\frac{\partial u}{\partial x}+\frac{\partial v}{\partial y}=0 \qquad (2.24a)$$

and

$$u\frac{\partial u}{\partial x} + v\frac{\partial u}{\partial y} = -\frac{1}{\rho}\frac{\partial p}{\partial x} + \nu\frac{\partial^2 u}{\partial y^2} \tag{2.24b}$$

inside the boundary layer $0 \le y \le \delta(x)$ and $0 \le x \le \ell$.

(iii) Stokes' Equation

When the viscous forces are dominant, the Reynolds number (Re = $\frac{\text{inertial forces}}{\text{viscous forces}}$; see Sect. 1.1) is very small, i.e., the term $(\vec{v}\cdot\nabla)\vec{v}$ in Eq. (2.22) is negligible. As a result, the *Stokes equation* is *obtained* which holds for "creeping" flows (see Example 3.9 for an application):

$$\rho\frac{\partial \vec{v}}{\partial t} = -\nabla p + \mu\nabla^2\vec{v} \tag{2.25}$$

(iv) Euler's Inviscid Flow Equation

For frictionless flows $(\mu = 0)$, Eq. (2.22) reduces to:

$$\rho\frac{D\vec{v}}{Dt} = -\nabla \rho + p\vec{g} \tag{2.26}$$

which is the *Euler equation.* Although ideal fluids, i.e., inviscid flows, hardly exist, the second-order term also vanishes when $\nabla^2\vec{v} \approx 0$; for example, outside boundary layers as indicated with the velocity profile in Fig. 2.9. In fact, aerodynamics people employ Eq. (2.26) to find the pressure field around airfoils (see p_{outer} in Fig. 2.6).

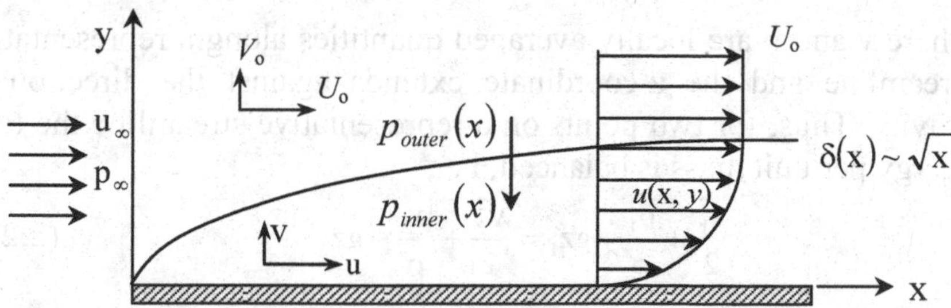

Fig. 2.6 Laminar flat-plate boundary layer

(v) Bernoulli's Equation

Equation (2.26) applied in 2-D to a representative streamline along coordinate "s" yields (see Fig. 2.7):

$$\frac{\partial v_s}{\partial t} + v_s \frac{\partial v_s}{\partial s} + \frac{1}{\rho}\frac{\partial p}{\partial s} + g \underbrace{\sin\theta}_{\partial z/\partial s} = 0 \tag{2.27a}$$

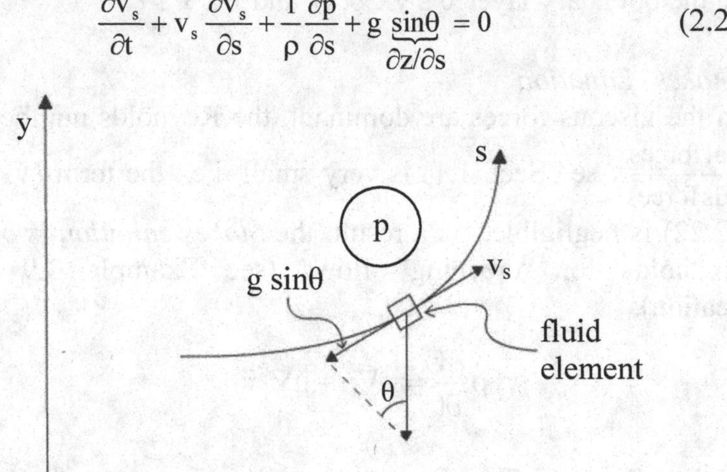

Fig. 2.7 A fluid element along a representative streamline

which leads to the *Bernoulli equation.* Multiplying Eq. (2.27a) through by ∂s and integrating, yields for *steady incompressible inviscid flows*:

$$\frac{v^2}{2} + \frac{p}{\rho} + gz = C \tag{2.27b}$$

where v and p are locally averaged quantities along a representative streamline and the z-coordinate extends against the direction of gravity. Thus, for two points on a representative streamline the total energy per unit mass is balanced, i.e.,

$$\frac{v_1^2}{2} + \frac{p_1}{\rho} + gz_1 = \frac{v_2^2}{2} + \frac{p_2}{\rho} + gz_2 \tag{2.28}$$

For example, for a given system where $v_2=0$ (e.g., point 2 of a streamline is on the front of a submerged body) and $g\Delta z \approx 0$, we have:

$$p_2 = p_1 + \frac{\rho v_1^2}{2} \tag{2.29}$$

where p_2 is the total or *stagnation point pressure*, p_1 is the *thermodynamic* (or a *static*) *pressure* at point 1; and $\rho v_1^2/2$ is the *dynamic pressure* at point 1. One application of Eq. (2.29) is the Pitot-static tube, which measures $\Delta p = p_2 - p_1$, so that $v_1 = \sqrt{2(p_2 - p_1)/\rho}$ (see Fig. 2.8).

An *extended, i.e., more realistic form of Bernoulli's equation* adds a frictional loss term to the R.H.S. of Eq. (2.28). For example, multiplying Eq. (2.28) through by ρ yields an energy balance per unit volume:

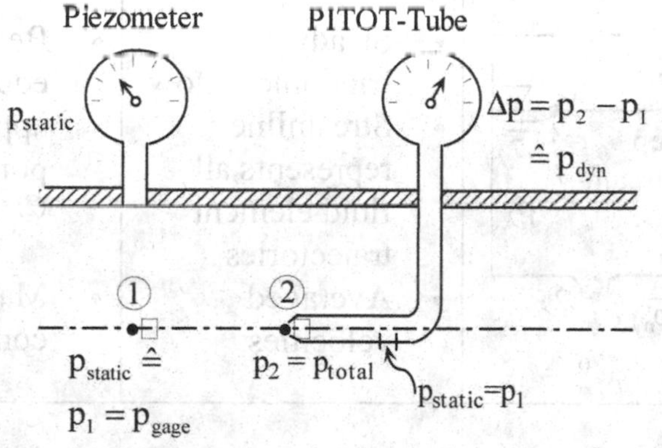

Fig. 2.8 Different manometers to measure different pressures

$$\underbrace{\frac{\rho}{2}v_1^2 + p_1 + \rho g z_1}_{\sim E_{total}} = \underbrace{\frac{\rho}{2}v_2^2}_{\sim E_{kin.}} + \underbrace{p_2}_{\sim E_{pr.}} + \underbrace{\rho g z_2}_{\sim E_{pot.}} + \underbrace{H_f}_{E_{loss}} \tag{2.30}$$

where H_f represents an energy loss between stations ① and ②.

Naturally, Eq. (2.30) can also be expressed in terms of heights (or "heads"):

$$\frac{v_1^2}{2g} + \frac{p_1}{\gamma} + z_1 = \frac{v_2^2}{2g} + \frac{p_2}{\gamma} + z_2 + h_f \tag{2.31}$$

The specific weight $\gamma \equiv \rho g$ and the frictional loss term $h_f \sim \tau_{wall} \sim \Delta p$ is usually expressed as a portion of the kinetic energy (see Chap. 3).

Example 2.9: Tank Drainage

Consider the efflux of a liquid through a smooth orifice of area A_0 at the bottom of covered tank (A_T). The depth of the liquid is y. The pressure in the tank exerted on the liquid is p_T while the pressure outside of the orifice is p_a. Develop an expression for $v_0(y)$.

Sketch	*Assumptions*	*Concept*
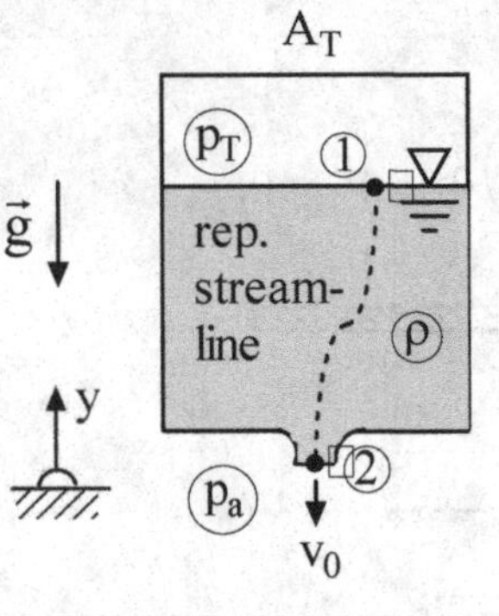 A_T; p_T; ①; $\vec{g}$; rep. stream-line; ρ; y; ②; p_a; v_0	• Steady frictionless flow • Streamline represents all fluid element trajectories • Averaged velocities	• Bernoulli's equation applied to points ① and ② • Mass conservation

Solution: At descending point ① and exit point ②, most of the information is given. Thus, Eq. (2.28) now reads:

$$\frac{v_1^2}{2} + \frac{p_T}{\rho} + gy = \frac{v_0^2}{2} + \frac{p_a}{\rho} + 0$$

The velocities are related via continuity (see Eq. (2.7)):

$$v_1 A_T = v_0 A_0$$

so that with $v_1 = v_0\, A_0 / A_T$,

$$v_0 = \left[\frac{\frac{2}{\rho}(p_T - p_a) + 2gy}{1 - \left(\frac{A_0}{A_T}\right)^2}\right]^{1/2}$$

Comment: Toricelli's law, $v = \sqrt{2gh}$ can be directly recovered when $\Delta p \approx 0$ and $A_T >> A_O$. Clearly, the liquid level y and pressure drop $p_T - p_A$ are the key driving forces. The solution breaks down when $A_O \to A_T$.

Example 2.10: Poiseuille Flow

Considering steady laminar fully-developed (unidirectional) flow of an incompressible fluid in a pipe of length "L" and constant cross sectional area (radius r_0), i.e., Poiseuille flow. Establish reduced forms of the N–S equations and solve for the axial velocity u(r).

Sketch	*Assumptions*	*Concept*
p_1, p(x), $\vec{g}$, p_2, $r = r_0$, r, u(r), θ, $r = 0$, x, u_{max}, (ρ,μ), L	As stated, i.e., • $\partial / \partial t = 0$ (steady-state) • $\vec{v} \Rightarrow u$ (uni-directional) • $\partial / \partial x = 0$ (fully-developed) • $\partial / \partial \theta = 0$ (axisymmetric)	Reduced N-S eq'ns based on the *postulates*: $\vec{v} = (u,0,0)$, where $u = u(r)$ only; and $\nabla p \Rightarrow -\frac{\partial p}{\partial x} \hat{=} \frac{\Delta p}{L}$ $= \frac{p_1 - p_2}{L} = ¢$

Solution:

- Check continuity: $\nabla \cdot \vec{v} = 0$ can be reduced to $\frac{\partial u}{\partial x} = 0$.
- The Equation of Motion (Eq. (2.21)) reduces to:

$$0 = -\nabla p + \mu \nabla^2 \vec{v} \tag{E.2.10.1}$$

or with the *postulates* and App. A

$$0 = (\frac{\Delta p}{L}) + \mu \frac{1}{r} \frac{d}{dr}(r \frac{du}{dr}) \tag{E.2.10.2}$$

subject to the boundary conditions: $u(r = r_0) = 0$ (no slip) and at $r = 0,\ du/dr = 0$ (axisymmetry). Double integration and invoking the two boundary conditions yields:

$$u(r) = \frac{1}{4\mu}(\frac{\Delta p}{L})(r_0^2 - r^2) = u_{max}\left[1 - (\frac{r}{r_0})^2\right] \tag{E.2.10.3}$$

Comments: The solution (E.2.10.3) is the base case for steady laminar *internal* flows with numerous applications in terms of industrial pipe flows, such as pipe networks, heat exchangers and fluid transport, as well as idealized tubular flows in biomedical, chemical, environmental, mechanical and nuclear engineering. Equation (E.2.10.3) not only provides the parabolic velocity profile $u(r)$ but can also be used to calculate the volumetric flow rate Q (Eq. (2.5)), wall shear stress τ_w (Eq. (2.19)), pressure drop Δp (either via Eq. (2.5) or using Eq. (2.19)), head loss h_f (cf. Eq. (2.31b)), etc. Section 3.2 and Part B further illustrate the use of the Poiseuille flow solution (E.2.10.3).

2.5 Conservation Laws of Energy and Species Mass

Although many natural and industrial fluid flow problems are non-isothermal, fluid mechanics education typically only deals with constant-temperature flows, leaving thermal flows for separate thermodynamics and convection heat transfer courses. Species mass transfer is entirely left for chemical and biomedical engineers. Thus, we highlight the energy RTT plus the resulting heat transfer equation with an analogy to mass transfer, to lay the groundwork for some interesting engineering applications in Chaps. 3, 4, 7 and 9.

The first law of thermodynamics states that energy forms can be converted but the total energy is constant, i.e., conserved:

$$E_{total} = E_{kinetic} + E_{potential} + E_{internal} + \ldots = \text{constant} \tag{2.32}$$

2.5.1 Global Energy Balance

For *observation time* Δt the energy balance for *any* system can be written as:

$$\underbrace{\Sigma E_{in} - \Sigma E_{out}}_{\substack{\text{Net energy transfer by heat,} \\ \text{work, and/or mass flow in/out}}} = \underbrace{\Delta E_{system} := [\Delta kE + \Delta pE + \Delta U]_{system}}_{\substack{\text{Change of total energy inside the system} \\ \text{(e.g., kinetic, potential and internal energies)}}} \tag{2.33}$$

where the internal energy U is the sum of all microscopic energy forms, i.e., mainly due to molecular vibration.

On a rate basis:

$$\Sigma \dot{E}_{\text{in}} - \Sigma \dot{E}_{\text{out}} = \left.\frac{\mathrm{d}E}{\mathrm{d}t}\right|_{\text{system}} \approx \left.\frac{\Delta E}{\Delta t}\right|_{\text{system}} \tag{2.34}$$

For example, for a *closed system* with heat transferred to the system and work done by the system (see Fig. 2.9):

$$\dot{Q} - \dot{W} = \frac{\Delta E}{\Delta t} \tag{2.35}$$

or during Δt considering only an internal energy change from State 1 <initial> to State 2 <final>:

$$Q - W = \Delta U = U_2 - U_1 \tag{2.36}$$

Heat can be transferred via conduction, convection and/or thermal radiation; work done by the closed system, such as a piston-cylinder device, is typically boundary work:

$$W = \int_1^2 F ds := \int_1^2 p d\forall \tag{2.37}$$

For an *open system*, i.e., control volume, energy forms flowing in and out of the system have to be accounted for. Thus, Eq. (2.42) can be written for a fixed control volume with uniform streams entering and leaving (see Fig. 2.9b):

$$\left.\frac{\Delta E}{\Delta t}\right|_{c.\forall.} = \Sigma[\dot{Q}+\dot{W}]_{in} - \Sigma[\dot{Q}+\dot{W}]_{out} + \Sigma\left[\dot{m}\left(h+\frac{v^2}{2}+gz\right)\right]_{in}$$

$$-\Sigma\left[\dot{m}\left(h+\frac{v^2}{2}+gz\right)\right]_{out} \tag{2.38}$$

Where $h \equiv \tilde{u} + \frac{p}{\rho} = c_p T$ is the enthalpy per unit mass, combining internal energy, $\tilde{u} = c_v T$, and flow work due to pressure p moving specific volume v, i.e., pv; $e_{kin} = v^2/2$ is the kinetic energy, and $e_{pot} = gz$ is the potential energy per unit mass.

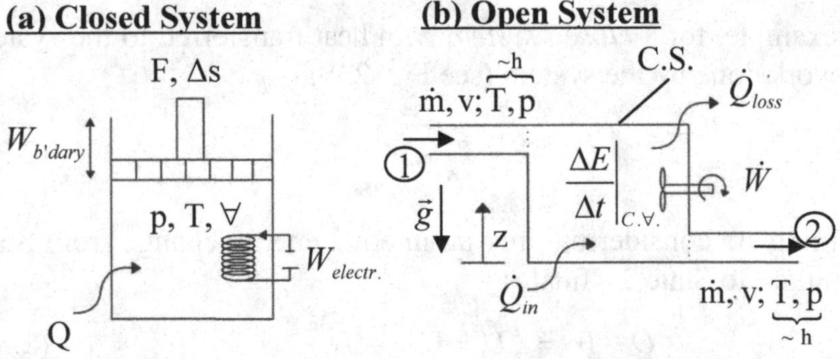

Fig. 2.9 Energy transfer for closed and open systems

As an aside, for steady, single inlet/outlet, frictionless flows without heat transferred and work done, the Bernoulli equation appears:

$$\left[\frac{p}{\rho}+\frac{v^2}{2}+gz\right]_{in}=\left[\frac{p}{\rho}+\frac{v^2}{2}+gz\right]_{out} \tag{2.39}$$

2.5.2 Energy Conservation in Integral Form

Taking $\boldsymbol{B}_N \equiv \mathrm{E}_{\text{total}}$ and hence $\beta \equiv \mathrm{e}_t$, the energy RTT reads [*Recall*: Eqs. (2.1) and (2.35)]:

$$\left.\frac{\mathrm{DE_t}}{\mathrm{Dt}}\right|_{\substack{\text{closed}\\ \text{system}}} \equiv \dot{\mathrm{Q}}-\dot{\mathrm{W}}=\frac{\partial}{\partial \mathrm{t}}\int_{\text{C.V.}}\rho \mathrm{e_t dV}+\int_{\text{C.S.}}\rho \mathrm{e_1}\vec{\mathbf{v}}\cdot d\vec{\mathbf{A}} \tag{2.40}$$

Typically, $\mathrm{E_{total}}=\mathrm{E_{kinetic}}+\mathrm{E_{internal}}:=\frac{\mathrm{m}}{2}|\vec{\mathbf{v}}|^2+\mathrm{m}\tilde{\mathrm{u}}$, i.e., the L.H.S. has to be separately expressed. The physical meaning of each term in Eq. (2.40) can be summarized as:

$$\begin{Bmatrix}\text{Time-rate-of-change}\\ \text{of } \mathrm{E_{total}} \text{ in material}\\ \text{volume moving with}\\ \text{the flow}\end{Bmatrix}=\begin{Bmatrix}\text{Rate-of-work}\\ \text{done on the C.V. by}\\ \text{surface forces plus}\\ \text{body forces}\end{Bmatrix}+\begin{Bmatrix}\text{Net heat flux}\\ \text{across the C.S}\\ \text{of the}\\ \text{control volume}\end{Bmatrix}$$

where the total heat flux $\vec{\mathbf{q}}_t=\vec{\mathbf{q}}_{\text{cond.}}+\vec{\mathbf{q}}_{\text{conv.}}+\vec{\mathbf{q}}_{\text{rad.}}$. In some cases, in order to complete the energy conservation law, a distributed internal heat source term, e.g., $\int_V \rho\hat{\mathrm{q}}_{\text{int}}\mathrm{dV}$, may have to be added. Finally, using Eq. (2.40), the energy RTT (see Sect. 2.2) reads for a stationary control volume:

$$\int_{C.V.} \rho(\vec{\mathbf{v}} \cdot \vec{\mathbf{f}}_b)dV + \int_{C.S.} (\vec{\mathbf{v}} \cdot \vec{\vec{\tau}})dA + \int_{C.S.} \vec{\mathbf{q}}_t \cdot d\vec{\mathbf{A}} + \int_{C.V.} \rho\hat{q}_{int}dV = \frac{\partial}{\partial t}\int_{C.V.} \rho e_t dV + \rho e_t \vec{\mathbf{v}} \cdot d\vec{\mathbf{A}} \tag{2.41}$$

2.5.3 Energy and Species Mass Conservation in Differential Form

Employing again the divergence theorem (see Eq. (2.9)), we can rewrite Eq. (2.41) as:

$$\rho\frac{De_t}{Dt} \equiv \rho\left[\frac{\partial e_t}{\partial t} + (\vec{\mathbf{v}} \cdot \nabla)e_t\right] = \rho(\vec{\mathbf{f}}_b \cdot \vec{\mathbf{v}}) + \nabla \cdot \left(\vec{\vec{T}} \cdot \vec{\mathbf{v}}\right) + \nabla \cdot q_t \tag{2.42}$$

where $e_t = \frac{1}{2}|\vec{\mathbf{v}}|^2 + \tilde{u} \cong \frac{1}{2}|\vec{\mathbf{v}}|^2 + c_v T$ and $\vec{\mathbf{q}}_t = -k\nabla T + \ldots$, where T is now the temperature.

A simpler derivation of the energy equation (2.42), resulting in a directly applicable form, starts with $h = \tilde{u} + p/\rho$ <enthalpy per unit mass> as the principal unknown, and considering $\vec{\mathbf{q}}$ <diffusive heat flux>, and $\nu\Phi$ <energy dissipation due to viscous stress>, we obtain:

$$\frac{\partial}{\partial t}(\rho h) + \nabla \cdot (\rho\vec{\mathbf{v}}h) = -\nabla \cdot \vec{\mathbf{q}} + \mu\Phi \tag{2.43}$$

With $dh \equiv c_p dT$, or simplified to $h = c_p T$ when c_p = constant, $\vec{\mathbf{q}} = k\nabla T$ after Fourier, and $\mu\Phi = \tau_{ij}\partial v_i/\partial x_j$, we obtain for thermal flow with constant fluid properties the *heat transfer equation*:

$$\frac{\partial T}{\partial t} + (\vec{\mathbf{v}} \cdot \nabla)T = \alpha\nabla^2 T + \frac{\mu}{\rho c_p}\Phi \tag{2.44}$$

where $\alpha = k/(\rho c_p)$ is the thermal diffusivity.

It is interesting to note that when contracting Eq. (2.44),

$$\frac{DT}{Dt} = \alpha\nabla^2 T + S_T \tag{2.45}$$

has the same form as the ***species mass transfer*** equation:

$$\frac{Dc}{Dt} = \mathcal{D}\nabla^2 c + S_c \tag{2.46}$$

where $\mathcal{D}$ is the binary diffusion coefficient and S_c denotes possible species sinks or sources. Clearly, momentum diffusivity ν (Eq. (2.22)), thermal diffusivity α (Eq. (2.44)) and mass diffusivity $\mathcal{D}$ (Eq. (2.46)) have the same dimensions [length^2/time].

Example 2.11: Thermal Pipe Flow ($q_{wall} = \cent$)

Consider Poiseuille flow (see Example 2.10) where a uniform heat flux, q_w, is applied to the wall of a pipe with radius r_0.

(A) Set up the governing equations for the fluid temperature assuming thermally fully-developed flow, i.e.,

$$\frac{T_w - T}{T_w - T_m} \equiv \Theta\left(\frac{r}{r_0}\right) \tag{E.2.11.1}$$

where $T_w(x)$ is the wall temperature, $T(r,x)$ is the fluid temperature, and $T_m(x)$ is the cross-sectionally averaged temperature, i.e.,

$$T_m = \frac{1}{\bar{u}A}\int_A uTdA \tag{E.2.11.2}$$

Note that $\Theta = \Theta(r)$ only, describing thermally fully-developed flow.

(A) Solve a reduced form of the heat transfer equation (Eq. 2.44) and develop an expression for the Nusselt number, defined as:

$$Nu = \frac{2r_0}{k}\frac{q_w}{T_w - T_m} := \frac{hD}{k} \tag{E.2.11.3a, b}$$

where k is the fluid conductivity and D is the pipe diameter.

Sketch	*Assumptions*	*Concept*
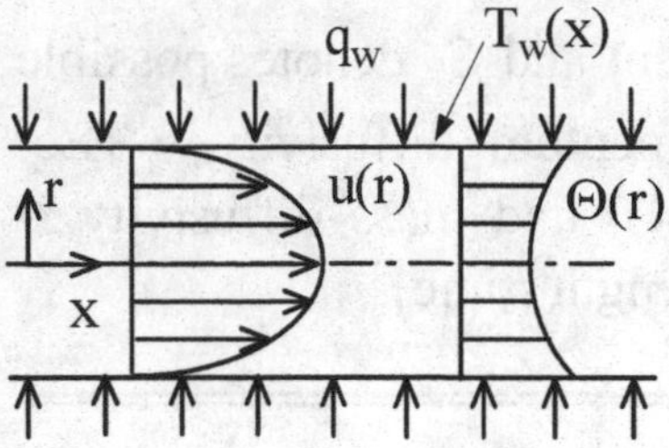 q_w $T_w(x)$ r u(r) $\Theta(r)$ x	As stated, i.e., • u(r) as in Eq. (E.2.10.3) • $\partial T / \partial t = 0$ <steady-state> • $(\vec{v} \cdot \nabla) T \Rightarrow u \partial T / \partial x$ • $\alpha \nabla T \Rightarrow \frac{\alpha}{r} \frac{\partial}{\partial r}(r \frac{\partial T}{\partial r})$	Reduced heat transfer equation (Eq. (2.44)) based on assumptions

Solution:

(A) With the result of Example 2.10 and the reduced heat transfer equation in cylindrical coordinates from App. A (see also list of assumptions), we have:

$$\frac{u(r)}{\alpha} \frac{\partial T}{\partial x} = \frac{1}{r} \frac{\partial}{\partial r}\left(r \frac{\partial T}{\partial r}\right) \tag{E.2.11.4}$$

(B) Employing the dimensionless temperature profile $\Theta(\frac{r}{r_0}) \equiv \Theta(\hat{r})$ given as Eq. (E.2.11.1), we can rewrite Eq. (E.2.11.4) as

$$-2 \frac{hD}{k}(1 - \hat{r}^2) = \frac{d^2 \Theta}{d\hat{r}^2} + \frac{1}{\hat{r}} \frac{d\Theta}{d\hat{r}} \tag{E.2.11.5}$$

Specifically,

- For $q_w = \text{¢}, \frac{\partial T}{\partial x} = \frac{dT_w}{dx} = \frac{dT_m}{dx} = \frac{2}{r_0} \frac{q_w}{\rho c_p \bar{u}} = \text{¢};$ (E.2.11.6a–c)
- As stated, $hD/k \equiv Nu_D := \frac{D}{k} \frac{q_w}{T_w - T_m} = \text{¢};$ (E.2.11.7a, b)
- And with $d\Phi / d\hat{r}$ being finite at $\hat{r} = 0$, we obtain

$$T - T_w = -(T_w - T_m)\mathrm{Nu}_D\left(\frac{3}{8} - \frac{\hat{r}^2}{2} + \frac{\hat{r}^4}{8}\right)$$

Now, by definition

$$T_w - T_m = \frac{2\pi}{\pi r_0^{\,2}\bar{u}}\int_0^{r_0}(T_w - T)\, u(r) r dr \qquad \text{(E.2.11.8)}$$

so that, when combining both equations and integrating, we have

$$1 = 4\mathrm{Nu}_D\int_0^1\left(\frac{3}{8} - \frac{\hat{r}^2}{2} + \frac{\hat{r}^4}{8}\right)(1-\hat{r}^2)\hat{r}\, d\hat{r}$$

from which we finally obtain:

$$\mathrm{Nu}_D = \frac{48}{11} = 4.36 \qquad \text{(E.2.11.9)}$$

Comment: It is interesting to note that for hydrodynamically and thermally fully-developed flow in a tube, subject to a constant wall heat flux, the Nusselt number (or the heat transfer coefficient) is constant. The same holds for the isothermal wall condition; however, the Nu-value is lower (see Kleinstreuer 1997; or Bejan 2002).

Example 2.12: Scale Analyses (for readers who like heat transfer)

Task A: Scale Natural Convection on a Vertical Wall

Consider steady laminar "natural convection" where, for example, air near a heated vertical wall rises against gravity. Within that thin thermal wall layer, the governing equations are (see Sketch):

$$\text{(air flow)} \quad \underbrace{u\frac{\partial u}{\partial x} + v\frac{\partial u}{\partial y}}_{\text{inertia}} = \underbrace{\nu\frac{\partial^2 u}{\partial y^2}}_{\text{friction}} + \underbrace{g\beta(T - T_\infty)}_{\text{buoyancy}} \qquad \text{(E.2.12.1)}$$

where β is the volumetric expansion coefficient, ($\beta \approx -\frac{1}{T}$ for air), T_∞ is the ambient air temperature far away from the heated wall and T is the actual air temperature within the thermal boundary layer. Thus, a second equation for T is necessary:

$$\text{(air temperature)} \qquad \underbrace{u\frac{\partial T}{\partial x} + v\frac{\partial T}{\partial y}}_{\text{heat convection}} = \underbrace{\alpha\frac{\partial^2 T}{\partial y^2}}_{\text{heat conduction}} \qquad \text{(E.2.12.2)}$$

Note: Equations (E.2.12.1) and (E.2.12.2) are special cases of Eqs. (2.23b) and (2.45b), respectively.

(i) Nondimensionalize Eq. (E.2.12.1), using $u \to \bar{u}$, $v \to \bar{v}$; $\partial x \to \ell$, $\partial y \to \delta$; and $\Delta T = T_w - T_\infty$ as scales, i.e., reference parameters, where l is the wall height, δ is the (variable) boundary-layer thickness, and T_W is the (high) wall temperature.

(ii) Form proportionalities between different terms in these equations, representing forces and fluxes, to generate dimensionless groups and functional dependencies for (x) and $\bar{u}(x)$.

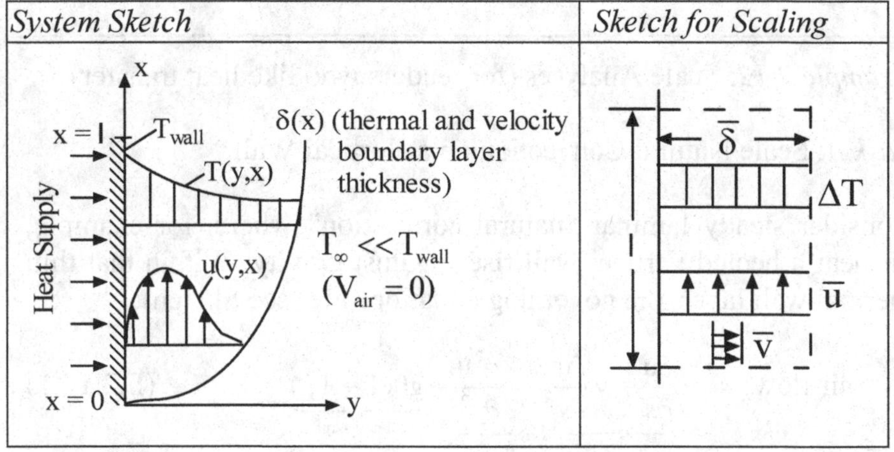

Solutions:

(*i*) Replacing the variables u, v, x, y, and T by $u = \hat{u}\,\bar{u}$, $v = \hat{v}\,\bar{v}$, $x = \hat{x}\ell, y = \hat{y}\,\delta$ and $T = \hat{T}\,\Delta T$ in Eqs. (E.2.13.1, 2), we obtain:

$$\left(\frac{\bar{u}^2}{\ell}\right)\bar{u}\frac{\partial \hat{u}}{\partial \hat{x}} + \left(\frac{\bar{u}\,\bar{v}}{\delta}\right)\hat{v}\frac{\partial \hat{u}}{\partial \hat{y}} = \left(\frac{\nu \bar{u}}{\delta^2}\right)\frac{\partial^2 \hat{u}}{\partial \hat{y}^2} + (g\beta\Delta T)$$

or, dividing through by $(\bar{u}^2/\ell)$ we obtain:

$$\bar{u}\frac{\partial \hat{u}}{\partial \hat{x}} + \underbrace{\underbrace{\left(\frac{\bar{v}}{u}\right)\left(\frac{\ell}{\delta}\right)}_{\varepsilon\cdot\varepsilon^{-1}}}_{O(1)}\frac{\partial \hat{u}}{\partial \hat{y}} = \underbrace{\underbrace{\frac{\nu}{\bar{u}l}}_{Re_l^{-1}}\underbrace{\left(\frac{\ell}{\delta}\right)^2}_{\varepsilon^{-2}}}_{O(1)}\frac{\partial^2 \hat{u}}{\partial \hat{y}^2} + \underbrace{\frac{g\beta\Delta T}{\bar{u}^2}\ell}_{\sim Ri(\text{Richardson number})} \qquad \text{(E.2.12.3)}$$

where by definition $\varepsilon < 1$, $Re_\ell >> 1$, and the Richardson number

$$Ri = \frac{g\beta\Delta T\ell^3}{\nu^2\left(\frac{u\ell}{\nu}\right)^2} \equiv \frac{Gr}{Re^2} \cong \frac{\text{Grashof \#}}{\text{Reynolds \# squared}} \qquad \text{(E.2.12.4)}$$

Clearly, we obtained a dimensionless PDE for buoyancy-driven flow plus a new dimensionless group, the Richardson number which encapsulates buoyancy, inertia and viscous forces.

(*ii*) Scale analysis of the first convection term in Eq. (E.2.12.2) balanced by the thermal diffusion term yields with $\delta T \to \Delta T = T_w - T_\infty$:

$$\bar{u}\frac{\Delta T}{\rho} \sim \alpha\frac{\Delta T}{\delta^2} \quad \text{or} \quad \bar{u} \sim \frac{\alpha\ell}{\delta^2} \qquad \text{(E.2.12.5)}$$

Recall from Eq. (E.2.12.1) that very near the vertical wall, where the inertia effects are negligible (i.e., $u < 1$, $\partial u/\partial x < 1$, and $v << 1$), upward buoyancy forces are counteracted by viscous forces, via:

$$g\beta\Delta T \sim \nu\frac{\bar{u}}{\delta^2} \qquad \text{(E.2.12.6)}$$

Eliminating $\bar{u}$ in Eq. (E.2.12.6) with Eq. (E.2.12.5) yields

$$\nu \frac{\alpha l}{\delta^4} \sim g\beta\Delta T$$

or

$$\frac{\delta}{\ell} \sim \left(\frac{g\beta\Delta T}{\alpha\nu} \ell^3 \right)^{-\frac{1}{4}} = Ra_{\ell}^{-1/4} \qquad \text{(E.2.12.7)}$$

where Ra = GrPr is the Rayleigh number, a ratio of buoyancy to thermal/viscous forces, while $Pr = \nu/\alpha$ is the Prandtl number, relating the fluid's kinematic viscosity and thermal diffusivity.

Replacing the wall or plate height ℓ by x, we can deduce that the boundary layer thickness (here $\delta = \delta_{th}$ for Pr = 1) varies as:

$$\delta \sim x^{1/4} \qquad \text{(E.2.12.8a)}$$

Combining (E.2.12.5) with (E.2.12.7) provides an expression for the vertical mean or reference velocity, i.e.,

$$\bar{u} \sim \sqrt{x} \quad \text{or} \quad u_{ref} = \sqrt{g\beta(T_w - T_\infty)\ell} \qquad \text{(E.2.12.8b,c)}$$

Task B: A Note on the Reynolds Number

The Reynolds number, being the ratio of inertial vs. viscous forces, is universally considered to be "always important". Actually, this is not always the case. For example, for Poiseuille flow (see Example 2.10) Re = 0 in light of the definition used in Example 1.1 because the inertia term $(\vec{v} \cdot \nabla)\vec{v}$ is identical to zero while the fluid flow is in dynamic equilibrium between the driving force (i.e., pressure gradient) and resistance (i.e., shear stress). Thus, on a case-by-case basis, some dimensionless groups may need a reinterpretation. For steady pipe flow we could rewrite the re-definition as the ratio of flow momentum to wall resistance.

$$Re_D = \frac{4\dot{m}}{\pi\mu D} := \frac{\dot{m}v}{\mu v D} \qquad \text{(E.2.12.9a, b)}$$

Alternatively, we could focus on the pipe entrance region in which $(\vec{v}\cdot\nabla)\vec{v} \Rightarrow \left(v_r \frac{\partial v_z}{\partial r} + v_z \frac{\partial v_z}{\partial z}\right)$ is non-zero and the conversion of U_{inlet} to a fully-developed $v_z(r) = 2v_{av}\left[1-\left(\frac{r}{R}\right)^2\right]$ takes place, where $Re_D < 2{,}000$.

2.6 Homework Assignments

Solutions to homework problems done individually or in, say, three-person groups should help to further illustrate fluid dynamics concepts as well as approaches to problem solving, and in conjunction with App. A, sharpen the reader's math skills (see Fig. 2.1). Note, there is no substantial correlation between good HSA results and fine test performances, just *vice versa*. Table 1.1 summarizes three suggestions for students to achieve a good grade in fluid dynamics – for that matter in any engineering subject. The key word is "independence", i.e., equipped with an equation sheet (see App. A), the student should be able: (i) to satisfactory answer all concept questions and (ii) to solve correctly all basic fluid dynamics problems.

The "Insight" questions emerged directly out of the Chap. 2 text, while some "Problems" were taken from lecture notes in modified form when using White (2006), Cimbala & Cengel (2008), and Incropera et al. (2007). Additional examples, concept questions and problems may be found in any UG fluid mechanics and heat transfer text, or on the Web (see websites of MIT, Stanford, Cornell, Penn State, UM, etc.).

2.6.1 Text-Embedded Insight and Problems

2.6.1 Provide quick answers (sketches, streamlines, math, physics) to the following counter-intuitive fluid-flow scenarios:

(a) Keeping the tailgate of a pick-up truck *up* reduces aerodynamic drag (Why?) and hence saves gasoline; although, most drivers intentionally keep it *down* and even install "airflow" nets to retain cargo when accelerating.
(b) Under otherwise identical conditions, it is easy to *blow* out a candle but nearly impossible to *suck* it out. Why?
(c) When bringing a spoon near a jet, e.g., faucet stream, it gets *sucked into* the stream. Try it out and explain!
(d) Chunks of metal are *torn out* from ship propellers at high speeds after a long period of time in operation. Why?
(e) The long hair of a girl driving a convertible is being pushed *into her face* rather than *swept back*. How come?
(f) Consider a humming bird being quasi-stationary in your moving car. What happens to it if you suddenly brake?

2.6.2 Consider the del-operator ∇ (see App. A.1.2):
(a) Explain $\nabla \cdot \vec{\mathbf{v}}$, $\nabla \vec{\mathbf{v}}$ and $\nabla \times \vec{\mathbf{v}}$, where $\vec{\mathbf{v}}$ is a velocity vector.
(b) Show that $\nabla \cdot \nabla = \nabla^2$, the Laplacian operator. (c) What is the physical meaning of $(\vec{\mathbf{v}} \cdot \nabla)\vec{\mathbf{v}}$ and how can the result (A.1.15) be interpreted?

2.6.3 In light of Sect. A.2.3, derive a finite-difference approximation for, say, $f'(x_i)$ of second-order accuracy.

2.6.4 The theorem of Green (or Gauss), known as the "divergence theorem" (see Eq. (A.3.1) in App. A) transforms a surface integral of a vector function into a volume integral and *vice versa*: (a) Would the conversion also work for a scalar function? (b) Provide a graphical illustration with physical explanation of Eq. (A.3.1).

2.6.5 Plot the error functions (A.3.3) and (A.3.4) and give two applications.

2.6.6 Classify (see Page A-15) the following ODEs and suggest best possible solution techniques:
(a) $y' = a(x) + b(x)y + c(x)y^2 + d(x)y^3$;

(b) $y' = a(x)y^n + b(x)y$

(c) $f(xy' - y) = g(y')$

(d) $x^2y'' + xy' - (x^2 + n^2)y = f(x)$

2.6.7 In Sect. 2.1, under "Fundamental Assumptions", it is stated that a fluid is a continuum when $\Delta\forall_{min} \leq 10^{-9}\,mm^3$, i.e., $L_{min} \leq 1\mu m$. What is the ratio of $\lambda_{fluid}/L_{system}$ for (a) air and (b) water when λ_{fluid} is an appropriate molecular length scale and $L_{system} \approx L_{min}$.

2.6.8 Can the Reynolds Transport Theorem (RTT) applied to quantities in all three tensor ranks, i.e., scalar vector and tensor? Give one example each!

2.6.9 Of interest is the conservation of *angular momentum* in terms of the RTT, i.e., $\frac{D}{Dt}\vec{\mathbf{H}}_{system} = \sum \mathbf{M}$. Here, $\vec{\mathbf{H}}_{system} = \vec{\mathbf{r}} \times m\vec{\mathbf{v}}$, is the angular momentum with m being the constant mass, $\sum \vec{\mathbf{M}}$ is the net moment applied to the system, and $\vec{r}$ is the moment radius (or arm). Express $\sum \vec{\mathbf{M}}$ for any control volume, i.e., stationary, moving, or deforming.

2.6.10 A horizontal fire-hose (nozzle diameter 6 cm) discharges typically 5 m^3/min of water ($\rho = 1{,}000$ kg/m^3). Find the force (and direction) to hold the fire-hose in place. Comment!

2.6.11 Consider a jet-plane with the engine mounted at the rear end discharging $\dot{m}_{gas} = 18$ kg/s of exhaust gases at v = 250 m/s relative to the plane. To shorten landing, a deflector-vane (called a "thrust reverser") is lowered into the path of the exhaust stream, which deflects the gases and hence aids in braking. Analyze the effect of the vane angle, θ, on the braking force, i.e., a horizontal $\theta = 0°$ (no effect) to a 180° (full effect). Discuss the $F_{brake}(\theta)$ plot!

2.6.12 A skydiver ($m_{total} = m_{person} + m_{gear}$) jumps out of a plane and at terminal velocity v_T he/she opens at t = 0 a parachute ($F_{drag} = kv^2$) and lands with a final velocity v_F. Show that $k = mg / v_F^2$ and derive an equation for *v*(*t*). Note that $v = v(t, v_T, v_F, g)$, but not m_{total}; why? Plot v(t) for typical parameter values and comment.

2.6.13 Consider a radial water turbine where water from a nozzle enters the impeller at its outer edge (diameter D) with velocity v (and mass flow rate $\dot{m}$) under angle θ w.r.t. the radial direction. The water leaves with v in radial direction. What are the resulting shaft torque and turbine power output when the maintained angular velocity is ω_0?

2.6.14 Based on the definition Eq. (2.16) for the correction factor α, compare $\alpha_{laminar}$ (Poiseuille flow) and $\alpha_{turbulent}$ (use the 1/7-law), and comment.

2.6.15 Provide a physical interpretation of Eq. (2.17) and sketch an example with all four terms appearing.

2.6.16 Employing the REV-approach of Sect. 2.3.3, derive the transient 3-D continuity equation in *cylindrical* coordinates. Note, for incompressible fluids $\nabla \cdot \vec{v} = 0$; does that also hold for transient flows?

2.6.17 List the Pros & Cons of the integral approach, say, using the RTT or lumped-parameter method, vs. the differential approach, i.e., solving reduced forms of the continuity and Cauchy or N–S equations. Give two examples each when to use which methodology.

2.6.18 Derive the Bernoulli equation two distinct ways, relying on first principles. Interpret each term of the resulting equation three ways.

2.6.19 Discuss with examples *boundary work vs. flow work*, and explain why on a differential bases we start with δW rather than dW?

2.6.20 Elaborate on the origin/derivation and similarities between Eqs. (2.45) and (2.46).

2.6.2 Additional Problems

2.6.21 Liquid flow velocity at a solid wall: (a) Empirical evidence shows that $\vec{v}_{fluid} = \vec{v}_{wall}$, e.g., if the wall/surface is stationary and impermeable, u = v = w = 0; however, wall-velocity slip may occur; but, when and why? (b) From the view-point of two reference frames, i.e., one fixed on the ground and the other fixed to the moving object (say, an airplane), what is $\vec{v}_{air}$ in the two reference-frame cases and what is $\vec{v}_{air}$ far away from the plane in both reference frames?

2.6.22 Oil at 60°C ($\rho = 864 kg/m^3$; $\mu = 72.5 \times 10^{-3} kg/m \cdot s$) is pushed ($\Delta p_{gage} = 1 atm$) between two horizontal parallel plates ($L = 1.5m, W = 0.75m$) a small distance h apart. Plot the Reynolds number, $Re_h = \rho u_{mean} h/\mu$, as a function of spacing h for $10\,\mu m \le h \le 1\,mm$, and comment.

2.6.23 A velocity field is given/measured as:

$$\vec{v} = (ax + b)i + (cx^2 - ay)j$$

where a–c are constants: (a) Classify this flow field; (b) check if the associated pressure field, $p = p(x, y)$, is a smooth function.
Hint: Cross-differentiation of the pressure gradients should generate the same results for "function smoothness", i.e.,

$$\frac{\partial}{\partial x}\left(\frac{\partial p}{\partial y}\right) = \frac{\partial}{\partial y}\left(\frac{\partial p}{\partial x}\right)$$

2.6.24 Compare "gravity-film flow" for (i) two vertical parallel walls a distance h apart; and (ii) a vertical wall with smooth falling film of thickness h. Assume constant pressure and no end effects. Plot both velocity profiles and discuss the impact of h on Q, the volumetric flow rate. Provide one sample application each.

2.6.25 Consider a vertical concentric shaft rotating and translating (d, ω_0, u_0) inside a stationary cylinder (i.e., housing of inner diameter D) with a lubricant (or slurry) of properties ρ and μ: (a) What industrial operation can be modeled with this shaft/pipe annulus? (b) Set up the reduced N–S equations plus B.Cs based on suitable assumptions and postulates. (c) Can you solve for the velocity field and draw a representative velocity profile? (d) Estimate the amount of lubricant which has to be constantly supplied.

2.6.26 Consider levitation of a sphere (D, ρ_{sp}) in an upward stream of air (ρ, μ, T): (a) What is the necessary air-velocity v(D) for different property ratios ρ_{sp}/μ_{air}; (b) Does your analysis hold for droplets as well?

Recall: $Re_D = \dfrac{\rho v D}{\mu} < 1.0$ for creeping flow analysis.

Chapter 3

Introductory Fluid Dynamics Cases

The previous chapter featured several examples to illustrate each topic introduced (or just reviewed) in Sects. 2.2–2.5. They included a few special cases of the conservation laws in both integral form [i.e., the simplified Reynolds Transport Theorem (RTT)] and differential form [i.e., the reduced Navier–Stokes (N–S) equations]. In this chapter, additional physical insight and a few new applications are provided, while problem-solving skills are honed in preparation for theories/models and applications in Parts B and C. Specifically, with Sects. 3.4–3.7 four optional topics are included to better prepare students for the subsequent chapters, industrial demands, and/or post-graduate courses.

3.1 Inviscid Flow Along a Streamline

As shown in Sect. 2.4.3, Bernoulli's equation, typically for steady, incompressible, inviscid flow along a representative streamline, is a special case of the momentum equation. Alternatively, it was also derived from the energy equation in Sect. 2.5, and it could be also derived with the Reynolds Transport Theorem applied to a streamtube (see HSAs in Sect. 2.6). Bernoulli's equation and simple algebra often suffice to explain complex phenomena, answer some of the questions stated in the introduction to Chap. 2 (Sect. 2.1), or compute conjugated velocity–pressure problems (see Examples 3.1C and D). Thus, the next four examples illustrate both the power and ease-of-use of the Bernoulli equation. For steady *inviscid* flow (see Eq. (2.39)):

C. Kleinstreuer, *Modern Fluid Dynamics: Basic Theory and Selected Applications in Macro- and Micro-Fluidics*, Fluid Mechanics and Its Applications 87,
DOI 10.1007/978-1-4020-8670-0_3, © Springer Science+Business Media B.V. 2010

$$\frac{p}{\rho g} + \frac{v^2}{2g} + z = \text{const.} \tag{3.1a}$$

or applied to two points on a representative streamline:

$$\frac{p_1}{\rho g} + \frac{v_1^{\ 2}}{2g} + z_1 = \frac{p_2}{\rho g} + \frac{v_2^{\ 2}}{2g} + z_2 \tag{3.1b}$$

Examples 3.1 A–D: Physical Insight Gained from Solving Bernoulli's Equation.

(A) Why do airplanes fly? Well, *they need a lift force larger than the plane+ passenger and baggage weights.*

Sketch:

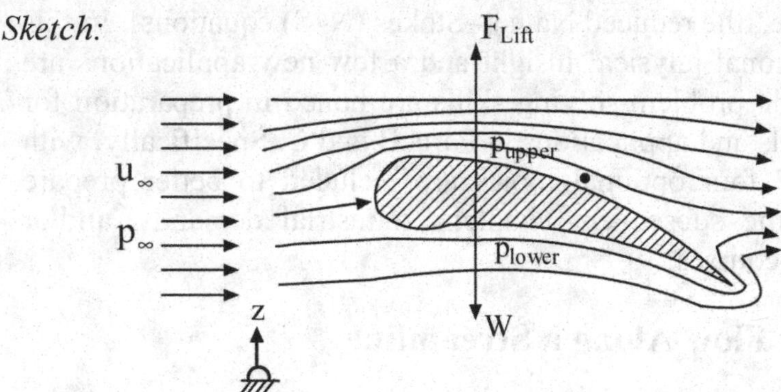

Rational:

- Bernoulli states $\frac{p}{\gamma} + \frac{v^2}{2g} + z = ¢$; here, with $z_\infty = z_{lower} \approx z_{upper}$:

$$p_e + \frac{\rho}{2} v_e^{\ 2} = p_u + \frac{\rho}{2} v_u^{\ 2} = p_\infty + \frac{\rho}{2} v_\infty^{\ 2} \tag{E.3.1.1}$$

- Observation
  Above the wing (or airfoil) streamlines cluster and the air velocity is very high so that with z ≈ constant, p_{upper} is low according to Bernoulli. The opposite occurs below the wing, i.e., $v_{lower} \downarrow p_{lower} \uparrow$. Hence, for an effective airfoil surface:

$$F_{lift} = (p_l - p_u) A_{surface} > W \tag{E.3.1.2}$$

(B) Why do roofs fly off houses during tornadoes? *Perhaps because the pressure field above the pitched roof is much lower than the atmospheric pressure in the attic.*

Sketch:

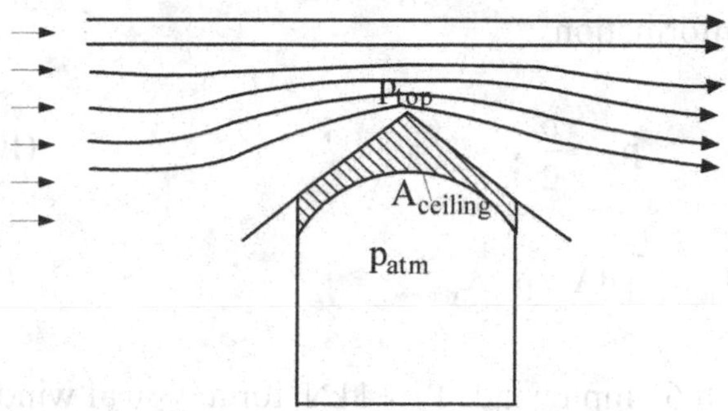

Rational:

- Similar to (A), the pressure field on the roof due to the locally very high wind velocity is very low so that the pressure force inside the house, $p_{atm} * A_{ceiling}$, blows the roof off the house.

(C) Estimate the wind force on a highrise window.

Sketch	*Assumptions*	*Concepts*
① ② F_R	• Steady incompressible flow on representative streamlines • No frictional effects • Constant velocity at Point ① and uniform pressure on window	• Bernoulli • Gage pressure $p_1 = 0$ • $\Delta z \approx 0$ • $v_z = 0$ at the stagnation point

Solution:

In general,

$$\frac{v_1^2}{2g}+\frac{p_1}{\rho g}+z_1=\frac{v_2^2}{2g}+\frac{p_2}{\rho g}+z_2 \qquad \text{(E.3.1.3a)}$$

With the given information,

$$p_2=\frac{\rho}{2}v_1^2 \qquad \text{(E.3.1.3b)}$$

and hence

$$F_R=\int p dA \approx p_2\, A_{\text{window}} \qquad \text{(E.3.1.4)}$$

For example, for a 65 mph wind, $F_R \approx 1\,\text{kN}$ for a typical window. Similarly, the form drag, part of the total drag on submerged bodies

$$F_{\text{Drag}}^{\text{total}}=F_{\text{net pressure}}^{\text{form}}+F_{\text{wall shear}}^{\text{friction}} \qquad \text{(E.3.1.5)}$$

can be estimated using Bernoulli's equation, i.e.,

$$F_{\text{Pressure}}=\Delta p A_{\text{projected}} \qquad \text{(E.3.1.6a)}$$

where $$\Delta p=p_{\text{stagnation}}-p_{\text{wake}} \qquad \text{(E.3.1.6b)}$$

(D) Consider an oscillating disk suspended on a fluid jet

A horizontal disk of mass M can only move vertically when a water jet (d_0, v_0) strikes the disk from below. Obtain a differential equation for the disk height h(t) above the jet exit plane when the disk is initially released at $H>h_0$, where h_0 is the equilibrium height. Find an expression for h_0, sketch h(t), and explain.

Sketch	*Assumptions*	*Concepts*
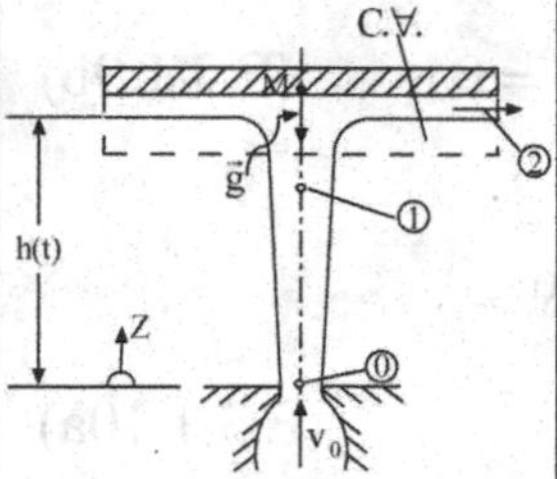	• Steady laminar frictionless flow represented by a streamline with points 0, 1, 2 • Near z = 0: $\rho=¢$, $v_0=¢$, and $p_0=p_1=p_2$ • Moving, accelerating, C.∀., i.e., $v_{rel}=v_{fluid}-v_{C.\forall.}$ and $a_{C.\forall.}=a_{disk}=d^2h/dt^2$ • Averaged velocities, i.e., $v_{C.\forall.}=dh/dt$; $v_2\approx 0$ • $M_{disk} \gg m_{fluid}\big\vert_{C.\forall.}$, and $A_1 > A_0$	• Bernoulli Equation • Continuity Equation • Momentum RTT

Solution:

- (Bernoulli) $\frac{p_0}{\rho}+\frac{v_0^2}{2}+gz_0=\frac{p_1}{\rho}+\frac{v_1^2}{2}+gz_1$ can be reduced to:

$$\frac{v_0^2}{2}=\frac{v_1^2}{2}+g\,h(t)\Rightarrow v_1=\sqrt{v_0^2-2gh} \qquad \text{(E.3.1.7a, b)}$$

- (Continuity) $v_0\,A_0=v_1\,A_1\Rightarrow A_1=A_0\,\frac{v_0}{v_1}$

- (Momentum RTT)

$$\overset{\approx 0}{\cancel{F_s}}+F_B-\int_{C.\forall.}a_{C.\forall.}\,\rho d\forall=\overset{0}{\cancel{\frac{\partial}{\partial t}\int_{C.\forall.}v\rho d\forall}}+\int_{c.s.}v\rho\vec{v}_{rel}\cdot d\vec{A} \qquad \text{(E.3.1.8a)}$$

can be reduced to:

$$-Mg-Ma=v_{rel}[-\rho(v_{rel}\,A_1)]+\overset{\approx 0}{\cancel{v_2}}\,\dot{m}_2 \qquad \text{(E.3.1.8b)}$$

with $v_{rel}=v_1-\frac{dh}{dt}$ and $a=\frac{d^2h}{dt^2}$, we have:

$$M\left(g + \frac{d^2h}{dt^2}\right) = \rho\left(v_1 - \frac{dh}{dt}\right)^2 A_1 \tag{E.3.1.9a}$$

Substituting v_1 and A_1 yields:

$$\ddot{h} - \frac{\rho\, v_0\, A_0}{M\sqrt{v_0^2 - 2gh}}\left(\sqrt{v_0^2 - 2gh} - \dot{h}\right)^2 + g = 0 \tag{E.3.1.9b}$$

Now, at equilibrium height, $h = h_0, \dot{h} = \ddot{h} = 0$. Thus,

$$-\rho\sqrt{v_0^2 - 2gh_0}\; A_0 v_0 + Mg = 0 \tag{E.3.1.10a}$$

or

$$h_0 = \frac{v_0^2}{2g}\left[1 - \left(\frac{gM}{\rho\, v_0^2\, A_0}\right)^2\right] := \frac{v_0^2}{2g}\left[1 - \left(\frac{Mg}{\dot{m} v_0}\right)^2\right] \tag{E.3.1.10b, c}$$

Graph:

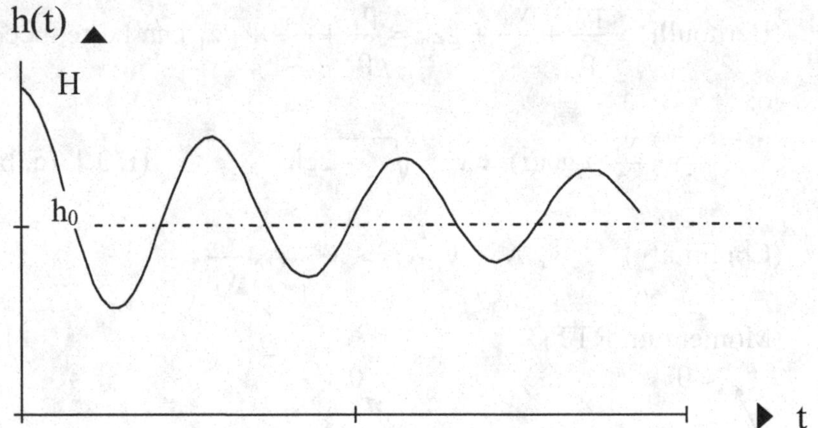

Comments:

(i) Although frictionless flow was assumed, the ODE for h(t) is nonlinear which implies oscillations as well as a decrease in amplitude.

(ii) This problem solution couples Bernoulli's equation with mass and momentum conservation and shows the interactive nature of fluid mechanics (see also Chap. 8).

(iii) The Bernoulli equation is even more powerful when extended to include frictional and form losses; for example, in pipe flow with entrance effects, changes in cross sectional area, valves, etc (see Eq. (2.31)). This is further discussed in Chap. 4.

3.2 Quasi-unidirectional Viscous Flows

Analytic solutions to unidirectional one- or two-dimensional flows are most desirable for a better understanding of the physics of fluid flow, i.e., "physical insight", as well as for benchmark computer model validations.

In unidirectional or *parallel flows*, the velocity vector has only one component, e.g., $\vec{\mathbf{v}} = (u, 0, 0)$, where in general u = u(y, z; t) and hence Stokes equation (Eq. 2.22) has to be solved. Unidirectional flow implies that the pressure gradient is constant, the flow is fully-developed, and the resulting momentum equation expresses a dynamic equilibrium (or force balance) at all times between a 1-D driving force and frictional resistance:

$$\underbrace{\frac{\partial p}{\partial x} \text{ and/or } \rho g_x}_{\text{driving forces}} \sim \underbrace{\frac{\partial \tau_{xy}}{\partial y}}_{\text{resistance}} \tag{3.2a, b}$$

3.2.1 Steady 1-D Laminar Incompressible Flows

For steady 1-D incompressible flow u = u(y), or $v_z = v_z(r)$ in cylindrical coordinates, as already discovered when discussing Couette and Poiseuille flows in Examples 2.8 and 2.10. In terms of the reduced x-momentum of the Navier–Stokes (N–S) equations, where u = u(y) only and $\frac{\partial p}{\partial x} = \text{constant}$, we have:

$$0 = -\left(\frac{\partial p}{\partial x}\right) + \mu \frac{d^2 u}{dy^2} + \rho g_x \tag{3.3a}$$

Rewriting Eq. (3.3a) with $-\frac{\partial p}{\partial x} \approx \frac{\Delta p}{l}$ as

$$\frac{d^2u}{dy^2} = \frac{1}{\mu}\left[\left(\frac{\Delta p}{l}\right) - \rho g_x\right] = \not{c} \tag{3.3b}$$

it is apparent that steady laminar 1-D parallel flows can be described by a second-order ODE (see App. A)

$$u'' = K \tag{3.3c}$$

subject to two boundary conditions, e.g., "no-slip" and symmetry.

In general, Table 3.1 summarizes for any fluid flow system, necessary *assumptions and their consequences.* In order to understand and appreciate the "differential approach" for solving modern fluid dynamics problems, solutions of Eq. (3.3c) are discussed in subsequent examples. They should illustrate the following:

- There is no mystery to setting up and finding exact (or useful approximate) solutions to reduced forms of the N–S equations.
- These results (see Examples 3.2–3.5) provide some interesting insight to the physics of fluid flow. They form base-case solutions to a family of engineering applications, such as film coating, internal flows and lubrication, as well as non-Newtonian fluid flows, such as exotic oils, blood, paints, polymeric liquids, etc. (see also Sect. 6.3).

For all practical purposes, the generic steps for *setting up differential analysis problems* in fluid mechanics and solving the resulting differential equation, include:

(i) Clever placement of the appropriate coordinate system into the system sketch is important, which also reveals the principal flow direction and hence identifies the momentum component of interest.

(ii) The velocity vector $\vec{v}$ and pressure gradient ∇p are the key unknowns; thus, based on the given flow system and with the help of Table 3.1, *postulates* for $\vec{v}$ and ∇p have to be provided first. Specifically which $\vec{v}$-component and ∇p-component are non-zero and what are their functional dependence can be determined via the stated assumptions, boundary conditions, and check of the continuity equation (see Examples 2.9–2.11).

(iii) Once functional postulates are determined, the non-zero component of the N–S equations can be deduced and the resulting ODE integrated, subject to appropriate boundary conditions.

Table 3.1 Flow assumptions and their impact

Flow assumption	Explanation	Examples
• Time-dependence	$\frac{\partial}{\partial t} = 0$ implies steady-state; but, flow is transient when $\vec{\mathbf{v}} = \vec{\mathbf{v}}(t)$	Transient flow examples: pulsatile flow; fluid-structure vibration; flow start-up/shut-down; etc.
• Dimensionality	Required number of space coordinates	1-D: Couette flow; Poiseuille flow; Thin-film flow 2-D: Boundary – layer flow; Pipe- entrance flow 3-D: Everything real …
• Directionality	Required number of velocity components: $\vec{v} = (u, v, w)$ or $\vec{v} = (v_z, v_r, v_\theta)$	Usually same as "Dimensionality" with some exceptions: For example, for the rotating parallel disk (or viscous clutch) problem, $\vec{\mathbf{v}} = (0, v_\theta, 0)$ where $v_\theta = v_\theta(r, \theta)$, i.e., unidirectional; but, the system is 2-D
• Development phase	$\frac{\partial v}{\partial s} = 0$: fully-developed flow ($s \triangleq$ axial coordinate)	"Fully-developed" … implies no velocity profile changes in that direction, say, s

• Symmetry	$\frac{\partial}{\partial n} = 0$: midplane (n $\hat{=}$ normal coordinate) $\frac{\partial}{\partial \theta} = 0$: axisymmetry	Self-explanatory
• FlowRegime: $\left\langle \begin{array}{l} \text{Laminar} \rightarrow \text{Re}_{\max} < \text{Re}_{critical} \\ Turbulent \rightarrow \text{Re} > \text{Re}_{critical} \end{array} \right.$ where $\text{Re}_{critical} \approx \begin{cases} 2{,}000 \textit{ for pipe flow} \\ 5 \times 10^5 \textit{ for flat plate } B-L \textit{ flow} \end{cases}$		

Note: The *Equation Sheet* (see App. A) should be the best and only information source (other than data tables and data charts) accompanying all solution procedures.

Example 3.2: Film Coating

Consider "film coating", i.e., a liquid fed from a reservoir forms a film pulled down on an inclined plate by gravity. Obtain the velocity profile, flow rate, and wall shear force.

Sketch

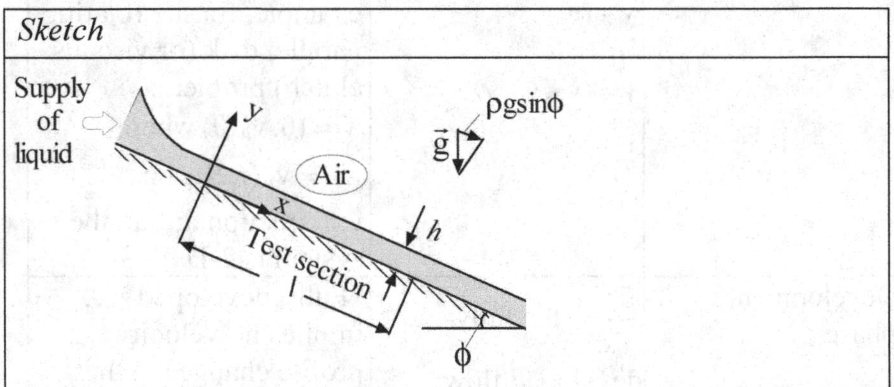

Assumptions	*Concepts*
• Steady laminar fully-developed flow • Negligible air–liquid shear stress • Constant h, p, ρ and μ • Neglect end effects, i.e., h = const. • Re ≤ 20 to avoid film surface rippling	• Reduced N–S eqn. $\vec{v} = (u, 0, 0)$; $u = u(y)$ only; $\nabla p = 0$ • gravity, ρg sin ϕ, drives the film flow

Solution:

Concerning the Solution Steps (i)–(iii) just discussed, an inclined x–y coordinate system is attached to the plate, i.e., the x-momentum equation is of interest (Note, the y- and z- momentum equations are both zero). Based on the sketch and assumptions, $\vec{v} = (u, 0, 0)$, $\nabla p \approx 0$ (thin-film condition, or no air-pressure variations), and $\partial / \partial t = 0$ (steady-state). With $v = w = 0$, continuity $\nabla \cdot \vec{v} = 0$ indicates that:

$$\frac{\partial u}{\partial x} + 0 + 0 = 0 \qquad \text{(E.3.2.1)}$$

which implies fully-developed (or parallel) flow; thus, u = u(y) only.

Consulting the equation sheet (App. A) and invoking the postulates u = u(y), v = w = 0, and $\partial p / \partial x = 0$, we have with the body force component $\rho g \sin \phi$,

$$0 = \mu \frac{d^2 u}{dy^2} + \rho g \sin \phi \qquad \text{(E.3.2.2)}$$

subject to u(y = 0) = 0 <no-slip> and $\tau_{\text{interface}} = \mu \left.\frac{du}{dy}\right|_{y=h} \approx 0$ which implies that at y = h → du/dy = 0 <zero velocity gradient>. Double integration of Eq. (E.3.2.2) in the form $u'' = K$ (see App. A), yields after invoking the two B.C.s:

$$u(y) = -\frac{\rho g h^2 \sin \phi}{2\mu} [\left(\frac{y}{h}\right)^2 - 2y/h] \qquad \text{(E.3.2.3a)}$$

Clearly,

$$u(y=h)=u_{max}=\frac{\rho g h^2}{2\mu}\sin\phi \tag{E.3.2.3b}$$

and the average velocity

$$u_{av}=\frac{1}{A}\int_A \vec{\mathbf{v}}\cdot d\vec{\mathbf{A}} := \frac{1}{h}\int_0^h u\,dy = \frac{\rho g h^2 \sin\phi}{3\mu} \tag{E.3.2.4a}$$

Hence,

$$u_{av}=\frac{2}{3}u_{max} \tag{E.3.2.4b}$$

In order to determine the film thickness, we first compute the volumetric flow rate Q, which is usually known.

$$Q=\int \vec{\mathbf{v}}\cdot d\vec{\mathbf{A}} := b\int_0^h u\,dy := u_{av}(hb) \tag{E.3.2.5a}$$

where b is the plate width. Thus,

$$Q=\frac{\rho\, g\, b \sin\phi}{3\mu}h^3 \tag{E.3.2.5b}$$

from which

$$h=\left(\frac{3\,\mu\, Q}{\rho\, g\, b \sin\phi}\right)^{1/3} \tag{E.3.2.5c}$$

The shear force (or drag) exerted by the liquid film onto the plate is

$$F_s=\int_A \tau_{yx}\,dA = b\int_0^l\left(-\mu\left.\frac{du}{dy}\right|_{y=0}\right)dy \tag{E.3.2.6a}$$

Thus,

$$F_s=\rho\, g\, b\, h\, l \sin\phi \tag{E.3.2.6b}$$

which is the x-component of the weight of the whole liquid film along $0 \le x \le l$.

Graph:

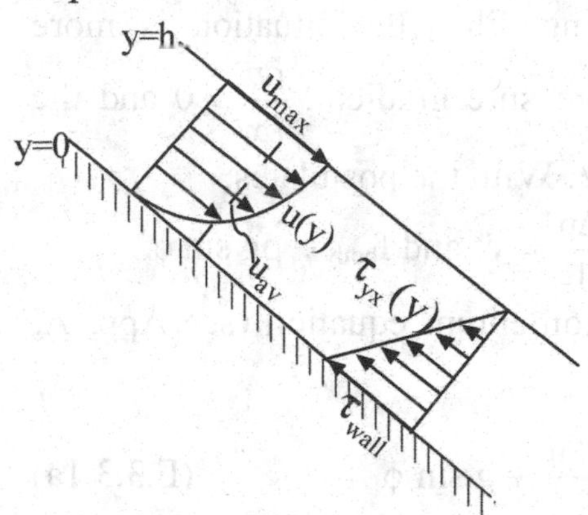

Comments:

- $u_{average}$ can be used to compute the film Reynolds number, $Re_h = \frac{u_{av}\, h}{\nu}$, and check if $Re_h < Re_{critical} \approx 20$
- Given Q (or $\dot{m} = \rho Q$), the coating thickness can be estimated

Note: The case of wire coating, i.e., falling film on a vertical cylinder is assigned as a homework problem (see Sect. 3.8).

Example 3.3: Flow Between Parallel Plates

Consider viscous fluid flow between horizontal or tilted plates a small constant distance h apart. For the test section $0 \le x \le l$ of interest, the flow is fully-developed and could be driven by friction when the upper plate is moving at $u_0 = ¢$, or by a constant pressure gradient, dp/dx, and/or by gravity (see Couette flow, Example 2.8)

Sketch	*Assumptions*	*Concepts*
$\vec{g}$, p(x), y, x, x=0, h, ρ,μ, u_0, φ, x=l	• Steady laminar unidirectional flow • Constant h, $\partial p/\partial x$, and fluid properties	• Reduced N-S equation where $\vec{v} = (u, 0, 0)$ and $-\frac{\partial p}{\partial x} \approx \frac{\Delta p}{l} = ¢$ • u = u(y) only

Solution: Although this problem looks like a simple case of a moving plate on top of a flowing film, the situation is more interesting due to the additional pressure gradient, $\frac{\partial p}{\partial x} \gtreqless 0$ and the boundary condition $u(y = h) = u_0 = ¢$. With the postulates:

$v = w = 0, \quad u = u(y)$ only, $-\frac{\partial p}{\partial x} \approx \frac{\Delta p}{l} = ¢$, and $f_{body} = \rho g \sin \phi$,

continuity is fulfilled and the x-momentum equation (see App. A, Equation Sheet) reduces to:

$$0 = \frac{1}{\rho}\left(\frac{\Delta p}{l}\right) + \nu \frac{d^2u}{dy^2} + g \sin \phi \tag{E.3.3.1a}$$

or

$$\frac{d^2u}{dy^2} = \frac{-1}{\mu}\left[\left(\frac{\Delta p}{l}\right) + \rho g \sin \phi\right] = ¢ \tag{E.3.3.1b}$$

subject to $u(y = 0) = 0$ and $u(y = h) = u_0$. Again, we have to solve an ODE of the form $u'' = K$. Introducing a dimensionless pressure gradient

$$P \equiv -\frac{h^2}{2\mu u_0}\left[\left(\frac{\Delta p}{l}\right) + \rho g \sin \phi\right] \tag{E.3.3.2}$$

we can write the solution u(y), known as Couette flow, in a more compact form, i.e.,

$$\frac{u(y)}{u_0} = \frac{y}{h} - P\left(\frac{y}{h}\right)\left(1 - \frac{y}{h}\right) \tag{E.3.3.3}$$

Graph ($\phi = 0$):

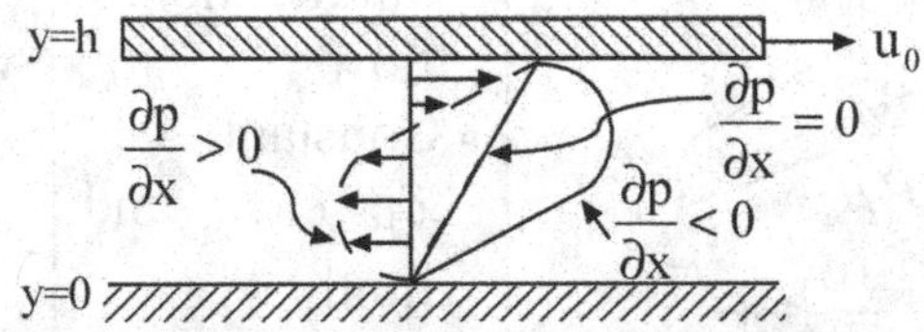

Comments:

- The pressure gradient, P or $\frac{\partial p}{\partial x}$, greatly influences u(y). Clearly, $\frac{\partial p}{\partial x}<0$ implies a "favorable" and $\frac{\partial p}{\partial x}>0$ an "adverse" pressure gradient
- For $u_0=0$, we have flow between parallel plates, i.e.,

$$u(y)=+\frac{h^2}{2\mu}\left[\left(\frac{\Delta p}{l}\right)+\rho g\sin\phi\right]\left[\frac{y}{h}-\left(\frac{y}{h}\right)^2\right] \qquad \text{(E.3.3.4)}$$

Example 3.4: Steady Laminar Fully-Developed Flow in a Pipe: Poiseuille Flow Revisited and Extended

Sketch	*Assumptions*
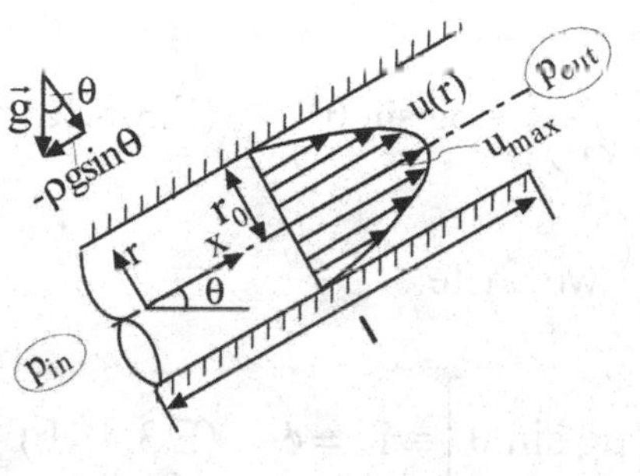	• As stated • Constant $-\frac{\partial p}{\partial x}\approx\frac{\Delta p}{l}$ Note: $\frac{\partial p}{\partial x}\approx\frac{p_1-p_2}{x_1-x_2}$ or $-\frac{\partial p}{\partial x}=\frac{p_{in}-p_{out}}{x_1-0}=\frac{\Delta p}{l}$ Constant fluid properties ρ and μ
Concepts	
• Reduced N–S equations in cylindrical coordinates • $\frac{\partial}{\partial t}=0$ <steady state> • $\frac{\partial}{\partial \theta}=0$ <axisymmetric> • $\frac{\partial}{\partial x}=0$ <fully developed>	

Poiseuille flow is the base case of all laminar internal flows. Thus, the following results will be frequently used in subsequent sections and chapters.

Postulates: $\vec{v} = (u, 0, 0)$; $u = u(r)$ only; $\nabla p \to \frac{\partial p}{\partial x} = ¢$

Boundary Conditions: The obvious one is $u(r = r_0) = 0$ and the second one could be $u(r = 0) = u_{max}$; however, u_{max} is unknown. Thus, we use symmetry on the centerline, i.e.,

$$\left.\frac{du}{dr}\right|_{r=0} = 0$$

Continuity:

$$\frac{\partial u}{\partial x} + 0 + 0 = 0 = \succ \frac{\partial u}{\partial x} = 0 < \text{fully-developed flow confirmed} >$$

x-momentum:

$$0 = -\frac{\partial p}{\partial x} + \mu\left[\frac{1}{r}\frac{\partial}{\partial r}\left(r\frac{\partial u}{\partial r}\right)\right] - \rho g \sin\theta \tag{E.3.4.1a}$$

With $u = u(r)$ only and $-\partial p/\partial x = \Delta p/l$, we write:

$$\frac{1}{r}\left[\frac{d}{dr}\left(r\frac{du}{dr}\right)\right] = \frac{1}{\mu}\left[\left(\frac{-\Delta p}{l}\right) + \rho g \sin\theta\right] \equiv K = ¢ \tag{E.3.4.1b}$$

or after separation of variables

$$\int d\left(r\frac{du}{dr}\right) = K\int r\,dr + C_1 \tag{E.3.4.2a}$$

so that

$$\frac{du}{dr} = \frac{K}{2} + \frac{C_1}{r} \tag{E.3.4.2b}$$

A second integration yields:

$$u = \frac{K}{4} r^2 + C_1 \ln r + C_2 \tag{E.3.4.2c}$$

From the first B.C. $$0 = \frac{K}{4} r_0^2 + C_1 \ln r_0 + C_2 \tag{E.3.4.3a}$$

while the second B.C. [see Eq. (E.3.4.2b)] yields

$$0 = 0 + \frac{C_1}{r}$$

which forces C_1 to be zero. From Eq. (E.3.4.3a) we have

$$C_2 = -\frac{K}{4} r_0^2 \tag{E.3.4.3b}$$

so that with $K \equiv \frac{1}{\mu}\left(\frac{-\Delta p}{l} + \rho g \sin\theta\right)$, we obtain:

$$u(r) = \frac{K r_0^2}{4}\left[1 - \left(\frac{r}{r_0}\right)^2\right] \tag{E.3.4.4}$$

For a horizontal pipe, $\theta = 0$ and hence

$$u(r) = \frac{r_0^2}{4\mu}\left(\frac{\Delta p}{l}\right)\left[1 - \left(\frac{r}{r_0}\right)^2\right] = u_{max}\left[1 - \left(\frac{r}{r_0}\right)^2\right] \tag{E.3.4.5a, b}$$

Notes:

- The average velocity, $u_{av} \equiv \bar{u} = \frac{1}{A}\int u(r) dA$ where $dA = 2\pi r dr < \text{cross - sectional ring} >$, so that

$$\bar{u} = \frac{1}{2\mu}\left(\frac{\Delta p}{l}\right)\int_0^{r_0}\left[1 - \left(\frac{r}{r_0}\right)^2\right] r dr = \frac{r_0^2}{8\mu}\left(\frac{\Delta p}{l}\right) \tag{E.3.4.6a, b}$$

- The volumetric flow rate, $Q = u_{av} A$, where $A = \pi r_0^2$, can be used to calculate the necessary pressure drop to maintain the flow:

$$Q = \frac{\pi r_0^4}{8\mu}\left(\frac{\Delta p}{l}\right) \tag{E.3.4.7a}$$

so that with a given Q-value (or Reynolds number)

$$\Delta p = \frac{8\,\mu\,Q}{\pi\,r_0^4}\,l \tag{E.3.4.7b}$$

- The pressure drop $\Delta p = p_{in} - p_{out}$ is positive while the pressure gradient

$$\frac{\partial p}{\partial x} < 0$$

- By inspection of Eq. (E.3.4.5), $u(r = 0) = u_{max}$, where

$$u_{max} = \frac{r_0^2}{4\,\mu}\left(\frac{\Delta p}{l}\right) = \frac{1}{2}\,u_{av} \tag{E.3.4.8}$$

- The wall shear stress, i.e., the frictional impact of viscous fluid flow on the inner pipe surface, is:

$$\tau_{wall} \equiv \tau_{rx}\Big|_{r=r_0} = -\mu \left.\frac{du}{dr}\right|_{r=r_0} \tag{E.3.4.9a}$$

$$\tau_w = \frac{r_0}{2}\left(\frac{\Delta p}{l}\right) \tag{E.3.4.9b}$$

Thus, a second expression for the pressure drop (or gradient) is found:

$$\Delta p = \frac{2\tau_w}{r_0}\,l \tag{E.3.4.10}$$

Equation (E.3.4.10) will be most valuable in solving industrial pipe-flow problems in conjunction with the extended Bernoulli equation (see Sect. 4.2).

Examples 3.5: Parallel Flows in Cylindrical Tubes

Case A: Consider flow through an annulus with tube radius R and inner concentric cylinder aR, where a < 1, creating a ring-like cross-sectional flow area.

Sketch	*Assumptions*	*Concept*
l; p_{in}; $v_z(r)$; p_{out}; r; R; z; aR; ℄	• Steady laminar axisymmetric flow • Constant $\partial p/\partial z$ and constant ρ and μ	• Reduced N–S equations in cylindrical coordinates

Postulates: $\vec{v} = (0, 0, v_z)$; $\nabla p \Rightarrow \frac{\partial p}{\partial z} \approx -\frac{\Delta p}{l} = ¢$; $v_z = v_z(r)$ only

$$\frac{\partial}{\partial t} = 0 < \textit{steady-state} >; \qquad \frac{\partial}{\partial \phi} = 0 < \textit{axisymmetric} >$$

Continuity Equation: $0 + 0 + \frac{\partial v_z}{\partial z} = 0$, i.e., fully-developed flow

Boundary Conditions: $v_z(r = aR) = 0$ and $v_z(r = R) = 0$

Solution: Of interest is the z-momentum equation, i.e., with the stated postulates (see App. A, Equation Sheet):

$$0 = -\frac{\partial p}{\partial z} + \mu\left[\frac{1}{r}\frac{\partial}{\partial r}\left(r\frac{\partial v_z}{\partial r}\right)\right] \tag{E.3.5.1a}$$

Thus, with $P \equiv \frac{1}{\mu}\left(\frac{-\Delta p}{l}\right)$

$$\frac{d}{dr}\left(r\frac{dv_z}{dr}\right)=P\cdot r \tag{E3.5.1b}$$

and hence after double integration,

$$v_z(r)=\frac{P}{4}r^2+C_1\ln r+C_2 \tag{E.3.5.2}$$

Invoking the BCs,

$$0=\frac{P}{4}(aR)^2+C_1\ln(aR)+C_2$$

and

$$0=\frac{P}{4}R^2+C_1\ln R+C_2$$

yields

$$v_z(r)=\frac{PR^2}{4}\left[1-\left(\frac{r}{R}\right)^2-\frac{1-a^2}{\ln(1/a)}\ln\left(\frac{R}{r}\right)\right] \tag{E.3.5.3}$$

and

$$\tau_{rz}=\mu\frac{dv_z}{dr}:=\frac{\mu P}{2}R\left[\left(\frac{r}{R}\right)-\frac{1-a^2}{2\ln(1/a)}\left(\frac{R}{r}\right)\right] \tag{E.3.5.4a, b}$$

Notes:

- For Poiseuille flow, i.e., no inner cylinder, the solution is (see Example 3.4):

$$v_z(r)=\frac{R^2}{4\mu}\left(\frac{\Delta p}{l}\right)\left[1-\left(\frac{r}{R}\right)^2\right] \tag{E.3.5.5}$$

This solution is not recovered when letting $a \to 0$ because of the prevailing importance of the ln-term near the inner wall.

- The maximum annular velocity is not in the middle of the gap $aR \le r \le R$, but closer to the inner cylinder wall, where the velocity gradient is zero and hence

$$\tau_{rz}\big|_{r=bR}=0$$

This equation can be solved for b so that $v_z(r = bR) = v_{max}$.

- The average velocity is $v_{av} = \int v_z(r)dA$, where dA = $2\pi r dr$ $\langle$cross-sectional ring of thickness dr$\rangle$, so that

$$v_{av} = \frac{R^2}{8\mu}\left(\frac{\Delta p}{l}\right)\left[\frac{1-a^4}{1-a^2} - \frac{1-a^2}{\ln(1/a)}\right] \tag{E.3.5.6}$$

and hence

$$Q = v_{av}[\pi R^2(1-a^2)] := \frac{\pi R^4}{8\mu}\left(\frac{\Delta p}{l}\right)\left[(1-a^4) - \frac{(1-a^2)^2}{\ln(1/a)}\right] \tag{E.3.5.7}$$

- The *net* force exerted by the fluid on the solid surfaces comes from two wall shear stress contributions:

$$F_s = \left(-\tau_{rz}\big|_{r=aR}\right)(2\pi aRl) + \left(\tau_{rz}\big|_{r=R}\right)(2\pi Rl) \tag{E.3.5.8a}$$

$$\therefore \quad F_s = \pi R^2 \Delta p(1-a^2) \tag{E.3.5.8b}$$

Case B: Consider Case A but now with $\frac{\partial p}{\partial z} \equiv 0$ and the inner cylinder rotating at angular velocity ω_0 = ¢ <cylindrical Couette flow>; in general, the outer cylinder could rotate as well, say with ω_1 = ¢.

Sketch	*Assumptions*	*Concepts*
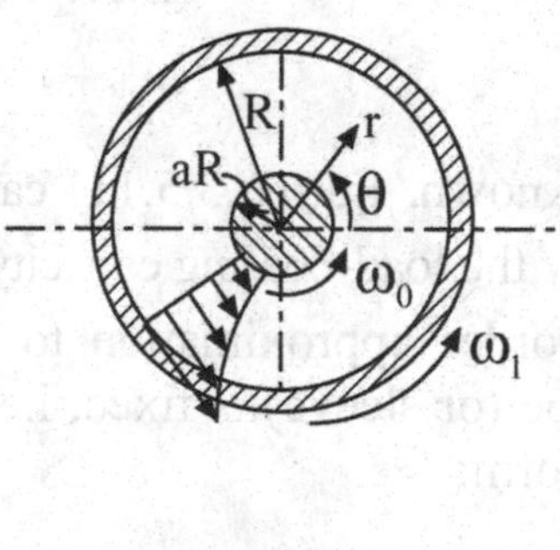	• Steady laminar axisymmetric flow • Long cylinders, i.e., no end effects • Small ω's to avoid Taylor vortices	• Reduced N-S equations in cylindrical coordinates • Postulates: $\vec{v} = (0, v_\theta, 0)$ $\frac{\partial p}{\partial \theta} = \frac{\partial p}{\partial z} = 0$

Solutions: With $v_r = v_z = 0$; $\frac{\partial}{\partial t} = \frac{\partial}{\partial \theta} = 0$; and $v_\theta = v_\theta(r)$ only (see Continuity and BCs) we reduce the θ-component of the Navier–Sokes equation (see Equation Sheet) to:

$$0 = 0 + \mu\left[\frac{\partial}{\partial r}\left(\frac{1}{r}\frac{\partial}{\partial r}(rv_\theta)\right)\right] \quad \text{(E.3.5.9)}$$

subject to

$$v_\theta(r = aR) = \omega_0(aR) \quad \text{and} \quad v_\theta(r = R) = \omega_1 R\,.$$

Again, as in simple Couette flow after start-up, the moving-wall induced frictional effect propagates radially and the forced cylinder rotations balanced by the drag resistance generate an equilibrium velocity profile. Double integration yields:

$$v_\theta(r) = C_1 r + \frac{C_2}{r} \quad \text{(E.3.5.10a)}$$

where

$$C_1 = \frac{\omega_1 R^2 - \omega_0(aR)^2}{R^2 - (aR)^2} \text{ and } C_2 = \frac{a^2R^4(\omega_0 - \omega_1)}{R^2 - (aR)^2} \quad \text{(E.3.5.10b, c)}$$

Notes:

- The r-momentum equation reduces to:

$$-\frac{v_\theta^2}{r} = -\frac{1}{\rho}\frac{\partial p}{\partial r} \quad \text{(E.3.5.11)}$$

  Thus, with the solution for $v_\theta(r)$ known, Eq. (E.3.5.11) can be used to find $\partial p / \partial r$ and ultimately the load-bearing capacity.
- Applying this solution as a first-order approximation to a journal bearing where the outer tube (or sleeve) is fixed, i.e., $\omega_1 \equiv 0$, we have in dimensionless form:

$$\frac{v_\theta(r)}{\omega_0 R} = \frac{a^2}{1 - a^2}\left(\frac{R}{r} - \frac{r}{R}\right) \quad \text{(E.3.5.12)}$$

- The torque necessary to rotate the inner cylinder (or shaft) of length l is

$$T = \int (aR)dF := (aR)\int_0^l \tau_{r\theta}\big|_{r=aR}\, dA \qquad \text{(E.3.5.13)}$$

where $dA = \pi(aR)dz$ and $\tau_{r\theta}\big|_{r=aR} = \mu\left[r\frac{d}{dr}\left(\frac{v_\theta}{r}\right)\right]_{r=aR}$.

Thus with:

$$\tau_{surface} \equiv \tau_{r\theta}\big|_{r=aR} = 2\mu\frac{\omega_0 R^2}{R^2-(aR)^2} \qquad \text{(E.3.5.14)}$$

$$T = \tau_{surf} A_{surf}(aR) := 4\pi\mu(aR)^2\, l\, \frac{\omega_0}{1-a^2} \qquad \text{(E.3.5.15)}$$

Graph:

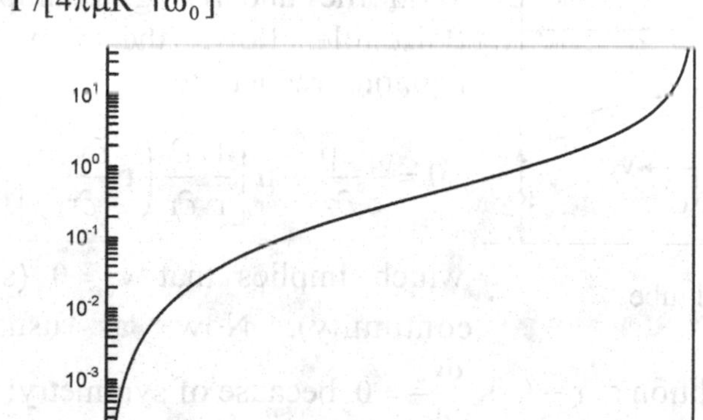

Comments:

- An electric motor may provide the necessary power, $P = T\omega_0$, which turns into thermal energy which has to be removed to avoid overheating.
- The Graph depicts the nonlinear dependence of T(a) for a given system. As the gap between rotor (or shaft) and stator widens, the wall stress increases (see Eq. (E.3.5.14)) as well as the surface area and hence the necessary torque.

3.2.2 Nearly Parallel Flows

Steady laminar flows in conduits with *slightly non-parallel* walls (or plates) have some practical importance. Examples include tapered tubes, cone-plate viscometers, slider bearings, converging/diverging slit flows, etc. For such cases, the key assumption is that of "uni-directional" flow (as in Poiseuille or Couette flows); although, there is a *small* second velocity component, which is first ignored. Then, the slight geometric changes, and hence more realistic flow fields, are incorporated by invoking the "no-slip" condition at the converging or diverging wall. To illustrate the two-step procedure, let us consider a mildly tapered pipe where the radius changes as (Fig. 3.1):

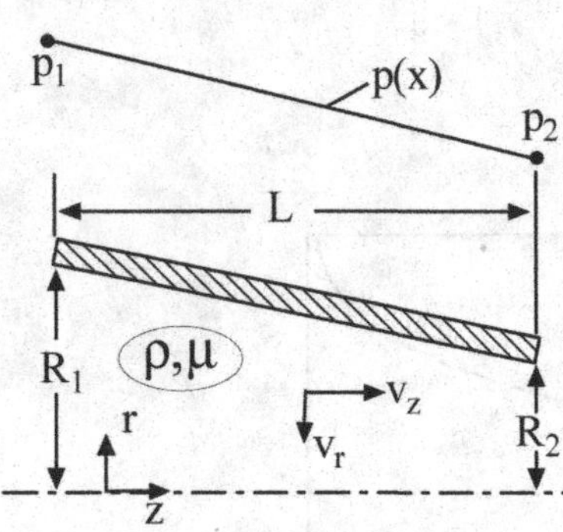

Fig. 3.1 Tapered tube

$$R(z) = R_1 - \frac{\Delta R}{L} z \tag{3.4}$$

With the underlying assumption of Poiseuille flow, the z-momentum equation reduces to:

$$0 = -\frac{\partial p}{\partial z} + \mu \left[\frac{1}{r} \frac{\partial}{\partial r} \left(r \frac{\partial v_z}{\partial r} \right) \right] \tag{3.5}$$

which implies that $v_r = 0$ (see also continuity). Now, as usual, the boundary condition at r = 0 is $\frac{dv_z}{dr} = 0$ because of symmetry; but, the no-slip condition at the tube wall reads:

$$v_z[r = R(z)] = 0 \tag{3.6}$$

which introduces the unique tube geometry and 2-D flow pattern. The solution is:

$$v_z = v_z(r, z) = v_{max} \left[1 - \left(\frac{r}{R(z)} \right)^2 \right] \tag{3.7}$$

Checking the continuity equation for axisymmetric flow:

$$\frac{1}{r}\frac{\partial}{\partial r}(r\,v_r) + \frac{\partial v_z}{\partial z} = 0 \tag{3.8}$$

we see that $v_r \neq 0$ because $v_z = v_z(z, r)$. In fact, Eq. (3.8) can be employed to find an expression for $v_r(r, z)$ considering that $v_r(r = 0) = 0$ or $v_r[r = R(z)] = 0$ (see HWA in Sect. 3.8).

Additional examples of nearly unidirectional flows include lubrication and stretching flows as given in Sect. 4.3; also discussed in Papanastasiou (1994), Kleinstreuer (1997), Middleman (1998) and Panton (2005), among others.

3.3 Transient One-Dimensional Flows

Time-dependent viscous flows occur in nature (e.g., blood flow and respiratory airflow as well as tidal motion and wind pattern) as well as in industry (e.g., flow-induced vibration, flow start-up or shut-down, pressure waves, etc.). As such flow phenomena are even more interesting than steady parallel flows, the necessary inclusion of the time dimension render the mathematics involved a bit more complicated, i.e., instead of ODEs, partial differential equations (PDEs) have to be solved. The reward is that more realistic fluid flow problems can be solved and some of these basic transient flow solutions provide new physical insight to more complex flow phenomena.

Of the two start-up problems considered, the first one (after Stokes) is a suddenly accelerated plate above which a body of fluid is set into motion due to frictional effects. The resulting expression for the thickness of that region of influence indicates the existence of a boundary layer, some 60 years later fully described by Prandtl. The second problem is a suddenly applied (constant) pressure gradient for tubular flow which, after some time, establishes itself into Poiseuille flow (see Example 3.4).

3.3.1 Stokes' First Problem: Thin Shear-Layer Development

Consider a horizontal plate or wall carrying a stagnant body of fluid (i.e., $u = 0$ when $t \leq 0$ for all y; see Fig. 3.2). Suddenly,

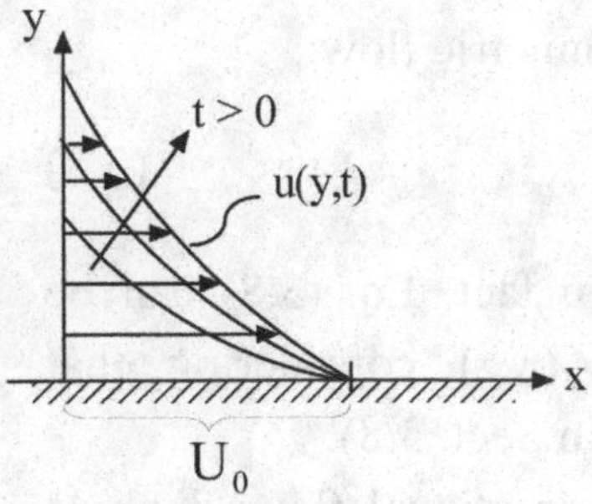

Fig. 3.2 Changing velocity profile with time inside thin shear-layer

the solidsurface attains (at y = 0) a finite velocity, i.e., $u = U_0$ when t > 0. Recalling that $v_{wall} = v_{fluid}$ (no-slip condition), this plate motion sets up, within a growing layer, parallel flow of the viscous fluid, i.e., u = u(y, t). The atmospheric pressure is constant everywhere. Thus, with the postulates $\vec{v} = [u(y, t), 0, 0]$ and $\nabla p = 0$ (3.9a, b) we can reduce the x-momentum equation to be:

$$\frac{\partial u}{\partial t} = \nu \frac{\partial^2 u}{\partial y^2} \tag{3.10}$$

Equation (3.10) is known as the transient one-dimensional *diffusion equation* [cf. Eq. (2.25)]. In the present case, it describes "momentum diffusion" normal to the axial parallel flow induced by the wall motion. As implied, the associated initial/boundary conditions are:

$$u(t \le 0, y) = 0, \text{ but } u(t > 0; y = 0) = U_0; \text{ for } y \to \infty, u = 0.$$

Because the evolution of u(y) with time shows similar profiles (Fig. 3.2), the independent variables y and t can be combined in conjunction with the fluid viscosity ν (see App. A). Thus, for the new *dimensionless* variable

$$\eta = \eta(y, t; \nu) \tag{3.11a}$$

we demand formally

$$[\eta] \hat{=} y^a t^b \nu^c = [1] \tag{3.11b}$$

or by simple inspection with a = 1

$$\eta = \frac{y}{t^{0.5}\nu^{0.5}} \hat{=} [1] \tag{3.11c}$$

For convenience, the dimensionless independent variable can be written as:

$$\eta = \frac{y}{2\sqrt{\nu t}} \tag{3.11d}$$

It is apparent that

$$u(y, t) \sim f[\eta(y, t)] \tag{3.12a}$$

where $f(\eta)$ is a dimensionless dependent variable. To turn the proportionality into an equation, we utilize the plate speed U_0, so that

$$u(y, t) = U_0 f(\eta) \tag{3.12b}$$

Now, with Eqs. (3.11d) and (3.12b), the governing PDE (3.10) can be transformed into an ODE for $f(\eta)$, i.e.,

$$f'' + 2\eta f' = 0 \tag{3.13}$$

subject to

$$f(\eta = 0) = 1 \text{ and } f(\eta \to \infty) \to 0 \tag{3.14a, b}$$

The solution is $f(\eta) = 1 - \text{erf}(\eta)$, where $\text{erf}(\eta)$ is the error function (App. A), so that

$$\frac{u}{U_0} = 1 - \frac{2}{\sqrt{\pi}} \int_0^{\eta} \exp(-\eta^2) d\eta \tag{3.15}$$

When plotting Eq. (3.15), it turns out that for $\eta = 2.0$ the moving-plate effect on the fluid body peters out, i.e., $f(\eta = 2) \equiv \frac{u}{U_0} \approx 0.01$. This implies that the region of frictional influence, i.e., $0 \leq y \leq \delta$, can be estimated from $y(\eta = 2) = \delta$ as

$$\delta \approx 4\sqrt{\nu t} \tag{3.16a}$$

Replacing t in terms of the plate travel time, i.e., $t = \frac{x}{U_0}$, Eq. (3.16a) can be written as:

$$\delta \approx 4\sqrt{\nu \frac{x}{U_0}} \tag{3.16b}$$

which can also be expressed as:

$$\frac{\delta}{x} \approx \frac{4}{\sqrt{Re_x}} \tag{3.16c}$$

where $Re_x = U_0 x / \nu$ is the local Reynolds number and δ is the extent of the shear layer in which u = u(y). Outside the shear layer, i.e., $y \geq \delta$, $u = 0$ in this case.

Note, δ(x) is fundamental to laminar thin-shear-layer (TSL), or boundary-layer (B-L) theory (see Sect. 5.2).

Akin to Stokes First Problem discussed here is his second problem solution, that of an oscillating flat plate. Even more complicated is laminar flow generated by start-up of a rotating disk in a reservoir of a viscous fluid. It is three-dimensional because fluid exits radially the finite disk (v_r-component) because of the centrifugal force; this vanishing fluid is constantly replaced by swirling, incoming fluid (v_θ *– and* v_z *– components*). Setting up these and other transient flow problems is part of the HWAs in Sect. 3.8.

3.3.2 Transient Pipe Flow

One of the most famous examples of transient internal flow is pulsatile flow in a tube, e.g., blood flow in a straight artery, first solved analytically in 1955 by Womersley as discussed in Nichols & O'Rourke (1998). An industrial application, i.e., sudden start-up of fluid flow, is given in Example 3.6.

Example 3.6: Start-up Flow in a Tube

Consider a viscous fluid at rest in a horizontal tube when suddenly a constant pressure gradient, Δp/L, is applied. For example, a valve connecting a pipe to a reservoir is suddenly opened. Find an expression for the resulting u(r, t).

Sketch	*Assumptions*	*Concepts*
p(x); L; r; z	• Pressure gradient $-\frac{\partial p}{\partial x} \approx \frac{\Delta p}{L} = ¢$ at all times • Transient laminar 1-D flow • Constant fluid properties	• Reduced N–S equations • Superposition of steady and transient contributions

Solution: Postulate that the actual velocity u(r, t) can be decomposed into a steady-state part and a transient part, i.e.,

$$u(r,t) = u(r)\big|_{ss} + u(r,t)\big|_{tr} \tag{E.3.6.1}$$

With $\vec{v} = [u(r,t); 0; 0]$ and $-\nabla p \rightarrow \frac{\Delta p}{L} = ¢$ the governing momentum equation is

$$\frac{\partial u}{\partial t} = \frac{1}{\rho}\left(\frac{\Delta p}{L}\right) + \frac{\nu}{r}\frac{\partial}{\partial r}\left(r\frac{\partial u}{\partial r}\right) \tag{E.3.6.2}$$

subject to

$$u(r, t=0) = 0; \;\; u(r = r_0, t) = 0, \text{ and } \left.\frac{\partial u}{\partial r}\right|_{r=0} = 0 \tag{E.3.6.3a–c}$$

Clearly, the steady-state part, u_{ss}, is the Poiseuille-flow solution, i.e.,

$$u_{ss}(r) = u_{max}\left[1 - \left(\frac{r}{r_0}\right)^2\right] \tag{E.3.6.4a}$$

where

$$u_{max} = \frac{1}{4\mu}\left(\frac{\Delta p}{L}\right) \tag{E.3.6.4b}$$

Knowing $u_{ss}(r)$ and employing the dimensionless variables

$$\hat{u} = \frac{u_{tr}}{u_{max}}, \quad \hat{r} = \frac{r}{r_0}, \quad \text{and} \quad \hat{t} = \frac{\nu t}{r_0} \tag{E.3.6.5a–c}$$

Equation (E.3.6.2) can be transformed to the well-known form:

$$\frac{\partial \hat{u}}{\partial \hat{t}} = \frac{1}{\hat{r}} \frac{\partial}{\partial \hat{r}} \left(\hat{r} \frac{\partial \hat{u}}{\partial \hat{r}} \right) \tag{E.3.6.6}$$

subject to

$$\hat{u}(\hat{t}=0) = 1 - \hat{r}^2; \quad \hat{u}(\hat{r}=1) = 0, \quad \text{and} \quad \left.\frac{\partial \hat{u}}{\partial \hat{r}}\right|_{\hat{r}=0} = 0 \tag{E.3.6.7a–c}$$

The solution is an infinite series in $\hat{r}$ times a decaying exponential function in $\hat{t}$, i.e.,

$$\hat{u} \sim \sum \text{fct}(\hat{r}) * e^{-\hat{t}} \tag{E.3.6.8}$$

Graph:

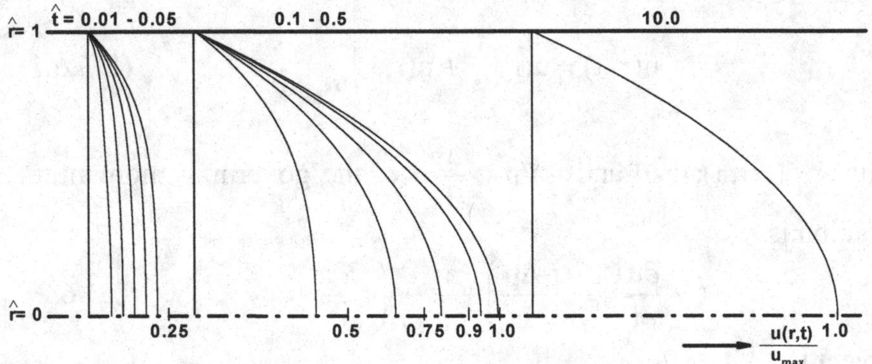

Comments:

The final solution for the axial tubular velocity $u(\hat{r}, \hat{t})$ is graphed for $0 \le \hat{r} \le 1$ and $0 \le \hat{t} \le 10.0$. When at $\hat{t} \approx 10.0$, $u_{tr} \to 0$ and $u = u_{ss}$, i.e., Poiseuille flow has been established.

It is interesting to note that the suddenly elevated tube-inlet pressure starts the core-fluid off almost uniformly (for $0.01 \le \hat{t} < 0.05$) and then, after $\hat{t} = 0.2$, in conjunction with the no-slip condition a parabolic velocity profile forms.

3.4 Simple Porous Media Flow

In numerous natural and industrial processes, a fluid flows (or just seeps or migrates) through a *fully-saturated* porous medium. Examples include blood flow through tissue, groundwater flow through geologic media, oil or steam dispersion through sand and porous rock, fluid flow through a container packed with spheres, pellets, or granular material (with upward flow known as fluidized beds), moisture migration through a porous composite, mixture flow across membranes or filters, coolant flow through microchannels, etc. In any case, as indicated in Fig. 3.3, the local velocity field through pores, capillaries, fissures, microchannels and/or around packed spheres, pellets, cylinders, fibers, cells or granular material is very complicated. For that reason, a volume-averaged, i.e., superficial velocity

$$\langle \vec{u} \rangle = \frac{1}{\forall} \int \vec{u} d\forall = \frac{Q}{A} \tag{3.17a, b}$$

is introduced, where the length-scale of the control volume, $\forall^{1/3}$, is smaller than the characteristic length of the system, say, channel height or pipe diameter; but, also $\forall^{1/3} >> d_p$, i.e., the pore (or pellet) diameter, so that we have $d_p << \forall << L_{system}$. Note, the volume integral extends over the volume space occupied by the fluid. Clearly, $\langle \vec{u} \rangle$ is a function of the driving forces (e.g., pressure gradient and/or gravity), fluid properties (i.e., viscosity and density) as well as the porous medium structure and material. As with all complicated fluid flow problems, possible solutions start with dimensional analysis and experimental investigations. Specifically, based on Darcy's observations in 1856, the uniform superficial 1-D velocity of

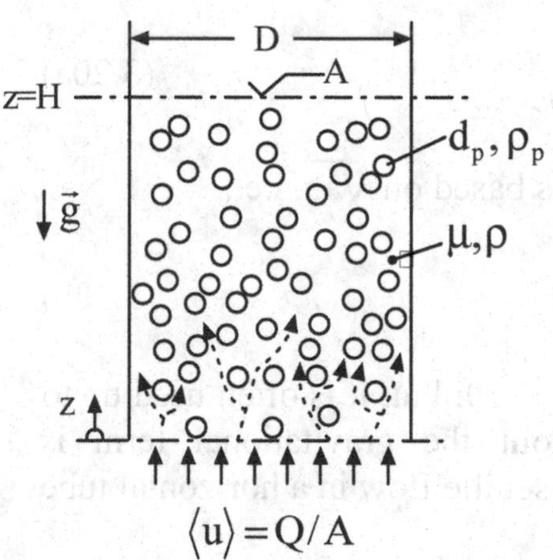

Fig. 3.3 Porous medium column

a viscous fluid through a homogeneous porous medium is (Fig. 3.3):

$$\langle u \rangle = -\frac{æ}{\mu}\left(\frac{dp}{dz} + \rho g_z\right) \tag{3.18a}$$

Where æ $[L^2]$ is the permeability, i.e., $\sqrt{æ}$ [L] is a length-scale representative of the effective pore diameter. Note, $æ/\mu \equiv K$ is the "hydraulic conductivity". Not surprising, æ has been correlated to the porosity ε, for example by Ergun (1952) as:

$$æ = \frac{\varepsilon^3 d_p^2}{C(1-\varepsilon)^2} \tag{3.18b}$$

where C = 150 ~ 180.

Assuming negligible gravitational effects, Eq. (3.18a) can be used to determine the key driving force, i.e., the pressure gradient as:

$$\frac{dp}{dz} = \frac{C\mu\langle u \rangle}{d_p^2}\frac{(1-\varepsilon)^2}{\varepsilon^3} \tag{3.19}$$

The porosity, or void fraction, is a volume ratio, i.e., $\varepsilon = \forall_{void}/\forall_{total} \approx \forall_{particles}/\forall_{total}$. It can be employed to construct a particle Reynolds number; for example, for a uniformly packed bed of spheres,

$$Re_p = \frac{\rho\langle u \rangle d_p}{(1-\varepsilon)\mu} \tag{3.20a}$$

Alternatively, the Reynolds number is based on $\sqrt{æ}$, i.e.,

$$Re = \frac{\langle u \rangle\sqrt{æ}}{\nu} \tag{3.20b}$$

In any case, Eq. (3.18a) holds for Re < 10; but, it is often used up to Re ≈ 10. When Eq. (3.18a) without the gravitational term is compared with Eq. (E.3.4.6b) for Poiseuille flow in a horizontal tube of radius R,

$$v_{av} = \frac{R^2}{8\mu}\left(-\frac{dp}{dx}\right) \tag{3.21}$$

the similarity is obvious. Thus, it is conducive that a homogeneous porous medium could be modeled as an assemblage of tiny straight tubes (i.e., pores) with laminar fully-developed flow of a viscous fluid.

Because of the obvious shortcomings of Eq. (3.18a), e.g., $\mathrm{Re} \le O(1)$, 1-D flow only, the no-slip condition cannot be enforced, local variations in the flow field and porous material structure are averaged out, several additions to Eq. (3.18a) have been proposed. The two most famous ones are the Brinkman (1947)-Forchheimer (1901) extensions to Darcy's law, where in 3-D:

$$\nabla p = -\frac{\mu}{æ}\vec{\mathbf{u}} - \underbrace{C_F \frac{\rho \vec{\mathbf{u}}|\vec{\mathbf{u}}|}{\sqrt{æ}}}_{\text{Forchheimer term}} + \underbrace{\hat{\mu}\nabla^2 \vec{\mathbf{u}}}_{\text{Brinkman term}} \tag{3.22}$$

Here, $\vec{\mathbf{u}} \mathrel{\hat{=}} \langle \vec{\mathbf{u}} \rangle$, C_F is a dimensionless form-drag constant which depends on the characteristics of the porous medium and the bounding walls, if any; typically, $C_F \approx 0.55$. The effective viscosity $\hat{\mu} \approx \mu$; however, the ratio $\hat{\mu}/\mu$ can depend again on the nature of the porous medium.

When comparing Eq. (3.22) with the Navier–Stokes equation (2.22), it is apparent that the nonlinear Forchheimer term relates to the inertia term. It is a necessary addition for porous media flow with Re > 10, i.e., when the pore flow is still laminar but the form drag posed by the porous material structure becomes important. The second-order Brinkman term relates to the viscous drag term. It is necessary to enforce no-slip boundary conditions, i.e., the term is significant in thin wall shear layers, typically of thickness $\sqrt{æ} << L_{\text{system}}$.

Quite frequently, channel flow interacts with (saturated) porous medium flow (see Fig. 3.4) where at the interface between fluid layer and medium the magnitude and gradient of the

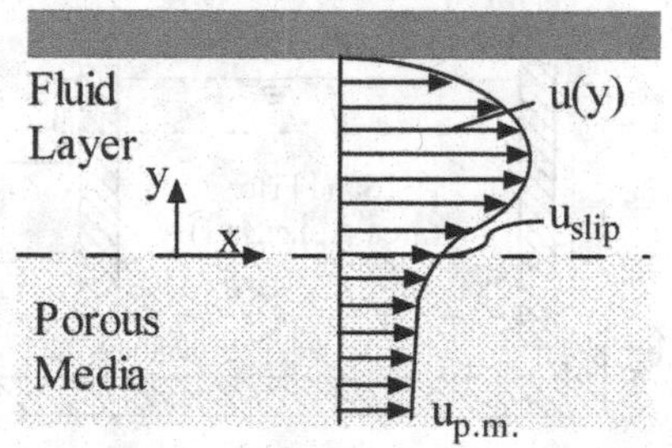

Fig. 3.4 Unidirectional channel flow with porous medium flow

velocity have to match. As a first approximation, Beavers & Joseph (1967) suggested:

$$\left.\frac{\partial u}{\partial y}\right|_{y=0} = \frac{\alpha}{\sqrt{æ}}\left(u_{slip} - u_{p.m.}\right) \tag{3.23}$$

where $u_{slip} = u(y=0)$ and $u_{p.m.}$ is the averaged superficial velocity well below the interface. More realistic interface conditions were given by Ochoa-Tapia & Whitaker (1995):

$$u = v \tag{3.24a}$$

and

$$\frac{1}{\varepsilon}\frac{du}{dy} - \frac{dv}{dy} = \frac{\beta}{\sqrt{æ}}u \tag{3.24b}$$

where $u \hat{=} u_{slip}$, $v \hat{=} u_{p.m.}$, and α, β are measured coefficients.

Example 3.7: Darcy's Experiment

The given Sketch depicts the basic set-up of Darcy's experiment. In terms of the form factor $\phi \equiv A/l$ of the porous slab, find the flow rate for the given data, i.e., A = 10 m^2, l = 1.5 m, h = 1.0 m, and the Darcy coefficient $k \equiv \gamma K = 0.5\,\text{cm/s}$; $\gamma \equiv \rho g$. Derive also Darcy's Law.

Sketch	*Assumptions*
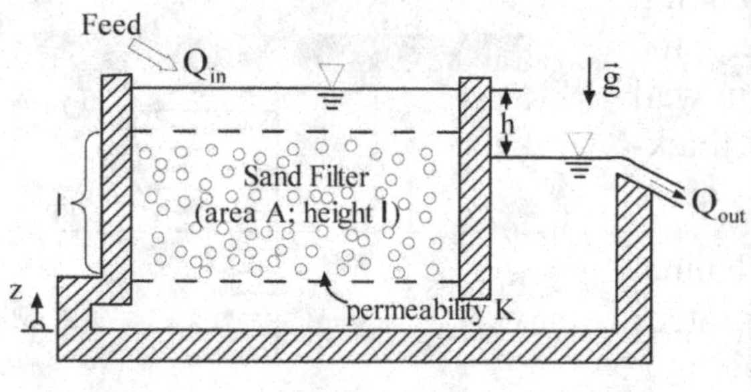 Feed; Q_{in}; $\vec{g}$; h; l; z; Sand Filter (area A; height l); permeability K; Q_{out}	• Steady laminar flow with $Re \le 1.0$ • Height h kept constant • Constant properties • Homogeneous filter without wall effects

Solution:

- Based on dimensional analysis we can state:

$$Q = Q(\underbrace{æ,\ \mu}_{\sim k \text{ or } K};\ v;\ A;\ \underbrace{h,\ l}_{\sim \Delta p}) \tag{E.3.7.1}$$

- From laboratory observations:

$$\frac{Q}{A} \sim \frac{h}{l} \text{ or } \frac{Q}{A} \sim \frac{\Delta p}{\rho g L} \quad \text{after Bernoulli}$$

Darcy employed a coefficient of proportionality k, so that

$$Q = \left(k\frac{h}{l}\right)A = vA \tag{E.3.7.2}$$

That coefficient depends on both the porous matrix as well as fluid properties, i.e.,

$$k = æ\rho g/\mu = \gamma K \tag{E.3.7.3a, b}$$

where æ is the permeability, $\gamma = \rho g$ the specific weight, and K the hydraulic permeability.

From these findings the previous equations can be formulated, i.e.,

$$v \equiv \langle u \rangle = -K\frac{\partial p}{\partial z} \tag{E.3.7.4}$$

and as a 3-D extension:

$$\nabla p = -\frac{\mu}{æ}\vec{v} \tag{E.3.7.5}$$

For the present problem,

$$\phi \equiv \frac{A}{l} := 6.6\,\text{m} \text{ and } Q = kh\phi := 0.033\,\text{m}^3/\text{s}$$

Note: Typical Darcy coefficients, note $[k] \mathrel{\hat{=}} \frac{\text{length}}{\text{time}}$, and average grain sizes are given below for different soil types.

Soil type	Clean gravel	Fine gravel	Coarse sand	Fine sand	Clay
Grain size [mm]	4~7	2~4	0.5	0.1	0.002
k-range [cm/s]	2.5~4.0	1.0~3.5	0.01~1.0	0.001~0.05	10^{-9}~10^{-6}

Example 3.8: Evaluation of the Hydraulic Conductivity

As discussed, a possible porous-medium model is a structure with n parallel tubes of diameter d, representing straight pores or capillaries. Assuming steady laminar fully-developed flow in horizontal tubes, find an expression for K= æ/μ.

Sketch	*Concept*
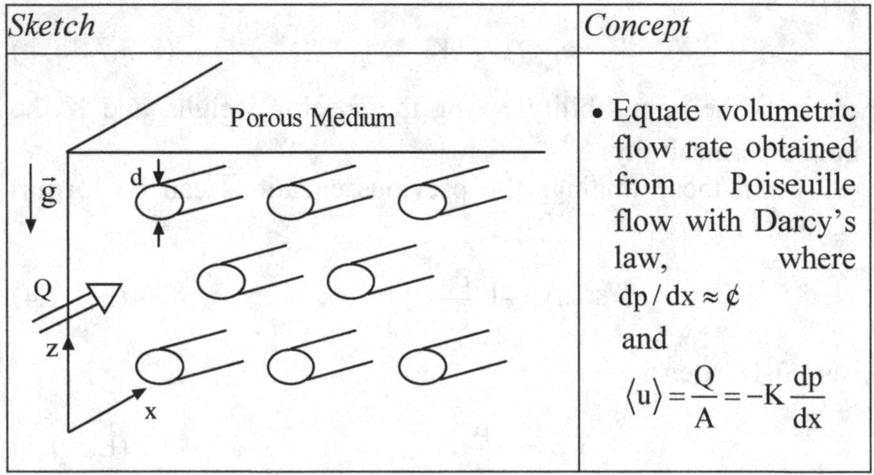	• Equate volumetric flow rate obtained from Poiseuille flow with Darcy's law, where $dp/dx \approx ¢$ and $\langle u \rangle = \frac{Q}{A} = -K\frac{dp}{dx}$

Solution: Using Eq. (E.3.4.7a) to obtain the flow rate per pore or tube, we have:

$$\frac{Q}{n} = v_{av} A_{tube} = -\frac{\pi d^4}{128\mu}\left(\frac{dp}{dx}\right) \tag{E.3.8.1a}$$

As indicated in Fig. 3.3, $\langle u \rangle = Q/A := \varepsilon v_{av}$, so that $A = \frac{n}{\varepsilon} A_{tube}$. Hence,

$$\frac{Q}{n} = \langle u \rangle \frac{1}{\varepsilon} A_{tube} = -K\left(\frac{dp}{dx}\right)\frac{d^2\pi}{4\varepsilon} \quad \text{(E.3.8.1b)}$$

so that by inspection

$$K = \frac{d^2\varepsilon}{32\mu} \quad \text{(E.3.8.2)}$$

where $0 < \varepsilon < 1.0$.

Comment: Once the average pore diameter has been estimated, setting $\varepsilon = 0.5$ and the fluid viscosity known, K and æ can be calculated and hence suitable porous medium flow analyses can be carried out.

Example 3.9: Creeping Flow in a Channel Filled with a Porous Medium

A slab of homogeneous porous medium of thickness h, bound by impermeable channel walls, represents a good base case for axial flow through porous insulation, a filter, a catalytic converter, tissue, i.e., extracellular matrix, etc.

Sketch	*Assumptions*	*Concepts*
y, h/2, μ, æ, Q, x	• Creeping flow, i.e., $\langle u \rangle \equiv u$, where u(y) only • $\partial p / \partial x = \not{c}$ • Constant properties • Symmetry	Use of the Darcy-Brinkman equation to invoke no-slip at channel walls

Solution: The 1-D form of Eq. (3.22) without the "high-speed" Forchheimer term reads:

$$\frac{dp}{dx} = -\frac{\mu}{æ}u + \mu\frac{d^2u}{dy^2} = ¢ \qquad (E.3.9.1a)$$

or

$$\frac{d^2u}{dy^2} - \frac{1}{æ}u = \frac{1}{\mu}\left(\frac{dp}{dx}\right) = ¢ \qquad (E.3.9.1b)$$

subject to $u\left(y = \frac{h}{2}\right) = 0$ and $\frac{du}{dy} = 0$ at $y = 0$.

The homogeneous solution to Eq. (E.3.9.1b) can be found on page 130 of Polyanin & Zaitsev (1995) or other math handbooks:

$$u(y) = -\frac{æh}{\mu}\left(\frac{dp}{dx}\right)\left[1 - \frac{\cosh\left(y/\sqrt{æ}\right)}{\cosh\left(h/\sqrt{æ}\right)}\right] \qquad (E.3.9.2)$$

An *effective* hydraulic conductivity can be estimated, given the channel flow rate per unit depth, i.e.,

$$\hat{Q} = 2\int_0^{h/2} u\,dy := -\frac{æh}{\mu}\left(\frac{dp}{dx}\right)\left[1 - \frac{2\sqrt{æ}}{h}\tanh\left(\frac{h}{2\sqrt{æ}}\right)\right] \qquad (E.3.9.3a)$$

Defining K_{eff} as $\bar{u}/(dp/dx) = -\frac{\hat{Q}/h}{dp/dx}$ we obtain:

$$\frac{K_{eff}}{æ/\mu} \equiv \frac{K_{eff}}{K} = 1 - \frac{2\sqrt{æ}}{h}\tanh\left(\frac{h}{2\sqrt{æ}}\right) \qquad (E.3.9.3b)$$

Graphs:

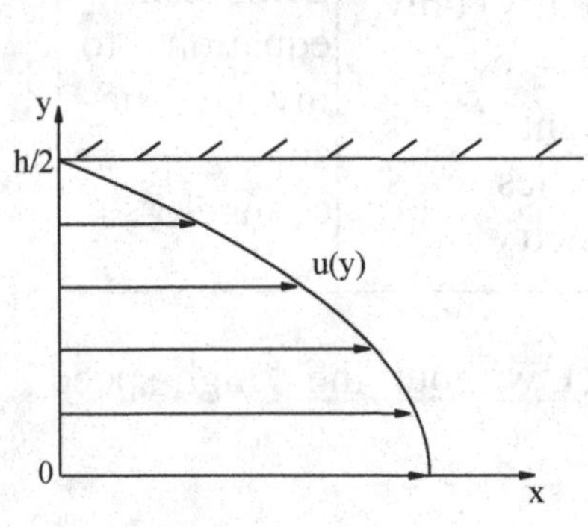

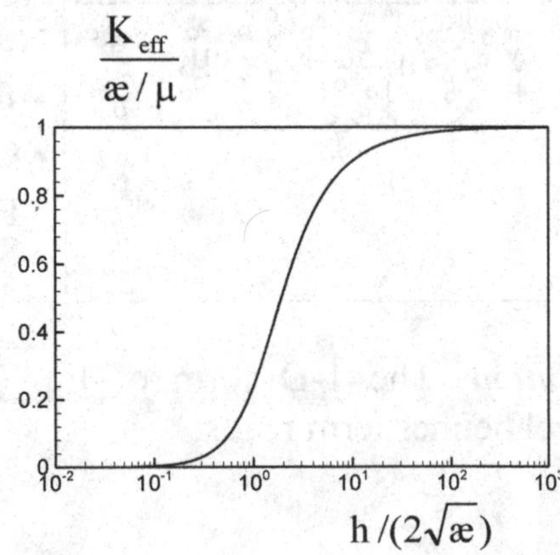

Comments: The first graph depicts a u(y)-profile determined by the $(1-\cosh y)$ function (i.e., compare Eq. (E.3.9.1b) with Eq. (E.3.4.5) for Poiseuille flow). The second graph shows the expected nonlinear increase of the nondimensionalized K_{eff} with channel height h.

Example 3.10: Radial Flow Through a Porous-Walled Tube

Consider pressure-driven flow in radial direction through a porous tubular wall. Applications include radial flow in a porous pipe, ultrafiltration, tubular filter or membrane, seepage into a lymph vessel, etc. For the given system (see Sketch), find the radial p(r) distribution and v(r) profile as well as the added mass flow rate over the tube length L.

Sketch	*Assumptions*	*Concepts*
p_0; $v_0=Q/A_{surf}$; $r=R_2$; $r=R_1$; p_i; r; z; $\Rightarrow \dot{m}$; $r=0$; $z=0$; $z=L$	• Steady radial seepage in $R_1 \le r \le R_2$ • Constant pressures and properties • No gravity effect	• Darcy's law: $\vec{v}=-\frac{\kappa}{\mu}\nabla p$ where $\vec{v}=(0,v,0)$ and $v=v(r)$ • Mass balance

Solution: Writing Eq. (3.22) in reduced form (see Assumptions and Concept):

$$\vec{v}=-\frac{æ}{\mu}\nabla p \qquad \text{(E.3.10.1)}$$

and taking the divergence of Eq. (E.3.10.1) we have with $\nabla \cdot \vec{v} = 0$ (continuity equation for incompressible fluids)

$$0 = -\frac{æ}{\mu}\nabla^2 p \qquad \text{(E.3.10.2)}$$

where ∇^2 is the Laplace operator (App. A). For our 1-D case in cylindrical coordinates, Eq. (3.10.2) is simply

$$\frac{1}{r}\frac{d}{dr}\left(r\frac{dp}{dr}\right) = 0 \qquad \text{(E.3.10.3)}$$

subject to $p(r = R_1) = p_i$ and $p(r = R_2) = p_o$. Double integration leads to

$$p(r) = C_1 \ln r + C_2$$

And finally for $p_o > p_i$:

$$\frac{p - p_i}{p_o - p_i} = \frac{\ln(r/R_1)}{\ln(R_2/R_1)} \qquad \text{(E.3.10.4)}$$

Now, with Darcy's law in the r-direction, i.e.,

$$v = -\frac{æ}{\mu}\left(\frac{dp}{dr}\right) \qquad \text{(E.3.10.5a)}$$

we obtain with the given p(r) and $\Delta p \equiv p_o - p_i$

$$v(r) = -\frac{æ}{\mu}\frac{\Delta p}{\ln(R_2/R_1)}\frac{1}{r} \qquad \text{(E.3.10.5b)}$$

A radial mass balance provides the added mass flow rate

$$\Delta\dot{m} = -\rho v\big|_{r=R_1} A_{surface} \qquad \text{(E.3.10.6a)}$$

where $A_{surface} = 2\pi R_1 L$, so that

$$\Delta\dot{m} = \frac{2\pi æ\rho\Delta p}{\mu \ln(R_2/R_1)} L \qquad \text{(E.3.10.6b)}$$

Graph:

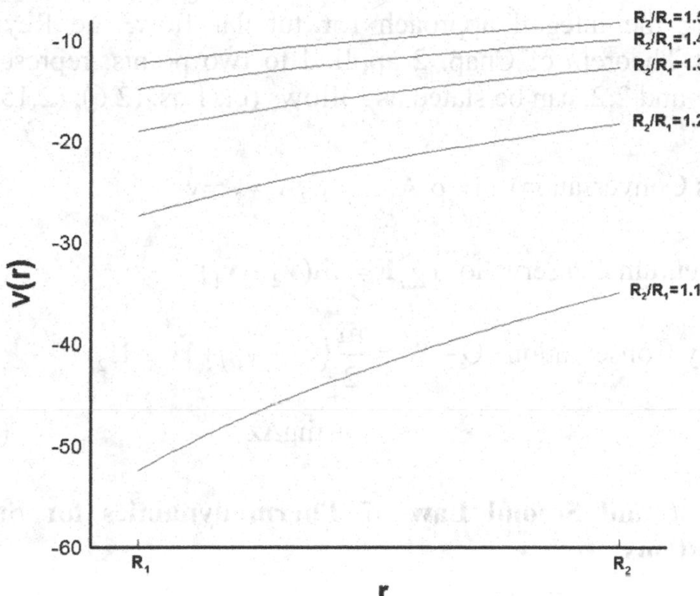

Comments: The v(r)-function (E.3.10.5b) is hyperbolic, i.e., for a given Δp, æ, μ and geometry, v decreases inversely with r. As expected radial mass influx increases with wall permeability, tube length, and pressure difference.

3.5 One-Dimensional Compressible Flow

As we recall from the Big Bang Theory, everything is compressible, e.g., the universe was once a pea-size thing. Other examples include compression of rock formations during an earthquake, transient fluid flow resulting in vibrations and noisy pipe systems (i.e., water hammer), and of course gases when the Mach number, $M \equiv v_{fluid} / c_{sound}$ exceeds 0.3 (see Fig. 1.3). Specifically, the importance of airflow applications for $M > 0.3$ has spawned separate engineering programs focusing on, say air-compressor design and aerodynamics, including supersonic/hypersonic airflows, propulsion, aircraft design, and space vehicles.

Our scope of flows with compressibility effects is restricted to steady 1-D internal gas flows with area-averaged velocities. Thus, employing the integral approach for tubular flow, the Reynolds Transport Theorem of Chap. 2 applied to two points, representing Sects. 2.1 and 2.2, can be stated as follows (cf. Eqs. (2.6), (2.15) and (2.38))

$$\text{(Mass Conversation)}\ \dot{m} = \rho_1 A_1 v_1 = \rho_2 A_2 v_2 := \text{¢} \tag{3.25}$$

$$\text{(Momentum Conservation)} \sum \vec{\mathbf{F}} = \dot{m}(\vec{\mathbf{v}}_2 - \vec{\mathbf{v}}_1) \tag{3.26}$$

$$\text{(Energy Conservation)}\ \dot{Q} - \dot{W} = \frac{\dot{m}}{2}\left(v_2^2 - v_1^2\right) + \dot{H}_1 - \dot{H}_2 + \dot{m} g \Delta z \tag{3.27}$$

3.5.1 First and Second Law of Thermodynamics for Steady Open Systems

Describing air with the Ideal Gas Law

$$p = \rho R T, \tag{3.28a}$$

recalling that the change in enthalpy rate at constant heat capacity is:

$$\Delta H \equiv \dot{H}_2 - \dot{H}_1 = \dot{m}\Delta h = \dot{m} c_p (T_2 - T_1), \tag{3.28b}$$

and using

$$\frac{c_p}{c_v} \equiv k \ \text{with}\ R = c_p - c_v, \tag{3.28c, d}$$

we can rewrite the energy equation (3.27) on a rate basis with $\Delta z \approx 0$ as:

$$\frac{\dot{Q} - \dot{W}}{\dot{m}} = \frac{1}{2}\left(v_2^2 - v_1^2\right) + \frac{k}{k-1}\left[\left(\frac{p_2}{\rho_2}\right) - \left(\frac{p_1}{\rho_1}\right)\right] \tag{3.29}$$

While Eq. (3.29) represents the first law of thermodynamics, the second one can be stated in form of a balance, where $S_{gen} > 0$:

$$\sum S_{in} - \sum S_{out} \pm \sum \frac{Q}{T_{ambient}} + S_{gen} = \Delta S\big|_{C.\forall.} \equiv S_2 - S_1 \qquad (3.30)$$

where Q is the heat transferred w.r.t. an ambient temperature source. $S_{gen} = \Delta S_{total} = \Delta S_{system} + \Delta S_{surrounding}$ is always a positive quantity (or zero under idealized conditions). For isentropic processes, i.e., internally reversible and adiabatic,

$$\Delta S_{syst.} = 0 \quad \text{or} \quad S_2 = S_1 \qquad (3.31a, b)$$

For a closed stationary system of fixed mass (see Sect. 1.3), we write the first law on a differential basis as:

$$\delta Q - \delta W = dU \qquad (3.32)$$

With $dS = \frac{\delta Q}{T}\Big|_{reversible}$ and boundary work $\delta W = pd\forall$ we obtain Gibbs first relation:

$$TdS = dU + pd\forall \text{ or per unit mass } Tds = du + pdv \qquad (3.33a, b)$$

where $v \equiv \frac{\forall}{m}$ is the specific volume.

With the definition of the enthalpy per unit mass, $h \equiv u + pv$, we obtain

$$dh = du + pdv + vdp \qquad (3.34)$$

so that Gibbs' second relation reads:

$$Tds = dh - vdp \qquad (3.35a)$$

or

$$ds = \frac{dh}{T} - \frac{v}{T}dp \qquad (3.35b)$$

where the specific volume is also $v \equiv 1/\rho$.

For example, air regarded as an ideal gas ($p = RT/v$) provides p-v and h-T relationships so that Eq. (3.35b) can be expressed as:

$$ds = c_p \frac{dT}{T} - R\frac{dp}{p} \tag{3.36}$$

Integration with $c_p \approx \bar{c}_p = \text{¢}$ yields

$$\Delta s = c_p \ln\left(\frac{T_2}{T_2}\right) - R\ln\left(\frac{p_2}{p_1}\right) \tag{3.37}$$

Similarly, Eq. (3.35b) can be expressed with $du = \bar{c}_v dT$ as:

$$ds = c_v \frac{dT}{T} + R\frac{dv}{v} \tag{3.38a}$$

so that after integration

$$\Delta s = c_v \ln\left(\frac{T_2}{T_2}\right) + R\ln\left(\frac{v_2}{v_1}\right) \tag{3.38b}$$

For isentropic flow, i.e., s = constant, $\delta Q = 0$ <adiabatic> and hence $\Delta s = 0$. Using this in Eqs. (3.37) and (3.38b) yields:

$$\frac{T_2}{T_1} = \left(\frac{p_2}{p_1}\right)^{(k-1)/k} \quad \text{and} \quad \frac{p_2}{p_1} = \left(\frac{\rho_2}{\rho_1}\right)^k \tag{3.39/40}$$

where $k \equiv c_p / c_v$, $R \equiv c_p - c_v$, and $v \equiv 1/\rho$.

Example 3.11: Bernoulli's Equation Revisited

Show that for adiabatic frictionless (i.e., isentropic) flow the Bernoulli equation, derived in Sect. 3.2 from the momentum equation, is identical to the energy equation.

Solution: Equation (3.27) with $\dot{Q} = \dot{W} = 0$ can be written for two points on a horizontal streamline as:

$$h_1 + \frac{u_1^2}{2} = h_2 + \frac{u_2^2}{2} = \text{¢} \tag{E.3.11.1}$$

Recall from Sect. 3.2 that for steady inviscid flow, the Euler equation can be written as

$$u du + \frac{dp}{\rho} = 0 \qquad \text{(E.3.11.2a)}$$

or in integrated form as the Bernoulli equation, i.e.,

$$\frac{u^2}{2} + \int \frac{dp}{\rho} = ¢ \qquad \text{(E.3.11.2b)}$$

Now, for isentropic flow, i.e., zero entropy changes, Eq. (3.35a) reduces to ($v = 1/\rho$):

$$0 = dh - v dp \quad \text{or} \quad dh = \frac{dp}{\rho} \qquad \text{(E.3.11.3a, b)}$$

Using (E.3.11.3b) in (E.3.11.2a) we obtain

$$u du + dh = 0 \qquad \text{(E.3.11.4)}$$

which is exactly the differential form of the energy equation (E.3.11.1).

3.5.2 Sound Waves and Shock Waves

Analogous to the ripple effect, i.e., gravity waves, created by a pebble thrown into a lake, sound travels through any medium as small-amplitude pressure disturbances. Such a weak pressure pulse is called a *sound wave*. Its speed, c, can be obtained by applying the 1-D mass and momentum RTT to a control volume moving with a tiny sound wave propagating through an undisturbed medium, such as air, water, or steel (see Fig. 3.5). For steady uniform flow of a compressible fluid, we consider the differential changes (see RTT mass and momentum balances) for a very small control volume.

Mass Conservation: $$\int_{C.S.} \rho \vec{v}_{rel} \cdot d\vec{A} = 0$$

$$(-\rho c A) + [(\rho + d\rho)(c - du)]A = 0$$

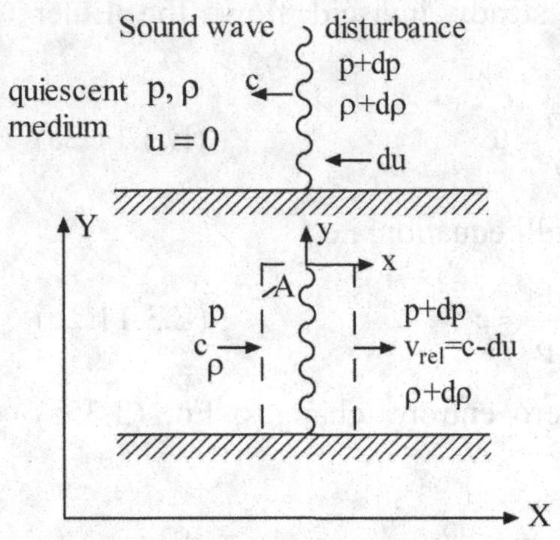

Fig. 3.5 Sound wave viewed from inertia frame X–Y and moving with x–y coordinates

Neglecting ($d\rho du$) as a higher-order term, we obtain

$$du = c\frac{d\rho}{\rho}$$

Now du is expressed with a *Momentum Balance*:

$$\sum F_x = \int_{C.S.} u\rho \vec{v}_{rel} \cdot d\vec{A}$$

$$pA - (p + dp)A = c(-\rho cA) + (c - du)[(\rho + d\rho)(c - du)A]$$

$$\therefore \quad du = \frac{dp}{\rho c}$$

so that after equating both results

$$c^2 = \frac{dp}{d\rho} \tag{3.41a}$$

Clearly, for incompressible fluids, $d\rho = 0$ and hence $c \to \infty$.

Assuming that for $f \leq 18{,}000\,\text{Hz}$ the sound wave propagates isentropically, i.e., reversible pressure pulse and no heat loss, $p = p(\rho, s)$ only. This implies that

$$c = \sqrt{\left.\frac{\partial p}{\partial \rho}\right|_{s=\not c}} \tag{3.41b}$$

and with Eq. (3.40) we have $p/\rho^k = \text{const}$, i.e.,

$$\frac{dp}{d\rho} = k\frac{p}{\rho} := c^2 \tag{3.42a}$$

so that for an ideal gas,

$$c = \sqrt{kRT} \tag{3.42b}$$

Hence, the Mach number, an indicator of compressibility effects (see Sect. 2.4) can be written as:

$$M \equiv \frac{v}{c} = v/\sqrt{kRT} \tag{3.43}$$

Notes

- *Sound speed examples*: $c_{air} = 340$ m/s. $c_{water} = 1,450$ m/s, and $c_{steel} = 5,000$ m/s.
- *Mach number ranges*: $0 < M_{subsonic} < 0.9 < M_{transonic} < 1.2 < M_{supersonic} < 5 < M_{hypersonic}$
- *The Mach cone*: Figure 3.6 depicts sound waves propogating every Δt-seconds from a point noise source which has different velocities, i.e., $0 \le v < c$ and $v > c$.

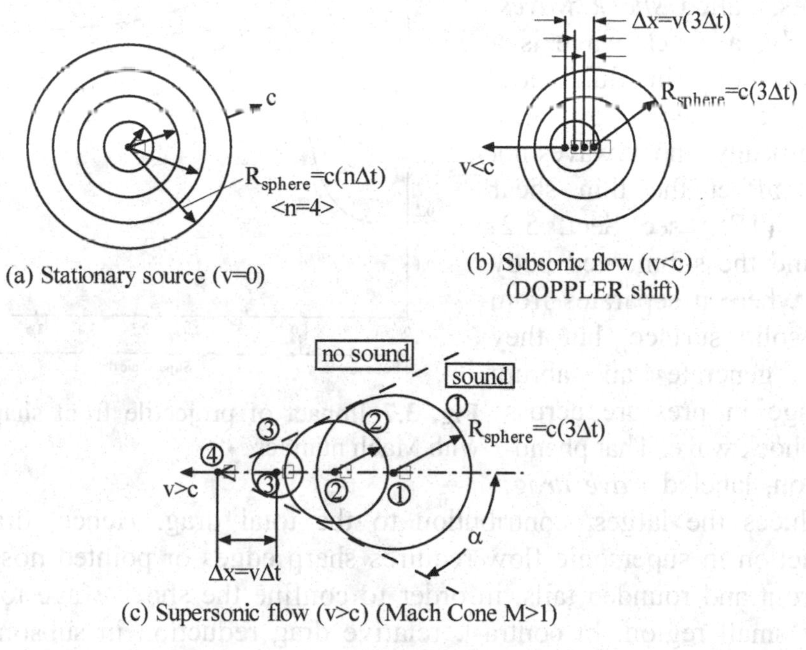

Fig. 3.6. Sound wave propagations

With respect to Fig. 3.6c, the point noise source, e.g., a fighter jet, moves to the left with supersonic speed, i.e., the jet is always ahead of the ever-expanding, spherical sound waves it generates. The resulting Mach cone is tangent to each sphere, i.e.,

$$\sin\alpha = \frac{c(n\Delta t)}{v(n\Delta t)} = \frac{c}{v} \tag{3.44a}$$

or

$$\alpha = \sin^{-1}(1/M) \tag{3.44b}$$

Outside the cone is the "zone of silence", i.e., it is dead quiet inside the jet because the created sound (wave) cannot reach the escaping jet.

Such sound waves, often called Mach waves, occur on supersonic aircrafts with a needle-like nose or wings with razor-sharp leading edges. In contrast, blunt noses or wings going v > c generate large-amplitude waves, called *shock waves*. Clearly, a shock wave is a discontinuity in fluid flow variables (see Fig. 3.5). Specifically, shock waves not only affect the thin shear layer (TSL; see Sect. 5.2) around the submerged body and where it separates from the solid surface, but they also generate an abrupt change in pressure across the shock wave. That phenomenon, labeled *wave drag*, produces the largest contribution to the total drag. Hence, drag reduction in supersonic flow requires sharp edges or pointed noses in front and rounded tails, in order to confine the shock wave to a very small region. In contrast, relative drag reduction in subsonic flow is achieved when a streamlined round nose and gradually pointed tail can be designed (see Fig. 3.7).

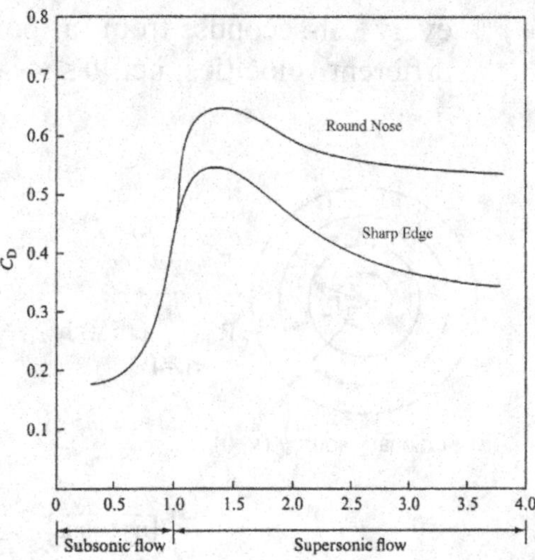

Fig. 3.7 Impact of projectile front shape with Mach number

Example 3.12: Compressibility Factor

Compare incompressible vs. compressible stagnation-point flow of an ideal gas and express the difference in terms of the compressibility factor

$$\beta \equiv \frac{p_s - p_0}{(p_s - p_0)\big|_{\rho=\not{c}}} = \frac{p_s - p_0}{\frac{\rho_0}{2}u_0^2}$$

as a function of Mach number $M_0 = u_0 / c$ and heat capacity ratio $k = c_p / c_v$.

Sketch	*Assumptions*	*Concepts*
u_0, p_0, ρ_0, T_0, (S)	• Steady isentropic flow, i.e., frictionless adiabatic	• Bernoulli's Equation • Ideal Gas Law • Mach # Correlation

Solution:

(A) *Incompressible Flow*: Applying Bernoulli between points 0 and s with $\Delta z = 0$ and $u_s = 0$ at the stagnation point,

$$p_s = p_0 + \frac{\rho}{2}u_0^2 \tag{E.3.12.1}$$

Thus, with $\rho_0 = \rho_s = \rho$ for incompressible flow (see given β - definition):

$$\beta = \frac{p_s - p_0}{\rho u_0^2 / 2} \equiv 1$$

To express the stagnation point temperature, we use (see ideal gas):

$$\frac{T_s}{T_0} = \frac{p_s}{p_0} \tag{E.3.12.2a}$$

so that with Eq. (E.3.12.1):

$$T_s = T_0(p_s / p_0) = T_0\left[1 + \frac{u_0^2}{2}\left(\frac{\rho}{p_0}\right)\right] \tag{E.3.12.2b}$$

Employing the ideal gas law in the form $\rho/p_0 = (RT_0)^{-1}$, we rewrite Eq. (E.3.12.2b) as:

$$T_s = T_0 + \frac{u_0^2}{2R} \tag{E.3.12.2c}$$

(B) Compressible Flow: For this situation, Bernoulli's equation reads:

$$\int \frac{dp}{\rho} + \frac{u^2}{2} = ¢ \tag{E.3.12.3a}$$

Expressing ρ as $\rho(p) = (p/C)^{1/k}$, $C \equiv p^{1/k}/\rho$, we can integrate the first term, i.e.,

$$\int \frac{dp}{\rho} = \frac{k}{k-1}\left(\frac{p}{\rho}\right) + \text{constant}$$

Hence, Eq. (E.3.12.3a) now reads:

$$\frac{k}{k-1}\frac{p}{\rho} + \frac{u^2}{2} = ¢ \tag{E.3.12.3b}$$

Applying Eq. (E.3.12.3b) to points ⓪ and ⓢ yields:

$$\frac{k}{k-1}\left(\frac{p_s}{\rho_s}\right) = \frac{k}{k-1}\left(\frac{p_0}{\rho_0}\right) + \frac{u_0^2}{2} \tag{E.3.12.3c}$$

Expressing Eq. (3.12.3c) again in terms of T_s with the equation of state $p/\rho = RT$, we have

$$T_s = T_0 + \frac{k-1}{k}\frac{u_0^2}{2R} \tag{E.3.12.4a}$$

Comparing (E.3.12.4a) with (E.3.12.2c) it is already evident that the second-term coefficient (k − 1)/k (≈0.3 for air) captures the difference between incompressible and compressible flows. In order to express β in terms of the Mach number, we use Eq. (3.42b) which provides $kR = c^2 / T_0$ and $u_0^2 / c^2 \equiv M_0^2$, i.e.,

$$T_s = T_0 \left[1 + \frac{1}{2}(k-1) M_0^2 \right] \qquad \text{(E.3.12.4b)}$$

Now, with Eq. (3.39) we obtain

$$p_s = p_0 \left(\frac{T_s}{T_0} \right)^{\frac{k}{k-1}} = p_0 \left(1 + \frac{k-1}{2} M_0^2 \right)^{\frac{k}{k-1}} \qquad \text{(E.3.12.5a)}$$

For subsonic flow $\frac{k-1}{2} M_0^2 < 1$ and Eq. (E.3.12.5a) can be expanded with the binominal theorem to (see App. A):

$$\frac{p_s}{p_0} = \left(1 + \frac{k-1}{2} M_0^2 \right)^{\frac{k}{k-1}} \approx 1 + \frac{k}{2} M_0^2 + \frac{k}{8} M_0^4 + \frac{k(2-k)}{48} M_0^6 + \ldots \text{(E.3.12.5b)}$$

In order to form the compressibility factor, we rearrange (E.3.12.5b) as

$$p_s - p_0 = \frac{k}{2} p_0 M_0^2 \left[1 + \frac{1}{4} M_0^2 + \frac{2-k}{24} M_0^4 + \ldots \right] \qquad \text{(E.3.12.5c)}$$

However, by definition the factor

$$\frac{k}{2} p_0 M_0^2 = \frac{k}{2} p_0 \frac{u_0^2}{k p_0 / \rho_0} = \rho_0 \frac{u_0^2}{2}$$

so that

$$p_s - p_0 = \frac{\rho_0}{2} u_0^2 \left[1 + \frac{1}{4} M_0^2 + \frac{2-k}{24} M_0^4 + \ldots \right] \qquad \text{(E.3.12.5d)}$$

Finally,

$$\frac{p_s - p_0}{\frac{\rho_0}{2} u_0^2} \equiv \beta = 1 + \frac{1}{4} M_0^2 + \frac{2-k}{24} M_0^4 + \ldots$$

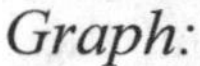
Graph:

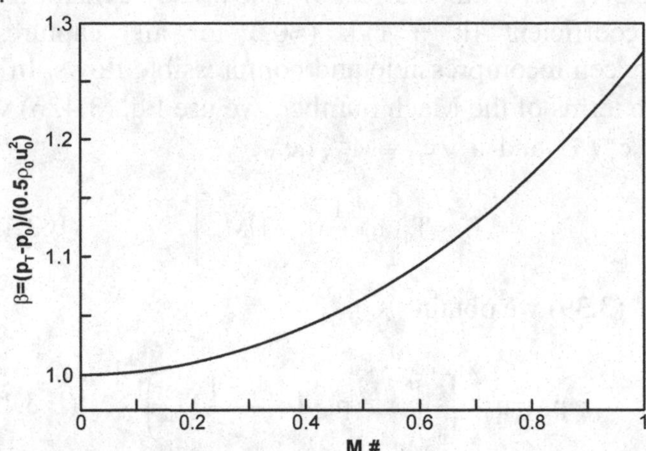

Comment: The graph $\beta(M)$ indicates that for $M < 0.3$ in air (being an ideal gas), the error in assuming incompressible flow is less than 2%; however, at $M = 1$ the error is 28%.

3.5.3 Normal Shock Waves in Tubes

As mentioned, shock waves are large-amplitude disturbances where over a very short distance, say 10^{-4} mm, the Mach number and all gas properties change. Such sudden changes can be oriented normal or angled (i.e., oblique) to the flow direction. Examples include exit flow in a rocket or jet engine nozzle as well as airflow ahead of a bullet in a gun barrel, around a supersonic aircraft or a large explosion source.

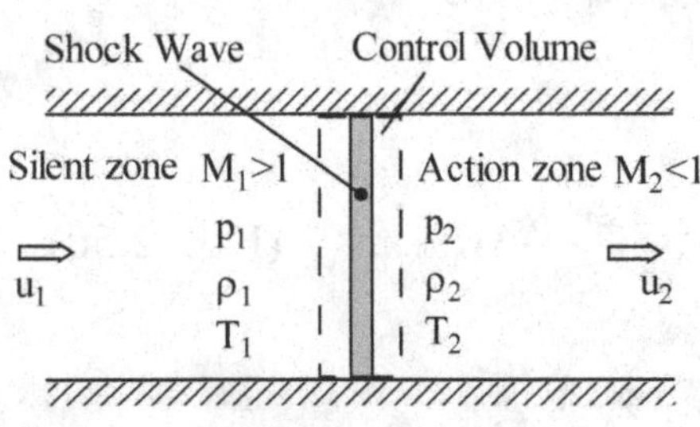

Fig. 3.8 Stationary normal shock wave in a tube

- Continuity: $\dot{m}_1 = \dot{m}_2$

$$\rho_1 u_1 = \rho_2 u_2$$

with $u = cM = M\sqrt{kRT}$

$$\rho_1 M_1 \sqrt{T_1} = \rho_2 M_2 \sqrt{T_2}$$

and with $\rho_1/\rho_2 = (p_1/p_2)(T_2/T_1)$ from the equation of state, we obtain

$$p_1 M_1 \sqrt{T_1} = p_2 M_2 \sqrt{T_2} \tag{3.45}$$

For a mathematical description, we assume the *normal* shock wave in a tube to be a "zero-thickness discontinuity" where viscous dissipation and heat transfer effects are negligible. The goal is to relate downstream Mach number, pressure and temperature to the corresponding upstream values (see Fig. 3.8).

- Energy Equation (E.3.11.1):

$$h_1 + \frac{u_1^2}{2} = h_2 + \frac{u_2^2}{2} = h_s$$

where $h_s = c_p T_s$ is the stagnation enthalpy (see Example 3.12). Thus, by rewriting Eq. (E.3.12.4b) we can relate upstream and downstream quantities as:

$$T_1\left(1 + \frac{k-1}{2} M_1^2\right) = T_2\left(1 + \frac{k-1}{2} M_2^2\right) \tag{3.46}$$

- Momentum Equation (3.26): $F_1 - F_2 = \rho_1 A u_1 (u_2 - u_1)$

or

$$p_1 - p_2 = \rho_1 u_1 u_2 - \rho_1 u_1^2 \tag{3.47}$$

However, with continuity $\rho_1 u_1 = \rho_2 u_2$, Mach number definition $M = u/\sqrt{kRT}$, and equation of state $\rho RT / p \equiv 1$, we have

$$p_1\left(1 + kM_1^2\right) = p_2\left(1 + kM_2^2\right) \tag{3.48}$$

The three equations (Eqs. (3.45), (3.46), and (3.48)) can be employed to correlate specific quantities before and after the shock wave. For example, focusing on the Mach numbers and pressures, we can obtain:

$$M_2 = \left\{\frac{M_1^2 + 2/(k-1)}{[2k/(k-1)]M_1^2 - 1}\right\}^{1/2} \tag{3.49}$$

and

$$\frac{p_2}{p_1} = \frac{2k}{k+1} M_1^2 - \frac{k-1}{k+1} \tag{3.50}$$

For air (k = 1.4) the dependencies of $M_2(M_1)$ and $p_2/p_1(M_1)$ are depicted in Fig. 3.9.

Equation (3.50) can be used to estimate the shock strength, i.e., the pressure increase

$$\frac{p_2 - p_1}{p_1} = \frac{p_2}{p_1} - 1 := \frac{2k}{k+1}\left(M_1^2 - 1\right) \tag{3.51}$$

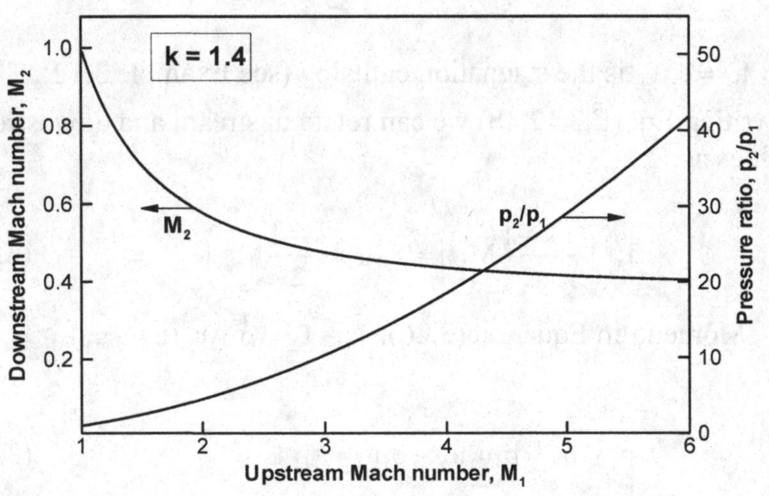

Fig. 3.9 Velocity and pressure changes across a normal shock wave

---

Example 3.13: Shock-Wave Induced Velocity

A normal shock wave travels through stagnant air at v = 700 m/s, Assuming T_{air} = 15°C, k = 1.4, and R = 287 $m^2/(s^2K)$, find the induced velocity $\Delta v = v_2 - v_1$.

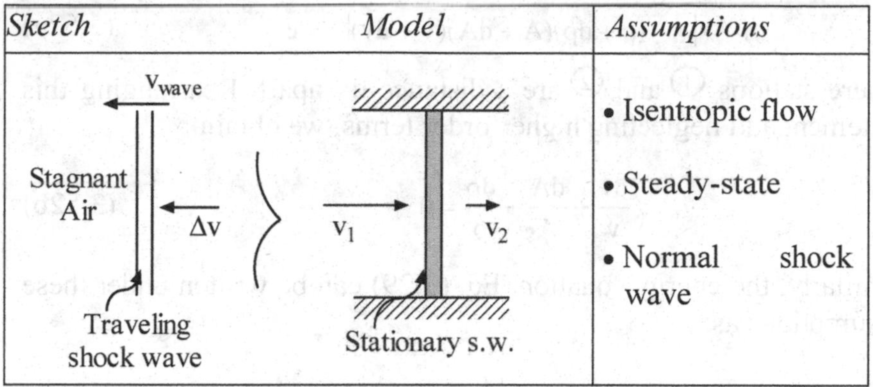

Solution: Using Eq. (3.49) where $M_1 = \frac{v_1}{c} = \frac{v_1}{\sqrt{kRT_1}} := 2.06$,

$$M_2 = \left(\frac{M_1^2 + 5}{7M_1^2 - 1} \right)^{1/2} := 0.567$$

To obtain v_2, i.e., $v_2 = M_2 c_2 = M_2 \sqrt{kRT_2}$, we need T_2. Employing Eq. (3.46)

$$T_2 = \frac{T_2 \left(M_1^2 + 5\right)\left(7M_1^2 - 1\right)}{36M_1^2} := 500\,\text{K}$$

We now can calculate

$$v_2 = 254\,\text{m/s} \text{ and } \Delta v = 254 - 700 = -446\,\text{m/s}$$

Comments: A downstream flow with $v_2 = 254\,\text{m/s}$ is observed from the stationary shock wave where the upstream airflow approaches at $v_1 = 700\,\text{m/s}$. The induced velocity of 446 m/s is in the same direction as the propagating shock wave (see Sketch).

3.5.4 Isentropic Nozzle Flow

Considering steady isentropic uniform flow in a nozzle or diffuser, we know from continuity, $\dot{m} = ¢$, that

$$\rho A v\big|_{①} = (\rho + d\rho)(A + dA)(v + dv)\big|_{②} = ¢ \tag{3.52a}$$

where stations ① and ② are a distance dx apart. Rearranging this statement and neglecting higher-order terms, we obtain

$$\frac{dv}{v} + \frac{dA}{A} + \frac{d\rho}{\rho} = 0 \tag{3.52b}$$

Similarly, the energy equation, Eq. (3.29) can be written under these assumptions as:

$$\frac{v^2}{2} + \frac{k}{k-1}\frac{p}{\rho} = ¢ \text{ OR } \frac{1}{2}(v + dv)^2 + \frac{k}{k-1}\frac{p + dp}{\rho + d\rho}$$

Again, rearranging, neglecting higher-order terms and assuming small-amplitude, moderate-frequency waves so that

$$\frac{p}{\rho^k} = ¢ \quad \text{or} \quad \frac{dp}{d\rho} = k\frac{p}{\rho} \tag{3.53a, b}$$

we obtain:

$$v dv + k\frac{p}{\rho}\frac{d\rho}{\rho} = 0 \tag{3.54}$$

Eliminating $d\rho/\rho$ with Eq. (3.54), we have with $c^2 = k\frac{p}{\rho}$ and M=v/c:

$$\left(M^2 - 1\right)\frac{dv}{v} = \frac{dA}{A} \tag{3.55}$$

Equation (3.55) for steady uniform isentropic flow in an area-changing conduit, A = A(x) lends itself to the following observations:

(a) Nozzle, i.e., accelerating, flows:

- If M > 1 and A increases, then dv > 0
- If M < 1 and A decreases, then dv > 0

(b) Diffuser, i.e., decelerating, flows:

- If $M > 1$ and A decreases, then $dv < 0$
- If $M < 1$ and A increases, then $dv < 0$

In summary, (see Fig. 3.10) supersonic flow accelerates in a diverging section and decelerates in a converging section – quite the opposite to subsonic flow where compressibility effects are negligible.

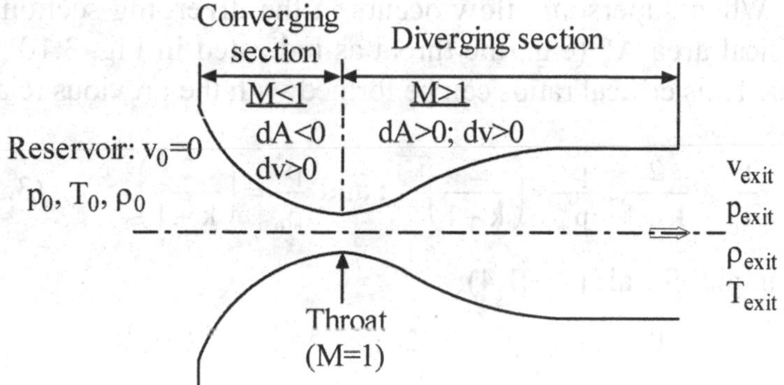

Fig. 3.10 Supersonic converging-diverging nozzle

As the discussion of the conservation-of-mass statement, Eq. (3.55), indicates, supersonic nozzle flow can generate a number of interesting flow situation, largely dependent on the type of nozzle and the magnitude of the exit pressure.

Considering representative streamline with a point in the reservoir ($v_0 = 0$) and any nozzle section (see Fig. 3.10), the energy equation (E.3.11.1) can be written as:

$$c_p T_0 = \frac{v^2}{2} + c_p T \qquad (3.56a)$$

Again, with $v = Mc$, $c = \sqrt{kRT}$, $c_p = c_v + R$ and $k = c_p / c_v$, we obtain:

$$\frac{T_0}{T} = 1 + \frac{k-1}{2} M^2 \qquad (3.56b)$$

Using the previously derived thermodynamic relationships for isentropic flow (see Eqs. 3.39 and 3.40), we also have

$$\frac{p_0}{p}=\left(1+\frac{k-1}{2}M^2\right)^{\frac{k}{k-1}} \tag{3.57}$$

and

$$\frac{\rho_0}{\rho}=\left(1+\frac{k-1}{2}M^2\right)^{\frac{k}{k-1}} \tag{3.58}$$

When supersonic flow occurs in the diverging section, there is a critical area A* (e.g., the throat as indicated in Fig. 3.10) where $M \equiv 1.0$. Thus critical ratios can be formed with the previous relations:

$$\frac{T^*}{T_0}=\frac{2}{k+1};\ \frac{p^*}{p_0}=\left(\frac{2}{k+1}\right)^{\frac{k}{k-1}};\ \text{and}\ \frac{\rho^*}{\rho_0}=\left(\frac{2}{k+1}\right)^{\frac{1}{k-1}} \tag{3.59a–c}$$

For example, for air (k = 1.4):

$$T^*=0.8333T_0,\ p^*=0.5283p_0,\ \text{and}\ \rho^*=0.6340\rho_0$$

To obtain the associated mass flow rate, we rewrite the conservation law

$$\dot{m}=\rho A v \tag{3.60}$$

with $\rho=\frac{p}{RT}$ and $v=M\sqrt{kRT}$ as well as Eqs. (3.57 and 3.58) as:

$$\dot{m}=p_0 A\sqrt{\frac{k}{RT_0}}M\left(1+\frac{k-1}{2}M^2\right)^{\frac{k+1}{2(1-k)}} \tag{3.61a}$$

or with M = 1 in A = A*

$$\dot{m}=p_0 A^*\sqrt{\frac{k}{RT_0}}\left(\frac{k+1}{2}\right)^{\frac{k+1}{2(1-k)}} \tag{3.61b}$$

The maximum mass flow rate, called “choked flow”, is $\dot{m}_{choked}=\rho^* v^* A^*$, or with $A^*=A_{exit}$ of a converging nozzle and M = 1,

$$\dot{m}_{choked}=p_0 A_e\sqrt{\frac{k}{RT_0}}\left(\frac{k+1}{2}\right)^{\frac{k+1}{2(1-k)}} \tag{3.61c}$$

If the back-pressure drops below p*, the flow remains choked.

Clearly, the mass flow rate of a given (ideal) gas depends only on the reservoir conditions p_0 and T_0 as well as the critical nozzle area A* where M = 1.0. Forming a ratio of Eqs. (3.61b) and (3.61a) yields an expression for A*, i.e.,

$$\frac{A^*}{A} = M\left[\frac{k+1}{2+(k-1)M^2}\right]^{\frac{k+1}{2(1-k)}} \tag{3.62}$$

Example 3.14: Converging–Diverging Nozzle Flow

Consider ideal gas flow in a converging-diverging nozzle (A_{throat} =10 cm^2 and A_{exit} =40 cm^2), fed by a reservoir (T_0 =20°C, p_0 = 500 kPa absolute). Determine the nozzle exit pressures such that M=1 in $A_{throat} \equiv A^*$. Specifically, a varying receiver pressure, p_r, can produce different mass flow rates and throat conditions (see Sketch)

Sketch:

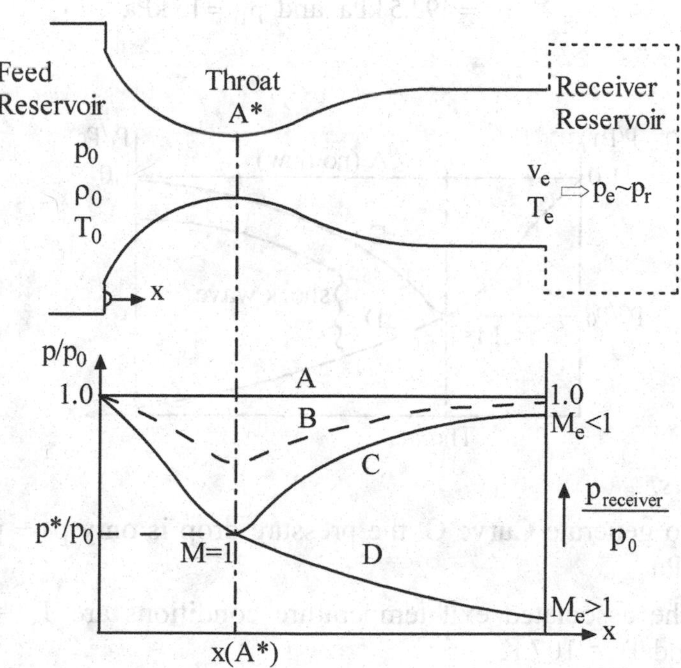

Note: Receiver (or back) pressure p_r can be regulated via a valve and vacuum pump

Solution: Clearly, when $p_r = p_0$ (see Curve A), no flow can occur. If p_r is slightly lower than p_0, subsonic flow occurs (see Curve B). Lowering p_r further results in a pressure distribution $p(x)/p_0$ (see Curve C) where M = 1 in the throat. Now, a second particular p_r – value generates again subsonic flow with M = 1 in the throat, i.e., Curve D, because there are two solutions to Eq. (3.57) for M_{exit} or for that matter to Eq. (3.62).

Given $A_{exit} / A_{throat} \equiv A / A^* = 40/10 = 4$ and using Eq. (3.62), two M_e – values can be obtained via trial-and-error, i.e., $M_e \approx 0.147$ and $M_e \approx 2.94$. Employing Eq. (3.57), the corresponding exit pressures are

$$p_e \equiv p_r \hat{=} p_c = 0.985 p_0 \text{ and } p_D = 0.0298 p_0$$

or with p_0 = 500 kPa

$$p_c = 492.5\,\text{kPa} \text{ and } p_D = 15\,\text{kPa}$$

Graph:

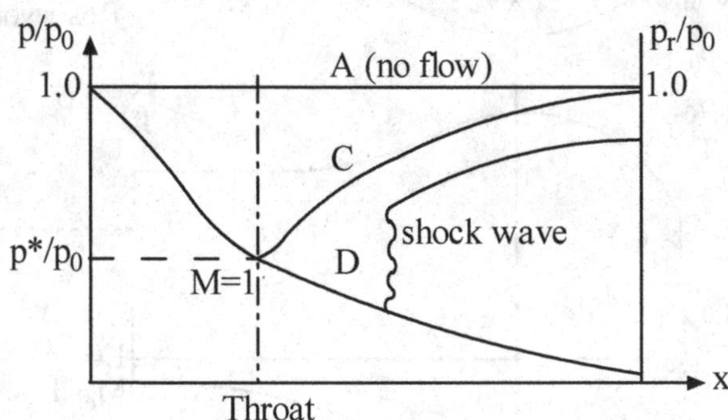

Comments:

- To generate Curve C, the pressure drop is only $p_0 - p_c$ =7.5 kPa
- The associated exit temperature conditions are T_c =292 K and T_D= 107 K

- That allows computation of the gas exit velocities to

$$v_c = M_c\sqrt{kRT_c} = 50\ m/s$$

and

$$v_D = M_D\sqrt{kRT_D} = 610\ m/s$$

- Any outlet pressure in the range of $p_c < p_r < p_D$, generating pressure distributions $p(x)/p_0$ between Curves C and D, produce shock waves inside or outside the diverging part of the nozzle and the flow is generally non-isentropic.

3.6 Forced Convection Heat Transfer

Figure 3.11 summarizes the interactions between fluid mechanics and heat transfer. Specifically, in Sect. 1.2 the heat flux vector, i.e. Fourier's law, was introduced as:

$$\vec{q} = -k\,\nabla T \tag{3.63}$$

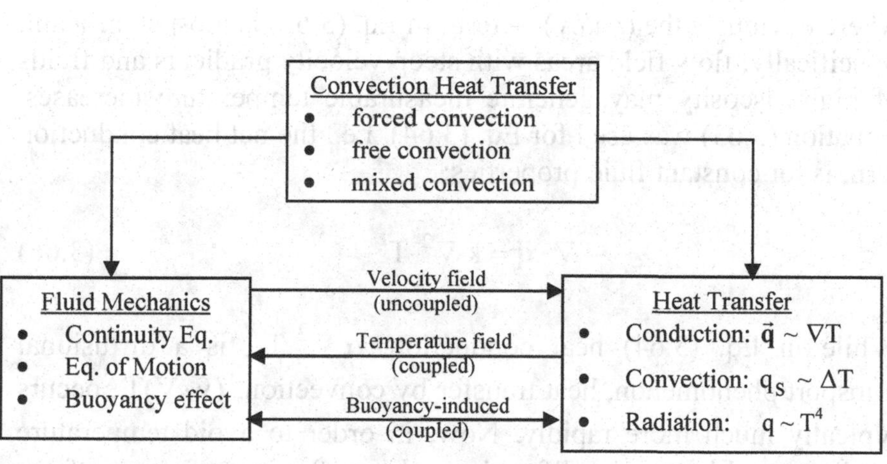

Fig. 3.11 Convection heat transfer components

where k is the (isotropic) thermal conductivity. Then, in Sect. 2.5.3 the heat transfer equation

$$\frac{\partial T}{\partial t} + (\vec{\mathbf{v}} \cdot \nabla) T = \alpha \nabla^2 T - \frac{\mu}{\rho c_p} \Phi \pm S_{heat} \tag{3.64}$$

was derived, where $\alpha \equiv k/(\rho\, c_p)$ is the thermal diffusivity, $\mu\Phi = \tau_{ij}\, \partial v_i / \partial x_j$ is the viscous dissipation function, and S_{heat} is a possible source or sink of energy, e.g., due to chemical reaction. In rectangular coordinates we have:

$$\Phi = 2\left[\left(\frac{\partial u}{\partial x}\right)^2 + \left(\frac{\partial v}{\partial y}\right)^2 + \left(\frac{\partial w}{\partial z}\right)^2\right] + \left[\left(\frac{\partial u}{\partial y} + \frac{\partial v}{\partial x}\right)^2 + \left(\frac{\partial v}{\partial z} + \frac{\partial w}{\partial y}\right)^2 + \left(\frac{\partial w}{\partial x} + \frac{\partial u}{\partial z}\right)^2\right] - \frac{2}{3}\left(\frac{\partial u}{\partial x} + \frac{\partial v}{\partial y} + \frac{\partial w}{\partial z}\right)^2 \tag{3.66}$$

where obviously the $(\partial u/\partial y)^2$ – term in Eq. (3.65) is most significant. Specifically, flow field areas with steep velocity gradients and fluids of high viscosity may generate measurable temperature increases. Equation (3.63) was used for Eq. (3.64), i.e., the net heat conduction term is for constant fluid properties:

$$-\nabla \cdot \vec{\mathbf{q}} = k\, \nabla^2 T \tag{3.66}$$

While in Eq. (3.64) heat conduction, $\alpha \nabla^2 T$, is a diffusional transport phenomenon, heat transfer by convection, $(\vec{v} \cdot \nabla)T$, occurs typically much more rapidly. Now, in order to avoid temperature gradients and hence simplify things, the surface heat flux, q_s, from a hot surface of temperature T_s into a moving stream with reference temperature T_{ref} can be based on the temperature difference $\Delta T = T_s - T_{ref}$, rather than Eq. (3.63), i.e., as attributed to Newton:

$$q_s = h(T_s - T_{ref}) \tag{3.67}$$

Here, h is the convective heat transfer coefficient, $T_{reference}$ is either T_{mean}, the cross-sectionally averaged fluid temperature, i.e., $T_m(x)$, or T_∞, as in thermal boundary-layer theory (Sect. 5.2), or the fluid bulk temperature $T_b = (T_{in} + T_{out})/2$, with T_∞ and T_b being constant. As always, the heat flow rate is then:

$$\dot{Q} = A_s q_s \tag{3.68}$$

3.6.1 Convection Heat Transfer Coefficient

The heat transfer coefficient h encapsulates all possible system parameters, such as temperature difference, thermal boundary-layer thickness Reynolds number, fluid Prandtl number, and wall geometry.

(a) Newton's law of cooling

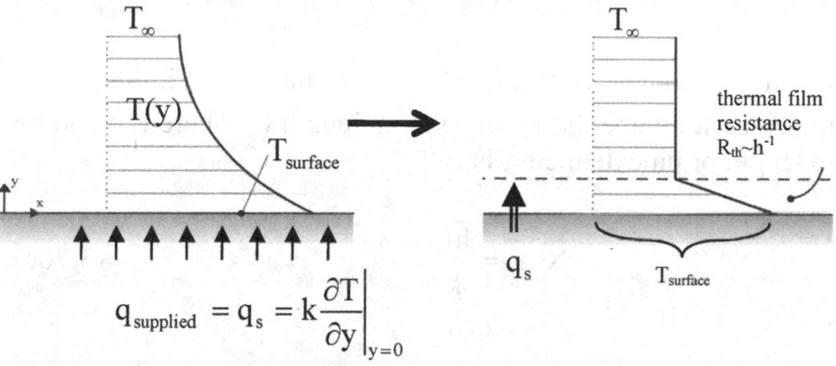

(b) Thermal boundary-layers

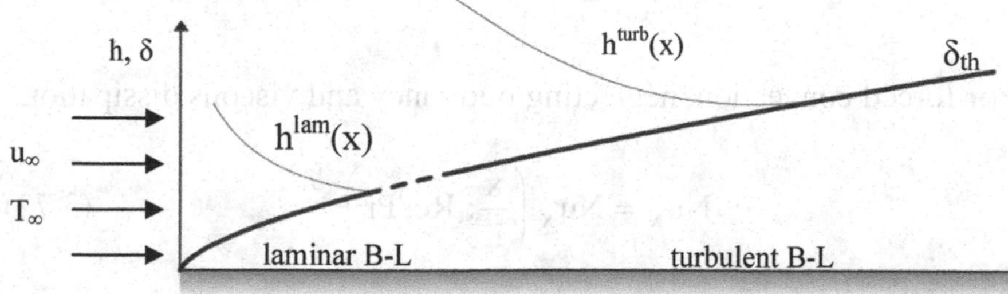

Fig. 3.12 Dependence of heat transfer coefficient on flow regime

Clearly, h is not a property such as k; but, it is a convenient artifact to calculate q_s or T_s, and ultimately $\dot{Q}$ which also greatly depends on the heat transfer area A_s. For example, for boundary layer flow:

$$h = \frac{q_s}{T_s - T_\infty} = \frac{-k}{T_s - T_\infty} \left. \frac{\partial T}{\partial y} \right|_{y=0} \tag{3.69a, b}$$

Typical h-values range from 20–300 for gases to 5,000–50,000 $W/(m^2 \cdot °C)$ for liquid metals. Figure 3.12 visualizes Eq. (3.69).

The Nusselt Number Equation (3.69a) can be non-dimensionalized by inspection, using the axial coordinate x as a length scale, i.e.,

$$\frac{h\,x}{k} \equiv Nu_x = \frac{q_s\,x}{k(T_s - T_\infty)} \tag{3.70a, b}$$

where Nu_x is known as the local Nusselt number. Similarly, the average Nusselt number based on system length L, where L could be a plate length or pipe diameter, is:

$$\overline{Nu_L} = \frac{\bar{h}L}{k} \tag{3.70c}$$

where

$$\bar{h} = \frac{1}{L}\int_0^L h(x)dx \tag{3.71}$$

For forced convection, neglecting buoyancy and viscous dissipation,

$$Nu_x = Nu_x\left(\frac{x}{L}; Re, Pr\right) \tag{3.72}$$

In summary, the main objective is to find the temperature field $T = T(x, y, z; t)$ and then obtain h (or Nu_x) in order to calculate the surface heat flux or temperature. The following solution steps are for constant-property fluids, i.e., one-way coupled problems:

- Solve, subject to appropriate boundary conditions, a reduced form of Eq. (3.64) after securing a computed (or measured) velocity function (see, for example, Sect. 3.2).
- Calculate the wall temperature gradient and obtain, via Eq. (3.69), h(x) and Nu(x).

Reynolds–Colburn Analogy Note that, as an *alternative approach*, Reynolds and Colburn established an analogy between heat and momentum transfer. It is based on the similarity between dimensionless temperature and velocity profiles in boundary layers (see Sect. 3.5 HWAs):

$$\frac{1}{2} C_f(x) = St_x \, Pr^{2/3} \quad \text{for} \quad 0.6 < Pr < 60 \tag{3.73}$$

where the skin friction coefficient and Stanton number are:

$$C_f = \frac{2\tau_{wall}}{\rho \, u_\infty^2} \quad \text{and} \quad St_x = Nu_x/(Re_x \, Pr) = h(x)/(\rho \, c_p \, u_\infty) \tag{3.74a–c}$$

Clearly, once the wall shear stress of a thermal boundary-layer problem is known, Nu(x) or h(x) can be directly obtained.

Example 3.15: Simple Couette Flow with Viscous Dissipation

As an example of planar lubrication with significant heat generation due to oil-film friction, consider simple thermal Couette flow with adiabatic wall and constant temperature of the moving plate.

Sketch	*Assumptions*	*Approach*
y=d; u_o; $\left(\frac{\partial p}{\partial x}=0\right)$; u(y); $T_s=T_o$; y; $q_s=0$; x	• Steady laminar 1-D flow • $\nabla p = 0$; u_0 and d are constant	• Reduced N–S equations and HT eqs. • Constant thermal wall cond.

Solution:

Based on the postulates $\vec{v} = [u(y), 0, 0]$ and $\nabla p \equiv 0$, the Navier–Stokes equations reduce to:

$$0 = 0 \quad \langle\text{continunity}\rangle$$

and

$$0 = \frac{d^2u}{dy^2} \quad \langle\text{x-momentum}\rangle \tag{E.3.15.1a}$$

subject to $u(y = 0) = 0$ and $u(y = d) = u_0$. Thus,

$$u(y) = u_0 \frac{y}{d} \tag{E.3.15.1b}$$

The heat transfer equation (3.64) with $\Phi = \left(\frac{\partial u}{\partial y}\right)^2$ from Eq. (3.65) reduces to:

$$k \frac{d^2T}{dy^2} = -\mu \left(\frac{u_0}{d}\right)^2 \tag{E.3.15.2}$$

subject to $\left.\frac{dT}{dy}\right|_{y=0} = 0$ and $T(y = d) = T_0$

Double integration yields:

$$T(y) = T_0 + \frac{\mu u_0^2}{2k}\left[1 - \left(\frac{y}{d}\right)^2\right] \tag{E.3.15.3}$$

At the plate surface $q_s = q(y = d) = -k \left.\frac{\partial T}{\partial y}\right|_{y=d}$ we have:

$$q_s = \mu \frac{u_0^2}{d} \tag{E.3.15.4}$$

Comments: Clearly, as μ and u_0 increase and the spacing decreases, q_s shoots up. For simple Couette flow du/dy evaluated at y = d is equal to u_0 / d so that $q_s = u_0\, \tau_{wall}$ here, which is a simple example of the heat transfer and momentum transfer relation (see Reynolds–Colburn analogy). Of interest would be the evaluation of the mean fluid temperature, $T_m = \frac{1}{\dot{m}} \int_A \rho\, u\, T\, dA$, to estimate h from $q_s = h(T_0 - T_m)$.

Example 3.16: Reynolds–Colburn Analogy Applied to Laminar Boundary-Layer Flow

Consider a heated plate of length L and constant wall temperature T_w, subject to a cooling air-stream (u_∞, T_∞). Find a functional dependence for $q_w(x)$.

Sketch	*Assumptions*	*Approach*
	• Thermal Blasius flow (see Sect. 5.2) • Constant properties	• Reynolds–Colburn analogy

Solution: Rewriting Eq. (3.73) we have:

$$\frac{q_w(x)}{\rho\, c_p\, u_\infty (T_w - T_\infty)} = C_f\, Pr^{-2/3} \qquad \text{(E.3.16.1)}$$

From Example 5.1 we can deduce that

$$C_f \sim Re_x^{-1/2} \qquad \text{(E.3.16.2)}$$

where actually $C_f = 0.664/\sqrt{Re_x}$ as shown in Sect. 5.2. Now, with everything else being constant

$$q_w(x) \sim \frac{K}{\sqrt{x}}, \quad K = ¢ \qquad \text{(E.3.16.3a, b)}$$

Graph:

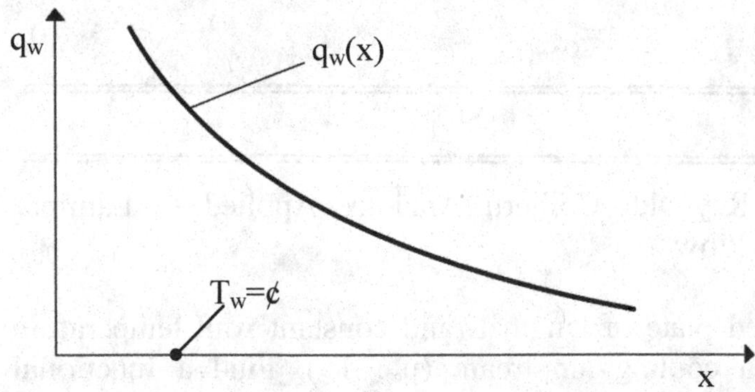

Comment: The wall heat flux from the plate surface decreases nonlinearly with plate distance because of the increasing Re(x), or better, the larger thermal boundary-layer thickness, $\delta_{th}(x)$, and hence milder wall temperature gradients (see Eq. 3.63).

Example 3.17: Lubricated Shaft Rotation with Heat Generation

Consider "cylindrical" Couette flow where, somewhat similar to Example 3.15, the rotating shaft (R_i, ω_i) is adiabatic and the stationary housing (R_0, T_0) is isothermal. In light of viscous dissipation, find T(r) as well as T_{max} at $r = R_i$ and $\hat{Q}_{wall}(r = R_0)$.

Sketch	*Assumptions*	*Approach*
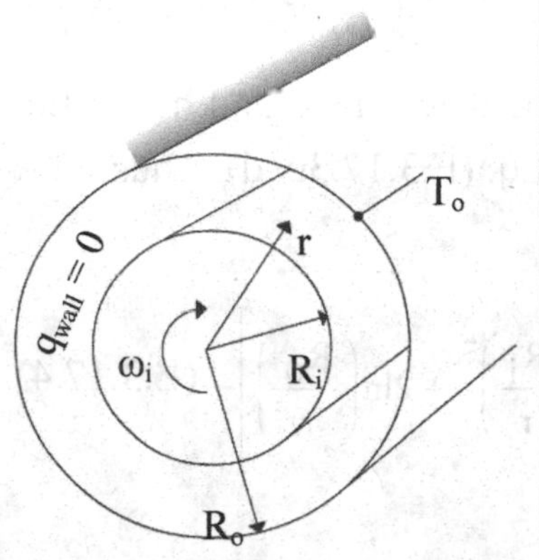 T_o, r, $q_{wall} = 0$, ω_i, R_i, R_o	• Steady laminar 1-D axisymmetrical flow • No gravity or end effects • Constant properties	• Reduced Θ-momentum equation • Postulate $v_\theta = v_\theta(r)$ only

Solution: Based on the postulate and assumptions, the Continuity Equation is satisfied and the Θ-momentum equation in cylindrical coordinates reduces to (see Equation Sheet in App. A):

$$\frac{d}{dr}\left(\frac{1}{r}\frac{d}{dr}(r\, v_\theta)\right) = 0 \qquad \text{(E.3.17.1a)}$$

subject to

$$v_\theta\,(r = R_i) = (\omega\, R)_i \quad \text{and} \quad v_\theta(r = R_0) = 0 \qquad \text{(E.3.7.1b, c)}$$

Double integration and invoking the B.C.s yields:

$$v_\theta(r) = \frac{\omega_i R_i (R_0/R_i)^2}{\left(\frac{R_0}{R_i}\right)^2 - 1}\left[\frac{R_i}{r} - \frac{r}{R_i}\right] \qquad \text{(E.3.17.2)}$$

The heat transfer equation (see Equation Sheet in App. A) reduces to:

$$0 = \frac{k}{r}\frac{d}{dr}\left(r\frac{dT}{dr}\right) - \mu\,\Phi \qquad \text{(E.3.17.3a)}$$

where

$$\Phi = \left(\frac{dv_\theta}{dr} - \frac{v_\theta}{r}\right)^2 \qquad \text{(E.3.17.3b)}$$

and as stated:

$$\left.\frac{dT}{dr}\right|_{r=R_i} = 0;\ T(r = R_0) = T_0 \qquad \text{(E.3.17.3c,d)}$$

With $v_\theta(r)$ given, Eq. (E.3.17.3b) can be determined and hence Eq. (E.3.17.3a) can be integrated subject to Eq. (E.3.17.3c, d). Thus,

$$T(r) = T_0 + \frac{\mu}{4k}\left[\frac{2\omega_i R_i}{1-\left(\frac{R_i}{R_0}\right)^2}\right]^2\left[\left(\frac{R_i}{R_0}\right)^2 - \left(\frac{R_i}{r}\right)^2 + 2\ln\left(\frac{R_0}{r}\right)\right] \qquad \text{(E.3.17.4)}$$

Now, either by inspection of (E.3.17.4) or setting dT/dr to zero, T_{max} occurs at $r = R_i$. In dimensionless form,

$$\frac{T_{max} - T_0}{\frac{\mu}{4k}\left[\frac{2\omega_i R_i}{1-\left(\frac{R_i}{R_0}\right)^2}\right]^2} = \left[\left(\frac{R_i}{R_0}\right)^2 - 1 + 2\ln\left(\frac{R_0}{R_i}\right)\right] \qquad \text{(E.3.17.5)}$$

The wall heat transfer rate per unit length is:

$$\hat{Q}_{wall} = (2\pi R_0)\, q(r = R_0);$$

where $q(r = R_0) = -k \left.\frac{dT}{dr}\right|_{r=R_0}$ (E.3.17.6a, b)

Hence,

$$\hat{Q}_w = \frac{4\pi\mu(\omega R)_i^2}{1-\left(\frac{R_i}{R_0}\right)^2} \tag{E.3.17.6c}$$

Graphs:

(a) Velocity profiles

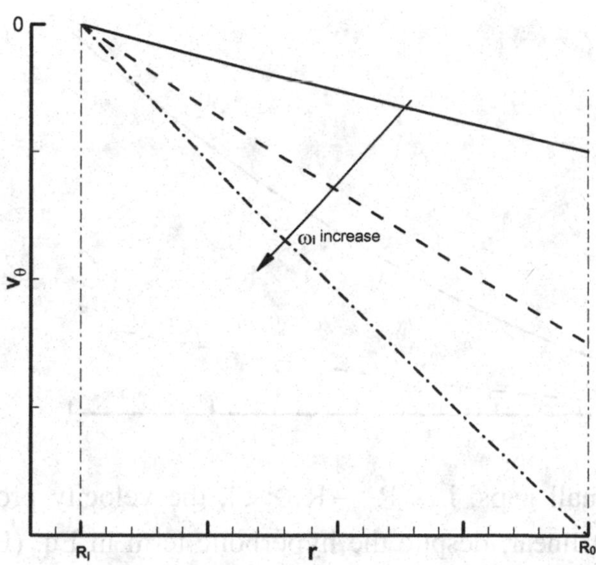

(b) Temperature profiles

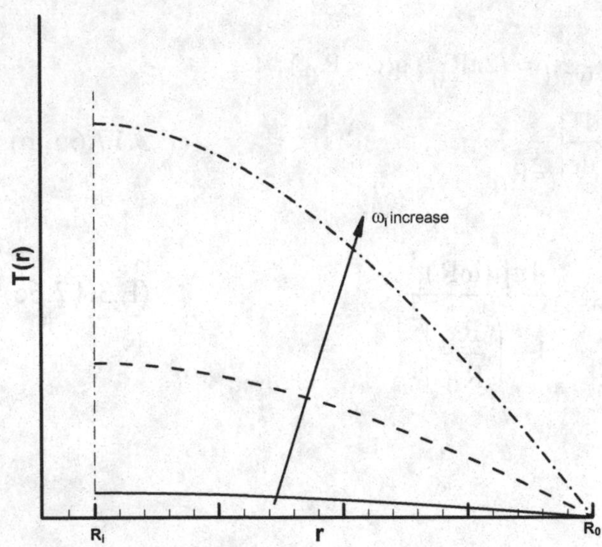

(c) Wall heat flow rate

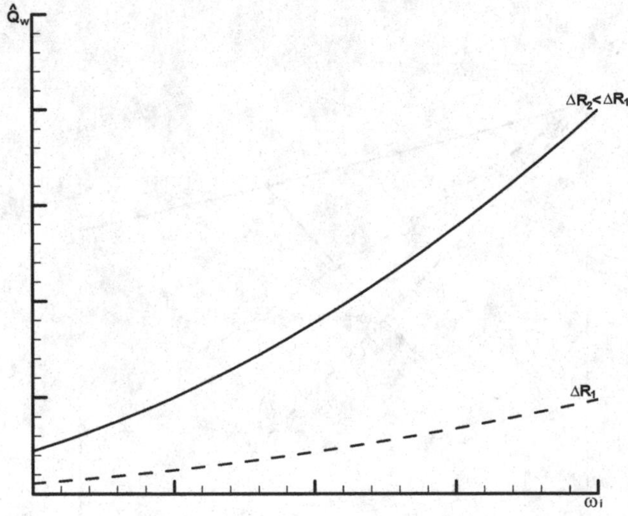

Comments:

- For small gaps, i.e., $R_o - R_i << 1$, the velocity profiles are almost linear, despite the hyperbolic term in Eq. (E.3.17.2). Clearly, with $\Delta R << 1$, $v_\theta(r)$ is "linearized".
- This is not the case for T(r) due to the strong viscous (heating effect (see Graph b).

- As expected, $\hat{Q}_w(\omega_i)$ decreases with a strong nonlinear influence of the gap size (see Graph c)

3.6.2 Turbulent Pipe Flow Correlations

The boundary-layer results of Sect. 3.6.1 can be directly related to internal flows, especially fully-developed flow in circular tubes, noting that:

$$C_f = \frac{f}{4} \tag{3.75}$$

The (Fanning) friction factor C_f is given as Eq. (3.74a) and the (Darcy) friction factor f is defined as:

$$f \equiv \frac{2(\Delta p / L)D}{\rho u_m^2} \tag{3.76}$$

where $\Delta p / L$ is the constant pressure gradient, D is the tube diameter, ρ is the fluid density, and u_m is the mean (or average) velocity. Clearly, for Poiseuille flow

$$f_{laminar} = \frac{64}{Re_D} \tag{3.77}$$

However, for turbulent pipe flow semi-equational correlations are necessary (see also Moody chart in App. B). For example, Petukhov et al. (1970) suggested

$$f_{turbulent} = (0.79 \ln Re_D - 1.64)^{-2} \tag{3.78}$$

covering the range $3{,}000 < Re < 5 \times 10^6$.

Again, connecting turbulent momentum transfer in pipes to convection heat transfer, Eqs. (3.73) and (3.74a–c) are employed, i.e.,

$$\frac{C_f}{2} = \frac{f}{8} = St\,Pr^{2/3} = \frac{Nu_D}{Re_D\,Pr^{1/3}} \tag{3.79}$$

and combined with Eq. (3.78). The result is:

$$\mathrm{Nu_D} = 0.023\,\mathrm{Re_D^{4/5}}\,\mathrm{Pr}^{1/3} \tag{3.80}$$

which is very similar to the empirical Dittus–Boelter equation (see Incropera et al. 2007) for $0.6 \leq \mathrm{Pr} \leq 160$ and $\mathrm{Re_D} > 10^4$, when assuming $L/D \geq 10$. For a broader range of Pr # and Re # applications, Gnielinski (1976) postulated:

$$\mathrm{Nu_D} = \frac{(f/8)(\mathrm{Re_D} - 10^3)\,\mathrm{Pr}}{1 + 12.7(f/8)^{1/2}(\mathrm{Pr}^{2/3} - 1)} \tag{3.81}$$

which holds for $0.5 \leq \mathrm{Pr} \leq 2{,}000$ and $3{,}000 \leq Re_D \leq 5 \times 10^6$.

Example 3.18: Turbulent Thermal Pipe Flow

Consider fully-developed hot air flow in a metal tube ($\dot{m} = 0.05$ kg/s; $L = 5$ m; $D = 0.15$ m). Because of the cold ambient ($T_\infty = 0°C$, $h_{\mathrm{ambient}} = 6\ \frac{W}{m^2K}$), the air inlet temperature $T_1 = 103°C$ drops to $T_2 = 77°C$ at the tube outlet. Find the heat loss and surface temperature at x = L.

Sketch	*Assumptions*	*Concepts*
T_∞, h_{amb}, $\dot{m}_{air}$, T_1, D, T_2, o, x, x=L	• Steady-state • Air ≙ Ideal gas • Constant properties and parameters • Tube wall effects negligible	• Energy balance $\dot{Q} = \dot{m}c_p(T_1 - T_2)$ • Radial heat loss resistance in series • $\mathrm{Nu_D}$-correlation Eq. (3.80)

Solution:

- Properties (App. B) for $T_{ref} \approx 360\,\mathrm{K}$, Pr = 0.7, c_p = 1,010J/kg·K, k = 0.03 W/m·K, $\mu = 208.2 \times 10^{-7}$ N·s/m^2

- Reynolds number $Re_D = \frac{4\dot{m}}{\mu\pi D} := 20{,}384$ (i.e., turbulent)
- Heat loss over entire tube: $\dot{Q} = \dot{m}c_p(T_1 - T_2)$ $\dot{Q} = 1313$ J/s
- **Tube surface temperature T_s at x = L (see thermal resistance network):**

$q_s(x = L) \Rightarrow$
(Radial heat flux)

$T(x{=}L){=}T_2$ $T_{wall}{\equiv}T_s$ $T_{ambient}{=}T_\infty$

$R_{inner} = \frac{1}{h_{inner}}$ $R_{outer} = \frac{1}{h_{ambient}}$

Hence,

$$q_s(x = L) = \frac{T_2 - T_\infty}{\frac{1}{h_i} + \frac{1}{h_{amb.}}} \qquad \text{(E.3.18.1)}$$

where $h_i = h(x = L) = Nu_D \frac{k}{D}$ and $Nu_D = 0.023\, Re_D^{4/5}\, Pr^{1/3}$.

Finally, with Nu_D=57.22 and hence $h_i = 11.44 \frac{W}{m^2 \cdot K}$, so that:

$$q_s(x = L) = 305 J/(m^2 \cdot s) \qquad \text{(E.3.18.2)}$$

Now, applying again

$$q_s = \frac{T_2 - T_s}{1/h_i} \qquad \text{(E.3.18.3)}$$

we obtain

$$T_s = 77°C - \frac{305}{11.44} = 50.34°C \qquad \text{(E.3.18.4)}$$

3.7 Entropy Generation Analysis

3.7.1 Background Information

As mentioned in Sect. 3.5.1, the second law of thermodynamics is the "increase of entropy" principle for any real process, i.e.,

$$\Delta S_{total} = \Delta S_{system} + \Delta S_{surrounding} \equiv S_{gen} > 0 \tag{3.82}$$

While an entropy value S [kJ/K] indicates the degree of molecular material *randomness*, $S_{gen} > 0$ is necessary for a process to proceed or a device to work. The source of entropy change is heat transferred. As Clausius stated: $dS > \delta Q / T$, implying all irreversibilities are contributing, e.g., due to heat exchange with internal and/or external sources as well as internal friction (or viscous effects) and net influx of entropy carried by fluid streams (see Eq. (3.30)). The inequality (Eq. 3.82) can be recast as an entropy "balance" by adding S_{gen} on both sides, i.e.,

$$\underbrace{\sum_{in} ms - \sum_{out} ms + \sum \frac{Q}{T_{ambient}}}_{\Delta S_{surrounding}} + \underbrace{S_{gen}}_{\substack{\text{system}\\ \text{irreversibilities}}} = \Delta S\big|_{C.\forall.} = \underbrace{(ms)_{final} - (ms)_{initial}}_{\Delta S_{system}} \tag{3.83}$$

Clearly, the larger S_{gen} the more inefficient a process, device or system is, i.e., S_{gen} is equivalent to "amount of waste generated". In convection heat transfer this "energy destruction" appears as viscous dissipation and random disorder due to heat input:

$$S_{gen} \sim \mu\Phi \text{ and } k(\nabla T)^2 \tag{3.84a}$$

or

$$S_{gen}^{total} = S_{gen}^{friction} + S_{gen}^{thermal} \tag{3.84b}$$

3.7.2 Entropy Generation Derivation

For optimal system/device design it is important to find for a given objective the *best possible system geometry and operational conditions* so that, S_{gen} is a minimum. Thus, within the framework of convection heat transfer with Newtonian fluids, it is of interest to derive an expression for

$$S_{gen} = S_{gen}(\text{thermal}) + S_{gen}(\text{friction}) \qquad (3.85)$$

Clearly, Eq. (3.85) encapsulates the irreversibilities due to heat transfer ($S_{gen,thermal}$) and viscous fluid flow ($S_{gen,friction}$).

Considering a point (x,y,z) in a fluid with convective heat transfer, the fluid element dx–dy–dz surrounding this point is part of thermal flow system. Thus, the small element dx–dy–dz can be regarded as an open thermodynamic system, subject to mass fluxes, energy transfer, and entropy transfer interactions that penetrate the fixed control surface formed by the dx–dy–dz box of Fig. 3.13.

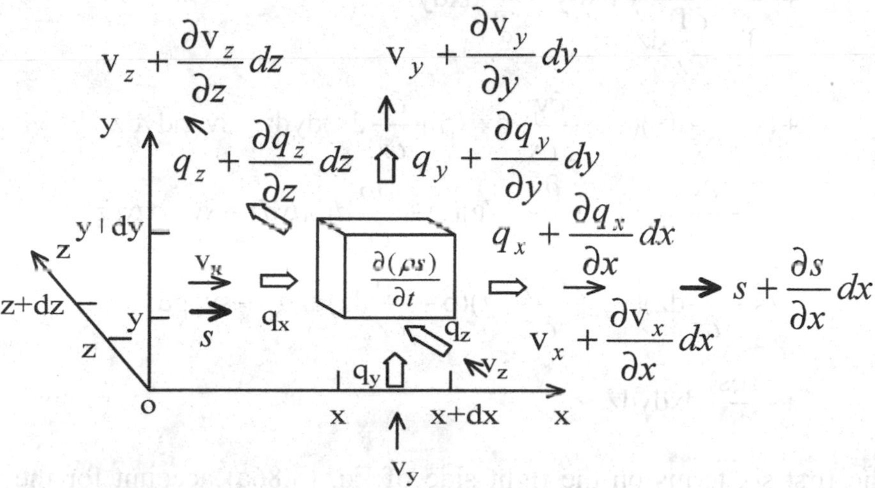

Fig. 3.13 The local generation of entropy in a flow with a viscous fluid and conductive heat transfer

Hence, the local volumetric rate of entropy generation (S_{gen} in $[\frac{kW}{m^3K}]$) is considered inside a viscous fluid with convective heat transfer without internal heat generation. The second law of thermodynamics for the dx–dy–dz box as an open system experiencing fluid flow and conductive heat transfer then reads, based on the Clausius definition $dS = \left.\frac{\delta Q}{T}\right|_{reversible}$ and Fig. 3.13:

$$S_{gen}dxdydz = \frac{q_x + \frac{\partial q_x}{\partial x}dx}{T + \frac{\partial T}{\partial x}dx}dydz - \frac{q_x}{T}dydz$$

$$+\frac{q_y + \frac{\partial q_y}{\partial y}dy}{T + \frac{\partial T}{\partial y}dy}dxdz - \frac{q_y}{T}dxdz$$

$$+\frac{q_z + \frac{\partial q_z}{\partial z}dz}{T + \frac{\partial T}{\partial z}dz}dxdy - \frac{q_z}{T}dxdy$$

$$+(s+\frac{\partial s}{\partial x}dx)(v_x + \frac{\partial v_x}{\partial x}dx)(\rho + \frac{\partial \rho}{\partial x}dx)dydz - sv_x\rho dydz$$

$$+(s+\frac{\partial s}{\partial y}dy)(v_y + \frac{\partial v_y}{\partial y}dy)(\rho + \frac{\partial \rho}{\partial y}dy)dxdz - sv_y\rho dxdz$$

$$+(s+\frac{\partial s}{\partial z}dz)(v_z + \frac{\partial v_z}{\partial z}dz)(\rho + \frac{\partial \rho}{\partial z}dz)dxdy - sv_z\rho dxdy$$

$$+\frac{\partial(\rho s)}{\partial t}dxdydz \qquad (3.86a)$$

The first six terms on the right side of Eq. (3.86a) account for the entropy transfer associated with heat transfer. Combining terms 1 and 2, 3and 4, 5 and 6 and dividing by dxdydz and taking the limit, the former six terms in Eq. (3.86a) can be reduced to:

$$\frac{T\frac{\partial q_x}{\partial x} - q_x\frac{\partial T}{\partial x}}{T(T+\frac{\partial T}{\partial x}dx)} + \frac{T\frac{\partial q_y}{\partial y} - q_y\frac{\partial T}{\partial y}}{T(T+\frac{\partial T}{\partial y}dy)} + \frac{T\frac{\partial q_z}{\partial z} - q_z\frac{\partial T}{\partial z}}{T(T+\frac{\partial T}{\partial z}dz)} =$$

$$\frac{1}{T}(\frac{\partial q_x}{\partial x} + \frac{\partial q_y}{\partial y} + \frac{\partial q_z}{\partial y}) - \frac{1}{T^2}(q_x\frac{\partial T}{\partial x} + q_y\frac{\partial T}{\partial y} + q_z\frac{\partial T}{\partial z}) \quad (3.86b)$$

Terms 7 to 12 in Eq. (3.86a) represent the entropy convected into and out of the system, while the last term is the time rate of entropy accumulation in the dx–dy–dz control volume. Decomposing and combining the last seven terms as well as considering in the limit, the last seven terms can be rearranged as:

$$\rho(\frac{\partial s}{\partial t}+v_x\frac{\partial s}{\partial x}+v_y\frac{\partial s}{\partial y}+v_z\frac{\partial s}{\partial z})$$

$$+s\left[\frac{\partial \rho}{\partial t}+v_x\frac{\partial \rho}{\partial x}+v_y\frac{\partial \rho}{\partial y}+v_z\frac{\partial \rho}{\partial z}+\rho(\frac{\partial v_x}{\partial x}+\frac{\partial v_y}{\partial y}+\frac{\partial v_z}{\partial z})\right] \quad (3.86c)$$

Combining Eqs. (3.86b, c), the local rate of entropy generation becomes:

$$S_{gen}=\frac{1}{T}(\frac{\partial q_x}{\partial x}+\frac{\partial q_y}{\partial y}+\frac{\partial q_z}{\partial z})-\frac{1}{T^2}(q_x\frac{\partial T}{\partial x}+q_y\frac{\partial T}{\partial y}+q_z\frac{\partial T}{\partial z})$$

$$+\rho(\frac{\partial s}{\partial t}+v_x\frac{\partial s}{\partial x}+v_y\frac{\partial s}{\partial y}+v_z\frac{\partial s}{\partial z})$$

$$+s\left[\frac{\partial \rho}{\partial t}+v_x\frac{\partial \rho}{\partial x}+v_y\frac{\partial \rho}{\partial y}+v_z\frac{\partial \rho}{\partial z}+\rho(\frac{\partial v_x}{\partial x}+\frac{\partial v_y}{\partial y}+\frac{\partial v_z}{\partial z})\right] \quad (3.87)$$

Note that the last term of Eq. (3.87) (in square brackets) vanishes identically based on the mass conservation principle (e.g., Bejan, 1995). Specifically, for homogeneous fluids:

$$\frac{D\rho}{Dt}+\rho\nabla\cdot\mathbf{v}=0 \quad (3.88)$$

where D/Dt is the substantial (material or Stokes) derivative:

$$\frac{D}{Dt}=\frac{\partial}{\partial t}+v_x\frac{\partial}{\partial x}+v_y\frac{\partial}{\partial y}+v_z\frac{\partial}{\partial z} \quad (3.89)$$

while $\mathbf{v}$ is the velocity vector (v_x, v_y, v_z). In vector notation the volume rate of entropy generation can be expressed as:

$$S_{gen}=\frac{1}{T}\nabla\cdot\mathbf{q}-\frac{1}{T^2}\mathbf{q}\cdot\nabla T+\rho\frac{Ds}{Dt} \quad (3.90)$$

According to the first law of thermodynamics, the rate of change in internal energy per unit volume is equal to the net heat transfer rate by conduction, plus the work transfer rate due to compression, plus the work transfer rate per unit volume associated with viscous dissipation, i.e.,

$$\rho \frac{D\tilde{u}}{Dt} = -\nabla \cdot \mathbf{q} - p(\nabla \cdot \mathbf{v}) + \mu\Phi \tag{3.91}$$

Writing the Gibbs relation $du = Tds - pd(1/\rho)$ and using the substantial derivative notation Eq. (3.89), we obtain:

$$\rho \frac{Ds}{Dt} = \frac{\rho}{T}\frac{Du}{Dt} - \frac{p}{\rho T}\frac{D\rho}{Dt} \tag{3.92}$$

Combining Eq. (3.92) with $\rho \frac{Ds}{Dt}$ given by Eq. (3.90) and $\rho \frac{Du}{Dt}$ given by Eq. (3.91), the volumetric entropy generation rate can be expressed as:

$$S_{gen} = -\frac{1}{T^2}\mathbf{q} \cdot \nabla T + \frac{\mu}{T}\Phi \tag{3.93}$$

If the Fourier law of heat conduction for an isotropic medium applies, i.e.,

$$\mathbf{q} = -k\nabla T \tag{3.94}$$

the rate of volumetric entropy generation (S_{gen}) in three-dimensional Cartesian coordinates is then (Bejan, 1996):

$$\begin{aligned} S_{gen} \equiv S_G = {} & \frac{k}{T^2}\left[(\frac{\partial T}{\partial x})^2 + (\frac{\partial T}{\partial y})^2 + (\frac{\partial T}{\partial z})^2 \right] \\ & + \frac{\mu}{T}\left(2\left[(\frac{\partial u}{\partial x})^2 + (\frac{\partial v}{\partial y})^2 + (\frac{\partial w}{\partial z})^2 \right] + (\frac{\partial u}{\partial y} + \frac{\partial v}{\partial x})^2 \right. \\ & \left. + (\frac{\partial u}{\partial z} + \frac{\partial w}{\partial x})^2 + (\frac{\partial v}{\partial z} + \frac{\partial w}{\partial y})^2 \right\} \end{aligned} \tag{3.95}$$

where u, v and w are velocity vector in x, y, and z direction, respectively; T is the temperature, k is the thermal conductivity and μ is the dynamic viscosity.

Specifically, the dimensionless entropy generation rate induced by fluid friction and heat transfer can be defined as follows:

$$S_{G,F} = S_{gen}(\text{frictional}) \cdot \frac{kT_0^2}{q^2} \tag{3.96a}$$

where
$$S_{gen}(\text{frictional}) = \frac{\mu}{T}\left\{2\left[(\frac{\partial u}{\partial x})^2 + (\frac{\partial v}{\partial y})^2 + (\frac{\partial w}{\partial z})^2\right] + (\frac{\partial u}{\partial y} + \frac{\partial v}{\partial x})^2 + (\frac{\partial u}{\partial z} + \frac{\partial w}{\partial x})^2 + (\frac{\partial v}{\partial z} + \frac{\partial w}{\partial y})^2\right\} \tag{3.96b}$$

while for the thermal entropy source,

$$S_{G,T} = S_{gen}(\text{thermal}) \cdot \frac{kT_0^2}{q^2} \tag{3.97a}$$

where
$$S_{gen}(\text{thermal}) = \frac{k}{T^2}\left[(\frac{\partial T}{\partial x})^2 + (\frac{\partial T}{\partial y})^2 + (\frac{\partial T}{\partial z})^2\right] \tag{3.97b}$$

Finally,

$$S_{G,total} = S_{gen}\frac{kT_0^2}{q^2} = S_{G,F} + S_{G,T} \tag{3.98}$$

where T_0 is the fluid inlet temperature and q is the wall heat flux.

Example 3.19: Thermal Pipe Flow with Entropy Generation

Deriving the irreversibility profiles for Hagen–Poiseuille (H–P) flow through a smooth tube of radius r_0 with uniform wall heat flux q [W/m^2] at the wall, the velocity and temperature for fully-developed regime are given by:

$$u = 2U[1 - (\frac{r}{r_0})^2] \tag{E.3.19.1}$$

and

$$T - T_s = -\frac{qr_0}{k}[\frac{3}{4} - (\frac{r}{r_0})^2 + \frac{1}{4}(\frac{r}{r_0})^4] \tag{E.3.19.2}$$

Sketch	*Assumptions*	*Conceptions*
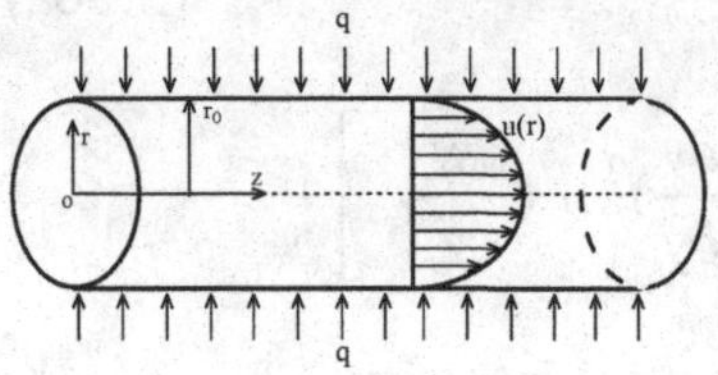	• Fully-developed H-P flow • Constant properties and parameters	• Volumetric entropy generation rate Eq. (3.95) • Thermal entropy generation • Frictional entropy generation

Solution:

The wall temperature $T_s = T(r = r_0)$ can be obtained from the condition

$$\frac{\partial T}{\partial x} = \frac{dT_s}{dx} = \frac{2q}{\rho c_p U r_0} = \text{constant} \qquad \text{(E.3.19.3)}$$

$$\frac{\partial T}{\partial r} = \frac{q}{k}[2\frac{r}{r_0} - (\frac{r}{r_0})^3] \qquad \text{(E.3.19.4)}$$

$$\frac{\partial u}{\partial r} = \frac{-4Ur}{r_0^2} \qquad \text{(E.3.19.5)}$$

Hense, the dimensionless entropy generation for fully-developed tubular H–P flow can be expressed as:

$$S_{gen}\frac{kT_0^{\,2}}{q^2} = \frac{4k^2}{(\rho c_p \overline{U} r_0)^2}\frac{T_0^{\,2}}{T^2} + (2R - R^3)^2\frac{T_0^{\,2}}{T^2} + \frac{16kT_0^{\,2}\mu\overline{U}^2}{q^2 T r_0^{\,2}}R^2$$

$$= \underbrace{\frac{16}{Pe^2}(\frac{T_0}{T})^2 + (2R - R^3)^2\frac{T_0^{\,2}}{T^2}}_{\text{heat transfer}} + \underbrace{\phi\frac{T_0}{T}R^2}_{\text{fluid friction}} \qquad \text{(E.3.19.6)}$$

with

$$Pe = Re \cdot Pr = \frac{2r_0 \rho c_p U}{k} \qquad \text{(E.3.19.7)}$$

and

$$R = \frac{r}{r_0}, \text{ and } \phi = \frac{16kT_0\mu U^2}{q^2 r_0^{\,2}} \qquad \text{(E.3.19.8a, b)}$$

Here, R is the dimensionless radius, T_0 is the inlet temperature which was selected as the reference temperature. On the right side of the Eq. (E.3.19.6), the first term represents the entropy generation by axial conduction, the second term is the entropy generated by heat transfer in radial direction, and the last term is the fluid friction contribution. Parameter ϕ , Eq. (3.19.8b), is the irreversibility distribution ratio ($\frac{S_{gen}(fluid\ \ friction)}{S_{gen}(heat\ \ transfer)}$).

Graph:

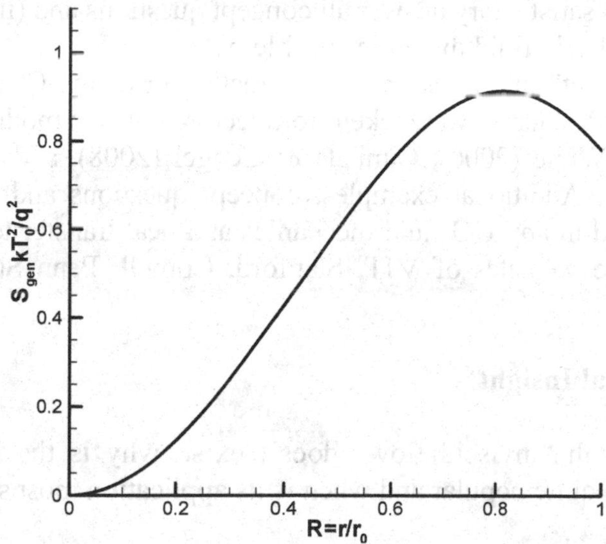

Comments:

As expected, according to Eq. (E.3.19.6), at the center point, i.e., R = 0, only the first term in the right side contributed to the dimensionless entropy generation rate; however, for Pe >> 1, the irreversibility due to axial conduction is negligible in the fully developed range. In contrast, in the wall region both thermal and frictional effects produce entropy with a maximum at $R \approx 0.8$ generated by dominant heat transfer induced entropy generation.

3.8 Homework Assignments

Solutions to homework problems done individually or in, say, three-person groups should help to further illustrate fluid dynamics concepts as well as approaches to problem solving, and in conjunction with App. A, sharpen the reader's math skills (see Fig. 2.1). Note, there is no substantial correlation between good HSA results and fine test performances, just *vice versa*. Table 1.1 summarizes three suggestions for students to achieve a good grade in fluid dynamics – for that matter in any engineering subject. The key word is "independence", i.e., equipped with an equation sheet (see App. A), the student should be able: (i) to satisfactory answer all concept questions and (ii) to solve correctly all basic fluid dynamics problems.

The "Insight" questions emerged directly out of the Chap. 3 text, while some "Problems" were taken from lecture notes in modified form when using White (2006), Cimbala & Cengel (2008), and Incropera et al. (2007). Additional examples, concept questions and problems may be found in any UG fluid mechanics and heat transfer text, or on the Web (see websites of MIT, Stanford, Cornell, Penn State, UM, etc.).

3.8.1 Physical Insight

3.8.1 Although "inviscid flow" does't exist, why is the Bernoulli equation still quite popular and when is its application most suitable?

3.8.2 In Example 3.1A, why is the air-velocity above the airfoil (or airplane wing) so much faster than below?

3.8.3 Explain the differences between thermodynamic (or static) pressure, dynamic pressure, and total (or stagnation) pressure. Draw a pressure probe which simultaneously can measure stagnation and static pressures.

3.8.4 It is desired to measure the total drag on an airplane whose cruising velocity is 300 mi/h. Mary and Berry suggest a one-twentieth scale model test in a wind tunnel at the same air pressure and temperature to determine the prototype drag coefficient. Is that feasible? Note: $a_{sound} \approx 750\,\text{mph}$

3.8.5 Consider the efflux of a liquid through a smooth orifice of area A_0 at the bottom of a covered tank (A_T). The depth of the liquid is *y*. The pressure in the tank exerted on the liquid is p_T while the pressure outside of the orifice is p_a. (A) Find the efflux velocity v_0. (B) Assuming $p_T = p_a$ <open tank>, estimate under what geometric condition (A_0 / A_T) can the non-steady term

$$\int_1^2 \frac{\partial v}{\partial t} ds \approx \frac{dv_1}{dt} y$$

be neglected, assuming small accelerated tank draining.

3.8.6 Consider a slider valve (h $\hat{=}$ partial opening) in pipe flow (d $\hat{=}$ diameter; v $\hat{=}$ mean velocity and ρ, μ are fluid properties). Find the key dimensionless groups by *inspection* (or dimensional analysis).

3.8.7 Draw carefully velocity profiles in a pipe's entrance and fully-developed regions for: (a) $Re_D \approx 1{,}800$ and (b) $Re_D \approx 18\,000$. Comment!

3.8.8 Consider steady flow (Re = 600) over a backward-facing step. Draw carefully three velocity profiles, i.e., well before and right after the step as well as downstream of the step.

3.8.9 From a math viewpoint, why is the assumption of "fully-developed" flow so important (if it is applicable for a given case)?

3.8.10 Revisiting Sect. 3.2.2, develop a criterion which sets the limit for the "nearly-parallel-flow" assumption.

3.8.11 List examples of transient flows in industry and nature. Why are Stokes' first and second Problem solutions so important and for which applications do they serve as math models?

3.8.12 What type of "2-D" problems described by PDEs can be transformed to ODEs with suitable "combined variables" (see Sect.3.3.1 and App. A)?

3.8.13 In order to solve "flow through porous media" *directly*, what kind of information would you need, say, for a packed bed?

3.8.14 In light of Eqs. (E.3.4.7a), (E.3.8.1), (E.3.18.1) and (3.68), it is evident that the flow rate of fluids or heat can be expressed as the ratio of "driving force"/resistance, e.g.,

$$Q = \begin{cases} \dfrac{\Delta p}{\sum R_{fluid}} \text{.................................for momentum transfer} \\ \\ \dfrac{\Delta T}{\sum R_{thermal}} \text{..............................for heat transfer.} \end{cases}$$

Write a report on the derivations and multitude of applications of such correlations. Is there something similar for species mass transfer?

3.8.15 Compressibility effects for gases may become important for Mach numbers $Ma \geq 0.3$. Provide a rational for this limit.

3.8.16 Discuss in terms of physical characteristics and math descriptions water waves, sound waves, and shock waves.

3.8.17 On the molecular level, why do frictional effects inside a system/fluid and heat transfer to the system/fluid increase the system's entropy?

3.8.18 How do heat conduction and convection heat transfer differ? How do $T_{surface}$, T_{wall}, T_{mean}, T_{bulk} and T_{∞} differ?

3.8.19 The convective heat transfer coefficient $h = \frac{q_{wall}}{T_{wall} - T_{ref}}$ is considered by some researchers as an artifact which masks the actual physics of convection heat transfer. Discuss this topic.

3.8.20 Derive Eq. (3.73) and comment on the advantages of heat-momentum transfer and heat-mass transfer analogies.

3.8.21 Revisit Example 3.17 and discuss the case of high-speed spindle rotation, i.e., $\omega_i >> 1$ and $q_{wall} \neq 0$.

3.8.2 Problems

3.8.22 An inverted U-tube acts as a water siphon (see Schematic). The bend in the tube is 1 m above the water surface; the tube outlet is 7 m below the water surface. The water issues from the bottom of the siphon as a free jet at atmospheric pressure. Determine (after listing the necessary assumptions) the speed of the free jet and the minimum absolute pressure of the water in the bend (see point A). Note: $p_{atm} = 1.01 \times 10^5$ N/m^2 and $\rho = 999$ kg/m^3.

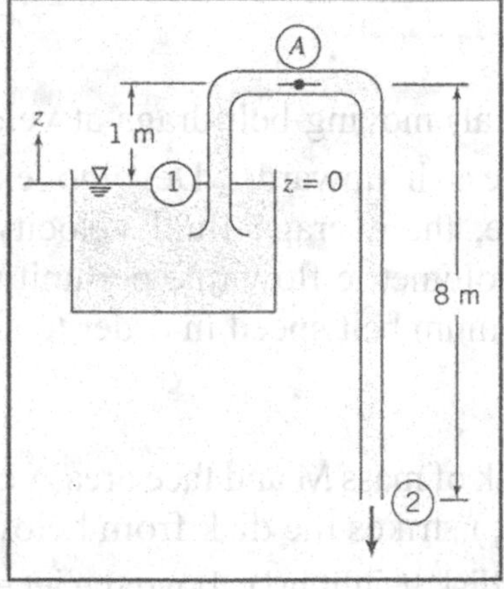

3.8.23 Consider steady fully-developed airflow in a smooth tube where a Pitot-static pressure arrangement measures p_{static} and

$p_{stagnation}$ as shown. Estimate: (a) the centerline velocity; (b) the volumetric flow rate; and (c) the wall shear stress.

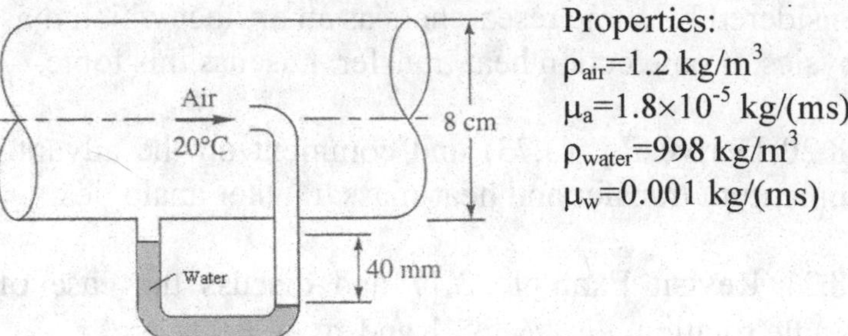

3.8.24 Draw very carefully a pressure-driven axial velocity profile, $v_z(r)$, in an annulus, i.e., a ring-like gap formed by a cylinder placed concentrically in a pipe. Is the profile symmetric with respect to the gap's center? If not, why not?

3.8.25 Consider a viscometer consisting of two concentric cylinders (R_1 and R_2) where the inner one is fixed and the outer one rotates with ω_0 = const. The gap, $\Delta R = R_2 - R_1$, is filled with a viscous (unknown) fluid. Solve for the velocity profile in the annular gap $R_1 \leq r \leq R_2$ and find an expression for the shear stress at the surface of the inner cylinder.

3.8.26 A (wide, vertical) moving belt drags at velocity v_0 a viscous fluid layer of thickness h upwards. Develop expressions for the film's velocity profile, the average fluid velocity, the shear stress distribution, and the volumetric flow rate per unit width. What is the condition for the minimum belt speed in order to achieve net upward flow?

3.8.27 A horizontal disk of mass M and face area A can move vertically when a water jet (d, v_0) strikes the disk from below. Obtain a differential equation for the disk height h(t) above the jet exit plane when the disk is initially released at $H > h_0$, where h_0 is the equilibrium height. Find an expression for h_0, sketch h(t), and explain.

3.8.28 Consider two inclined (angle θ) parallel plates a distance d apart where the upper plate moves at $u_0 = \not{c}$ and a constant pressure gradient is applied to the viscous fluid (ρ, μ).

(A) Derive an expression for the velocity and graph typical profiles for different dp/dx – values. Evaluate τ_{wall} and graph $\tau_{yx}(y)$.
(B) What are the conditions for simple Couette flow and Poiseuille flow?

3.8.29 The axial velocity of an incompressible fluid flowing between parallel plates is u(y) = fct. (Ay, By^2).

(a) Determine v(x, y).
(b) When and where is u = u(x, y).
(c) Draw a velocity profile for each case.

3.8.30 Consider a uniformly packed bed (or layer) of porous material, e.g., spherical particles. Develop an equation for the pressure drop

$$\frac{dp}{dz} = \text{fct}(U, \varepsilon, d_p, \text{and } \mu)$$

where U is Darcy's velocity, ε is the porosity (or void fraction), d_p is the mean particle diameter, and μ is the fluid viscosity. Plot $\frac{dp}{dz}(\varepsilon, d_p)$ with parameter values so that the Reynolds number remains

$$Re = \frac{\rho U d_p}{(1-\varepsilon)\mu} < 10$$

3.8.31 Develop a criterion for "fluidization", i.e., particle suspension, in a vertical porous bed. Thus, of interest is the superficial gas velocity, $U_{fluidize}$, which expands the bed and levitates all particles. Plot $U_{fluidize}(d_p, \varepsilon)$ for bed height H = 1m and $\varepsilon_{rest} = 0.35$ while $\varepsilon_{expanded} = 0.42$, $d_p = 50\mu m$, and $U_{initial} = 10^{-6}\, m/s$. Take $\rho_p = 1{,}650\ kg/m^3$, $\rho_{gas} = 3.5\ kg/m^3$, and $\mu_{gas} = 2\times 10^{-5}\ kg/m\cdot s$. In more general terms, which dimensionless groups (which depend on

$U_{fluidize}$) and ratios would form a basic criterion for nearly uniform fluidization?

3.8.32 A fluid flows radially through a porous cylindrical shell (R_i, R_o, L) of permeability æ. The outside pressure $p_2 = 1.5p_1$, the inner tube pressure. Find the shell pressure distribution, the radial flow velocity, and the (axial) mass flow rate for an incompressible fluid (ρ, μ). List a biomedical or industrial application for this model.

3.8.33 Consider horizontal, boundary-layer type thermal flow through a porous slab heated from below.

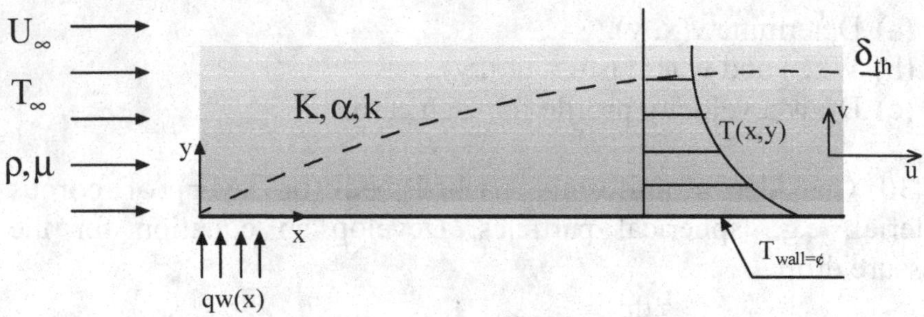

The known driving force is (proportional to) $dp/dx = -\mu U_\infty / K$ and it is assumed that $\delta_{th} << 1$. Within the B-L region:

$$\frac{\partial u}{\partial x} + \frac{\partial v}{\partial y} = 0 \qquad (33.1)$$

where $u = -\frac{K}{\mu}\frac{\partial p}{\partial x}$ and $v = -\frac{K}{\mu}\frac{\partial p}{\partial y}$ (33.2a, b)

and

$$u\frac{\partial T}{\partial x} + v\frac{\partial T}{\partial y} = \alpha\frac{\partial^2 T}{\partial y^2} \qquad (33.3)$$

$$qw \sim x - 1/2 \qquad (33.4)$$

(a) Derive these equations.
(b) Based on scale analysis of Eq. (33.3), show that

$$\frac{\delta_{th}}{x} \sim Pe_x^{-1/2}$$

where $Pe_x = U_\infty x/\alpha$ is the local Peclet number and that the local Nusselt number

$$Nu_x = \frac{q_w x}{k\alpha} \sim Pe_x^{1/2}$$

(c) Demonstrate that the PDE (33.3) can be transformed to the ODE

$$\theta'' + \frac{\eta}{2}\theta' = 0$$

subject to $\theta(0) = 0$ and $\theta(\infty) = 1$, where the combined variable $\eta = \frac{y}{x}Pe_x^{1/2}$ and $\theta = \frac{T - T_w}{T_\infty - T_w}$.

(d) Find $\theta(\eta)$, show that $Nu_x = 0.564\,Re_x^{1/2}$, plot Nu(x) for different U_∞ and α values, and comment!

3.8.34 Consider steady horizontal isentropic stagnation-point flow, e.g., an air stream at Ma = 0.7, density $\rho_0 = 1.8\ \text{kg/m}^3$ and temperature $T_0 = 75°\text{C}$ (R = 287 J/kg·K and $\chi = c_p/c_v = 1.4$) approaching a bluff body. Find the stagnation pressure, temperature, and density.

Hint: Derive first Bernoulli's equation for isentropic flow, i.e.,

$$\frac{\chi}{\chi - 1}\frac{p}{\rho} + \frac{v^2}{2} = \text{constant}$$

3.8.35 Nozzle geometry of a rocket engine: Find the throat and exit areas of a nozzle which should generate 30×10^3 N at z = 20km (when $p_{exit} = 5467\text{Pa}$), based on a stagnation pressure $p_0 = 10^3\,\text{kPa}$ and temperature $T_0 = 2{,}500\ \text{K}$. Assuming isentropic gas flow ($R = 280\ \text{m}^2/\text{s}^2\cdot\text{K}$ and $\chi = 1.4$).

Note: From isentropic flow tables for $\frac{p_{exit}}{p_{stagn.}} = 5.467 \times 10^{-3}$ we get $M_{exit} = 4.15$, $A_e/A^* = 12.2$, and $T_e/T_0 = 0.225$

3.8.36 Consider a Pitot-static tube with a U-tube mercury manometer which connects one arm to the total (or stagnation) point and the other to the static one. Inserted into an air stream ($p = 1.02\times10^5\,\mathrm{N/m^2}$, $T = 28°C, R = 287\,\mathrm{J/kg\cdot K}$, and $\chi = 1.4$) , find for $u_{air} = 50, 250$ and 420 m/s the Mach numbers and the associated manometer readings $h = \dfrac{(p_0 - p)}{\rho_{Hg}\cdot g}$, where $\rho_{Hg} = 13.6\times10^3\,\mathrm{kg/m^3}$.

Note: Ma < 0.3→ incompressible air; $0.3 \le Ma < 1.0$ → compressibility effects, and $Ma \ge 1.0$ → supersonic flow

3.8.37 Convection heat transfer: Consider a well-insulated tube ($D_i = 20$ mm , $D_o = 40$ mm) with an electric wall heater providing $q_{el} = 10^6\,\mathrm{W/m^2}$, in order to increase the average water temperature $T_i(x=0) = 20°C$ to $T_o(x=L) = 60°C$ when $\dot{m}_{water} = 0.1$ kg/s with $c_p = 4{,}179$ J/kg×K .

(a) Find the necessary tube length. (b) Assuming a constant convection coefficient $h = 1{,}500$ J/m²×K, find the inner-surface inlet and exit temperatures.

3.8.38 Consider a well-insulated counter-flow heat exchanger (i.e., a thin-walled, double pipe system) which connects a water heater to a shower. Specifically, cold water ($c_p = 4.18\,\mathrm{kJ/kg\cdot°C}$) enters at 15°C at a rate of 0.25 kg/s and is heated to 45°C by hot water ($c_p = 4.19\,\mathrm{kJ/kg\cdot°C}$) that enters at 100°C with 3 kg/s. Determine (a) the rate of heat transfer; and (b) the rate of entropy generation in the heat exchanger.

3.8.39 Consider a mixing chamber where liquid water with $\dot{m}_{water}$=2.5 kg/s at 200 kPa and 20°C is mixed with superheated steam at 200 kPa and 150°C, and the mixture leaves at 60°C and 200 kPa. The chamber loses heat at a rate of 1,200 kJ/min to the ambient ($T_{ambient} = 25°C$). Find $S_{total} = S_{gen.}$

References (Part A)

Beavers, G.S., Joseph, D.S., 1967, Journal of Fluid Mechanics, Vol. 30(1), pp. 197–207.

Bejan, A., 1995, *Convective Heat Transfer*, 2nd ed. Wiley, New York.

Bejan, A., 1996, *Entropy Generation Minimization: The Method of Thermodynamic Optimization of Finite-Size Systems and Finite-Time Processes*, CRC Press, Boca Raton, FL.

Bejan, A., 2002, International Journal of Energy Research, Vol. 26, pp. 545–565.

Brinkman, H.C., 1947, Physica, Vol. 13(8), pp. 447–448.

Cimbala, J.M., Cengel, Y.A., 2008, *Essentials of Fluid Mechanics: Fundamentals and Applications*, McGraw-Hill, New York, NY.

Ergun, S., 1952, Analytical Chemistry, Vol. 24(2), pp. 388–393.

Forchheimer, P., 1901, Zeitschrift Des Vereines Deutscher Ingenieure, Vol. 45, pp. 1782–1788.

Gnielinski, V., 1976, International Journal of Chemical Engineering, Vol. 16, pp. 359–368.

Incropera, F.P., DeWitt, D.P., Bergman, T.L., Lavine, A.S., 2007, *Introduction to Heat Transfer*, Wiley, New York, NY.

Kleinstreuer, C., 1997, *Engineering Fluid Dynamics*, Cambridge University Press, New York, NY.

Kleinstreuer, C., 2006, *Biofluid Dynamics – Principles and Selected Applications*, Taylor & Francis, Boca Raton, FL/London, New York.

Middleman, S., 1998, *An Introduction to Fluid Dynamics: Principles of Analysis and Design*, Wiley, New York.

Nichols, W.W., O'Rourke, M.F., et al., 1998, *McDonald's Blood Flow in Arteries: Theoretical, Experimental, and Clinical Principles,* Oxford University Press, New York.

Ochoa-Tapia, J.A. and Whitaker, S., 1995, International Journal of Heat and Mass Transfer, Vol. 34(14), pp. 2635–2646.

Panton, R.L., 2005, *Incompressible Flow*, Wiley, Hoboken, NJ.

Papanastasiou, T.C., 1994, *Applied Fluid Mechanics*, P T R Prentice Hall, Englewood Cliffs, NJ.

Petukhov, B.S., Zhilin, V.G., 1970, Journal of Engineering Physics, Vol. 19(3), pp. 1185–1194.

Polyanin, A.D., Zaitsev, V.F., 1995, *Handbook of Exact Solutions for Ordinary Differential Equations*, CRC Press, Boca Raton, FL.

White, F.M., 2006, *Viscous Fluid Flow*, McGraw-Hill, New York, NY.

Womersley, J.R., 1955, The Journal of Physiology, Vol. 127, pp. 553–563.

Part B

Conventional Applications

Part B: Conventional Applications

Chapter 4

Internal Flow

The classification of real flows as *internal* or *external*, i.e., depending on whether the fluid is forced through a conduit or past a submerged body, allows for system-tailored solution approaches. Basic examples are (internal) Poiseuille flow (see Sects. 2.4 and 3.2) and (external) Blasius flow (see Sects. 2.4.3 and 5.2). Open-*channel* flow forms a special (internal flow) category where three walls bound the fluid stream but the fourth (free) surface is typically the water air interface which can be smooth, rippled, wavy, or jumpy.

Alternatively, real flows can be grouped according to the *Reynolds-number range*, i.e., creeping flow, intermediate as well as laminar/turbulent boundary-layer flows, and very high-speed flows, such as supersonic and hypersonic. In any case, the macroscopic fluid-mass conservation condition for *internal* flows is:

$$\dot{m} = \int_A \rho\, v_n\, dA = (\rho\, v)_{av}\, A \overset{\rho = ⊄}{\downarrow}{=} \rho\, Q = ⊄ \text{ (constant)} \qquad (4.1a\text{–}d)$$

anywhere in the conduit. Clearly, implementation of Eq. (4.1) is not straight forward for external flows because of the usually unknown, varying “cross-sectional” area A. This can then be made up for by safely enlarging A such that the impact or perturbations of, say, the submerged body are not anymore registered (see Example 2.6).

4.1 Introduction

Circular pipes can withstand high pressures and hence are used to convey liquids, while non-circular ducts transport low-pressure gases, such as air in heating and cooling systems. In most internal

C. Kleinstreuer, *Modern Fluid Dynamics: Basic Theory and Selected Applications in Macro- and Micro- Fluidics*, Fluid Mechanics and Its Applications 87,
DOI 10.1007/978-1-4020-8670-0_4, © Springer Science+Business Media B.V. 2010

flow applications the *hydrodynamic entrance length* is approximately:

$$\frac{L_e}{D} = \begin{cases} 0.05\,\mathrm{Re} & \text{for laminar flow} \\ 1.36\,\mathrm{Re}^{1/4} \approx 10 & \text{for turbulent flow} \end{cases} \tag{4.2a, b}$$

For any given problem, L_e / D has to be checked to see if L_e is negligible when compared to the pipe length, and hence fully-developed flow can be assumed (see Sect. 3.2). To further simplify calculations and enlarge the range of arbitrary conduit applications, the hydraulic diameter is introduced:

$$D_h = \frac{4A}{P} \tag{4.3}$$

Equation (4.3) transforms non-circular ducts into pipes of equivalent diameter, where P is the perimeter. Most industrial flows, those requiring high fluid throughput, are turbulent, i.e.,

$$\mathrm{Re}_{D_h} = \frac{v_{av}\, D_h}{\nu} = \frac{4\dot{m}}{\pi\,\mu\, D_h} > 4{,}000 \tag{4.4a, b}$$

to be on the safe side, while for smooth pipes, ReD > 2,300.

For *single straight pipe* analysis, assuming unidirectional flow (i.e., u = u(r) only), geometric and kinematic *pipe-design* problems rely on the Moody chart (see App. B) and can be grouped as follows:

(i) Evaluate the necessary pump characteristics based on the computed *pressure drop* Δp (or head loss h_L) in order to convey a given maximum flow rate.
(ii) What is the pipe diameter D for a specified pressure drop, given pipe length and flow rate.
(iii) Determine the flow rate Q for a given pipe geometry (D, L, ε/D) and pressure drop, where ε/D is the relative surface roughness.

Problem types (ii) and (iii) require an iterative procedure because the Reynolds number, and hence the *friction factor f*, is not known.

In general,

$$h_{\substack{\text{total}\\ \text{losses}}} \equiv h_L = h_{\text{friction}} + h_{\text{form}} \tag{4.5a}$$

where

$$h_{\text{friction}} \equiv h_{\text{major}} = f\,\frac{v^2}{2g}\left(\frac{L}{D}\right) \tag{4.5b, c}$$

and

$$h_{\text{form}} \equiv h_{\text{minor}} = \Sigma\, K_L\left(\frac{v^2}{2g}\right)$$

so that with D = ¢:

$$h_L = \left[f\left(\frac{L}{D}\right) + \Sigma\, K_L\right]\frac{v^2}{2g} \tag{4.5d}$$

Here, $f = f(\text{Re}_D, \varepsilon/D)$ (see Moody chart) and loss coefficients K_L are tabulated for valves, elbows, pipe entrances/exits, etc., as given in any UG fluids text or engineering handbook.

For steady *laminar* incompressible fully-developed (i.e., Poiseuille) pipe flow, analytical expressions for all parameters can be derived while for isotropic *turbulent* flow the Moody chart or friction factor correlations (e.g., Haaland 1983) can be employed:

$$f^{-1/2} \approx -1.8\log\left[\frac{6.9}{\text{Re}_D} + \left(\frac{\varepsilon/D}{3.7}\right)^{1.11}\right] \tag{4.5e}$$

Pipe networks are either connected parallel pipes or pipes in series, typically "interrupted" by pumps and/or turbines. Thus, in addition to fluid-mass conservation for incompressible flow, i.e.,

$$\Sigma Q_{in} = \Sigma Q_{out} \tag{4.6}$$

the *extended* Bernoulli equation has to be considered:

$$\left(\frac{p}{\rho g}+\frac{\alpha v^2}{2g}+z\right)_{(1)} = \left(\frac{p}{\rho g}+\frac{\alpha v^2}{2g}+z\right)_{(2)} - h_{pump} + h_{turbine} + h_L \tag{4.7a}$$

Here, α is the "kinetic energy correction factor" which for laminar pipe flow is needed because $v^2/(2g)$ does not capture fully the kinetic energy of the flow. Specifically, $\alpha_{laminar} = 2.0$ and $\alpha_{turbulent} \approx 1.05$. The actual pump (P) and turbine (T) "heads" are:

$$h_{P,T} = \frac{P_{P,T}}{\dot{m}g} \tag{4.7b, c}$$

where $P = F \cdot v \equiv \dot{W}$ is the net power (or work per time) and $\dot{m}$ is the mass flow rate.

4.2 Laminar and Turbulent Pipe Flows

Steady fully-developed laminar flow problems with straight circular pipes can be solved analytically, even when basic heat transfer is included (see Sect. 3.6). However, for the entrance region, if at all important, and for noncircular ducts as well as turbulent flows, approximations and semi-empirical methods have to be employed, in case direct numerical solutions are not available or too costly.

4.2.1 Analytical Solutions to Laminar Thermal Flows

Based on the postulates $\vec{v} = [v_z \equiv u(r), 0, 0]$ and $-\partial p/\partial z = \Delta p / L = \not{c}$, the z-momentum of the Navier–Stokes equations in cylindrical coordinates for Poiseuille flow reduces to:

$$0 = \frac{\Delta p}{L} + \frac{\mu}{r}\frac{d}{dr}\left(\frac{1}{r}\frac{du}{dr}\right) + g_z \tag{4.8a}$$

subject to:

$$u(r = r_0) = 0 \quad \text{and} \quad \left.\frac{du}{dr}\right|_{r=0} = 0 \tag{4.8b, c}$$

The solution to Eqs. (4.8a–c) is the Poiseuille profile (see Example 3.4):

$$u(r) = u_{max}\left[1 - \left(\frac{r}{r_0}\right)^2\right] \tag{4.8d}$$

where $u_{max} = u(r = 0)$ contains the pressure gradient, viscosity, and gravity component.

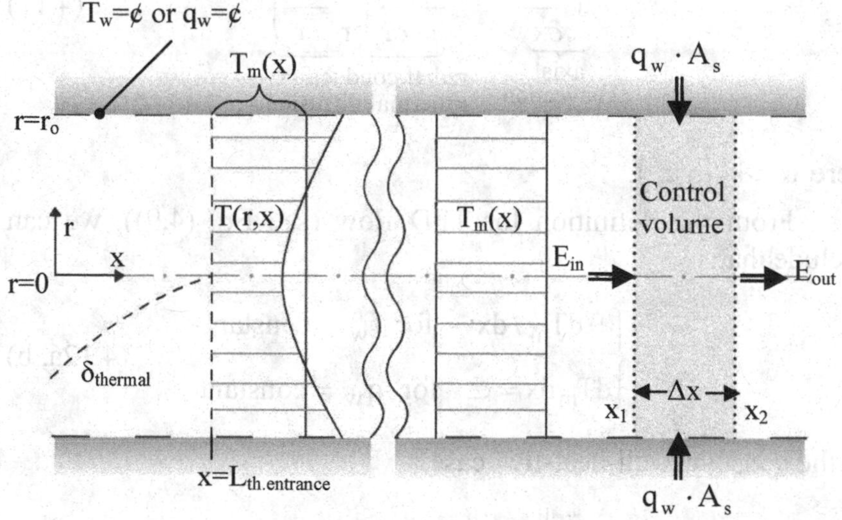

Fig. 4.1 Thermal pipe flow

When the pipe walls are heated or cooled, i.e., $T_{wall} \neq T_{fluid}$, *thermally fully-developed* (TFD) flow is assumed. That occurs not when $\partial T / \partial x = 0$; but, when the *dimensionless* temperature gradient is zero (see Fig. 4.1 and Example 2.11):

$$\frac{\partial}{\partial x}\left[\frac{T_w(x) - T(r, x)}{T_w(x) - T_m(x)}\right] \equiv \frac{\partial \theta}{\partial x} = 0 \tag{4.9}$$

where $T_{wall} \equiv T_{surface}$ for thin-walled (metallic) tubes.

The mean (or cross-sectionally averaged) temperature T_m is defined for $\rho = \not\subset$ as:

$$T_m = \frac{1}{u_{av}\, A} \int_A u(r)\, T(r, x) dA \tag{4.10}$$

Clearly, for u(r) see Eq. (4.8d) and T(r, x) can be obtained from the integration of the reduced form of the heat transfer equation (see Eq. (E2.11.4) or Eq. (3.64)):

$$\underbrace{u \frac{\partial T}{\partial x}}_{\substack{\text{axial} \\ \text{convection}}} = \underbrace{\frac{\alpha}{r} \frac{\partial}{\partial r}\left(\frac{1}{r} \frac{\partial T}{\partial r}\right)}_{\substack{\text{radial conduction} \\ \langle\text{thermal diffusion}\rangle}} \tag{4.11}$$

where $\alpha \equiv k/(\rho\, c_p)$.

From the definition for TFD flow (see Eq. (4.9)), we can conclude that:

$$\frac{\partial T}{\partial x} = \begin{cases} \theta\, dT_m / dx & \text{for } T_w = \text{constant} \\ dT_m dx = \not\subset & \text{for } q_w = \text{constant} \end{cases} \tag{4.12a, b}$$

For the constant-wall-heat-flux case,

$$q_w = k \left.\frac{\partial T}{\partial r}\right|_{r=r_0} = h(T_w - T_m) = \not\subset \tag{4.13a, b}$$

where $T_w = T_w(x)$ and $T_m = T_m(x)$ so that everywhere on the surface $\Delta T \equiv T_w - T_m$ = constant.

Equation (4.11) can be directly solved, typically using the boundary conditions (Fig. 4.1):

$$T(r = r_0) = T_w(x) \quad \text{and} \quad \left.\frac{\partial T}{\partial r}\right|_{r=0} = 0 \tag{4.14a, b}$$

Thus,

$$T(r, x) = T_w(x) - \frac{u_{max}}{\alpha}\left(\frac{dT_m}{dx}\right)\left(\frac{3r_0^2}{16} + \frac{r^4}{16r_0^2} - \frac{r^2}{4}\right) \tag{4.15}$$

In order to evaluate dT_m / dx and q_w, we use Eq. (4.13b) and equate:

$$\frac{q_w}{h} = T_m - T_w = \frac{11u_{max}}{96\alpha}\left(\frac{dT_m}{dx}\right)r_0^2 \tag{4.16}$$

An energy balance for the control volume $(\pi r_0^2)\Delta x$, see Fig. 4.1, yields:

$$E_{in} + q_w A_s = E_{out} \tag{4.17a}$$

Now, with $\dot{m} = \rho u_{av}(\pi r_0^2)$

$$\dot{m} c_p T_m|_{x_1} + q_w(2\pi r_0 \Delta x) = \dot{m} c_p T_m|_{x_2}$$

and in the limit:

$$(2\pi r_0)q_w = \dot{m} c_p \frac{dT_m}{dx}$$

so that

$$q_w = \frac{1}{2} \rho \, c_p \, r_0 \frac{dT_m}{dx} = \frac{T_m - T_w}{h} \qquad (4.17b, c)$$

which, with Eq. (4.16), yields for the heat transfer coefficient:

$$h = \frac{48}{22} \frac{k}{r_0} \qquad (4.18a)$$

Hence, the Nusselt number is:

$$Nu_D \equiv \frac{hD}{k} = 4.364 \qquad (4.18b)$$

Note, a numerical solution of Eq. (4.11) for the *isothermal* wall case (see Eq. (4.12a)) leads to:

$$Nu_D = 3.658 \qquad (4.18c)$$

Example 4.1: Derivation of the Frictional Loss Term, h_f, for Steady Laminar Flow in Smooth Pipes

Sketch	*Assumptions*	*Concept*
L, τ_w, r, x, D, τ_w, (1), (2), $\vec{g}$, z	• As stated, with $L >> L_{entrance}$ • Poiseuille flow • Constant cross sectional pipes • Constant properties and velocity average	• Combine Poiseuille flow solution with extended Bernoulli

Solution Steps:

(A) Check entrance length and assume that $Re_D < 2{,}300$

- Clearly, for any given data set $L_{entrance} = 0.05\,Re_D\,D$ should be much less than L_{pipe}.

(B) Poiseuille flow

- $\vec{v} = [u(r), 0, 0]$ and $\partial p / \partial x = -\frac{\Delta p}{L} = \text{constant}$

$$\bullet\, v = \frac{1}{A}\int u(r)\,dA = \frac{1}{\pi r_0^2}\int u_{max}\left[1-\left(\frac{r}{r_0}\right)^2\right](2\pi\, r\, dr)$$

$$= \frac{1}{2}u_{max} = \frac{r_0^2}{8\mu}\left(\frac{\Delta p}{L}\right) = Q/(\pi r_0^2) \qquad \text{(E.4.1.1a)}$$

$$\bullet\, \tau_{wall} = \mu \left.\frac{du}{dr}\right|_{r=r_0} = 4\mu\, v / r_0 = \frac{r_0}{2}\left(\frac{\Delta p}{L}\right) \qquad \text{(E.4.1.1b)}$$

(C) Extended Bernoulli equation applied to the pipe centerline from points ①–②

$$\bullet \left(\frac{p}{\rho g} + \frac{v^2}{2g} + z\right)_{①} = \left(\frac{p}{\rho g} + \frac{v^2}{2g} + z\right)_{②} + h_{loss} \qquad \text{(E.4.1.2)}$$

$$\bullet\, h_{loss} \equiv h_L = \underset{\langle major \rangle}{h_{friction}} + \underset{\langle minor \rangle}{h_{form}} \qquad \text{(E.4.1.3a)}$$

- With $\Delta z = 0$, $v_1 = v_2$

$$\text{and } h_{\text{form}} \approx 0, \; h_{\text{friction}} = h_f = \frac{\Delta p}{\rho g} \tag{E.4.1.3b}$$

(D) Combine (B) and (C) results

- From $v = \frac{r_0^2}{8\mu}\left(\frac{\Delta p}{L}\right); \quad r_0 = D/2$ (E.4.1.4a)

we have

$$\Delta p = 32\mu\, v\, L / D^2 \tag{E.4.1.4b}$$

As a result,

$$h_f = \frac{32\mu\, L\, v}{\rho\, g\, D^2} := \frac{128\mu\, L\, Q}{\pi\, \rho\, g\, D^4} \tag{E.4.1.5a, b}$$

Now, defining the friction factor f (Darcy–Weisbach) as:

$$f \equiv \frac{8\tau_w}{\rho\, v^2} := \frac{2D}{\rho\, v^2}\left(\frac{\Delta p}{L}\right) \tag{E.4.1.6a}$$

so that

$$\Delta p = f\, \frac{\rho\, v^2}{2}\left(\frac{L}{D}\right) \tag{E.4.1.6b}$$

and hence we can express the frictional loss also as:

- $h_f = \frac{\Delta p}{\rho g} = f\, \frac{v^2}{2g}\left(\frac{L}{D}\right)$ (E.4.1.7)

(E) Notes

- With $\tau_w = 4\mu\, v / r_0, \; f_{\text{laminar}} = \frac{32\mu v}{r_0\, \rho\, v^2} = \frac{64}{\text{Re}_D}$ (E.4.1.8, 9)

- From a 1-D force balance for any fully-developed flow regime (see momentum RTT) we obtain (with $v_{in} = v_{out}$):

$$\sum F_x = \underbrace{\Delta p(\pi r_0^2)}_{F_{net,pressure}} - \underbrace{\rho g(\pi r_0^2)L \sin\phi}_{W_x} - \underbrace{\tau_w(2\pi r_0^2)}_{F_{viscous}} \qquad \text{(E.4.1.10)}$$

so that with $L \sin\phi = -\Delta z$:

$$h_f = \Delta z + \frac{\Delta p}{\rho g} = \frac{2\tau_w}{\rho g}\frac{L}{r_0} = \frac{4\tau_w}{\rho g}\left(\frac{L}{D}\right) \qquad \text{(E.4.1.11)}$$

or with $f \equiv 8\tau_w \mid (\rho v^2)$ we have again:

$$h_f = f\frac{v^2}{2g}\left(\frac{L}{D}\right) \qquad \text{(E.4.1.12)}$$

Graph:

- On a log-log graph, we have the analytical and measured $f(Re_D)$ function:

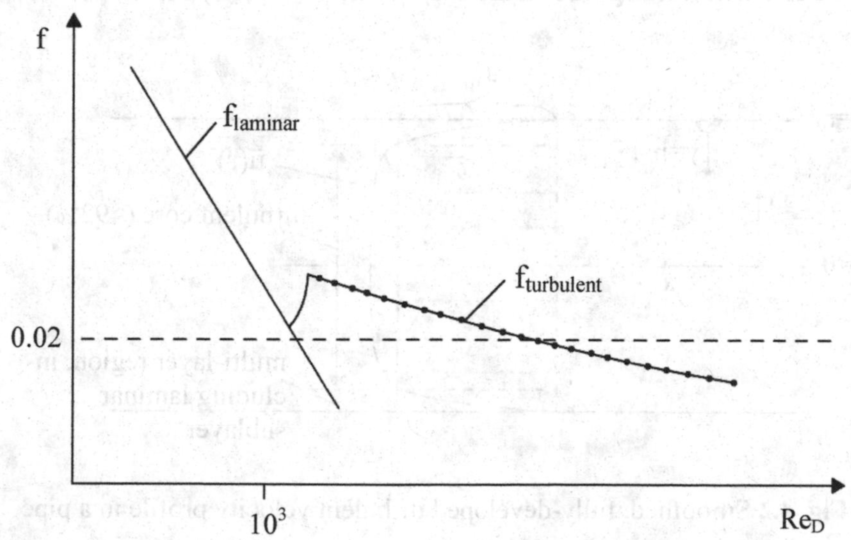

Commen:t

- Clearly, the h_f-correlation holds for both laminar *and* turbulent pipe flows, where

$$f_{laminar} = 64/\sqrt{Re_D} \quad \text{and} \quad f_{turbulent} = f\left(Re_D, \frac{\varepsilon}{D}\right)$$

Turbulent friction factor values are obtainable, for example, via Eq. (4.5e), or from the Moody chart (see App. B)

4.2.2 Turbulent Pipe Flow

Isothermal Flow The hydrodynamic entrance length for turbulent flow, say,
$Re_D > 4{,}000$, is:

$$L_e = 1.36D\,Re_D^{1.4} \tag{4.19a}$$

as confirmed experimentally, or simply:

$$L_e \approx 10D \tag{4.19b}$$

Thus, $L_{e/tub} << L_{e/laminar}$ (see Eq. (4.2)) because of the rapid fluid-element mixing due to the 3-D *random* velocity fluctuations.

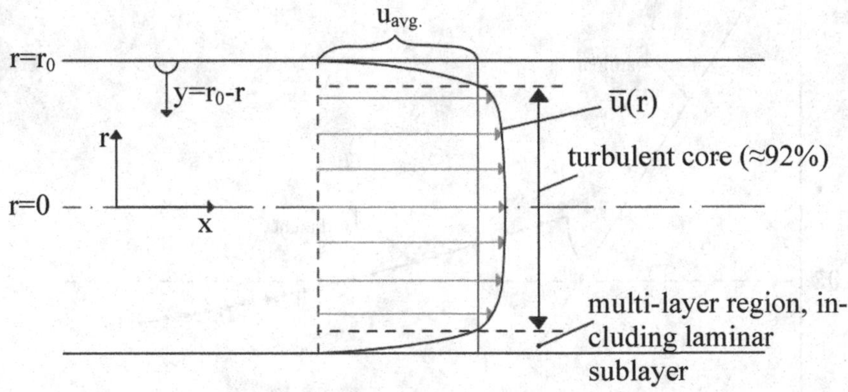

Fig. 4.2 Smoothed, fully-developed turbulent velocity profile in a pipe

While for Poiseuille flow $u(r) \sim r^2$ and $u_{av} = 0.5\,u_{max}$, in turbulent pipe flow $\bar{u}(r) \sim r^{1/7}$ and u_{av} is much closer to u_{max} (see Fig. 4.2). The actual turbulent velocity profile exhibits random fluctuations (i.e., u', v', w'), where in the x-direction:

$$u_{instantaneous} = \bar{u}_{time-averaged} + u'_{random} \tag{4.20a}$$

Clearly, time-smoothing (or averaging) yields:

$$\bar{u} = \frac{1}{\Delta t}\int_{t_1}^{t_2} u\,dt \quad \text{and} \quad \int_{t_1}^{t_2} u'\,dt = 0 \tag{4.20b, c}$$

The multi-layer region near the pipe wall consists of the laminar sublayer at the wall where $Re_D \to 0$ and $u \sim y$, a buffer layer and an overlap layer which connects to the turbulent core, which makes up about 92% of the cross section. Various semi-empirical velocity profiles have been postulated for turbulent flow in *smooth* pipes (see Schlichting & Gersten 2003), where the 1/7-law and a natural-log law are the favorite ones. For example,

$$\frac{\bar{u}}{u_{max}} = \left(\frac{y}{r_0}\right)^{1/n} \tag{4.21a}$$

where $\bar{u}$ is the time-averaged axial velocity, while $y = r_0 - r$. Via integration we obtain:

$$\frac{u_{av}}{u_{max}} = \frac{2n^2}{(n+1)(2n+1)} \tag{4.21b}$$

The coefficient, $n = n(Re_D)$ is in the range $6 \le n \le 10$ for $4{,}000 \le (Re_D) \le 3 \times 10^6$ with $n = 7$ being typical, i.e., $Re_D = 10^5$ and then $u_{av} / u_{max} = 0.817$ (see App. B for $n = n(Re_D)$).

In terms of "inner variables" (see Sect. 5.2), i.e., $u^+ = u/u_\tau$ ($u_\tau = \sqrt{\tau_w/\rho}$ being the "friction velocity") and $y^+ = u_\tau y/\nu$, we can obtain:

$$u^+ = 8.74(y^+)^{1/7} \tag{4.22}$$

where in the laminar sublayer $u^+ = y^+$. Borrowed from turbulent boundary-layer theory (Sect. 5.2) and empirically adjusted to pipe flow, there are the "log-laws", where the laminar sublayer $u^+ = y^+$ for $0 \leq y^+ \leq 5$ is connected via the buffer layer to the turbulent core region with $u^+ = u^+(\ln y^+)$ -functions (see Sect. 5.2 and Schlichting & Gersten 2003; among others).

Example 4.2: Turbulent Friction Factor Derivation for Flow in Smooth Pipes

Sketch	*Assumptions*	*Concepts*
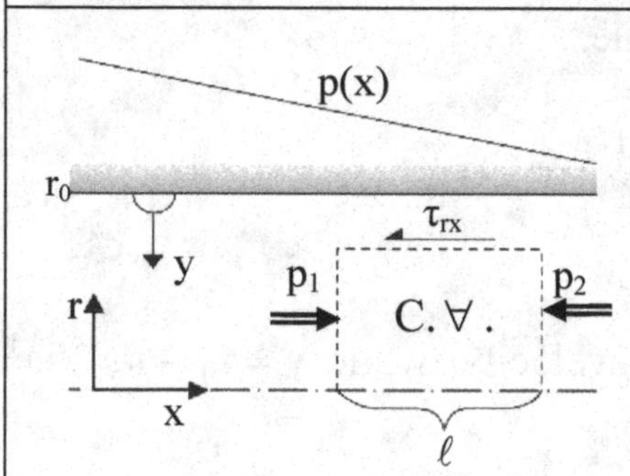 p(x), r_0, τ_{rx}, y, r, x, p_1, p_2, C.∀., ℓ	• Steady fully-developed turbulent flow in a smooth pipe with constant fluid properties	• Force balance for C.∀. $\hat{=} (r^2\pi)\,\ell$ • friction loss expression, i.e., $h_f \sim \Delta p \sim \tau_{wall}$ • Use of Eq. (4.22)

Solution:

- A force balance (see Sect. 2.4, Momentum RTT) on the C.∀., $\pi r^2 \ell$, yields:

$$\sum F_x = \Delta p(\pi r^2) - \tau_{rx}(2r\pi\ell) = \dot{m}(v_2 - v_1) = 0 \quad \text{(E.4.3.1a, b)}$$

which, evaluated at the wall, results in

$$\Delta p(\pi r_0^2) - \tau_w (2r_0 \pi \ell) = 0$$

so that

$$\tau_{rx} = \tau_w \frac{r}{r_0} \tag{E.4.3.2}$$

Note, Eq. (E.4.3.2) indicates that τ(r) is linear in pipe flow for *any* fully-developed flow regime. It turns out that is the case for non-Newtonian fluids as well (see Sect. 6.3).

- Expressing Δp from Eq. (E.4.3.1b) in h_f of Eq. (4.5d) with $\sum K_L = 0$, we obtain:

$$\Delta p - 2\ell\tau_w / r_0$$

and hence

$$h_f = \frac{\Delta p}{\rho g} = \frac{2\ell\tau_w}{\rho g r_0} = f \frac{v^2}{2g}\left(\frac{\ell}{2r_0}\right)$$

from which

$$f = \frac{8\tau_w}{\rho v^2} \tag{E.4.3.3a}$$

or with $\tau_w = \rho u_\tau^2$,

$$f = 8\left(\frac{u_\tau}{v}\right)^2 \tag{E.4.3.3b}$$

- Equation (4.22) can be evaluated at the centerline, where $y^+ = r_0 u_\tau / \nu$, so that we can obtain an expression for the elusive friction velocity as:

$$u_\tau = \left(\frac{u_{max}}{8.74}\right)^{7/8} \left(\frac{\nu}{r_0}\right)^{1/8} \tag{E.4.3.4}$$

Also,

$$u_{av} \equiv v = \frac{1}{A} \int u \, dA = 0.817 \, u_{max}$$

so that Eq. (E.4.3.4) can now be rewritten as:

$$u_\tau = \left(\frac{v}{7.1406}\right)^{7/8} \left(\frac{2\nu}{D}\right)^{1/8}$$

and hence

$$f = 0.3164 \, Re_D^{-1/4} \tag{E.4.3.5}$$

Note, Eq. (E.4.3.5), known as the Blasius correlation, differs significantly from $f_{laminar} = 64 \, Re_D^{-1/2}$. More extensive expressions taking pipe wall roughness into account is given as Eq. (4.5e). Alternatively, the Moody chart can be used (see App. B).

Example 4.3: Given h_f (or Δp) for Turbulent Pipe Flow, as well as D and L, Find the Volumetric Flow Rate Q When D = 0.3 m, L = 100 m, $\frac{\varepsilon}{D} = 2 \times 10^{-4}$, $h_f = 8\text{m}$, and $\nu_{oil} = 2 \times 10^{-5} \, \text{m}^2/\text{s}$.

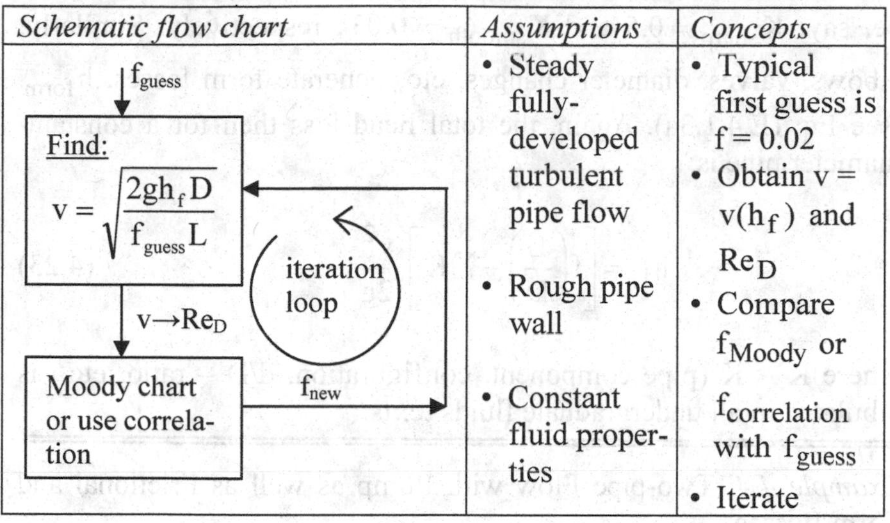

Schematic flow chart	Assumptions	Concepts
(flow chart)	• Steady fully-developed turbulent pipe flow • Rough pipe wall • Constant fluid properties	• Typical first guess is f = 0.02 • Obtain $v = v(h_f)$ and Re_D • Compare f_{Moody} or $f_{correlation}$ with f_{guess} • Iterate

Solution:

- Recall, $Q = v(\pi r_0^2)$, where v is unknown; but, v can be estimated with f_{guess}
- Wild guess: f = 0.004 and then

$$v = [2gh_f D/(f_{guess}\, L)]^{1/2} := 3.43 \frac{m}{s}$$

and hence $Re_D = \frac{vD}{\nu} = 5.1465 \times 10^4$ <turbulent flow>

- With $Re_D = 5 \times 10^4$ and $\frac{\varepsilon}{D} = 2 \times 10^{-4}$

$$f_{new} = 0.022$$

- Now v = 4.63 m/s and $Re_D = 6.94 \times 10^4$
- Finally, f = 0.021 and ultimately $Q = 0.335\ m^3/s$

Clearly, a sharp pipe inlet, creating locally a re-circulation zone (i.e., vena contracta) creates a larger head loss than a well rounded inlet. Specifically, in Eq. (4.23) the loss coefficients would be, say, $K_{sharp} = 0.5$ and $K_{smooth} = 0.03$, respectively. Similarly, elbows, valves, diameter changes, etc. generate form losses, h_{form} (see Eq. (E.4.1.3a). Again, the total head loss then for a constant-diameter pipe is:

$$h_L = \left[f\left(\frac{L}{D}\right) + \sum K \right] \frac{v^2}{2g} \tag{4.23}$$

where K = K (pipe component, configuration, d/D – ratio, etc.) is tabulated in all undergraduate fluids texts.

Example 4.4: Two-pipe Flow with Pump as well as Frictional and Form Losses

Given two water reservoirs ($H_A = 5\text{m}$ and $H_B = 15\text{m}$) connected by two parallel wrought-iron pipes ($D_1 = 0.05\text{m}$, $\varepsilon_1 = 0.04\text{mm}$; $L_1 \approx 15\text{m}$, and $D_2 = 0.10\text{m}$, $\varepsilon_2 = 0.06\text{mm}$; $L_2 \approx 15\text{m}$ with pump (10 kW; $\eta = 0.8$) to convey the water (20°C; $\rho = 988\ \text{kg/m}^3$ and $\mu = 10^{-3}$ kg/(m·s)). From reservoir A to reservoir B. Take $K_{bend} \approx 0.3$ and find Q_1 and Q_2!

Sketch	*Assumptions*
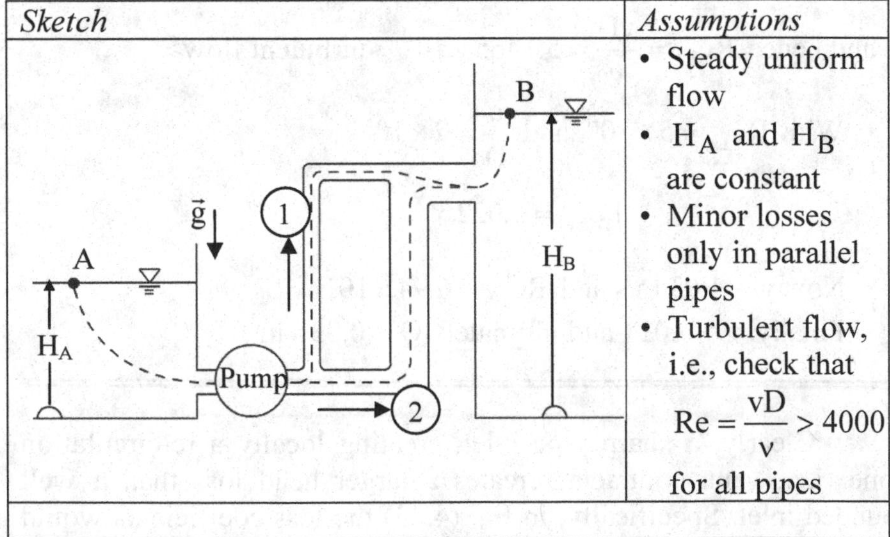 	• Steady uniform flow • H_A and H_B are constant • Minor losses only in parallel pipes • Turbulent flow, i.e., check that $Re = \frac{vD}{\nu} > 4000$ for all pipes

Concepts:

- Extended Bernoulli with $h_{pump} = \frac{\dot{W}_{electr} \cdot \eta}{\rho g Q}$ (see Eq. (4.7b))
- $h_L = h_{L,1} + h_{L,2}$ because Δp is the same for each pipe
- Total losses $h_L = h_{friction} + h_{form}$
- Friction factors from Eq. (4.5e) or Moody chart
- Flow rate $Q = Q_1 + Q_2$
- Solve system of n equations with n unknowns simultaneously. Alternatively, guess f_i and use Moody chart or correlation

Solution: With $v_A = v_B = 0$, $p_A = p_B$ and $\Delta z = H_B - H_A$ the extended Bernoulli equation yields:

$$h_{pump} = H_B - H_A + h_L = \frac{\dot{W}_{el} \cdot \eta}{\rho g Q} \qquad \text{(E.4.4.1a, b)}$$

where

$$h_{L,i} = \left[\left(f \frac{L}{D}\right)_i + \sum K_i\right] \frac{v_i^2}{2g}; \quad i = 1, 2 \qquad \text{(E.4.4.2a, b)}$$

Employing Eq. (4.5e):

$$f_i \approx \left\{-1.8 \log\left[\frac{6.9}{Re_i} + \left(\frac{\varepsilon / D_i}{3.7}\right)^{1.11}\right]\right\}^{-2} \qquad \text{(E.4.4.3)}$$

with

$$Re_i = \frac{v_i D_i}{\nu} \qquad \text{(E.4.4.4)}$$

$$v_i = Q_i / \left(\frac{\pi D_i^2}{4}\right) \qquad \text{(E.4.4.5a)}$$

and

$$Q = Q_1 + Q_2 \tag{E.4.4.5b}$$

Guess $f_1 = 0.02$ and $f_2 = 0.018$, which leads with

$$\frac{\varepsilon_1}{D_1} = 8 \times 10^{-4} \quad \text{and} \quad \frac{\varepsilon_2}{D_2} = 6 \times 10^{-4} \quad \text{using (E.4.4.3) to:}$$

$$Re_1 = 3 \times 10^5 \quad \text{and} \quad Re_2 = 8 \times 10^5 .$$

Now v_1 and v_2 can be computed as well as Q and h_L (see Eqs. (E.4.4.5a, b and (E.4.4.2a, b)). Clearly, Eq. (E.4.4.1b) has to be fulfilled, which requires h_L and Q have to match! If not, a new f-value has to be assumed and the calculations have to be repeated. The final result is $Q_1 = 0.0124$ kg/s and $Q_2 = 0.066$ kg/s.

Non-isothermal Turbulent Pipe Flow Again, the thermal entrance length has to be compared with the actual pipe length to assure that $L_{e,\text{thermal}} << L$ and hence fully-developed flow can be assumed. Now the Reynolds-Colburn analogy (see Eq. (3.87)) can be employed for turbulent flow where

$$c_f(x) = 2St_x \; Pr^{2/3} \equiv \frac{2\tau_w}{\rho v^2} \equiv \frac{f}{4} \tag{4.24a–c}$$

For smooth tubes (Eq. (E.4.3.5)):

$$f_1 = 0.3164 \, Re_D^{-0.25} \quad \text{when } 4 \times 10^3 \leq Re_D \leq 10^5 \tag{4.25a}$$

For higher flow rates based on measurements:

$$f_2 = 0.184 \, Re_D^{-0.2} \quad \text{when } 2 \times 10^4 \leq Re_D \leq 10^6 \tag{4.25b}$$

Now, with $St_x \equiv Nu_x/(Re_x\ Pr)$, where x is replaced by D, and employing, for example, the empirical f_2-expression we obtain:

$$\overline{Nu_D} = 0.023\ Re_D^{0.8}\ Pr^{1/3} \tag{4.26}$$

Equation (4.26) holds for $0.7 < Pr < 160$, $Re_D > 10^4$ and $\ell / D > 60$ and matches the empirical formula of Dittus & Boelter (1930) very well, as already discussed in Chap. 3.

In order to solve some practical convection heat transfer problems, e.g., basic heat exchanger performance, we recall the global energy balance (see Eq. (2.47)) for an open system with inlet ① and outlet ②:

$$\dot{Q}_{in} - \dot{W}_{out} = \dot{m}\{(h_2 - h_1) + 0.5(\alpha_2 v_2^2 - \alpha_1 v_1^2) + g(z_2 - z_1)\ \tag{4.27}$$

For a horizontal pipe of length L with specific inlet/outlet enthalpies $h_{1,2} = c_p\ T_{m1,2}$ and power preformed $\dot{W}_{out} = 0$, we have:

$$\dot{Q}_{1-2} \equiv \dot{Q}_L = \dot{m} c_p (T_{m,2} - T_{m,1}) \tag{4.28}$$

which holds for either of the two thermal wall conditions, i.e., $q_w = \not\subset$ or $T_w = \not\subset$. An alternative expression for the heat flow rate eminates from Eq. (3.84), i.e., based on Newton's law of cooling:

$$\dot{Q}(x) = h_x (2 r_0\ \pi\ x)\, [T_w(x) - T_m(x)] \tag{4.29}$$

Equating (4.28) and (4.29) yields in the limit, i.e., $\Delta T_m / \Delta x \rightarrow dT_m / dx$:

$$\frac{dT_m}{dx} = \frac{\pi\ D}{c_p\ \dot{m}}\ h_x (T_w - T_m) \tag{4.30}$$

Now, integrating from the pipe inlet $T_m(x=0) = T_{m,1}$ to any axial position x, we obtain with $\overline{h} = \frac{1}{L}\int_0^L h(x)\,dx$:

$$\ln \frac{T_m(x) - T_w}{T_{m,1} - T_w} = \frac{-\pi D \overline{h}}{c_p \dot{m}} x \qquad (4.31a)$$

or

$$T_m(x) = T_w + (T_{m,1} - T_w) \exp\left[\frac{-\pi D \overline{h}}{\dot{m} c_p} x\right] \qquad (4.31b)$$

For the uniform heat flux case with q_{wall} given, Eq. (4.28) can be rewritten for an arbitrary pipe section $\Delta x = x_2 - x_1$

$$\frac{q_w(\pi D \Delta x)}{\dot{m} c_p} = T_m(x_2) - T_m(x_1) \qquad (4.32a)$$

In the limit

$$\frac{dT_m}{dx} = \frac{q_w(\pi D)}{\dot{m} c_p} \qquad (4.32b)$$

or

$$T_m(x) = \frac{\pi D q_w}{\dot{m} c_p} \qquad (4.32c)$$

Figures 4.3a and b depict the two classic cases, i.e., T_w = constant and q_w = constant.

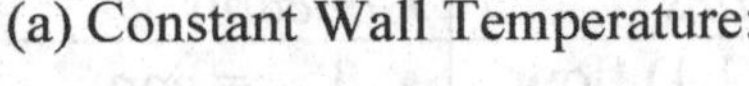

(b) Constant Wall Heat Flux:

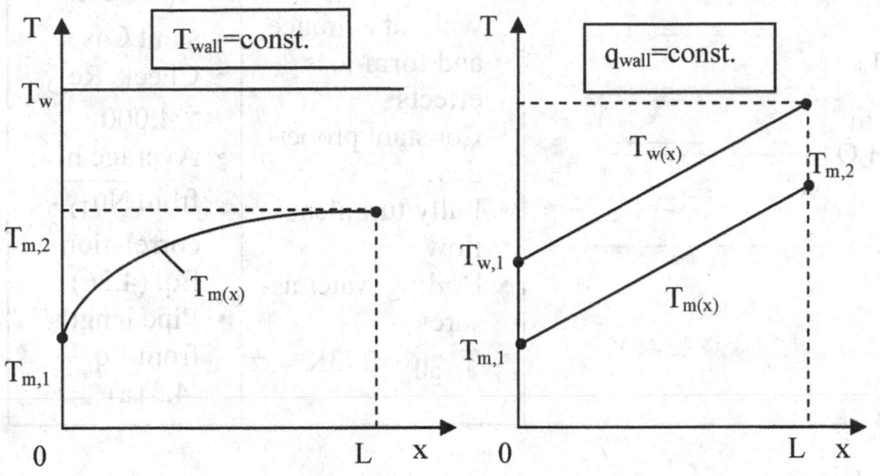

Fig. 4.3a, b Axial wall and mean temperature changes

In summary, Eqs. (4.28)–(4.32) can be selectively used depending on the given thermal boundary condition and if T_m, T_w, $\dot{Q}$ or q_w is of interest. It should be noted that the equations also hold for *developing* flows, provided that the local heat transfer coefficient, h_x, or its tube-averaged approximation is being employed.

Example 4.5: Single-Tube Heat Exchanger

Consider a smooth curved pipe of unknown length (D = 12.5 mm) immersed in a boiling-water reservoir ($T_R = 373K$, $p_R = p_{atm}$), where the tube water ($\dot{m} = 0.0576$ kg/s, $T_{in} = 300K$) should be heated to $T_{out} = 360K$. Determine the required tube length and total heat transfer rate. The property data for reference temperature $T_{ref} = 0.5\,(T_{in} + T_{out}) = 330K$ are: $\rho = 984\ kg/m^3$, $\mu = 4.9 \times 10^{-4}\ N \cdot s/m^2$, $c_p = 4184\ J/(kg \cdot K)$, k = 0.65 W/(m·K); and Pr = 3.15.

Sketch	*Assumptions*	*Concepts*
$T_{m,1}$ $\dot{m}$ H_2O $T_{m,2}$	• Steady 1-D flow without entrance and form-loss effects • Constant properties • Fully turbulent flow • Boiling water assures $T_{wall} = 373K = \not\subset$	• T_w = constant Case • Check $Re_D > 4{,}000$ • Average h from $\overline{Nu_D}$ -correlation, Eq. (4.26) • Pipe length from Eq. (4.31a)

Solution:

- The total heat flow rate (Eq. (4.28)) is:

$$\dot{Q}_{total} = \dot{m}\, c_p[T_m(x = L) - T_m(x = 0)] \tag{E.4.5.1}$$

$$= 0.0576 \cdot 4184(360 - 300)$$

$$\underline{\dot{Q}_{total} = 14.46\ kJ/s \triangleq 14.46\ kW}$$

- Equation (4.31a) can be rewritten as:

$$L = \frac{\dot{m}\, c_p}{\bar{h}(\pi D)} \ln\left[\frac{T_m(L) - T_w}{T_m(0) - T_w}\right] \tag{E.4.5.2}$$

where $\bar{h}$ is unknown,

- Equation (4.26) contains $\bar{h}$, i.e.,

$$\overline{Nu_D} \equiv \frac{\bar{h}\, D}{k} = 0.023\, Re_D^{0.8}\ Pr^{0.33} \tag{E.4.5.3}$$

where $Re_D = \frac{4\,\dot{m}}{\pi\,\mu\,D} := 12 \times 10^3$, i.e., turbulent flow occurs.

Hence,

$$\bar{h} = 3.47\,\frac{k\,W}{m^2 \cdot K}$$

Now, Eq. (E.4.5.2) can be evaluated with $T_w = 373K$ to obtain the pipe length:

$$L = 3.05\ m$$

Comments:
- Figure 4.3a depicts the pipe-water heating process in detail. Note that Eq. (E.4.5.1) is a global energy balance.
- The pipe length is based on an average heat transfer coefficient (Eq. (E.4.5.3)) and hence would be incorrect if the pipe entrance effect would be significant.
- For a constant-wall-heat-flux ease, say, with q_w, D and L given, as well as $T_m(x)$ and $T_w(x)$ measured, where $0 \le x \le L$, $h_x = h = \not\subset$ can be readily obtained from Eq. (4.29) and then $\dot{Q}_{total} = q_w(\pi\,D\,L)$.

Example 4.6: Basic Shell-and-Tube Heat Exchanger

Consider cooling of hot oil ($\dot{m}_{oil} = 5.5\ kg/s$; $T_{\dot{m}} = 380K$, $T_{out} = 320K$; $c_p = 2.122 kJ/(kg \cdot K)$) by water ($T_{in} = 280K$, $\dot{m}_{water} = 5.65 kg/s$; $c_p = 4.18 kJ/(kg \cdot K)$)

System:	*Model*	*Assumptions*
Oil; Shell; Water; Tube; Oil; $(\dot{m}, h)_{O,in}$; IN; $(\dot{m}, h)_w$; C.∀.; OUT; $(\dot{m}, h)_{O,out}$ (a) System (b) Model	• Steady turbulent uniform streams • Constant properties • Negligible ΔE_{kin} and ΔE_{pot}	• No losses • $\Delta\dot{Q}_{oil} = \Delta\dot{Q}_{water}$ <heat from the oil received by the water>

Solution:

$$\Delta\dot{Q}_{oil} = \Delta\dot{Q}_{water}\,; \qquad \Delta\dot{Q} = \dot{m}\Delta h = \dot{m}\, c_p\, \Delta T$$

$$(\dot{m}c_p)_{oil}\,(T_{in} - T_{out})_{oil} = (\dot{m}c_p)_{water}\,(T_{in} - T_{out})_{water}$$

Thus,

$$T_{out,w} = T_{in,w} + \frac{(\dot{m}c_p)_0\,(T_{in} - T_{out})_0}{(\dot{m}\,c_p)_w}$$

$$\underline{T_{out,w}} = 280\text{K} + 30\text{K} = \underline{310\text{K}}$$

Comments: The basic "heat-exchange balance", $\Delta\dot{Q}_A = \Delta\dot{Q}_B$, holds for other configurations as well. Here are two basic examples depicted:

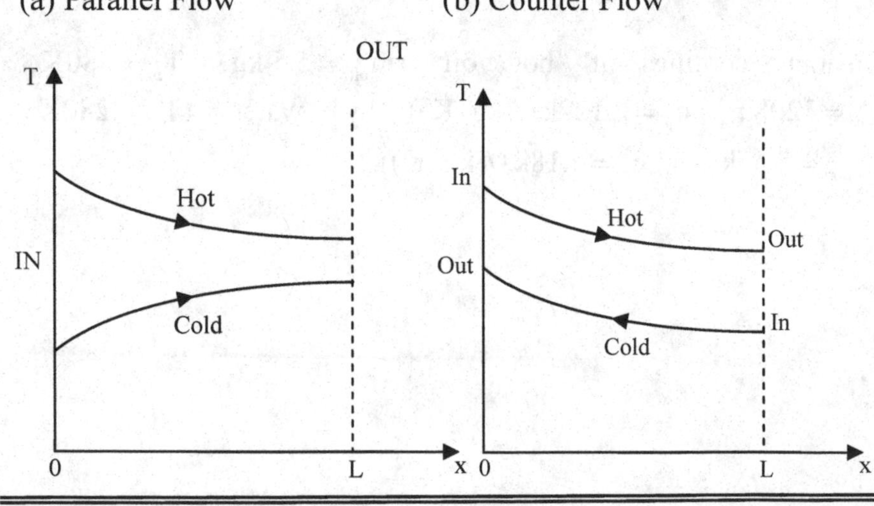

Clearly, in all cases $T_{fluid} = fct(x)$ which can be approximated by postulating:

$$\dot{Q} = (UA) \cdot \Delta T_{ln} \tag{4.33a}$$

where U is the "overall" heat-transfer coefficient, A(L) is the effective heat transfer area, and ΔT_{en} is the "log-mean" temperature difference based on the two inlet and outlet temperatures. Specifically,

$$\Delta T_{ln} = \frac{\Delta T_1 - \Delta T_2}{\ln\left(\frac{\Delta T_1}{\Delta T_2}\right)} \tag{4.33b}$$

where $\Delta T_1 = T_{wall} - T_{in}$ and $\Delta T_2 = T_{wall} - T_{out}$.

Typically, L_{pipe} required for shell-and-tube is 10× less than for the other configurations, i.e., parallel and counter-flow pipe arrangements as given in Example 4.6.

4.3 Basic Lubrication Systems

Lubrication is a subset of *tribology*, the theory of friction, lubrication, wear, damage and replacement of mechanical and biological systems, such as gliding/rotating machine parts or movable joints. In order to slide two solids relative to each other, a tangential resistance, i.e., a frictional force, F, has to be overcome:

$$F = f \cdot N \tag{4.34}$$

where N is the normal load, and f is the friction coefficient with a typical range of $0.01 < f < 1.0$.

Low *friction coefficients* can be achieved when the two solid body surfaces stay at all times separated via a suitable lubricant. Load-carrying lubrication relies on the *wedge-film* effect to work, where a minimum angled surface/plate velocity is required to generate a load-carrying pressure field (see Fig. 4.4a). Such a case of *hydrodynamic lubrication* greatly reduces wear and hence surface damage. In general, friction and wear should be considered as separate phenomena, as in the case of *"boundary" lubrication* when friction and wear are determined by the surface properties of the *solids and the chemical nature of the lubricant* rather than just its viscosity. Specifically, wear is the removal of one material by another due to sliding contact. Wear debris may ultimately cause device/machine (or human joint) failure. Other lubrication modes include squeeze-film and elasto-hydrodynamic lubrication, where in the latter case bounding material deformation, e.g., cartilage in joints, is considered as well (see Fig. 4.4a–c).

The basic mechanism of *squeeze-film* lubrication is shown in Fig. 4.4b while the human knee joint (Fig. 4.4c) is an illustrative example of coupled hydrodynamic lubrication with squeeze-film material deformation, labeled *elasto-hydrodynamic lubrication.* For example, for the problem of human-joint replacement the right selection of cup-and-ball material (e.g., polyethylene, metal alloys, or ceramics) and lubricant, as well as perfect sphericity and surface finish are important. Also of concern are the minimum sustained radial clearance and constant properties of the pseudo-synovial fluid in light of realistic load changes during walking, running, etc. In any case, problems may arise when particles are entrained due to wear, or boundary lubrication occurs when the fluid film breaks.

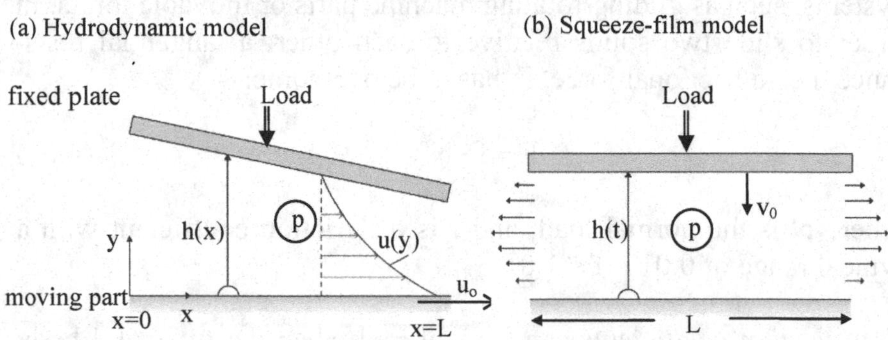

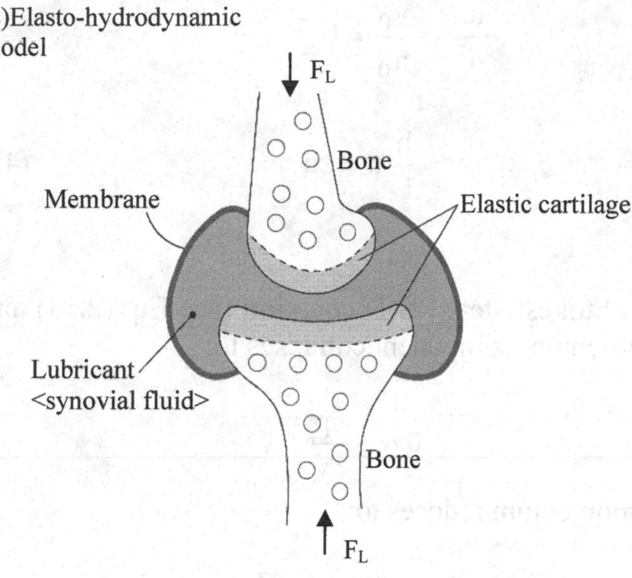

Fig. 4.4 Lubrication Models: (a) Hydrodynamic model; (b) Squeeze-film model; (c) Schematic of the knee (i.e., elasto-hydrodynamic model)

4.3.1 Lubrication Approximations

Preliminaries Underlying aspects of lubrication approximations are that geometrically h << L, or $\varepsilon = h/L \to 0$, the Reynolds number $Re = u_0 h_0 / \nu$ is small, allowing the use of Stokes equation (see Sect. 2.4), and the flows are laminar, quasi-steady and two-dimensional. Several lubrication approximations are introduced and illustrated, starting with simple slot flow and ending with the encompassing Reynolds lubrication equation in Sect. 4.3.2 (see Fig. 4.5).

For slot flow, the continuity equation

$$\frac{\partial u}{\partial x} + \frac{\partial v}{\partial y} = 0 \qquad (4.35a)$$

can be scaled, resulting in:

$$\frac{u_0}{L} + \frac{v_0}{h_0} = 0 \qquad (4.35b)$$

or

$$v = \frac{h}{L} u = \varepsilon u \qquad (4.35c, d)$$

where $\varepsilon = \frac{h}{L} << 1$.

Starting with Stokes' steady 2-D equation (see Eq. (2.34) and App. A), the y-momentum component collapses to:

$$0 = -\frac{\partial p}{\partial y} \qquad (4.36)$$

while the x-momentum reduces to:

$$0 = -\frac{dp}{dx} + \mu \frac{\partial^2 u}{\partial y^2} \qquad (4.37)$$

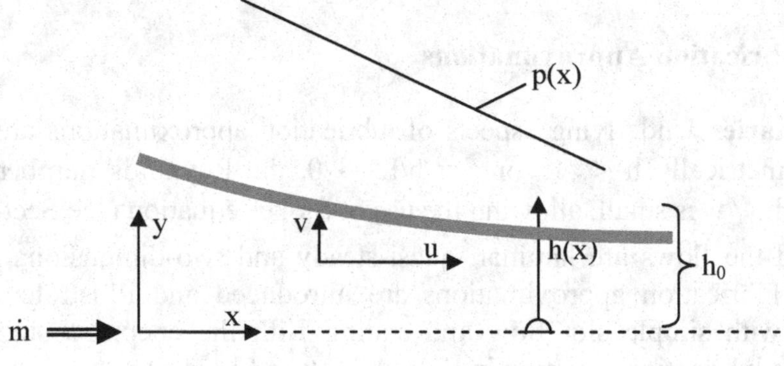

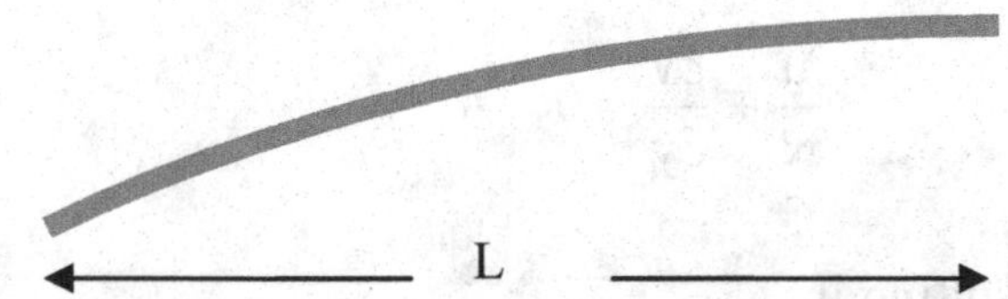

Fig. 4.5 Axisymmetric slot flow with mildly varying height

Slot Flow Considering slot flow with locally varying clearance h(x), we have as the solution to Eq. (4.37) (see Fig. 4.5 and Sect. 3.2):

$$u(x, y) = -\frac{h^2}{2\mu}\left(\frac{dp}{dx}\right)\left[1-\left(\frac{y}{h}\right)^2\right] \tag{4.38}$$

which is Poiseuille-type flow, depending on the local height h(x) and the local pressure gradient. Clearly, via the boundary condition u(y = h) = 0, the axial velocity is not only a function of y (as in Poiseuille flow) but also dependent on x, because h=h(x) as already discussed in Sect. 3.2. In order to complete the slot flow analysis, global continuity, $\dot{m} = \rho\bar{v}A = \rho Q$ = constant, requires for incompressible flow:

$$Q = \int_A v_n \, dA = 2\int_0^{h(x)} u\,dy := \frac{h^3}{2\mu}\left(\frac{\Delta p}{L}\right) \tag{4.39a–c}$$

Given the flow rate per unit depth, the quasi-constant pressure gradient, $\frac{dp}{dx} \cong -\frac{\Delta p}{L}$, can be eliminated in Eq. (4.38). Alternatively, Eq. (4.39c) delivers the overall pressure drop, i.e.,

$$p_1 - p_2 = 2\mu Q \int_0^L \frac{dx}{h^3} \tag{4.40}$$

It should be apparent that the governing equation (4.37) based on the lubrication approximation describes numerous other applications, e.g., flows in narrow channels, films, blood vessels, filters, porous media, etc.

Example 4.9: Sample Squeeze-Film System

Consider squeeze-film lubrication under a constant load L for different fluids, i.e., Newtonian, power-law, and viscoelastic.

Model Schematic	*Assumptions*	*Concepts*
L; z; p(r); v(z); r; h(t); $w = \dot{h}/2$; R; L	• Transient 2-D axisymmetric in- compressible flow • Gap height $h << R$ and $\dot{h} << \overline{v}$ • Normal stresses are negligible • Neglect end effects	• Reduced continu- ity and equation of motion • Insert appropriate rheology models for τ_{rz} • Average v(r, z, t) over z: $\overline{v} = \frac{2}{h}\int_0^{h/2} v\,dz$

Solution:

Based on the postulates $\vec{v} = [0;\, v = v(r, z, t);\, w = \dot{h}/2]$ and $\nabla p = dp/dr$, we reduce the continuity equation to:

$$\frac{1}{r}\frac{\partial(rv)}{\partial r} + \frac{\partial w}{\partial z} = 0 \qquad \text{(E.4.9.1)}$$

and the equation of motion to:

$$\frac{\partial}{\partial r}\,p(r, t) = \frac{\partial}{\partial z}\,\tau_{rz} \qquad \text{(E.4.9.2)}$$

Integration of Eq. (E.4.9.1) across the gap $-\frac{h}{z} \le z \le \frac{h}{2}$ requires:

$$\int dw = -\int \frac{1}{r}\frac{\partial(r\,v)}{\partial r}\,dz \qquad \text{(E.4.9.3)}$$

Specifically,

$$\int_{-h/2}^{h/2} dw = w\left(\frac{h}{2}\right) - w\left(-\frac{h}{2}\right) = \frac{\dot{h}}{2} - \left(-\frac{\dot{h}}{2}\right) = \dot{h} \qquad \text{(E.4.9.4a)}$$

and

$$-\int_{-h/2}^{h/2} \frac{1}{r}\frac{\partial(rv)}{\partial r}dz = -\frac{1}{r}\frac{d(r\bar{v})}{dr}\int_{-h/2}^{h/2} dz = -\frac{1}{r}\frac{d}{dr}(r\bar{v})h \qquad \text{(E.4.9.4b)}$$

so that

$$\dot{h} \equiv \frac{dh}{dt} = -\frac{1}{r}\frac{d}{dr}(rh\bar{v}) \qquad \text{(E.4.9.5a)}$$

where

$$\bar{v} = \frac{2}{h}\int_0^{h/2} v(r, z; t)\,dz \qquad \text{(E.4.9.5b)}$$

Before integrating Eq. (E.4.9.2) across the gap, τ_{rz} has to be defined.

(i) Newtonian fluid: $\tau_{rz} \approx \mu \frac{\partial v}{\partial z}$ (E.4.9.6)

Thus, Eq. (E.4.9.2) reads:

$$\mu \frac{\partial^2 v}{\partial z^2} = \frac{dp}{dr} \qquad \text{(E.4.9.7)}$$

subject to $v(z = \pm h/2) = 0$. Integration yields:

$$v = \frac{h^2}{2\mu}\left(\frac{dp}{dr}\right)\left[\left(\frac{z}{h}\right)^2 - \frac{1}{4}\right] \qquad \text{(E.4.9.8)}$$

Evaluating the average radial velocity as needed in Eq. (E.4.9.3), we have:

$$h\bar{v} = \frac{1}{\mu}\left(\frac{dp}{dr}\right)\int_0^{h/2}\left(z^2 - \frac{h^2}{4}\right)dz := \frac{-h^3}{12\mu}\frac{dp}{dr} \quad \text{(E.4.9.9)}$$

so that

$$\frac{1}{r}\frac{d}{dr}\left(r\frac{dp}{dr}\right) = \frac{12\mu}{h^3}\frac{dh}{dt} \quad \text{(E.4.9.10)}$$

subject to $p(r = R) = 0$ and $dp/dr = 0$. Thus, double integration yields:

$$p = \frac{3\mu R^2}{h^3}\dot{h}\left[\left(\frac{r}{R}\right)^2 - 1\right] \quad \text{(E.4.9.11)}$$

Now the load can be evaluated as:

$$L = \int_0^R p dA = 2\pi\int_0^R prdr := \frac{3\pi}{2}\mu R^4\ \dot{h}\big/h^3 \quad \text{(E.4.9.12)}$$

(ii) Power-law fluid: $$\tau_{rz} = K\left(\frac{\partial u}{\partial z}\right)^{n-1} \quad \text{(E.4.9.13)}$$

Starting over with Eq. (E.4.9.2) for this basic non-Newtonian fluid, the load is:

$$L = \frac{2\pi(2 + 1/n)^n}{n+3}K\,\mathrm{sgn}(\dot{h})|\dot{h}|^n\ h^{-(2n+1)}\ R^{n+3} \quad \text{(E.4.9.14)}$$

(iii) Linear viscoelastic fluid: $$\tau_{rz} = \int_{-\infty}^{t} G(t - t')\frac{\partial u(t')}{\partial z}dt' \quad \text{(E.4.9.15)}$$

leads to:

$$L = \frac{3\pi R^4}{2}\int_0^t Q(t - t')(\dot{h}/h^3)dt' \quad \text{(E.4.9.16)}$$

Graphs (for Case(i)):

(a) Profile of radial velocity component

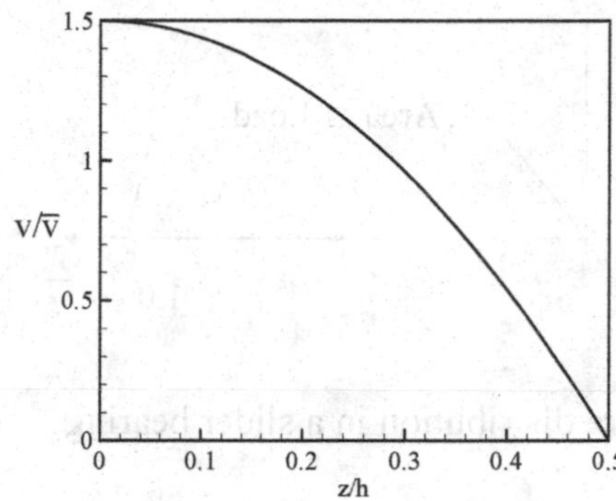

(b) Radial pressure distributions

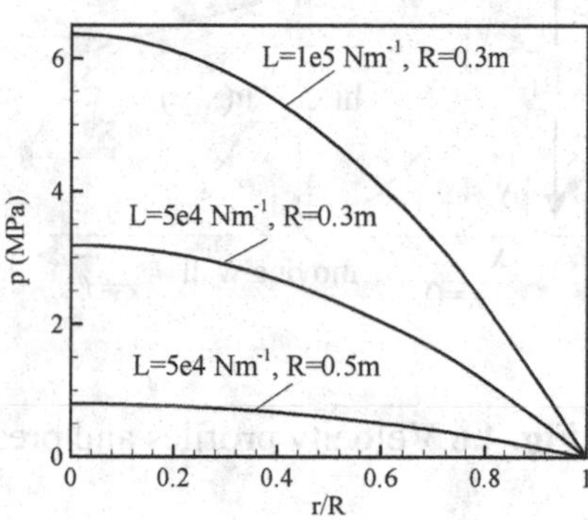

Comments:

As expected, the radial squeeze-film velocity v(z) is parabolic due to the approaching disks creating at time *t* with $\dot{h}_{disk}$ a pressure gradient dp/dr which is driving the fluid outwards. The pressure distribution (see Eq. (E.4.9.11)) clearly depends on the disk-approach speed $\dot{h}_{disk} = \dot{h}(L)$ and the disk radius. In turn, the load determines the disk speed (see Eq. (E.4.9.12)).

Hydrodynamic Lubrication As pointed out a few times, for fully developed internal flows (e.g., Couette and Poiseuille) the pressure gradient is constant and hence parabolic velocity profiles are ensured. When one stationary wall is slanted (or curved) and the other flat surface drags viscous fluid into the narrowing passage, the lubrication approximation holds. Specifically, in such planar bearings, the local pressure builds up, capable of sustaining a significant load exerted on the fixed wall, i.e., $L = p_{mean} \cdot A_{surface}$ (see Fig. 4.6).

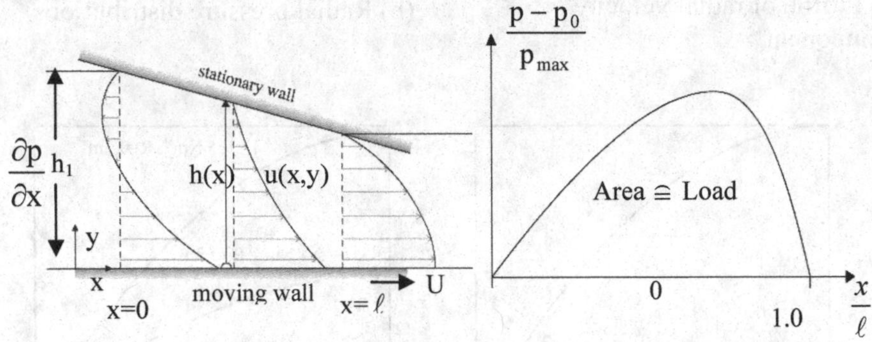

Fig. 4.6 Velocity profiles and pressure distribution in a slider bearing

Very similar to the slot-flow analysis, i.e., assuming a small, only slightly varying gap h(x), it can be shown that the viscous forces are dominant. Clearly,

$$v \ll u, \quad \partial^2 u/\partial x^2 \ll \partial^2 u/\partial y^2 \quad \text{and} \quad \partial p/\partial y \approx 0 \qquad (4.41a\text{–}c)$$

Specifically,

$$\frac{F_{inertia}}{F_{viscous}} = \frac{u\,\partial u/\partial x}{\nu\,\partial^2 u/\partial y^2} \sim \frac{U^2/\ell}{\nu\,U/h^2} = Re_\ell \left(\frac{h}{\ell}\right)^2 < 1 \qquad (4.41e\text{–}g)$$

Although $Re_\ell \equiv U\ell/\nu$ may be relatively high in lubrication, $Re_\ell (h/\ell)^2 = O(0.1)$ as shown. As a result, Stokes' equation (see Sect. 2.4) reduces to the Reynolds lubrication equation.

$$\nabla \cdot \vec{\mathbf{v}} = 0 \quad \text{and} \quad \nabla p = \mu\,\nabla^2\,\vec{\mathbf{v}} + \vec{\mathbf{g}} \qquad (4.42a, b)$$

For the given 2-D system (Fig. 4.6) these equations are simplified to (see Eq. (4.37)):

$$\frac{dp}{dx} = \mu \frac{d^2u}{dy^2} \tag{4.43a}$$

subject to:

$$u(y=0) = U; \quad u[y = h(x)] = 0 \tag{4.43b, c}$$

and

$$p(x = 0) = p, \text{ and } p(x = \ell) = p_0 \tag{4.43d, e}$$

Integration with respect to y yields:

$$u(x, y) = U\left(1 - \frac{y}{h(x)}\right) - \frac{[h(x)]^2}{2\mu}\left(\frac{dp}{dx}\right)\frac{y}{h(x)}\left(1 - \frac{y}{h(x)}\right) \tag{4.44}$$

where the pressure gradient is obtained from the (given) flow rate:

$$Q = \int_0^{h(x)} u(x, y)\, dy = -\frac{[h(x)]^3}{12\mu}\left(\frac{dp}{dx}\right) + \frac{hU}{2} \tag{4.45}$$

so that

$$p(x) = p_0 + 6\mu U \int_0^x \frac{dx}{[h(x)]^2} - 12\mu Q \int_0^x \frac{dx}{[h(x]^3} \tag{4.46}$$

Here h(x) is a design function where the stationary surface is usually slightly curved. Clearly, the last term in Eq. (4.46) indicates that (see also Eq. (E.4.9.11)):

$$p(x) \sim h^{-3}$$

which implies that very small gap sizes may produce very high pressure levels and hence load-carrying capacities.

4.3.2 The Reynolds Lubrication Equation

In case the upper slanted (or curved wall) moves downwards with a vertical velocity (see Example 4.9):

$$v(t) = -\frac{\partial h}{\partial t} \cong -V \tag{4.47}$$

Now, h = h(x, t) and the squeezing effect will produce also flow in the z-direction. Hence, with the following scales (see Fig. 4.6a):

$$x, z \sim L, \;\; y \sim h_0, \;\; t \sim L/U, \; u, w \sim U, \;\; v \sim \frac{h_0}{L} U, \;\; p \sim \mu \frac{uL}{h_0}$$

we have to solve:

$$0 = -\frac{\partial p}{\partial x} + \mu \frac{\partial^2 u}{\partial y^2} \tag{4.48}$$

and

$$0 = -\frac{\partial p}{\partial z} + \mu \frac{\partial^2 u}{\partial y^2} \tag{4.49}$$

Subject to the no-slip conditions at y = 0 and y = h(x, t). The results are (see HWA in Sect. 4.5):

$$u(x, y; t) = \frac{h^2}{2\mu}\left(\frac{\partial p}{\partial x}\right)\left[\left(\frac{y}{h}\right)^2 - \left(\frac{y}{h}\right)\right] + \left(1 - \frac{y}{h}\right) U \tag{4.50}$$

and

$$w(y, z; t) = \frac{h^2}{2\mu}\left(\frac{\partial p}{\partial z}\right)\left[\left(\frac{y}{h}\right)^2 - \left(\frac{y}{h}\right)\right] \tag{4.51}$$

This time we employ continuity on a differential bass, i.e.,

$$-\frac{\partial v}{\partial y}=\frac{\partial u}{\partial x}+\frac{\partial w}{\partial z} \tag{4.52}$$

Integrating Eq. (4.52), recalling that $V=\partial h/\partial t=\int_0^h \frac{\partial v}{\partial y}dy$ and applying Leibniz Rule (see App. A), we obtain:

$$\frac{\partial}{\partial x}\int_0^h u dy-\frac{\partial h}{\partial x}u(y=h)+\frac{\partial}{\partial z}\int_0^h w dy=-\int_0^h \frac{\partial v}{\partial y}dy=-V=-\frac{\partial h}{\partial t} \tag{4.53}$$

Inserting the profiles for u and w (see Eqs. (4.50) and (4.51)) into Eq. (4.53) yields the Reynolds equation for "lubrication" in a non-uniform, very narrow channel at small to moderate Reynolds numbers:

$$\frac{1}{\mu}\left[\frac{\partial}{\partial x}\left(h^3\frac{\partial p}{\partial x}\right)+\frac{\partial}{\partial z}\left(h^3\frac{\partial p}{\partial z}\right)\right]=6U\frac{\partial h}{\partial x}+12\frac{\partial h}{\partial t} \tag{4.54}$$

Example 4.10: Simple Slider Bearing with a Given Volumetric Flow Rate

Sketch	*Assumptions*	*Concepts*
fixed; p_1; $p_2=p_0$; $Q=\text{¢}$; h_1; $h(x)$; h_2; y; x; U; x=0; x=L	• Steady uni-directional laminar flow in a very small gap • Negligible inertia terms • No end effects • Constant fluid properties	• Reynolds lubrication equation • Conservation of mass • $\lvert h_2-h_1\rvert \ll L$

Solution: Again, the basic problem solution is a combination of Couette flow and Poiseuille flow (see Eq. (4.44)):

$$u(x, y) = U\left(1 - \frac{y}{h}\right) - \frac{h^2}{2\mu}\left(\frac{dp}{dx}\right)\frac{y}{h}\left(1 - \frac{y}{h}\right) \qquad \text{(E.4.10.1)}$$

where here

$$h \equiv h(x) = h_1 + \frac{h_2 - h_1}{L}\, x \qquad \text{(E.4.10.2)}$$

The flow rate Q is:

$$Q = \int_0^{h(x)} u(x, y)\, dy = \frac{h_1 h_2}{h_1 + h_2}\, U = \not\subset \qquad \text{(E.4.10.3)}$$

And from Eq. (4.45)

$$\frac{dp}{dx} = -\frac{12\mu}{h^3}\left(Q - \frac{hU}{2}\right) \qquad \text{(E.4.10.4)}$$

The maximum pressure occurs at an x-location where dp/dx = 0, i.e.,

$$h_{opt} = \frac{2Q}{U} = \frac{2h_1 h_2}{h_1 + h_2} = h_1 + \frac{h_2 - h_1}{L}\, x_{opt} \qquad \text{(E.4.10.5a)}$$

Thus,

$$x_{opt} = \frac{h_1 L}{h_1 + h_2} \qquad \text{(E.4.10.5b)}$$

The sustaining force per unit width is:

$$F = \int_0^L (p - p_0)\, dx \qquad \text{(E.4.10.6)}$$

where with Eq. (E.4.10.4) we obtain:

$$p(x) = p_0 + 6\mu U \int_0^x \frac{dx}{h^2} - 12\mu Q \int_0^x \frac{dx}{h^3} \qquad \text{(E.4.10.7)}$$

Tasks:

- Solve (E.4.10.7) and (E.4.10.6) with $h(x) = h_1 + \frac{h_2 - h_1}{L} x$
- Plot p(x) and indicate x_{opt} and p_{max} for a reasonable geometry (see Graph a)
- Plot Eq. (E.4.10.1) profiles (see Graph b)

Graphs:

(a) p(x)

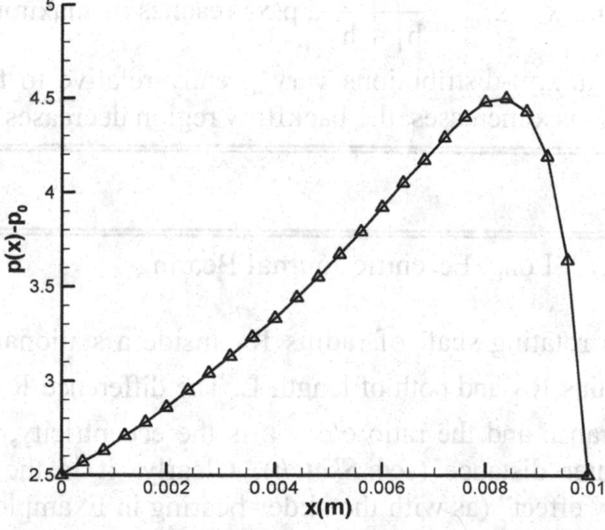

(b) u(x,y) in different location

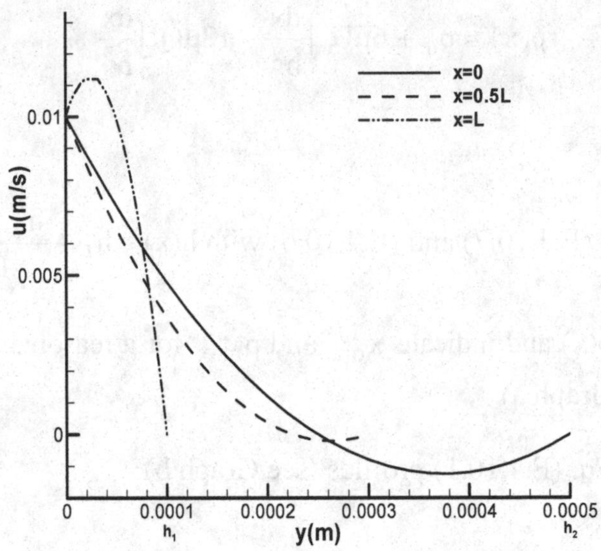

Comments:

(a) When $x = x_{opt} = \dfrac{h_1 L}{h_1 + h_2}$, p(x) reaches its maximum p_{max}.

(b) The u(x,y)-distributions vary greatly relative to the x position; as x increases, the backflow region decreases.

Example 4.11: Long, Eccentric Journal Bearing

Consider a rotating shaft of radius R_1 inside a stationary housing of inner radius R_2 and both of length L. The difference $R_2 - R_1 \equiv c$ is the clearance and the ratio $e/c \equiv \varepsilon$ is the eccentricity, where e is the off-center distance (see *Sketch*). Clearly, it is the "confined wedge-flow effect" (as with the slider bearing in Example 4.10) that generates the high, load-bearing pressure distribution. Again, the gap $h << 1$, where $h \approx c(1 + \varepsilon \cos \theta)$ which implies $h(\theta = 0) = h_{max} = c(1 + \varepsilon) = c + e$ and $h(\theta = \pi) = h_{min} = c - e$.

Derive an expression for the internal pressure in the form $\hat{p}(\theta)$, where $\hat{p} = \dfrac{p - p_0}{\mu\omega_0 (R_1/c)^2}$

Sketch	*Assumptions*	*Concepts*
θ=0, R_2, $0 \le z \le L$, h(θ), θ, R_1, e, ω_0, θ=π	• Steady laminar flow • Slow shaft rotation • $h/R_1 << 1$; $R_1/L << 1$ • $h = h(e, \theta)$ as given • Coordinate $x = R\,\theta$ • Constant properties • No end effects	• Reynolds lubrication equation (4.54) • Reynolds pressure conditions at $\theta = \theta_{cavity}$ $p = 0$ and $\dfrac{dp}{d\theta} = 0$

Solution:

Nondimensionalizations

$$\theta = \frac{x}{R_1},\ \hat{h} = \frac{h}{c},\ \hat{z} = \frac{z}{L/2},\ \text{and}\ \hat{p} = (p - p_0)/[\mu\omega_0 (R_1/c)^2]$$

Employing Eq. (4.54) with $(R_1/L)^2 \approx 0$ we have:

$$\frac{\partial}{\partial\theta}\left(\hat{h}^3 \frac{\partial \hat{p}}{\partial\theta}\right) = 6\,\frac{\partial \hat{h}}{\partial\theta} \qquad \text{(E.4.11.1)}$$

Integrating twice yields:

$$\hat{p}(\theta) = 6\int_0^{\theta} \frac{\hat{h} - C}{\hat{h}^3}\, d\theta \qquad \text{(E.4.11.2a)}$$

where the integration constant depends on the pressure boundary conditions assumed. For example, setting p = 0 and dp/dθ = 0 when a vapor cavity starts to form at θ = θcav, we get (Szeri 1980):

$$C = \hat{h}(\theta_{cav}) \tag{E.4.11.2b}$$

Typical θcav-values are 249.2° when ε = 0.1 and 219.7 when ε = 0.5. For the latter case, $\hat{p}(\theta)$ is graphed (see also Panton 2005).

Graph:

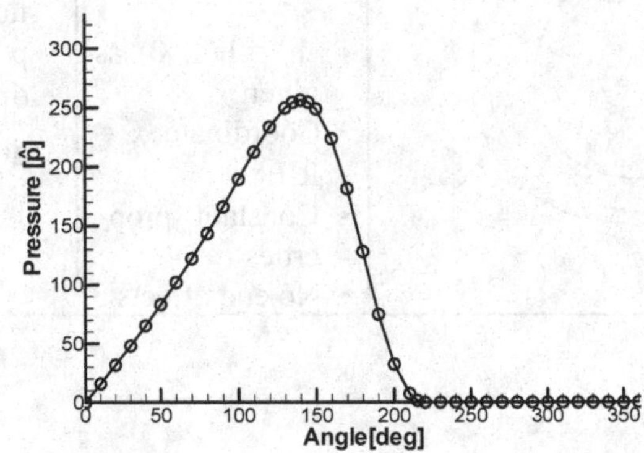

Comments:

- The pressure distribution in the journal bearing follows qualitatively that of planar bearings (see Example 4.10).
- Again, the magnitude of h and the pressure boundary conditions greatly determine $\hat{p}(\theta)$.

4.4 Compartmental Modeling

Most transport phenomena involving fluid–structure interaction, convective-diffusive mass transfer or modeling micro-scale devices coupled to macro-scale systems are rather complex. Hence, such problems have been traditionally analyzed with a lumped-parameter (or "black-box") approach, relying on the RTT for mass, momentum and energy transfer (see Sect. 2.2) or on coupled first-order rate equations for species-mass, i.e., material convection.

In general, a compartment is a "well-mixed box," e.g., a homogeneous isotropic region, an entity or device of constant volume, which has uniform inlet and outlet streams of different velocities, enthalpies and/or concentrations. Of interest are integral (or global) quantities, such as flow rates and exerted forces, or the time-rate-of change of species concentrations inside the compartment as well as species material transport, deposition and conversion. Examples include:

(i) Flow problems which can be solved with the mass/momentum Reynolds Transport Theorem (RTT), e.g., fluid-mass conservation in a pipe network or drag forces on submerged bodies (see Examples 2.2 and 2.6).
(ii) Energy/work exchange and heat transfer for closed and open systems employing the first law of thermodynamics via a "lumped-parameter" approach, or the energy RTT (see Examples 4.5 and 4.6).
(iii) Species-mass transfer in complex systems (e.g., the human body, part of the environment or a petroleum refinery) where the system is compartmentalized and transport phenomena described by *mass balances*, typically a coupled set of convective first-order rate equations.

While examples for Groups (i) and (ii) have been provided, compartmental analysis of complex species-mass transfer systems requires further discussion. Clearly, Group (iii) applications have been traditionally the domain of biomedical and chemical engineers, rather than mechanical engineers.

To begin with, a representative system is decomposed into a network of perfectly mixed, constant-volume compartments which are connected by ducts with negligible impact, i.e., no volume, dispersion and losses in the conduits. Figure 4.7a illustrates a biomedical example where the entire interactions in the human body between the flowing blood and the arterial walls, plus organs, labeled "tissue", are represented in a single unit comprising two well-mixed regions. The "injected tracer" may be a toxin or a therapeutic drug migrating into organ tissue which is governed by the overall species transfer parameter K. More generally, Fig. 4.7b and c show the schematics of compartments in parallel and in series, reminiscent of pipe networks.

Because of their stringent assumptions, compartment models have their limitations. In many applications the material in the compartment is not homogeneous. Furthermore, not all the inflowing/outflowing material plus material conversion due to chemical reaction can be fully accounted for. Examples include varying species concentrations, such as vertical nutrient concentrations in a lake, or an incomplete water balance of an estuary.

(a) Single-Compartment Model Application: Blood–Tissue Interaction

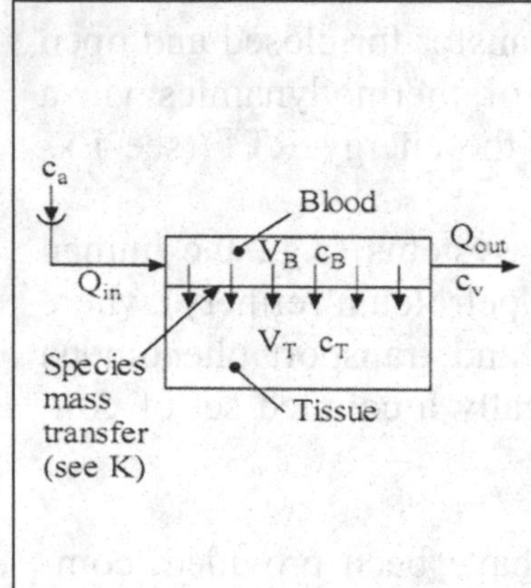

Notation

- $c_a \hat{=}$ artery injected tracer concentration
- $Q_{in} \hat{=}$ arterial blood stream
- $Q_{out} \hat{=}$ venous blood stream
- $\forall = V_B + V_T \hat{=}$ total compartment volume
- $c_v \hat{=}$ venous tracer concentration
- $K \hat{=}$ diffusion-convection coefficient

(b) Two Compartments in Parallel (c) Compartments in Series

$(Q = Q_1 + Q_2; R_{total}^{-1} = R_1^{-1} + R_2^{-1})$ $Q \hat{=}$ flow rate, $R \hat{=}$ flow resistance

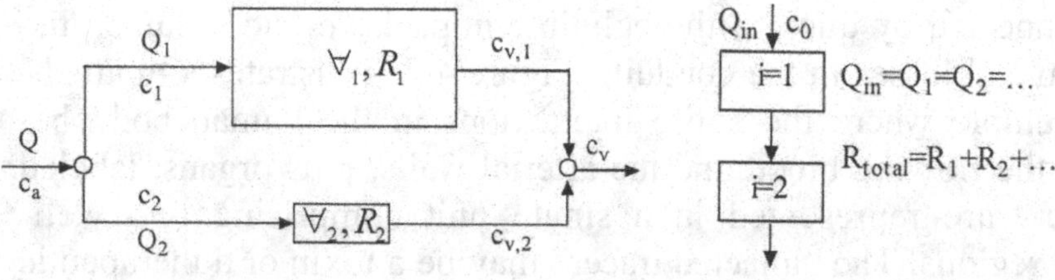

Fig. 4.7 Single- and multi-compartment models

4.4.1 Compartments in Parallel

In case a complex device or domain has to be subdivided, two compartments in parallel may form an appropriate model (see Fig. 4.7b). A fluid mass balance yields:

$$Q = Q_1 + Q_2 \tag{4.55}$$

while a species mass balance dictates:

$$QC_v = Q_1 C_1 + Q_2 C_2 \tag{4.56}$$

The transient 1-D mass transfer equation (see Eq. (2.55))

$$\frac{\partial c}{\partial t} + u\frac{\partial c}{\partial x} = D\frac{\partial^2 c}{\partial y^2} - kc^n \tag{4.57}$$

can be reduced for each compartment by neglecting diffusion and chemical reaction. Hence,

$$C_v(t) = \frac{Q_1}{Q_2} C_1 e^{-k_1 t} + \frac{Q_2}{Q} C_2 e^{-k_2 t} \tag{4.58}$$

where $k_i = (\kappa Q / \forall)_i$, with i = 1,2. Measuring Cv(t), Eq. (4.58) can be employed to find k_i or Qi/Q, where κ is the partition coefficient.

4.4.2 Compartments in Series

In general, the material conservation law for compartments in series (see Fig. 4.7c) can be expressed as:

$$\frac{dc_i}{dt} = \sum_{j=1}^{N} (q_{ji} - q_{ij}) + S_i(t) \tag{4.59}$$

where c_i is the amount of material per unit volume accumulating in compartment i, q_{ij} is the rate of material being transferred from compartment i to compartment j, and $S_i(t)$ is the rate of material injected from the system's exterior (or generated due to chemical reaction in side compartment i).

A constitutive relationship is necessary to express the transfer rates q_{ij} with the principal unknown c_i, i.e.,

$$q_{ij} = f(c_i) \tag{4.60a}$$

where a Taylor series expansion around point *c* yields:

$$f(c) = a_0 + a_1 c + a_2 c^2 + \ldots \tag{4.60b}$$

Recalling that *c* is positive and much less than unity, f(c) can be linearized to:

$$f(c) \approx a_1 c \tag{4.60c}$$

and hence

$$q_{ij} = k_{ij} c_i \tag{4.60d}$$

where k_{ij} are the rate constants.

Thus, for linear compartmental models with constant coefficients, Eq. (4.59) can be written for each compartment *i* as:

$$\frac{dc_i}{dt} = \sum_{j=1}^{N} (k_{ij}c_j - k_{ij}c_i) + S_i(t) \tag{4.61}$$

Equation (4.61) has a homogeneous solution of the form:

$$c_i(t) = \sum_{k=1}^{n} A_k e^{-a_k t} \tag{4.62}$$

where the coefficients A_k are proportional to k_{ij}, $c_i(t=0)$, and $S_i(t)$, while the eigen-frequencies $\alpha_k \sim k_{ij}$. The particular solution for $c_i(t)$ depends on the information for k_{ij}, $c_i(t=0)$, and $S_i(t)$. Linearity of Eq. (4.61) allows one to compute $c_i(t)$ in steps and then sum up all the (independent) solutions (see Example 4.13 as well as Hoffman 2001; among others).

Example 4.12: Simple Stability Analysis for a Single Compartment with Fluid Volume ∀(t) and Volume-Dependent Inflow and Outflow

Of interest is the equilibrium point, i.e., $\forall(t) = \forall_0 = \not\subset$, and an associated analysis of system stability. Applications include the human-body fluid and a hydro-electric reservoir.

Sketch	*Assumptions*	*Concept*
Q_{in} → [$\frac{d\forall}{dt}$] → Q_{out}	• Well-mixed compartment and homogenous fluid • Constant properties • Uniform inlet/ outlet streams	• First-order rate equation $d\forall/dt = Q_{in} - Q_{out}$

Solution:

- Fluid mass balance in terms of volume, i.e., with $m = \rho\forall$; $\rho =$ const.; and $\dot{\forall} = \forall / t = Q$:

$$\frac{d\forall}{dt} = Q_{in} - Q_{out} \tag{E.4.12.1}$$

where $Q = Q(\forall)$.

- Equilibrium solution:

$$\frac{d\forall}{dt} = 0, \quad \text{i.e.,} \quad Q_{in} = Q_{out} \qquad \text{(E.4.12.2a,b)}$$

Equations (E.4.12.2a, b) indicate that the amount of fluid in the compartment doesn't change, i.e., $\forall(t = t_0) = \forall_0$ and hence $Q_{in}(t_0) = Q_{out}(t_0)$.

Graph:

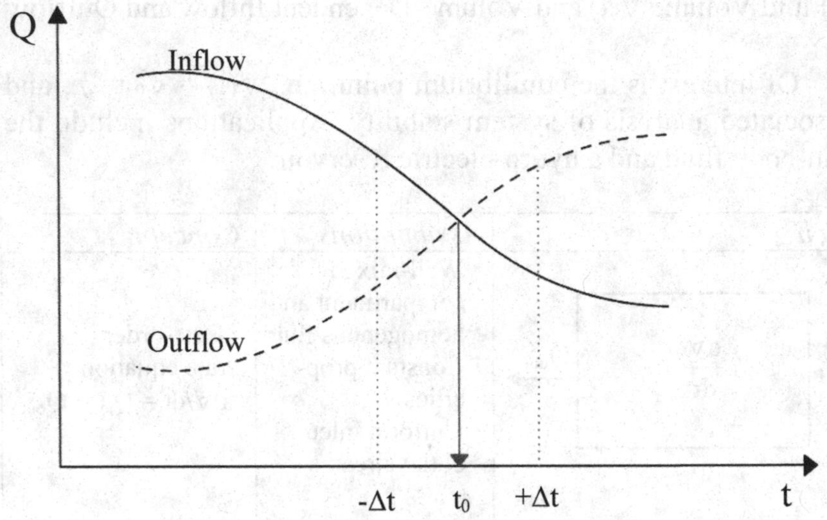

Comments:

- Using the plotted inflow/outflow functions as quantitative examples, small deviations from the equilibrium point ($\forall_0$) may lead to the following (see Graph):

Case (i): $\forall(t) = \forall_0 + \Delta\forall$, so that with $Q_{out}(t_0 + \Delta t) > Q_{out}(t_0)$ while $Q_{in}(t_0 + \Delta t) < Q_{in}(t_0)$. This implies that $Q_{in} - Q_{out} < 0$ and hence $\forall(t)$ decreases, i.e., $\forall(t)$ retreats back to the equilibrium volume $\forall_0$.

Case (ii): Perturbation of $\forall(t)$ to $\forall_0 - \Delta\forall$ yields $Q_{in}(t_0 - \Delta t) > Q_{out}(t_0 - \Delta t)$ and $d\forall/dt > 0$, i.e., the fluid volume increases towards $\forall_0$ again. In summary, the solution is *stable*.

- Different functions for Q_{in} and Q_{out} may prevent the solution to revert back to equilibrium, e.g., the compartment totally drains or overflows, and the system is unstable.

Example 4.13: Transient Species Concentrations in Open System with Linear Chemical Reaction

Consider a sudden pollutant input into a body of water with inflow and outflow streams, or a continuous flow chemical reactor. Set up the compartmental model and solve the system for two species input modes: impulse and step functions at time t = 0.

Sketch	*Assumptions*	*Concept*
IN Q_{in}, c_{in}; $\forall = ¢$; $c = c(t)$; $k = ¢$; $Q_{out} = Q_{in}$, $c_{out} = c(t)$ OUT	• Well-mixed compartment with $\forall$ = constant • Reaction rate is constant • Constant flow rate	• Species mass $m = \rho\forall$; $\forall$ = constant • Rate equation (4.59) with $\kappa = 1$

Solution: Equation (4.59) can be rewritten in terms of species mass, $m = \rho\forall$, as:

$$\frac{dm}{dt} = Q_{in}(t)\, c_{in}(t) - Q_{out}(t)\, c_{out}(t) \pm k(t)\, c(t)\, \forall(t) \quad \text{(E.4.13.1)}$$

In light of the assumptions (see *Sketch*), Eq. (E.4.13.1) reduces to:

$$\frac{dc}{dt} = \frac{Q}{\forall} c_{in}(t) - \frac{Q}{\forall} c(t) - k\, c(t) \quad \text{(E.4.13.2)}$$

Where $\forall/Q \equiv \tau$ is the "residence time", i.e., the theoretical duration that a fluid element (or particle) resides in the compartment.

Species Input ① (Impulse): At $t = 0$ a finite amount of, say, a pollutant is dumped into a lake (compartment), i.e., the lake water starts with an initial contaminant concentration $c_0 > 0$. No species enters the lake thereafter, i.e., $c_{in}(t) = 0$ because $\Delta t_{input} \approx 0$.

Species Input ② (Step Function): At $t = 0$ suddenly a fixed amount of a pollutant is discharged into the lake and that species input, $c_0 > 0$ and $c_{in}(t) = c_{in} = c_0$, stays constant.

The solutions to Eq. (E.4.13.2) for the two different input conditions are of the form of Eq. (4.62). Specifically, for Case ① we have with $c_0 > 0$ and $c_{in}(t) = 0$, while chemical reaction can be neglected, i.e., $k = 0$:

$$c(t) = c_0 \, e^{-t/\tau} \qquad \text{(E.4.13.3)}$$

And for Case ② with $c_{in}(t) = c_0$ = constant (see Polyaminal and Zaitsev 1995):

$$c(t) = \frac{c_0}{1 + k\tau}\left[1 - \exp(-t(k + \tau^{-1}))\right] \qquad \text{(E.4.13.4)}$$

Graph:

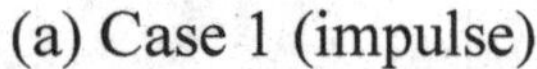
(a) Case 1 (impulse)

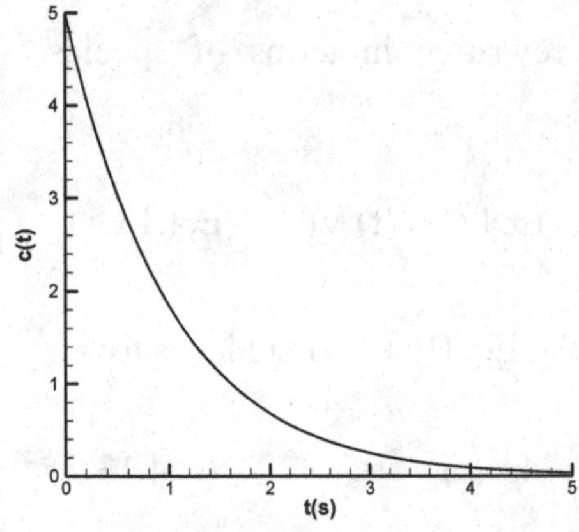

(b) Case 2 (step function)

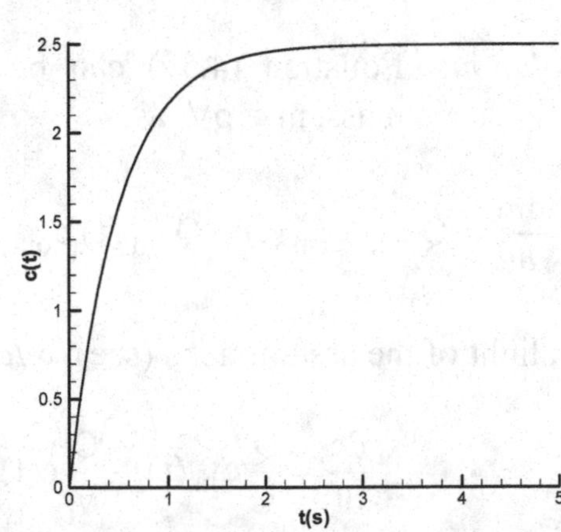

Comments:

- Clearly, with an initial amount of pollutant residing in the lake, i.e., $c(t=0)=c_0=5\,[M/L^3]$ in Case 1, and no additional input, the contaminant is (exponentially) washed out via Q_{out}. Note the pollutant residence time $\tau \sim Q_{out}^{-1}$.
- In Case 2, due to the constant species input (i.e., $c_{in}(t)=c_0$), an equilibrium concentration is reached due to the interplay of influx, efflux, and uptake.

Additional case studies using the compartmental transport approach may be found in UG heat/mass transfer textbooks as well as in Middleman (1972), Cooney (1976), Bird et al. (2002), Truskey et al. (2004), Kleinstreuer (2006) or Chandran et al. (2007), among other texts.

4.5 Homework Assignments

Solutions to homework problems done individually or in, say, 3-person groups should help to further illustrate fluid dynamics concepts as well as approaches to problem solving, and in conjunction with App. A, sharpen the reader's math skills (see Fig. 2.1). Unfortunately, there is no substantial correlation between good HSA results and fine test performances, just *vice versa.* Table 1.1 summarizes three suggestions for students to achieve a good grade in fluid dynamics – for that matter in any engineering subject. The key word is "independence", i.e., after studying the text and equipped with an equation sheet (see App. A), the student should be able: (i) to satisfactory answer all concept questions and (ii) to solve correctly all basic fluid dynamics problems.

The "Insight" questions emerged directly out of the Chap. 4 text, while some "Problems" were taken from lecture notes in modified, i.e., enhanced, form when using White (2006), Cimbala & Cengel (2008), and Incropera et al. (2007). Additional examples, concept questions and problems may be found in good UG fluid mechanics and heat transfer texts, or on the Web (see websites of MIT, Stanford, Cornell, Penn State, UM, etc.).

4.5.1 Text-Embedded Insight Questions and Problems

4.5.1 Derivations:

(a) Is there a theoretical back-up for the empirical correlations (Eq. 4.2a, b)?

(b) Derive Eq. (4.3) and list the advantages of working with the hydraulic diameter?

(c) What is the rational for Eq. (4.5d) and why is h_L so important?

(d) Show that $\alpha_{laminar} = 2.0$ for Eq. (4.7a).

(e) Derive Eqs. (4.12 a, b).

4.5.2 Why is $Nu|_{q_w=\not{c}} > Nu_D|_{T_w=\not{c}}$ (see Eq. (4.18b) vs. Eq. (4.18c))?

4.5.3 Discuss the pros and cons of turbulence in: (a) pipe flow and (b) tanks.

4.5.4 Explain Eq. (4.20c).

4.5.5 Turbulent pipe velocity profiles: (a) Plot Eq. (4.21a) for n=7 and 10, and discuss the pros and cons of this $\bar{u}(r)$ – profile correlation. (b) Derive Eq. (4.21b).

4.5.6 Develop an equation for: (a) the reactive force in slot flow (see Fig. 4.5); and (b) the load in squeeze-film lubrication (see Example 4.9).

4.5.7 Why is there possibly backflow in a slider bearing (see Fig. 4.6)?

4.5.8 Plot typical pressure distributions p(x) for hydrodynamic lubrication (see Eq. (4.46)) and resulting load capacities.

4.5.9 Derive Eq. (4.50) and discuss Eq. (4.54) with examples.

4.5.10 Derive Eq. (E.4.11.2a) and compute the load-carrying capacity of a typical journal bearing.

4.5.11 What is the difference between compartmental modeling (see Sect. 4.4) and the "lumped parameter" approach?

4.5.12 Justify the underlying assumptions for Eq. (4.58) and interpret / explain the partition coefficient $\kappa = k\forall / Q$.

4.5.13 Revisit Example 4.12 and run specific case studies, i.e., with realistic Q(t) functions (see Graph).

4.5.14 Determine the equilibrium conditions for Case 2 in Example 4.13.

4.5.2 Problems

4.5.15 Consider horizontal tubular flow ($D_1 \equiv D_{tube} = 5cm$, $\rho = 10^3$ kg/m^3) for which a Venturi meter ($D_2 \equiv D_{throat} = 3cm$, and $K = 0.98$) with a differential pressure gage measures $\Delta p = p_1 - p_2$

(a) Develop an equation for the mass flow rate $\dot{m}$ and plot $\dot{m}(\Delta p)$ for $1 \le \Delta p \le 10kPa$.

(b) Compute the average tube velocity for $\Delta p = 5kPa$.

4.5.16 A 50 hp air-compressor sucks in ambient air ($Q = 0.3m^3/s$, $p = 100kPa$, $\rho = 1.15kg/m^3$, $\mu = 1.8 \times 10^{-5} \frac{kg}{m \cdot s}$) through a long pipe ($D = 20cm$, $L = 20m$, $\varepsilon = 0.15mm$). What percentage of the compressor power is lost due to the (frictional) pipe resistance?

4.5.17 Consider two parallel-connected pipes, both of diameter D and of the same material but of lengths L and χL, where $1 < \chi < 4$. Find and plot the flow rate ratio Q_1 / Q_2 as a function of χ.

4.5.18 A typical industrial process is the discharge of a highly viscous liquid from a tank [D, $h(t = 0) = H$] via a horizontal long small pipe (d, L). Assuming laminar flow and negligible losses (e.g., pipe entrance, etc.), derive an ODE for h(t) which is the fluid depth. Solve for $t = t(h; L, d; D, H)$ and graph h(t).

4.5.19 Consider a basic slider-block as shown:

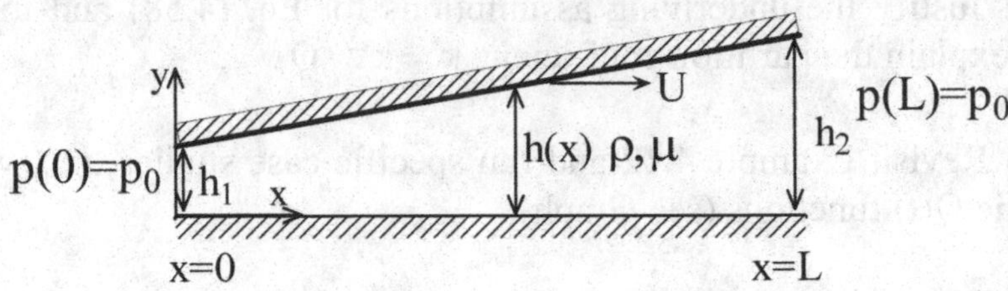

In order to apply Reynolds Lubrication Theory, we recall that $h_2 << L$ and $U = \not\subset$ so that due to the "wedge effect" a high-pressure in the lubrication layer generates a large lift, i.e., load-carrying capacity of the planar bearing.

(a) Find h(x), i.e., establish and reduce the Reynolds equation

$$\frac{\partial h}{\partial t} = \nabla \cdot [\frac{h^3}{12} \nabla p + \frac{h}{2} \vec{U}] \quad \text{to} \quad \frac{d}{dx}(h^3 \frac{dp}{dx}) = 6(1 - \frac{h_2}{h_1})$$

which is now in dimensionless form with U=1 (unity).

(b) Solve for $p(x) - p_0$ and dp/dx

(c) Plot $\frac{p(x) - p_0}{6(\frac{h_2 - h_1}{h_2 + h_1})}$, find the location for p_{max} and comment.

(d) Find the velocity field, $\vec{v} = \nabla p(\frac{y^2}{2\mu} - \frac{yh}{2\mu}) + \vec{U}(\frac{y}{h} - 1)$, and indicate how to compute the hydrodynamic forces, i.e.,

$$\vec{F} = -\int_A \hat{n}[p(x) - p_0] dA + \int_A \hat{n} \cdot \bar{\tau}_w dA$$

4.5.20 Consider a sphere of radius R approaching vertically under force F_z in a viscous fluid μ a horizontal wall. Note that the distance h(t,x), i.e., the sphere–wall gap function is small: $\varepsilon = \frac{h_0}{R} << 1$.

Thus, of interest is the film lubrication region. Specifically:

(a) Show that the Reynolds equation (see 4.5.19a) reduces to

$$12\mu\frac{\partial h}{\partial t}=\frac{1}{x}\frac{\partial}{\partial x}(xh^3\frac{\partial p}{\partial x})$$

and determine the gap function $h(x,t)=h_0(t)+f(x,R)$, where $h_0(t)$ is the smallest clearance.

The reference pressure at the film's outer boundaries, $\pm x_b$, is $p(x=\pm x_b)=p_b$

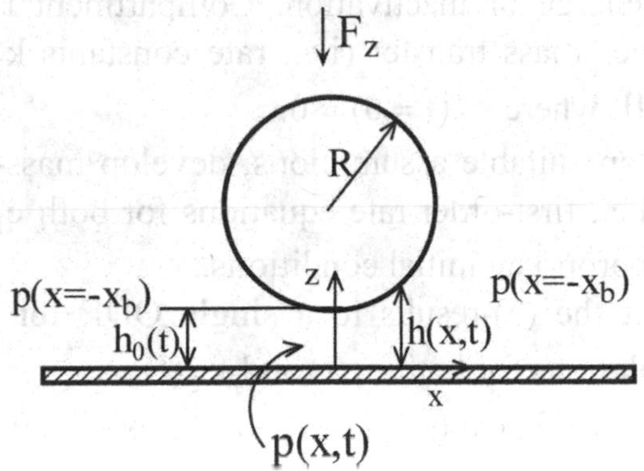

(b) Find $\partial p/\partial x=6\mu\frac{x}{h^3}\frac{dh}{dt}$ as well as an expression for $p(x)-p_b$.

(c) Assuming that only the film pressure contributes to the vertical force, develop an equation for

$$F_z=2\pi\int_0^{x_b}[p(x)-p_b]xdx$$

(d) With the approximation $\frac{x_b^2}{2h_0R}>>1$, show that

$F_z\approx 6\pi\mu R^2\frac{dh_0/dt}{h_0}$ and discuss various scenarios for

$$F_z\sim(h_0^{-1}\text{ and }\frac{dh}{dt}=const.,\text{ etc.})$$

(e) Given a constant F_z (e.g., the sphere's weight), find $h_0(t)$ subject to $h_0(t=0)=h_{initial}$, and plot dh_0/dt as well as $h_0(t)$ and comment.

(f) The (e)-result indicates that the sphere won't touch the wall within a finite time interval, which contradicts experimental observations. What happens?

4.5.21 Consider a two-compartment system. For example, drug injection (species mass M) into a main compartment I of volume $\forall_1$ and initial concentration $c_1(t=0) = c_0 = M/\forall_1$ and with drug clearance, i.e., discharge or inactivation. Compartment I intersects via two-way species mass transfer (i.e., rate constants k_1 and k_2) with compartment II, where $c_2(t=0) = 0$.

(a) Based on suitable assumptions, develop mass-balance equations, i.e., first-order rate equations for both $c_1(t)$ and $c_2(t)$, with appropriate initial conditions.

(b) Convert the (a)-results to a single ODE for $c_1(t)$ and then find solutions for both $c_1(t)$ and $c_2(t)$.

Note: $c_1(t) = c_0[\lambda_1 \exp(-\lambda_2 t) + (1-\lambda_1)\exp(-\lambda_3 t)]$, where λ_i 's are constants.

(c) Plot $c_1(t)$ for $0 \le t \le 100\,\text{h}$, $k_e = 3.8\times10^{-6}\,\text{s}^{-1}$, and constants $\lambda_1 = 0.4\text{s}^{-1}$, $\lambda_2 = 5\times10^{-5}\text{s}^{-1}$, and $\lambda_3 = 10^{-6}\text{s}^{-1}$.

(d) Determine the half-times of the drug $t_{1/2}$ which is defined as the time when $c = c_0/2$, representing the distribution phase (see λ_2) and the elimination phase (see λ_3).

(e) Calculate $\int_{t_0}^{t_e} c_1(t)dt$ and interpret the result.

Chapter 5

External Flow

5.1 Introduction

Internal *fully-developed* flows, such as Couette and Poiseuille, are completely dominated by viscous effects throughout the entire flow domain. In contrast, entrance flows and external flows feature highly viscous effects *confined* to rapidly growing "boundary layers" in the entrance region, or to thin shear layers along the solid surface. Clearly, the effect of no-slip wall velocity triggers an expanding *boundary layer* (Fig. 5.1). As demonstrated in Sect. 3.3.1 (Eq. (3.16a)), the boundary-layer thickness relates to the fluid's kinematic viscosity as (see also Eq. (5.8)):

$$\delta \sim \sqrt{\nu} \tag{5.1}$$

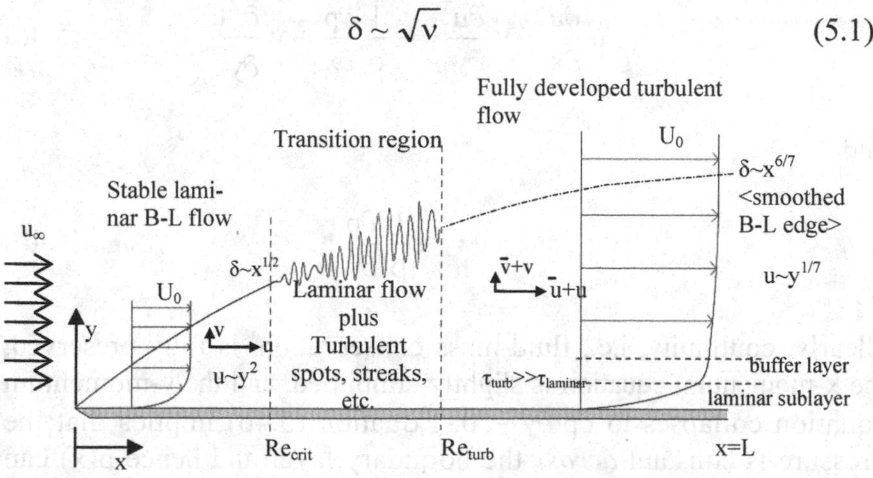

Fig. 5.1 Flat-plate laminar and turbulent boundary layers with velocity profiles

C. Kleinstreuer, *Modern Fluid Dynamics: Basic Theory and Selected Applications in Macro- and Micro- Fluidics*, Fluid Mechanics and Its Applications 87,
DOI 10.1007/978-1-4020-8670-0_5, © Springer Science+Business Media B.V. 2010

As the oncoming (free-stream) fluid particles are reaching the submerged body surface, say, a flat plate (see Fig. 5.1), they stick to the solid surface (see "no-slip condition" and Sect. 7.4). That affects the neighboring fluid layers and hence the frictional effect propagates outwards, i.e., normal to the solid surface, until at distance δ(x) the presence of the flat plate is not registered anymore. The basic hypothesis of Prandtl (1904) is that (see Fig. 5.1):

$$\frac{\delta}{L} \sim \frac{v}{u} << 1 \quad \text{and based on that} \quad \frac{\partial^2}{\partial x^2} << \frac{\partial^2}{\partial y^2} \qquad (5.2a, b)$$

which implies that axial momentum diffusion, $\nu\partial^2 u/\partial x^2$, is negligible. However, the convective term v $\partial u/\partial y$ has to be retained; because, although v is quite small, the normal gradient $\partial u/\partial y$ is large inside the boundary layer. Based on scale analysis and physical insight, the steady 2-D Navier–Stokes equations then reduce to (see Sect. 2.4 and HWA Sect. 5.5):

$$\frac{\partial u}{\partial x} + \frac{\partial v}{\partial y} = 0 \qquad (5.3)$$

$$u\frac{\partial u}{\partial x} + v\frac{\partial u}{\partial y} = -\frac{1}{\rho}\frac{\partial p}{\partial x} + \nu\frac{\partial^2 u}{\partial y^2} \qquad (5.4a)$$

and

$$0 = -\frac{1}{\rho}\frac{\partial p}{\partial y} \qquad (5.4b)$$

Clearly, continuity, i.e., fluid-mass conservation, is fully preserved, the x-momentum equation is slightly simplified, and the y-momentum equation collapses to $\partial p/\partial y = 0$. Equation (5.4b) implies that the pressure is constant *across* the boundary layer and hence p(x) can be determined with the Euler or Bernoulli equation just outside the boundary layer. Clearly, in reality there is no "edge"; but, for practical purposes we postulate that $\delta(x)$ has been reached when the ratio $u(y)/U_0$ differs less than 1% from unity there and all velocity gradients are basically zero (see Fig. 5.1).

5.2 Laminar and Turbulent Boundary-Layer Flows

While Prandtl's boundary-layer (B-L) theory will prevail as an educational tool to gain physical insight and explain special fluid flow phenomena, solutions to the boundary-layer equations outside aerodynamics have become largely superfluous. The reason is that nowadays commercial Navier–Stokes equation solvers, which even run on PCs, are being routinely employed to tackle simulation problems in academia and industry. In any case, the mother of all B-L flows is flow past a (razor-thin) horizontal flat plate of which only the upper half is usually shown (see Fig. 5.1).

5.2.1 Solution Methods for Flat-Plate Boundary-Layer Flows

Other than direct numerical solutions of Eqs. (5.3)–(5.4a, b), B-L flows are solved with the similarity analysis (for B-Ls with steady laminar self-similar velocity profiles) and the integral method (for both laminar and turbulent B-Ls).

Similarity Analysis for Laminar Boundary-Layer Flows One elegant solution of Eqs. (5.3)–(5.4a) relies on the *similarity* of the axial velocity profiles u(x, y) (see Kleinstreuer 1997; Sabersky et al. 1999; Panton 2005; among others). Employing the stream function approach, $u \equiv \partial\psi/\partial y$ and $v \equiv -\partial\psi/\partial x$, postulating a similarity variable $\eta = \eta(x, y)$ which combines both independent variables and setting $u/U_0 = f'(\eta)$, Eq. (5.3) is automatically fulfilled and the PDE (5.4) can be transformed into a third-order ODE for $f(\eta)$ as illustrated with a HWA (see Sect. 5.5). Specifically,

$$2f''' + ff'' = 0 \tag{5.5}$$

subject to

$$f'(0) = u(x, 0)/U_0 = 0\,,\ f'(\infty) = 1 \text{ and } f(0) = v(x, 0) = 0 \quad (5.6a\text{–}c)$$

where based on scaling:

$$\eta(x, y) = y\sqrt{\frac{U_0}{\nu x}} \quad \text{and} \quad \psi(x, y) = \sqrt{\nu\, U_0\, x}\, f(\eta) \qquad (5.7a, b)$$

The numerical (Runge–Kutta) solution of this system includes the following results:

$$\frac{\delta}{x} = \frac{5.0}{\sqrt{Re_x}} \quad \text{and} \quad C_f \equiv \frac{\tau_{wall}}{\rho/2\, U_0^2} = \frac{0.664}{\sqrt{Re_x}} \qquad (5.8, 5.9a, b)$$

where

$$Re_x = \frac{U_0\, x}{\nu} \quad \text{and} \quad \tau_w = \mu \left.\frac{\partial u}{\partial y}\right|_{y=0} \qquad (5.10, 5.11)$$

Equation (5.8) reveals that $\delta(x) \sim \sqrt{x}$ and that even for laminar B-L flow the Reynolds number has to be very large for hypothesis (5.2) to hold. Another interesting observation can be made when evaluating Eq. (5.4a) at the wall where $u = v = 0$, so that from Eq. (5.4a):

$$\left.\frac{\partial^2 u}{\partial y^2}\right|_{y=0} = \frac{1}{\mu}\frac{\partial p}{\partial x} \qquad (5.12)$$

Equation (5.12) indicates that the magnitude of the pressure gradient determines qualitatively the velocity profile near the wall. Clearly, for constant-pressure flow (Blasius 1908) $\partial^2 u/\partial y^2 = 0$ and hence the profile is linear at the wall. However, a negative (or favorable) pressure gradient causes a fuller profile because of the negative curvature, while a positive (or adverse) pressure gradient with positive curvature (i.e., $\partial^2 u/\partial y^2 > 0$) may generate boundary-layer separation and reverse flow.

When plotting $u(y) = U_0 df(\eta)/d\eta$ at any x-station, it could be well approximated by a parabolic function (or something locally similar, e.g., an exponential function or a sine- or tanh-function). Indeed, that idea of approximating the (unknown) u(x, y)-profile, subject to appropriate boundary conditions, is the kernel of the *momentum integral* relation (MIR) after Von Kármán (1921).

Integral Method for Laminar or Turbulent Boundary-Layer Flows For all kinds of similar or nonsimilar, laminar or turbulent boundary-layer type flows, the Von Kármán (1921) MIR may be most appropriate. The approximate solutions provide local as well as global system parameters such as the boundary-layer thickness, wall shear stress, flow rates, fluxes, drag coefficients and forces, in addition to realistic velocity profiles. Specifically, suitable velocity functions are postulated, subject to standard boundary conditions, and then momentum conservation is preserved. Clearly, the more accurate the assumed tangential/axial velocity profiles, the better are the system parameter results. In order to best illustrate the integral method, we focus on the steady laminar (or turbulent) 2-D flat-plate boundary-layer problem (see Fig. 5.1).

Along the boundary-layer edge Euler's equation holds so that:

$$-\frac{1}{\rho}\frac{\partial p}{\partial x} = U\frac{\partial U}{\partial x} \tag{5.13}$$

and hence Eq. (5.4a) can be rewritten, with $\mu\partial u/\partial y = \tau_{yx}$, as:

$$u\frac{\partial u}{\partial x} + v\frac{\partial u}{\partial y} = U\frac{\partial U}{\partial x} + \frac{1}{\rho}\frac{\partial \tau_{yx}}{\partial y} \tag{5.14}$$

From Eq. (5.3) we obtain:

$$v(x, y) = -\int \frac{\partial u}{\partial x}\, dy + f(x) \quad \text{or} \quad v = -\int_0^y \frac{\partial u}{\partial x}\, dy \qquad (5.15a, b)$$

which will be used to eliminate the normal (vertical) velocity v in Eq. (5.14).

Now, integrating Eq. (5.14) across the boundary layer yields:

$$\int_{y=0}^{y=\delta(x)} \left(u \frac{\partial u}{\partial x} + v \frac{\partial u}{\partial y} - U \frac{dU}{dx} \right) dy = \frac{1}{\rho} \int_0^{\delta} \frac{\partial \tau_{yx}}{\partial y}\, dy = -\frac{\tau_{wall}}{\rho} \qquad (5.16)$$

Integration by parts of the $v\partial u/\partial y$ – term with Eq. (5.15b) inserted results in:

$$-\int_0^{\delta} \left\{ \frac{\partial u}{\partial x} \int_0^{y} \frac{\partial u}{\partial x}\, dy \right\} dy = -\left[U \int_0^{\delta} \frac{\partial u}{\partial x}\, dy - \int u \frac{\partial u}{\partial x}\, dy \right] \qquad (5.17)$$

so that Eq. (5.16) finally reads:

$$\int_0^{\delta} \frac{\partial}{\partial x} [u(U-u)]\, dy + \frac{dU}{dx} \int_0^{\delta} (U-u)\, dy = \frac{\tau_w}{\rho} \qquad (5.18)$$

As mentioned, with the right choice of u(x, y) both sides of the momentum balance (5.18) have to be matched (see Example 5.1).

Example 5.1: Blasius Flow

Consider steady laminar high Reynolds-number flow past a horizontal flat plate with zero pressure gradient, i.e., given $u_\infty \equiv U = \text{constant}$. Employing the *momentum integral relation (MIR)* with a basic polynomial for u(x, y), find $\delta(x)$ and $\vec{v} = (u, v, 0)$.

Sketch	*Assumptions*	*Concepts*
	• As stated • Constant fluid properties • Mass-less plate with upper B-L only	• Integral boundary-layer equation (5.18) with $U = u_\infty = \not\subset$ • Postulate $u(x, y) = a + by + cy^2$

Solution: With $U = \not\subset$ and invoking Leibniz' rule, Eq. (5.18) reduces to:

$$\rho \frac{d}{dx} \int_0^\delta u(u_\infty - u)\, dy = \tau_w = \mu \left.\frac{\partial u}{\partial y}\right|_{y=0} \qquad \text{(E.5.1.1)}$$

where u_∞ is given and we employ the postulate:

$$u(x, y) = a + by + cy^2 \qquad \text{(E.5.1.2a)}$$

Clearly, the coefficients in the u(x, y) – profile function depend on x and the following three BCs determine the three coefficients a to c.

$$u(y = 0) = 0, \quad u(y = \delta) = u_\infty \text{ and } \left.\frac{\partial u}{\partial y}\right|_{y=\delta} = 0 \qquad \text{(E.5.1.2b–d)}$$

Note: A linear velocity-profile postulate would have missed incorporating the third B.C.
(E.5.1.2d) and hence would have generated an incorrect result.

Using Eq. (E.5.1.2a):

$$a = 0, \quad b = u_\infty / \delta \quad \text{and} \quad c = - u_\infty /(2\delta^2)$$

so that only δ(x) remains as an unknown:

$$u(x, y) = u_\infty \left[\frac{y}{\delta(x)} - \frac{1}{2}\left(\frac{y}{\delta(x)}\right)^2 \right] \tag{E.5.1.3}$$

Inserting Eq. (E.5.1.3) into Eq. (E.5.1.1), carrying out the integration and evaluating τ_{wall}, yields:

$$0.066\bar{6}\,\rho\, u_\infty^2 \frac{d\delta}{dx} = \mu \frac{u_\infty}{\delta} \tag{E.5.1.4a}$$

or

$$\delta \frac{d\delta}{dx} = K \equiv \frac{\mu}{0.066\bar{6}\,\rho\, u_\infty} \tag{E.5.1.4b}$$

subject to δ(x = 0) = 0. Thus,

$$\frac{\delta^2}{2} = Kx \tag{E.5.1.4c}$$

or

$$\frac{\delta(x)}{x} = \frac{5.477}{\sqrt{Re_x}} \tag{E.5.1.5}$$

Comments:

- $\delta(x) \sim \sqrt{x}$ is depicted in Fig. 5.1.
- Compared with the "exact" solution (5.8) this simple approximation is off by just under 10%.

5.2.2 Turbulent Flat-Plate Boundary-Layer Flow

When a critical (flat-plate) Reynolds number of about $Re_\ell = 5 \times 10^5$ has been reached, the boundary-layer flow becomes fully turbulent after undergoing transition as indicated in Fig. 5.1. The traditional approach for tackling turbulent flow problems is the numerical solution of the *Reynolds-averaged Navier–Stokes (RANS)* equations. Specifically, the *instantaneous* dependent variables, say v, (i.e., representing pressure, velocity, temperature, etc.) are decomposed according to Reynolds as:

$$v = \bar{v} + v' \,; \quad \bar{v} = \frac{1}{\Delta t}\int_{t_1}^{t_2} v \, dt \tag{5.19a, b}$$

where $\bar{v}$ is the time-averaged (i.e., smoothed) variable and v' is the randomly fluctuating part. This time-smoothing (or filtering) procedure applied to the N–S equations (see Sect. 2.4 and App. A) generates apparent (or Reynolds) stresses, $-\rho\overline{\vec{v}'\,\vec{v}'}$, caused by the 3-D velocity fluctuations $\vec{v}' = (u', v', w')$. The off-diagonal entries of the $\overline{\vec{v}'\,\vec{v}'}$ tensor, called a dyadic product, may cause large (turbulent) shear stresses while the sum of the diagonal

$$\frac{1}{2}(\overline{u'^2} + \overline{v'^2} + \overline{w'^2}) \equiv k = \frac{1}{2}\, \overline{v_i' \, v_i'} \tag{5.20}$$

is the turbulence kinetic energy. The RANS equations then read in index notation:

$$\frac{\partial \bar{v}_i}{\partial x_i} = 0 \tag{5.21a}$$

and

$$\frac{\partial \bar{v}_i}{\partial t} + \bar{v}_j \frac{\partial \bar{v}_i}{\partial x_j} = -\frac{1}{\rho}\frac{\partial \bar{p}}{\partial x_i} + \underbrace{\nu \frac{\partial^2 \bar{v}_i}{\partial x_j \, \partial x_j}}_{\sim \nabla\cdot\tau_{ij}^{lam}} - \underbrace{\frac{\partial}{\partial x_j}\left(\overline{v_i' \, v_j'}\right)}_{\sim \nabla\cdot\tau_{ij}^{turb}} \tag{5.21b}$$

The new (unknown) stress tensor

$\tau_{ij}^{turb} = -\rho\,\overline{v_i'\,v_j'}$ is expressed, following Boussinesq's idea, as:

$$\frac{\tau_{ij}^{turb}}{\rho} = \nu_{turb}\,\frac{\partial \overline{v}_i}{\partial x_j} \tag{5.22}$$

The *turbulent eddy viscosity* ν_t is a complex function of the turbulence kinetic energy, turbulence energy dissipation, turbulence time and length scales, velocity gradients, flow system characteristics, fluid viscosity, etc. Starting with Boussinesq (1877) and Kolmogorov (1942), numerous ν_t-models have been postulated (see Kleinstreuer 1997; Wilcox 1998; among others), each only applicable to a rather small class of (fully) turbulent flows. Clearly, the future belongs to *direct numerical simulation (DNS)* where the N–S equations are directly solved, i.e., without any turbulence modeling, on numerical mesh time and length scales smaller than those appearing in a given turbulent flow field.

In any case, engineers always found a way to bypass tasks which were mathematically or numerically difficult – as is the case in turbulence modeling. For example, in Sect. 4.2 fully-developed turbulent *pipe flow problems* were readily solved with the extended Bernoulli equation:

$$\frac{\Delta p}{\rho g} + \Delta z = h_f = f\left(\frac{L}{D}\right)\frac{v^2}{2g} \tag{5.23}$$

where the friction factor $f = f(Re_D, \varepsilon/D)$ encapsulates turbulence effects (see Moody Chart in App. B). Postulating semi-empirical turbulent velocity profiles directly bypasses any solution of the RANS equations. For example, the convenient power-law for pipe flow is (see Sect. 4.2):

$$\frac{\bar{u}}{u_{max}} = \left(1 - \frac{r}{R}\right)^{1/n} \quad \text{or} \quad \frac{\bar{u}}{u_\tau} = 8.74\left(\frac{yu_\tau}{\nu}\right)^{1/n} \qquad (5.24a, b)$$

where R is the pipe radius; n, typically 7, is a function of Re_D (see App. B or Schlichting & Gersten 2000); the friction velocity $u_\tau \equiv \sqrt{\tau_w/\rho}$, and the wall coordinate $y = R - r$.

For *turbulent boundary-layer flow* the linear laminar sublayer and the log-law with inner variables $u^+ \equiv \bar{u}/u_\tau$ and $y^+ \equiv yu_\tau/\nu$ are frequently employed:

$$u^+ = y^+ \quad \text{for} \quad 0 \le y^+ \le 5 \qquad (5.25a)$$

and

$$u^+ = \frac{1}{\kappa}\ln y^+ + B \quad \text{for} \quad 30 \le y+ \le 400 \qquad (5.25b)$$

where, based on measurements, $\kappa = 0.41$ and $B = 5.24$. In the buffer overlap region $5 < y^+ < 30$ another log-law is appropriate, if needed:

$$u^+ = 5.0\ln y^+ - 3.0 \qquad (5.25c)$$

Employing the integral method of Sect. 5.2.1, one turbulent boundary layer based on a logarithmic velocity profile can provide some good results within the range of $10^5 < Re_x < 10^9$ (see HWA in Sect. 5.5):

$$\frac{\delta(x)}{x} \approx 0.14\,Re_x^{-1/7} \qquad (5.26a)$$

$$\tau_w \approx 0.0125\rho U^2 Re_x^{-1/7} \qquad (5.26b)$$

$$c_f \approx 0.025\,Re_x^{-1/7} \qquad (5.26c)$$

$$F_D \approx 0.015\rho U^2 LW\,Re_x^{-1/7} \qquad (5.26d)$$

and

$$c_D \approx 0.030\,Re_x^{-1/7} \qquad (5.26e)$$

Example 5.2: Fully Turbulent Flow Characteristics in a Smooth Pipe

Water at 20°C flows through a horizontal pipe ($D = 0.1$ m, $Q = 4 \times 10^{-2}$ m³/s) driven by $\Delta p/\ell = 2.59$ kPa/m. Of interest is the "laminar sublayer thickness" δ_s; the ratio of centerline to average velocity, and, for radial position $r_p = 0.025$ m, the ratio of turbulent-to-laminar shear stresses, where $\tau_{laminar}$ is the laminar stress portion of the turbulent flow field at $r = r_p$.

Sketch

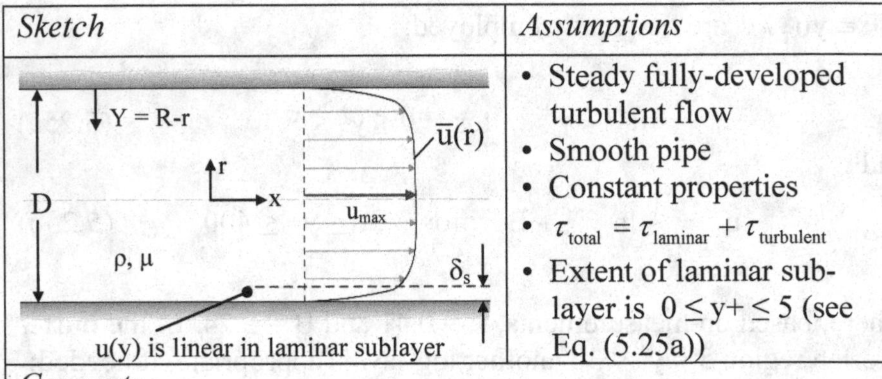

Assumptions

- Steady fully-developed turbulent flow
- Smooth pipe
- Constant properties
- $\tau_{total} = \tau_{laminar} + \tau_{turbulent}$
- Extent of laminar sub-layer is $0 \le y+ \le 5$ (see Eq. (5.25a))

Concepts

- Mass conservation $Q = \int_A \vec{\mathbf{v}} \cdot d\vec{\mathbf{A}} = u_{avg}\, A$
- Force balance (momentum RTT): $\Delta p \approx 4\tau_{wall}\ell / D$
- Turbulent velocity profiles

(i) Pipe flow: $\dfrac{\bar{u}}{u_{max}} = \left(1 - \dfrac{r}{R}\right)^{1/n}$; $n = n(Re)$;

Here, we have $Re_D = \dfrac{u_{avg}\, D}{\nu} = 5.07 \times 10^5$ and hence $n = 8.4$ (see App. B or Schlichting & Gersten 2000)

(ii) Pipe wall: $u^+ = y^+$ for $0 \le y^+ \le 5$ and friction velocity $u_\tau \equiv \sqrt{\tau_w / \rho}$

Properties: For water at 20°C, $\rho = 998\ \text{kg/m}^3$ and $\nu = 10^{-6}\ \text{m}^2/\text{s}$ (see App. B)

Solution:

(a) Sublayer thickness δ_s:

$$y^+(y = \delta_s) \equiv \frac{u_\tau\, \delta_s}{\nu} = 5.0 \qquad \text{(E.5.2.1a)}$$

$$\therefore \qquad \delta_s = 5\nu\sqrt{\tau_w/\rho}\,; \qquad \tau_w = \frac{D}{4}\left(\frac{\Delta p}{\ell}\right) \qquad \text{(E.5.2.1b, c)}$$

and hence

$$\tau_w = 64.8\ \text{N/m}^2 \quad \text{and} \quad u_\tau = 0.255\ \text{m/s}$$

so that

$$\underline{\underline{\delta_s = 0.02\ \text{mm}}}$$

(b) With $n \approx 8.4$ for $Re_D = 5.07 \times 10^5$ we have:

$$Q = v_{avg}(\pi R^2) = v_{max}\int_0^R \left(1 - \frac{r}{R}\right)^{1/8.4} (2\pi\, r\, dr) \qquad \text{(E.5.2.2a, b)}$$

or

$$\frac{u_{max}}{u_{avg}} = \frac{R^2}{2}\left[\int_0^R \left(1 - \frac{r}{R}\right)^{1/8.4} r\, dr\right]^{-1} := 1.186 \qquad \text{(E.5.2.3)}$$

while $u_{max} = 6.04$ m/s

(c) Anywhere in the pipe we have:

$$\tau_{rx} \equiv \tau_{total} = \tau_{turb} + \tau_{lam} \qquad \text{(E.5.2.4)}$$

From the momentum RTT for fully-developed flow:

$$\frac{\Delta p}{\ell} = \frac{2\tau_{rx}}{r} = \not{c}\text{, or with } \tau\left(r = \frac{D}{2}\right) = \tau_{wall} \qquad \text{(E.5.2.2a, b)}$$

$$\tau_{total} = \frac{2\tau_w}{D} r \quad \text{and hence} \qquad \text{(E.5.2.5c)}$$

$$\tau_{total}\left(r = r_p = 0.025 \text{ m}\right) = 32.4 \text{ N/m}^2$$

Recall: $\tau_{lam} = \mu \dfrac{d\bar{u}}{dr} := \dfrac{\mu\, u_{max}}{r\, R}\left(1 - \dfrac{r}{R}\right)^{\frac{1-n}{n}}$ (E.5.2.6a, b)

so that

$$\tau_{lam}(r = r_p) = (\nu\rho)\left.\frac{d\bar{u}}{dr}\right|_{r=r_p} := 0.0266 \frac{N}{m^2}$$

Finally, $\underline{\kappa} \equiv \dfrac{\tau_{turb}}{\tau_{lam}} = \dfrac{\tau_{total} - \tau_{lam}}{\tau_{lam}} := \underline{1220}$

Comments:

- The laminar sublayer, only 20 μm thick, exhibits a very steep velocity gradient.
- While $\dfrac{u_{max}}{u_{avg}} = 2.0$ for laminar flow because of the parabolic velocity profile, the ratio is only 1.186 for turbulent pipe flow.
- Midway between pipe centerline and wall, the turbulent shear-stress portion is 1,220 × $\tau_{laminar}$.

5.3 Drag and Lift Computations

Drag and lift computations are naturally most important in aero–dynamics as part of any aircraft design (see Bertin 2002; among others). In general, fluid flow around a submerged body, e.g., an airplane, a submarine, a race car and a cruising bird, or fixed objects such as weather balloons, commercial signs, buildings and large trees produce a resulting force. It can be decomposed into a component parallel to the object motion (or free-stream) called the *drag* and in the vertical direction, i.e., against gravity, called *lift*. Both the total drag and lift are due to frictional and net pressure effects and hence the surface force exerted by a fluid on the object is:

$$\vec{\mathbf{F}}_{\text{surface}} = \vec{\mathbf{F}}_{\text{pressure}} + \vec{\mathbf{F}}_{\text{viscous}} \tag{5.27a}$$

i.e.,

$$\vec{\mathbf{F}}_{s} = \oiint [-(p - p_{\infty})\hat{n} + \tau_{w}\,\hat{t}]\,dS \tag{5.27b}$$

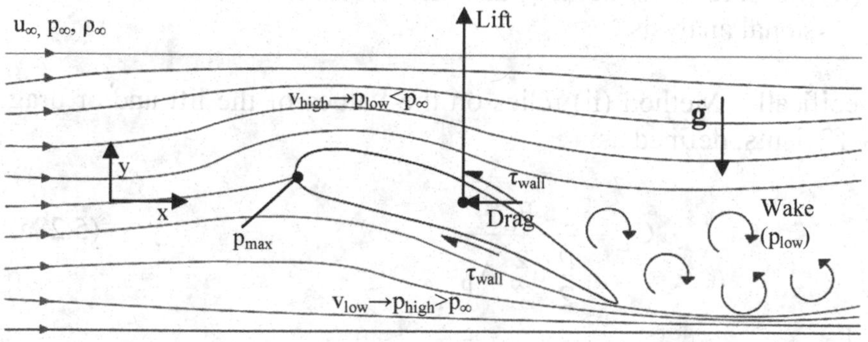

Fig. 5.2 Pressure and wall shear stress around an airfoil causing lift and drag

Here, $\hat{n}$ is the normal and $\hat{t}$ is the tangential unit vector. Assuming that the free-stream is parallel to the x-axis (see Fig. 5.2), then from Eq. (5.27a):

$$\vec{\mathbf{F}}_{Drag} \equiv \vec{\mathbf{F}}_{form} + \vec{\mathbf{F}}_{friction} = \hat{i}\, F_s \quad \text{and} \quad \vec{\mathbf{F}}_{Lift} = \hat{j}\, F_s \tag{5.27c, d}$$

Considering the streamlines around an idealized airplane wing (see Fig. 5.2), we detect accelerating airflow above the wing, which, according to Bernoulli's equation, translates into a low (i.e., suction) pressure while the opposite is true beneath the wing (see Example 3.1a). The difference, $\Delta p \cdot A_{horizontal}$, generates almost entirely the lift force, F_{lift}, in Eq. (5.27d). In contrast, the total drag results from the front-and-back pressure difference times projected vertical area, also called *form drag*, as well as the net wall shear stress on the object's surface area (see Eq. (5.27c) in conjunction with Eq. (5.27b)), called *frictional drag*.

In general, detailed p- and τ_w - distributions from numerical (CFD) simulations have to be known in order to then solve the closed surface integral (5.27a). Alternative approaches for finding $\vec{F}_s$ are:

(i) Analytic solutions, employing the momentum RTT, as discussed in Sects. 2.4 and 3.1
(ii) Measurements of drag and lift coefficients based on dimensional analysis

Specifically, Method (ii) relies on the values of the lift and/or drag coefficients, defined as:

$$C_L \equiv \frac{F_L}{\frac{\rho}{2} u_\infty^2 A_p} \tag{5.28a}$$

and

$$C_D \equiv \frac{F_D}{\frac{\rho}{2} u_\infty^2 A_p} \tag{5.28b}$$

Here u_∞ is the free-stream velocity, ρ is the fluid density, and A_p is the projected area, i.e., either the frontal (plane normal) or midplane area. Figures 5.3a, b depict typical force coefficients for an airfoil and a sphere, respectively.

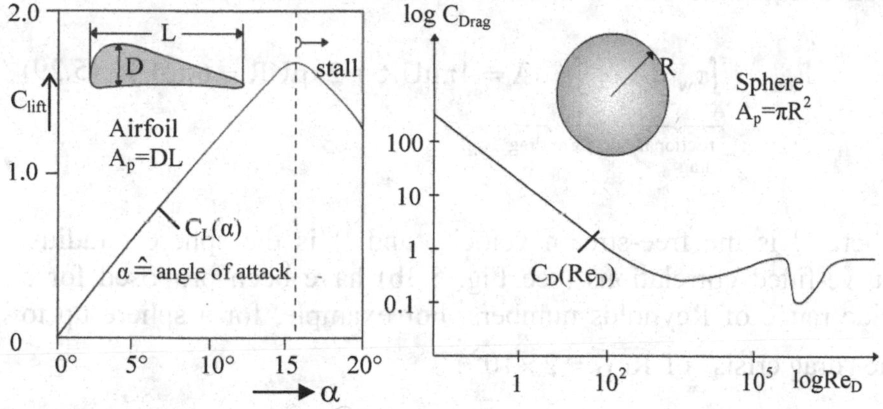

Fig. 5.3 (a) Airfoil lift coefficient as a function of inclined angle, where $C_L(\alpha_{critical})$ decreases because of upper boundary-layer separation, called "stall". (b) Average drag coefficient for a smooth sphere in uniform flow (see App. B)

The $C_D(Re_D)$ curve for spheres shown in Fig. 5.3b encapsulates different flow mechanisms:

- For Re ≤ 1 we are in the Stokes regime and $C_D = 24/Re_D$ without any flow separation which starts at $Re_D \approx 10$ and vortex shedding at $Re_D \approx 90$.
- At about $Re_D = 10^3$ the drag is 95% due to net pressure (i.e., form drag).
  Note: $C_D \sim v$ may decrease but $F_D \sim v^2$ increases.
- Between $10^5 < Re_D < 10^6$ the boundary layer becomes turbulent and flow separation is delayed from angle $\theta^{laminar}_{separation} \approx 80°$ to $\theta^{turbulent}_{separation} \approx 140°$
- At about $Re_D = 2 \times 10^5$, $C_D(Re_D)$ dips, i.e., a sudden drag reduction leading to instabilities at high speeds or, on a positive note, longer travel distances of, say, dimpled golf balls.

Because of the flow field complexities around submerged bodies, analytical solutions are not available. A famous exception is Stokes' equation for flow around a sphere when $\mathrm{Re_D} \leq 1.0$, i.e.,

$$F_{total} = \underbrace{\int_A \tau_w \, dA}_{\text{frictional drag}} + \underbrace{\int_A p \, dA}_{\text{form drag}} = 4\pi\mu UR + 2\pi\mu UR = 6\pi\mu UR \quad (5.29)$$

where U is the free-stream velocity and R is the sphere's radius. Curve-fitted correlations (see Fig. 5.3b) have been proposed for a wide range of Reynolds numbers. For example, for a sphere up to the "drag crisis" of $\mathrm{Re_D} \approx 2 \times 10^5$:

$$C_{D,sphere} \equiv \frac{F_{total}}{\frac{\rho}{2} U^2 A_{proj}} = \frac{24}{\mathrm{Re_D}} + \frac{6}{1+\sqrt{\mathrm{Re_D}}} + 0.4 \quad (5.30a)$$

while a simple drag coefficient for a cylinder in laminar cross-flow reads (White, 1974):

$$C_{D,cylinder} \approx 1 + \frac{10}{\mathrm{Re_D^{2/3}}} \quad (5.30b)$$

Example 5.3: Commercial Passenger Jet

Consider a cruising airplane ($m = 70 \times 10^3$ kg, $A_p = 150$ m^2, v = 350 mph) at 12 km altitude ($\rho_{air} = 0.312$ kg/m^3). What is the necessary angle of attack, lift-to-drag ratio, and what is the engine power requirement to maintain the crusing speed? How would you estimate the minimum take-off velocity?

Notes: For modern aircrafts, $C_{L,max} = 3.5$ <double-flapped wings> and $C_{D,cruising} \approx 0.03$.

Sketch	*Assumptions*	*Concepts*
F_L, F_D, α, v, ρ_{air}, two flaps	• Only the wings generate F_D and F_L • Steady (2-D) incompressible flow	• 1-D force balance • Power = $F_D \cdot v$ • Use of lift/drag coefficients

Solution: • Force coefficient $C_i = \frac{F_i}{\frac{\rho}{2} v^2 A_p}$; i = L or D

• Force balance: $F_{weight} = F_{Lift}$

• Power = $F_{Drag} \times$ velocity

• Angle-of-attack: $C_L = C_L(\alpha)$ where

$$C_L = \frac{F_L}{\frac{\rho}{2} v^2 A_p}, \quad F_L = W = m \cdot g = 686.7 \text{ kN}$$

With the conversion 1 mph ≙ 0.447 m/s and A_p = 150 m^2, we have C_L = 1.20 and then from Fig. 5.3a (or a handbook) $\alpha \approx 10°$.

• Lift-to-drag ratio and power requirement: $P = F_D \cdot v$ where

$$F_D = C_D A_p \frac{\rho}{2} v^2 := 17.2 \text{ kN}$$

Thus,

$$\kappa = \frac{F_L}{F_D} \approx 40 \quad \text{and} \quad P = 2{,}620 \text{ kW}$$

• Minimum take-off speed $v_{min} = \text{fct}(C_{L,max})$:
Take-off requires $F_L\big|_{max} \geq W$, i.e.,

$$C_{L,max}\, A_p \frac{\rho}{2} v_{min}^2 \geq mg \equiv W_{plane}$$

or

$$\underline{v_{min}} \geq \sqrt{\frac{2\,mg}{\rho\, C_{L,max}\, A_p}} := \underline{46.8\, m/s}$$

Comment: Clearly, a take-off speed exceeding v_{min}, say, v = 150 mph may be safer. Still, the lower v_{min} is for take-off and landing, the shorter the runway can be; however, the probability of a crash is greater.

Example 5.4: Turbulent Boundary-Layer Flows

Evaluate the B-L thicknesses above and below the water line and the maximum frictional drag for a 50 m long ship (10 m high) traveling at 15 mph. The kinematic viscosities are $\nu_{air} = 1.5 \times 10^{-5}\ m^2/s$ (at 20°C) and $\nu_{water} = 1.3 \times 10^{-6}\ m^2/s$ (at 10°C) for 5 m of ship surface below the waterline.

Sketch	*Assumptions*	*Concepts*
	• Fully turbulent flow for $0 \leq x \leq L$ • Constant pressure • Straight ship walls	• Use of correlations $\delta(x)$ and $c_f(x)$

Solutions:

From Sect. 5.2.2, for turbulent flow in the range $5 \times 10^5 < Re_x < 10^9$:

$$\frac{\delta(x)}{x} \approx 0.14\, Rc_x^{-1/7} \qquad \text{(E.5.4.1)}$$

and

$$c_f(x) = \frac{\tau_w(x)}{\frac{\rho}{2}U^2} \approx 0.025\, Re_x^{-1/7} \qquad \text{(E.5.4.2)}$$

Here, $Re_L\big|_{min} = UL/\nu_{air} := 2.22 \times 10^7$, while $Re_L\big|_{water} = 2.56 \times 10^8$, i.e., both are turbulent B-L flows. Hence,

$$\delta(x = L) = \begin{cases} 0.625 \text{ m} & \text{for air} \\ 0.44 \text{ m} & \text{for water} \end{cases} \qquad \text{(E.5.4.3a, b)}$$

Now, the drag force is:

$$F_D = \int_0^L \tau_w(x)\, h\, dx \qquad \text{(E.5.4.4a)}$$

or

$$\frac{F_D}{\frac{\rho}{2}U^2 Lh} = \int_0^L c_f(x)\, d\left(\frac{x}{L}\right) \approx 0.03\, Re_L^{-1/7} \qquad \text{(E.5.4.4b)}$$

$$\therefore \quad F_D = 0.03\, Re_L^{-1/7} \frac{\rho}{2} U^2\, Lh := 2.1 \times 10^4 \text{ N} \qquad \text{(E.5.4.4c)}$$

and hence

$$P_{ship} = F_D\, U \geq 1.4 \times 10^5 \text{ N} \frac{m}{s} \geq 0.14 \text{ MW} \qquad \text{(E.5.4.5)}$$

Comment: Clearly, the *actual* power required to maintain a speed of 15 mph through calm seas is larger because the air form drag were not considered.

5.4 Film Drawing and Surface Coating

Pressure, surface tension, viscous and gravitational forces play key roles in film drawing and surface coating. Typically, *nearly parallel flow* (see Sect. 3.2) and 1-D free-surface force balances are assumed to solve basic problems. Such simplifications entail a negligible normal velocity component and consequently $\partial u/\partial x \approx 0$ from continuity, and $\partial p/\partial y \approx 0$ from the y-momentum equation. Hence, the x-momentum equation (see also the boundary-layer assumption in Sect. 5.2) can be reduced to:

$$0 = -\frac{dp}{dx} + \mu\frac{\partial^2 u}{\partial y^2} + \rho g_x \tag{5.31}$$

where $-\,dp/dx = \Delta p/L =$ constant and $u = u(x, y)$ in case of slightly non-parallel geometric boundaries (see Sect. 4.3 and the lubrication assumption).

In case a meniscus forms between a rising liquid (on a vertical wall as shown in Example 5.5) and air, a 1-D force balance between surface tension and gravity yields (see Eq. (5.34)):

$$\frac{\sigma}{R} + \rho g h = 0 \tag{5.32}$$

where σ is the surface tension, $R \sim \frac{1}{h''}$ is the meniscus curvature, and $\rho g h$ is the fluid static pressure.

5.4.1 Drawing and Coating Processes

For certain drawing processes where an applied force pulls the film out of an extruder, the normal stresses are more important than the shear stresses and hence Eq. (5.31) has to be updated (see Example 5.7). Classical examples are the slider bearing (Example 4.10) and dip-coating (Example 5.6). Similar applications include film drawing, spin coating, calendaring (or extrusion coating), and sheet casting (see Fig. 5.4a–d). In most cases the material is non-Newtonian, such as polymers, melts, etc., and it changes its thermodynamic state from liquid (in the die) to solid (as the end product). Specifically, in *film drawing*, molten material is drawn out of a die and pulled up as a solid, thin sheet/film onto a large roll (see Fig. 5.4a). For *spin* coating (see Fig. 5.4b), a polymeric liquid is fed through the center of a fast-spinning disk, which then spreads radially and coats the disk surface. In polymer processing via *calendering* operation (see Fig. 5.4c), a relatively thick sheet is reduced to a final thickness determined by the clearance of the two counter-rotating rollers. In addition, such rollers can impart any surface pattern onto the sheet. *Film casting* is another extrusion process where a liquid film from a die is turned into a solid thin sheet via a cooled, rotating roller (see Fig. 5.4d).

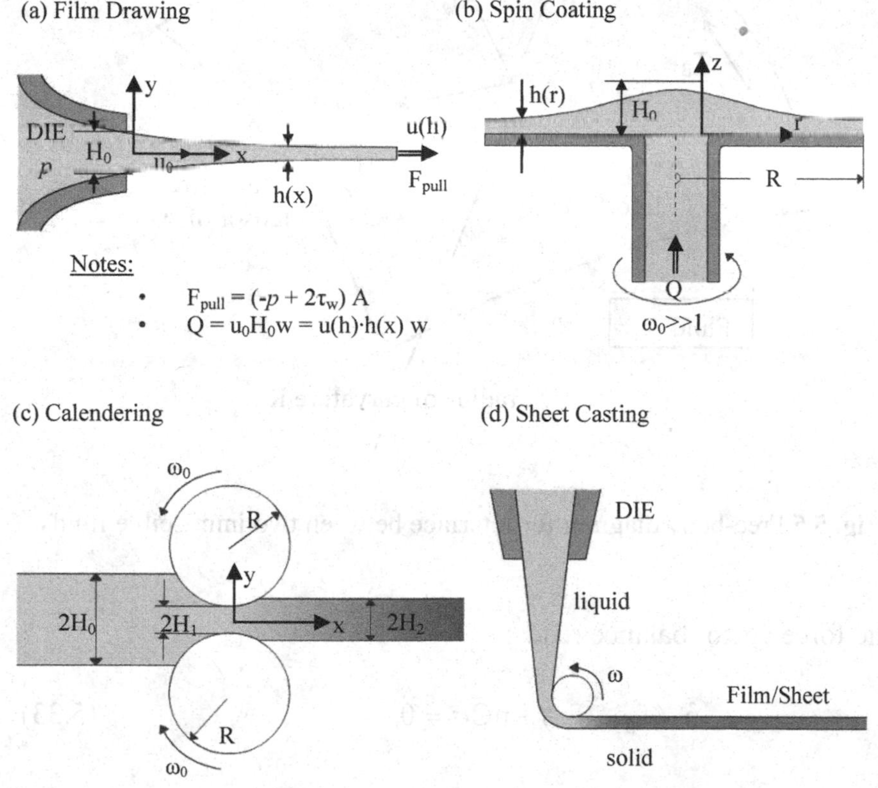

Fig. 5.4 Thin-film drawing and coating processes

5.4.2 Fluid-Interface Mechanics

Consider two immiscible, relatively moving fluids A and B, forming interface S of mean curvature $\overline{C}$. Ignoring surfactant and temperature gradients, the equilibrium forces are net total stress tensor T and surface tension $\sigma = F/l$ (Fig. 5.5).

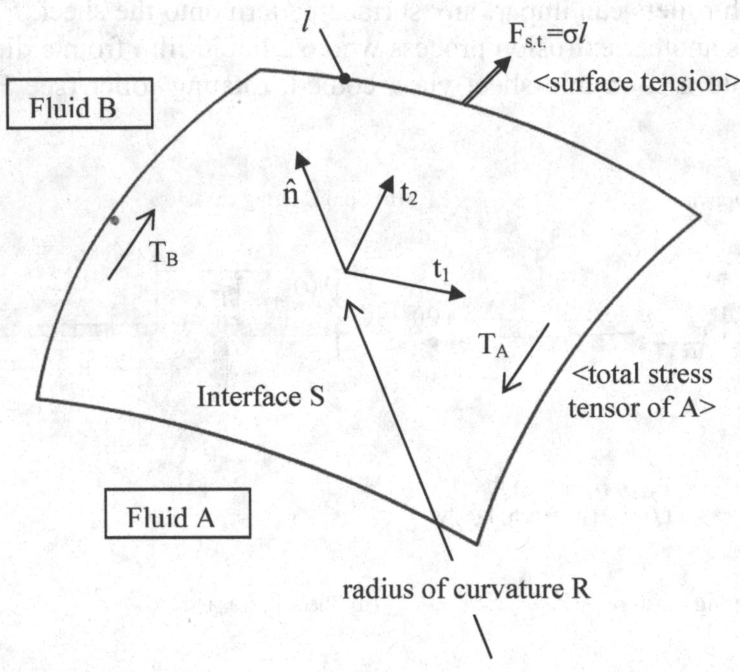

Fig. 5.5 Free-body diagram for interface between two immiscible fluids

The force vector balance reads:

$$\hat{n} \cdot (T_B - T_A) + \hat{n}\overline{C}\sigma = \vec{0} \qquad (5.33)$$

where $\overline{C} \equiv R^{-1}$ is the mean curvature of the interface S, σ is the tabulated surface tension, $T \equiv \vec{\vec{\tau}}_{total} = -p\hat{I} + \vec{\vec{\tau}}$, p is the thermodynamic pressure, $\hat{I}$ is the unit tensor, and $\vec{\vec{\tau}}$ combines the normal and tangential (i.e., shear) stresses. Under static, i.e. no-flow conditions, Eq. (5.33) reduces to the Young–Laplace equation:

$$-\Delta p + \overline{C}\sigma = 0 \tag{5.34}$$

where Δp is the static pressure difference between the two fluids. The mean surface curvature has to be determined based on the interface geometry:

$$\overline{C}\hat{n} = \frac{d\hat{t}}{ds} \tag{5.35}$$

where $\hat{n}$ and $\hat{t}$ are normal and tangent unit vectors and s is the interface arc length. Specifically, for a cylindrically symmetric surface described by h(x), with h' and h" being the contour derivatives:

$$\overline{C} = \frac{h''}{\sqrt{[1+(h')^2]^{3/2}}} \tag{5.36}$$

Equation (5.36) applied to standard gas–liquid configurations yields:

$$\overline{C} = \begin{cases} 0 & \text{for planar surfaces} \\ 2/R & \text{for spheres of radius R} \\ 1/R & \text{for cylinders of radius R} \end{cases}$$

Example 5.5: Meniscus at a Vertical Wall

Consider a gas-liquid interface with capillary rise h at a vertical wall forming the contact angle θ which is the inclination angle $\varphi = \frac{\pi}{2} - \theta$. Find h and θ.

Sketch	*Assumptions*	*Concepts*
	• Cylindrically symmetric surface • 1-D static analysis: Surface tension force balances gravity	• Young-Laplace equation with $\overline{C} = \frac{1}{R}$ and $\Delta p = -\Delta\rho g z$ from fluid statics

Solution: Equation (5.34) now reads

$$\Delta\rho g z + \overline{C}\sigma = 0 \tag{E.5.5.1}$$

and with Eq. (5.36)

$$\overline{C} \equiv \frac{1}{R} = \frac{z''}{[1+(z')^2]^{3/2}} = -\frac{\Delta\rho g}{\sigma} z = -Kz \tag{E.5.5.2}$$

subject to z(x=0) = 0 and z'(x=0) = tan φ = 0. Instead of solving this nonlinear ODE numerically, we start over with $\overline{C} = R^{-1}$, where the interface curvature is

$$\frac{1}{R} = \frac{d\phi}{ds} \tag{E.5.5.3}$$

Eliminating ds, we recall $\sin\phi = \frac{dz}{ds}$ so that

$$\frac{1}{R} = \frac{d\phi}{dz}\frac{dz}{ds} = \frac{d\phi}{dz}\sin\phi = -Kz$$

Integrating

$$\sin\phi d\phi = -Kzdz$$

yields:

$$\cos\phi = -\frac{K}{2}z^2 + C_1$$

where at z = 0 we have the incline angle $\phi = 0°$, i.e., $\cos\phi = 1$ and $C_1 = 1$. Thus,

$$\cos\phi - 1 = -2\sin^2\left(\frac{\phi}{2}\right) = -\frac{K}{2}z^2$$

or

$$z = \pm 2\sqrt{\frac{\sigma}{\Delta\rho g}}\sin\left(\frac{\phi}{2}\right) \tag{E.5.5.4}$$

Finally, z = h with contact angle θ, so that with $\phi = \frac{\pi}{2} - \theta$

$$h = 2\sqrt{\frac{\sigma}{\Delta\rho g}}\sin\left(\frac{\pi}{4} - \frac{\theta}{2}\right) \tag{E.5.5.5}$$

where $\theta < \frac{\pi}{2}$ for h to be positive.

Graph:

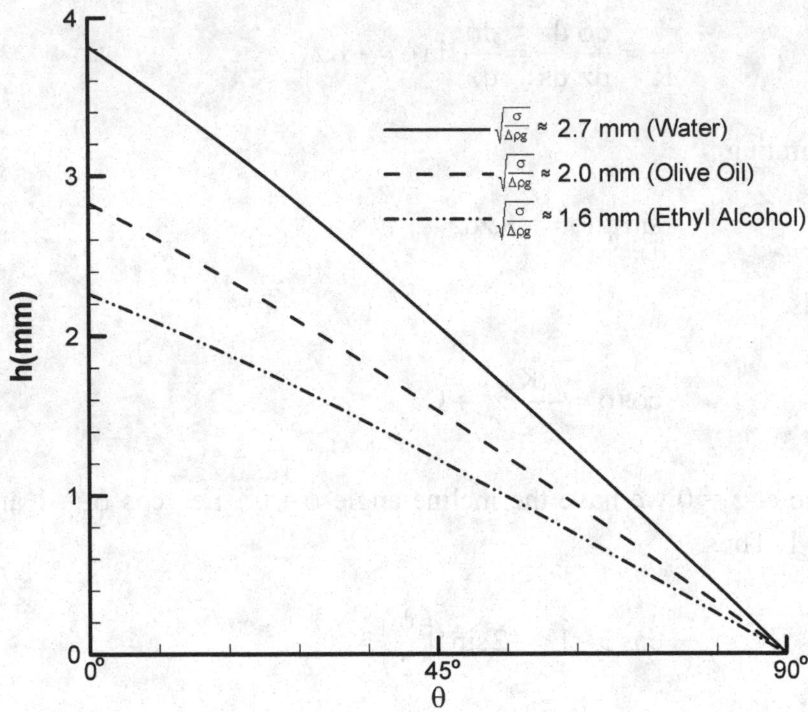

Comments:

- As expected $h \sim \sigma$, i.e., with increasing parameter $\sqrt{\dfrac{\sigma}{\Delta\rho g}}$ the (pulling) surface-tension force $\vec{F}_\sigma$ gets larger and hence h increases
- The magnitude of the vertical component of $\vec{F}_\sigma$ depends greatly on θ, i.e., $0° \leq \theta \leq 90°$, implying that for $\theta = 0°$, $F_{vertical} = F_{max}$ and hence $h = h_{max}$, while for $\theta = 90°$, $F_{vertical} = 0$.

Clearly, the key parameter in fluid-interface mechanics is the surface tension σ. Example 5.5 indicates that, if not known beforehand, σ-values can be obtained from a vertical-plate force balance, measuring the upward pulling force, the fluid densities, and the contact angle. Specifically, with respect to Fig. 5.6 the 1-D force balance can be formulated as follows.

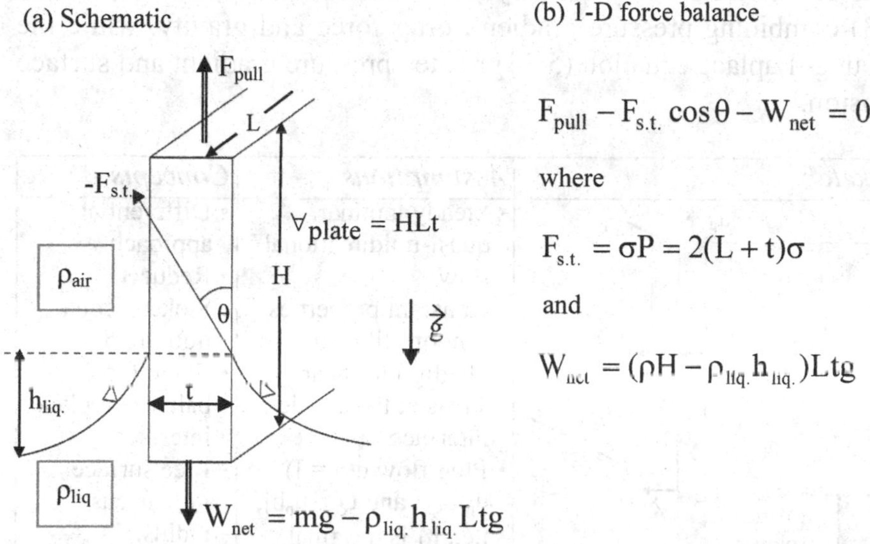

Fig. 5.6 System schematic plus force balance

Thus, ignoring the buoyance effect on the plate portion in the air,

$$F_{pull} = 2\sigma(L+t)\cos\theta + (\rho H - \rho_{liq.} h_{liq.})Ltg \tag{5.37}$$

With $h_{liq.}$ from the net-weight expression (see Fig. 5.6) and the plate geometry plus density known, only the pulling force and the contact angle have to be measured to compute the surface tension from Eq. (5.37).

Example 5.6: Surface Coating

In contrast to the falling film of constant thickness (see Example 3.2), we now consider more realistic surface coating. A thin plate is being submerged into a finite reservoir of a coating solution and then vertically withdrawn at a constant velocity.

The interplay of drag force, gravitational effect and surface tension is described by the Reynolds lubrication equation (see Sect. 4.3) combining pressure gradient, drag force and gravity, while the Young–Laplace equation (5.34) relates pressure gradient and surface tension.

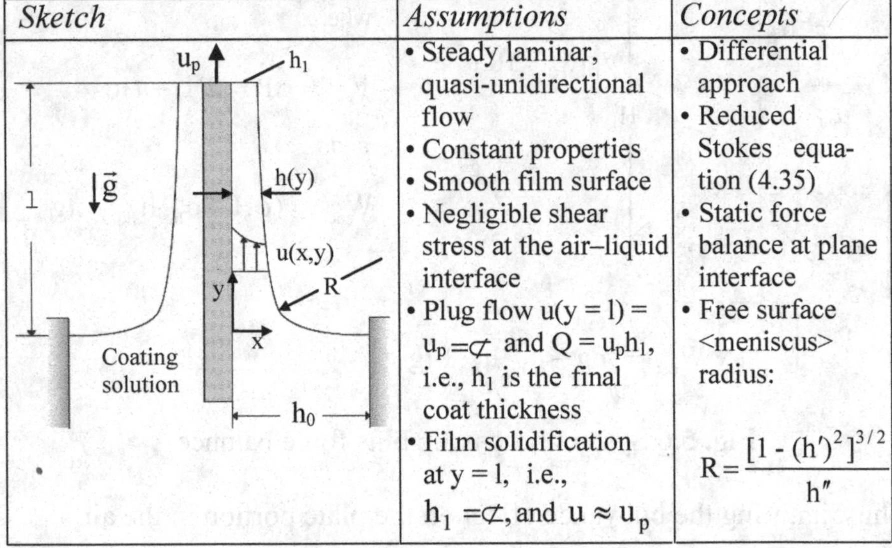

Sketch	*Assumptions*	*Concepts*
	• Steady laminar, quasi-unidirectional flow • Constant properties • Smooth film surface • Negligible shear stress at the air–liquid interface • Plug flow $u(y = l) = u_p = \not\subset$ and $Q = u_p h_1$, i.e., h_1 is the final coat thickness • Film solidification at $y = l$, i.e., $h_1 = \not\subset$, and $u \approx u_p$	• Differential approach • Reduced Stokes equation (4.35) • Static force balance at plane interface • Free surface <meniscus> radius: $R = \frac{[1-(h')^2]^{3/2}}{h''}$

Solution: Based on the stated assumptions, Stokes' equation reduces to:

$$-\frac{dp}{dy} + \mu \frac{d^2u}{dx^2} - \rho g = 0 \qquad \text{(E.5.6.1a)}$$

subject to:

$$u(x = 0) = u_p \quad \text{and} \quad \left.\frac{du}{dx}\right|_{x=h(y)} = 0 \qquad \text{(E.5.6.1b, c)}$$

which integrated yields:

$$u(x, y) = u_p + P[h(y)]^2 \left[\left(\frac{x}{h}\right)^2 - 2\left(\frac{x}{h}\right) \right] \qquad \text{(E.5.6.2a)}$$

where $$P \equiv \left(\frac{dp}{dy} + \rho g \right) \Big/ (2\mu) \qquad \text{(E.5.6.2b)}$$

Recall: A meniscus forms between a rising liquid on a vertical wall and air. A 1-D force balance between surface tension and gravity yields (see Eq. (5.34)):

$$\frac{\sigma}{R} + \rho g h = 0 \qquad \text{(E.5.6.3)}$$

where $\rho g h = p$ is the static pressure and meniscus radius $R = [1 - (h')^2]^{3/2} / h''$; so that for $(h')^2 << 1$ and after differentiation:

$$\frac{dp}{dy} = -\sigma \frac{d^3 h}{dy^3} \qquad \text{(E.5.6.4)}$$

Now, the flow rate per unit depth is

$$Q = \int_0^h u\, dx = -\frac{h^3}{3\mu}\left(\frac{dp}{dy} + \rho g \right) + u_p\, h = u_p\, h_1 \qquad \text{(E.5.6.5a–c)}$$

We substitute dp/dy and rearrange Eq. (E.5.6.5c) to:

$$\frac{h^3}{3}\left(\frac{d^3 h}{dy^3} - \frac{\rho g}{\sigma} \right) + \frac{\mu\, u_p}{\sigma}(h - h_1) = 0 \qquad \text{(E.5.6.6a)}$$

Introducing the capillary number $Ca = u_p\, \mu / \sigma$ and Stokes number $St = \mu\, u_p / (\rho g h_1^2)$, we recast Eq. (5.6.6a) in the form:

$$\frac{d^3h}{dy^3} - \frac{Ca}{h_1^2\, St} + \frac{3Ca}{h^2}\left(1 - \frac{h_1}{h}\right) = 0 \qquad \text{(E.5.6.6b)}$$

where Ca and St may vary within the limiting assumptions of laminar ripple-free film flow, i.e., $Ca \sim 10^{-5}$, and St is in the range of 0.1 to 1.0.

Numerical solution of Eq. (E.5.6.6b), subject to $h(y = l) = h_1, h(y = 0) = h_0$, and $dh / dy(y = l) = 0$, could be achieved with a Runge–Kutta ODE solver. Alternatively, applying Eq. (E.5.6.6b) to Eq. (E.5.6.4) yields:

$$\frac{dp}{dy} = -\sigma \frac{d^3h}{dy^3} = -\sigma \frac{Ca}{h_1^2\, St} + \sigma \frac{3Ca}{h^2}\left(1 - \frac{h_1}{h}\right)$$
$$= -\rho g + \frac{3u_p \mu}{h^2}\left(1 - \frac{h_1}{h}\right) \qquad \text{(E.5.6.7)}$$

Thus,

$$P \equiv \left(\frac{dp}{dy} + \rho g\right)/(2\mu) = \frac{3u_p}{h^2}\left(1 - \frac{h_1}{h}\right) \qquad \text{(E.5.6.8)}$$

We substitute P and rearrange Eq. (E.5.6.2a) to:

$$\frac{u(x, y)}{u_p} = 1 + \frac{3}{2}\left(1 - \frac{h_1}{h}\right)\left[\left(\frac{x}{h}\right)^2 - 2\left(\frac{x}{h}\right)\right] \qquad \text{(E.5.6.9)}$$

Now, for various y-stations, and hence assumed film thicknesses h in the range $h_1 \leq h \leq h_0$, we can plot Eq. (E.5.6.9).

Graphs:

(a) Vertical film velocity u (x,h)

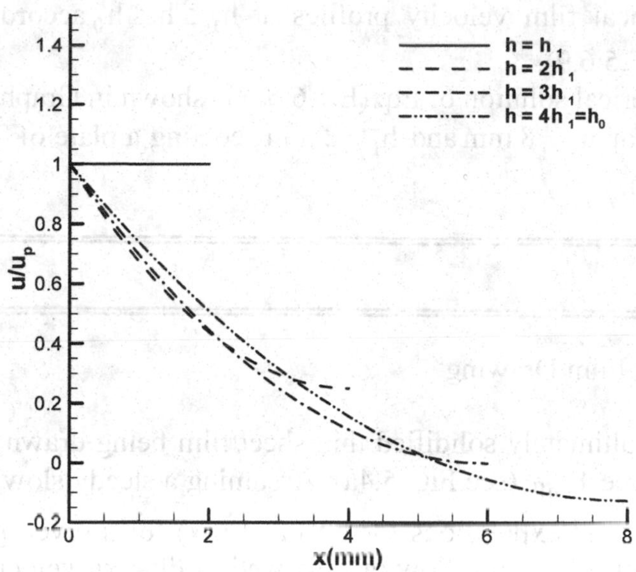

(b) Emerging film thickness h (y)

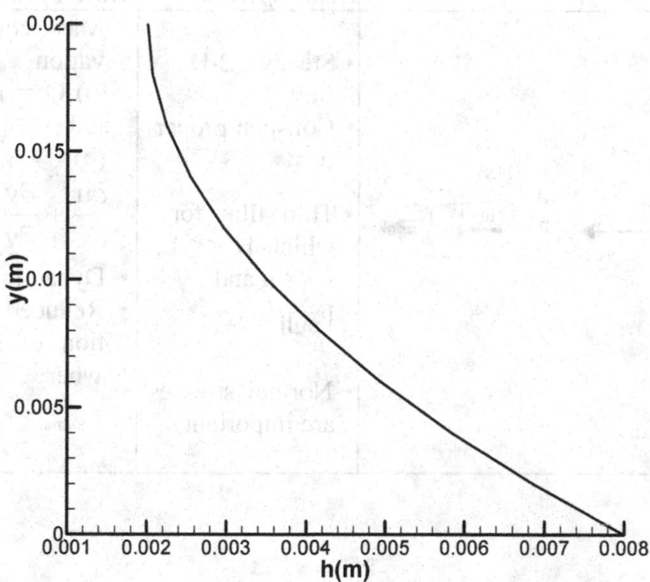

Comments:

- For various assumed h-values, i.e., y-stations, Graph (a) depicts the local film velocity profiles at $h_1 \le h \le h_0$ according to Eq. (E.5.6.9).
- Numerical solution of Eq. (E.5.6.6a) is shown in Graph (b) as h(y) for $h_0 = 8$ mm and $h_1 = 2$ mm, coating a plate of 20 mm length.

Example 5.7: Film Drawing

Consider an ultimately solidified thin sheet/film being drawn out of a die with force F_{pull} (see Fig. 5.4a). Assuming a steady slow extrusion process, find expressions for h(x) and u(x) for a given pulling force, volumetric material flow rate, as well as die-exit velocity and opening.

Sketch	*Assumptions*	*Concepts*
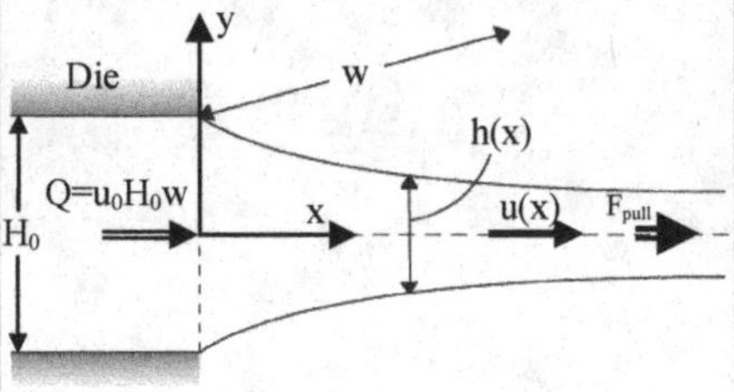 y, Die, w, h(x), $Q=u_0H_0w$, H_0, x, u(x), F_{pull}	• Steady 2-D flow • Constant properties • Thin film for which H << L, v << u and $F_{pull} = \not\subset$ • Normal stresses are important	• Mass conservation (a) Q = u(hw) and (b) $\frac{\partial u}{\partial x} + \frac{\partial v}{\partial y} = 0$ • Dynamic B. C. • Reduced Equation of Motion, where $\tau_{normal} >> \tau_{tangential}$

Solution:

- Continuity: $$\frac{\partial u}{\partial x}+\frac{\partial v}{\partial y}=0 \qquad \text{(E.5.7.1)}$$

- x-momentum: $$0=-\frac{\partial p}{\partial x}+\mu\left(2\frac{\partial^2 u}{\partial x^2}+\frac{\partial^2 u}{\partial y^2}\right) \qquad \text{(E.5.7.2)}$$

net normal stress changes

- y-momentum: $$0=-\frac{\partial p}{\partial y}+\mu\left(2\frac{\partial^2 v}{\partial x^2}+\frac{\partial^2 v}{\partial y^2}\right) \qquad \text{(E.5.7.3)}$$

- Free-surface boundary conditions

x-direction: $$\tau_{yx}\Big|_{y=\pm h/2}=0 \qquad \text{(E.5.7.4a)}$$

y-direction: $$[\ p+\tau_{yy}]\Big|_{y=\pm h/2}=0 \qquad \text{(E.5.7.4b)}$$

- Dynamic B. C.: $$\underbrace{-p+2\mu\frac{\partial u}{\partial x}}_{\text{pressure plus normal stress}}=\underbrace{\frac{F_{pull}}{h\,w}}_{\text{external force/area}} \qquad \text{(E.5.7.4c)}$$

With Eq. (E.5.7.4a) we assume:

$$\tau_{yx}=\mu\left(\frac{\partial v}{\partial x}+\frac{\partial u}{\partial y}\right)=0$$

and hence $\partial v/\partial x=\partial u/\partial y=0$, so that Eqs. (E.5.7.1, 2) reduce to:

$$\frac{\partial}{\partial x}\left(-p+2\mu\frac{\partial u}{\partial x}\right)=0 \qquad \text{(E.5.7.5)}$$

and

$$\frac{\partial}{\partial y}\left(-p + 2\mu \frac{\partial v}{\partial y}\right) = 0 \qquad \text{(E.5.7.6)}$$

Integrating Eq. (E.5.7.5) subject to B. C. (E.5.7.4c) yields:

$$-p + 2\mu \frac{\partial u}{\partial x} = \frac{F}{h\,w} \qquad \text{(E.5.7.7)}$$

Integrating Eq. (E.5.7.6) subject to B. C. (E.5.7.4b) yields:

$$-p + 2\,\mu \frac{\partial v}{\partial y} = 0$$

where $\frac{\partial v}{\partial y} = -\frac{\partial u}{\partial x}$ (see Eq. (E.5.7.1), so that $p = -2\mu \frac{\partial u}{\partial x}$ and hence Eq. (E.5.7.7) now reads:

$$\frac{du}{dx} = \frac{F}{4\mu\,h\,w} \qquad \text{(E5.7.8)}$$

In order to replace $\frac{du}{dx}$ we use $Q = U_0(H_0 w) = uhw = ⊄$ and differentiate the expression $uh = ⊄$:

$$u\frac{dh}{dx} + h\frac{du}{dx} = 0 \qquad \text{(E.5.7.9a)}$$

so that with Eq. (E.5.7.8):

$$\frac{Q}{h\,w}\frac{dh}{dx} + h\frac{F}{4\mu\,h\,w} = 0 \qquad \text{(E.5.7.9b)}$$

Hence,

$$\frac{dh}{dx} + \frac{F}{4\mu Q} h = 0 \qquad \text{(E.5.7.9c)}$$

subject to $h(x = 0) = h_0$.

Separation of variables and integration leads to:

$$h(x) = H_0 \exp\left[-\frac{F}{4\mu Q} x\right] \qquad \text{(E.5.7.10)}$$

The constant flow rate $u_0 H_0 = h u$ implies that

$$u(x) = U_0 \exp\left[\frac{F}{4\mu Q} x\right] \qquad \text{(E.5.7.11)}$$

Graphs:

(a) Film-thickness developments for different extrusion forces

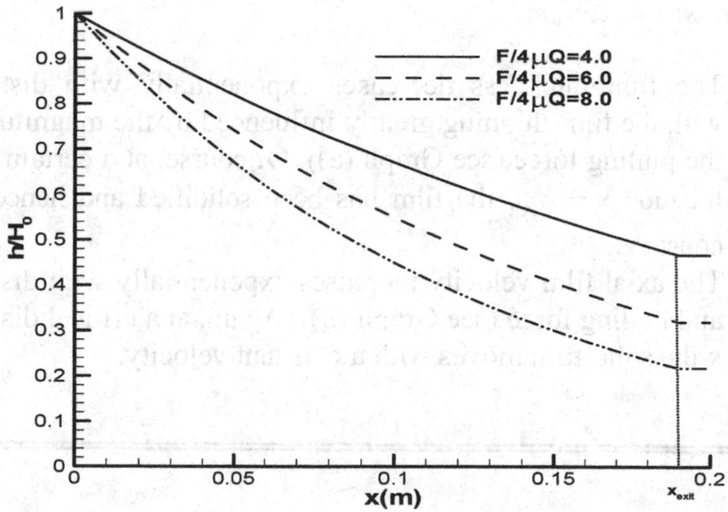

(b) Developing film velocity for different driving forces

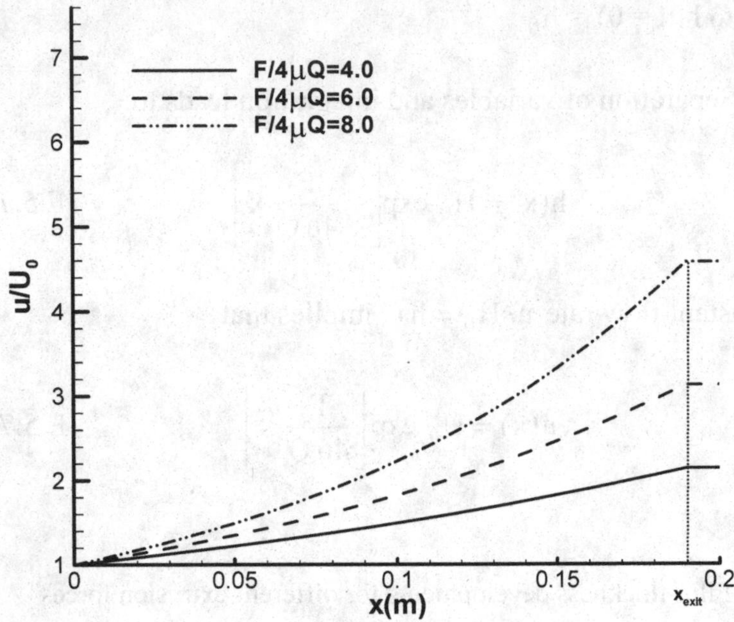

Comments:

- The film thickness decreases exponentially with distance, with the film-thinning greatly influenced by the magnitude of the pulling force (see Graph (a)). Of course, at a certain axial location $x = x_{exit}$ the film has been solidified and hence h = constant.
- The axial film velocity increases exponentially with distance and pulling force (see Graph (b)). Again, at a critical distance x the solid film moves with a constant velocity.

Example 5.8: Spin Coating

A steady, center-flow supplied material is radially spread and thinned on a horizontal, fast-spinning disk (see Fig. 5.4b). Assuming lubrication-type flow, derive the radial velocity profile $v_r(r, z)$ and film thickness h(r) during the material-application stage for a given material flow rate Q and disk angular velocity ω_0 (after Papanastasiou 1994).

Sketch	*Assumptions*	*Concepts*
	• Steady 2-D axi-symetric flow $\vec{v} = [v_r, v_\theta = r\omega_0, 0]$ • Constant properties • $v_r = v_r(r, z)$ where $\frac{\partial v_r}{\partial r} << \frac{\partial v_r}{\partial z}$ • Note: $\nabla p \approx 0$	• Differential approach • Reduced continuity and N-S equations (see Sect. 4.3)

Solution: Based on the stated assumptions, the 2-D continuity equation in cylindrical coordinates reduces to:

$$\frac{1}{r}\frac{\partial}{\partial r}(r\, v_r) = 0 \tag{E.5.8.1a}$$

which implies that

$$v_r = \frac{f(z)}{r} \tag{E.5.8.1b}$$

The r-momentum equation reduces to:

$$v_r \frac{\partial v_r}{\partial r} - \frac{v_\theta^2}{r} = \nu \frac{\partial^2 v_r}{\partial z^2} \tag{E.5.8.2a}$$

With the "lubrication assumption" of Sect. 4.3, i.e., $\partial v_r / \partial r << \partial v_r / \partial z$, and $v_\theta = r\,\omega_0$, Eq. (E.5.8.2a) becomes:

$$\frac{\partial^2 v_r}{\partial z^2} = -\frac{\omega_0^2}{\nu} r \qquad \text{(E.5.8.2b)}$$

subject to $v_r(z=0)=0$ (no slip) and $\left.\frac{\partial v_r}{\partial z}\right|_{z=h} = 0$ (no stress).

Double integration yields:

$$v_r(r, z) = \frac{\omega_0^2}{\nu}\left[h(r)\,z - \frac{z^2}{2}\right] r \qquad \text{(E.5.8.3)}$$

The film thickness h(r) can be obtained via the constant supply-rate condition:

$$Q = \bar{v} A = \int_0^{h(r)} v_r(r, z)\, dA\,; \quad dA = 2\pi r\, dz \qquad \text{(E.5.8.4a)}$$

which gives:

$$Q = \frac{2}{3}\frac{\pi\omega_0^2 r^2}{\nu} h^3 \qquad \text{(E.5.8.4b)}$$

so that

$$h(r) = \left(\frac{3\nu Q}{2\pi\,\omega_0^2\, r^2}\right)^{1/3} \qquad \text{(E.5.8.5a)}$$

or with $\frac{r}{R} \equiv \hat{r}$

$$\hat{h} = \frac{h}{R} = \left(\frac{3\nu Q}{2\pi\omega_0^2 R\hat{r}^2}\right)^{1/3} = K\left(\frac{r}{R}\right)^{-2/3} \qquad \text{(E.5.8.5b)}$$

Graph:

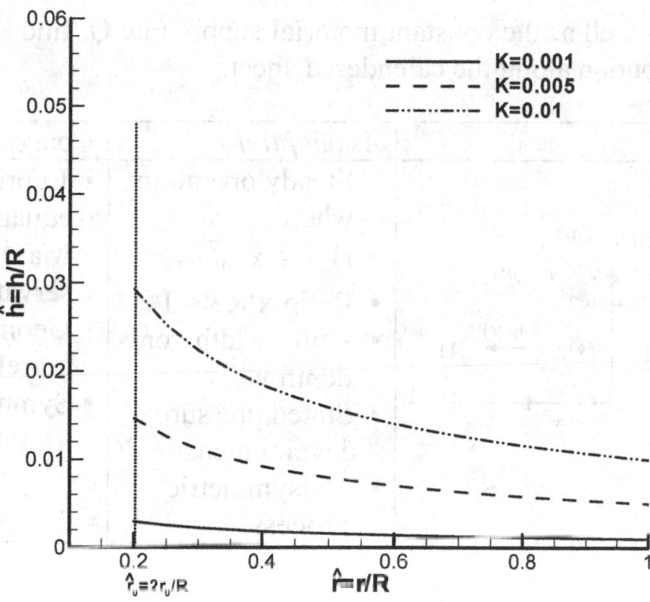

Comments:

The film thins in radial direction in balance with the two key parameters, i.e., material supply rate Q and disk rotation ω_0, where $h \sim (Q^{1/3}$ and $\omega_0^{-2/3})$. Note that the region $0 \le \hat{r} \le 0.2$ was not considered in this model.

Example 5.9: Calendering Operation

Consider a rolling process where a relatively thick sheet ($2H_0$) is pressed into a thin sheet of 2H (see Fig. 5.4c). While the material gage pressure is zero upstream of the rollers until $x = x_{in}$, it rapidly peaks inside the gap $x_{in} < x < x_{out}$ and then returns to atmospheric pressure for $x \ge x_{out}$ when the sheet takes on the final thickness 2H.

Thus, the focus is on h(x) for $-x_{in} \leq x \leq x_{out}$. Given the roller separation $2H_1$, the roller data (R, ω_0) and sheet thicknesses ($2H_0$ and 2H) as well as the constant material supply rate Q, find the pressure distribution along the calendered sheet.

Sketch	*Assumptions*	*Concepts*
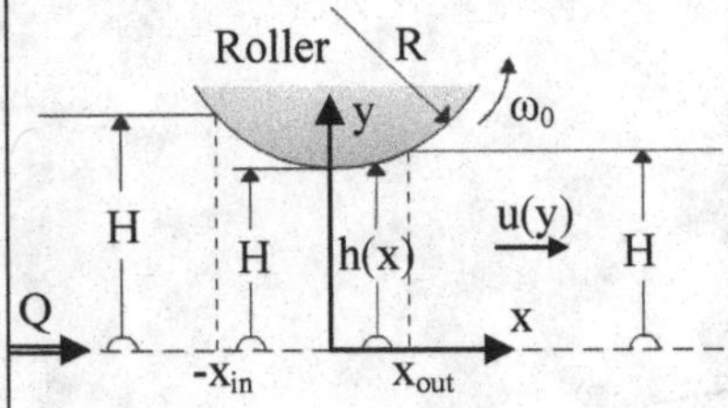	• Steady operation where $H(x \geq x_{out}) = \not\subset$ • Ratio x/R << 1 • Unit width or depth w • Stated pressure developments • Axisymmetric process	• Lubrication equation • Mass conservation • Geometric correlations • Symmetry

Solution:

- Inlet-to-outlet height between rollers is:

$$h(x) = (R + H_1) - \sqrt{R^2 - x^2} \qquad \text{(E.5.9.1a)}$$

With $\frac{x}{R} << 1$,

$$h(x) = (R + H_1) - R\left[1 - \left(\frac{x}{R}\right)^2\right]^{1/2} = H_1 + \frac{x^2}{2R} \qquad \text{(E.5.9.1b, c)}$$

Thus,

$$h(x = x_{in}) = H_0 = H_1 + \frac{x_{in}^2}{2R} \qquad \text{(E.5.9.2a)}$$

and

$$h(x = x_{out}) = H = H_1 + \frac{x_{out}^2}{2R} \tag{E.5.9.2b}$$

- The volumetric material supply rate is:

$$Q = \int_A u(y)dA = \bar{v}A = (R\,\omega_0)\,(2H_1 w) \tag{E.5.9.3}$$

- Thin-film equation (or lubrication theory):

$$0 = -\frac{dp}{dx} + \mu\frac{\partial^2 u}{\partial y^2} \tag{E.5.9.4a}$$

subject to:

$$u[y = h(x = 0)] = R\,\omega_0 \quad \text{and} \quad \left.\frac{\partial u}{\partial y}\right|_{y=0} = 0 \tag{E.5.9.4b, c}$$

Double integration yields:

$$u(x, y) = R\,\omega_0 - \frac{h^2}{2\mu}\left[\frac{dp(x)}{dx}\right]\left[1 - \left(\frac{y}{h}\right)^2\right] \tag{E.5.9.5}$$

Again, the pressure gradient is deduced from $Q = \not{c}$, i.e.,

$$Q = 2\int_0^h u\,dy = 2h\left[R\,\omega_0 - \frac{h^2}{3\mu}\left(\frac{dp}{dx}\right)\right]$$

$$= 2R\,\omega_0\,H_1 \tag{E.5.9.6a–c}$$

Now, solving for the pressure gradient:

$$\frac{dp}{dx} = \frac{3\mu\,R\,\omega_0(h - H_1)}{h^3} \tag{E.5.9.7}$$

where

$$h(x) = H_1 + \frac{x^2}{2R} \text{ for } -x_{in} \le x \le x_{out} \qquad \text{(E.5.9.8)}$$

and

$$p(x = x_{out}) = 0$$

Dimensionalization of Eq. (5.9.5) yields with $\hat{x} = \frac{x}{(2H_1R)^{1/2}}$, $\hat{y} = y/H_1$, $\hat{u} = u/U = u/R\omega_0$, and $\hat{h} = h/H_1$:

$$\hat{u}(\hat{x}, \hat{y}) = 1 - \frac{3\hat{x}^2}{2(1+\hat{x}^2)}\left[1 - \left(\frac{\hat{y}}{1+\hat{x}^2}\right)^2\right] \qquad \text{(E.5.9.9)}$$

Graph:

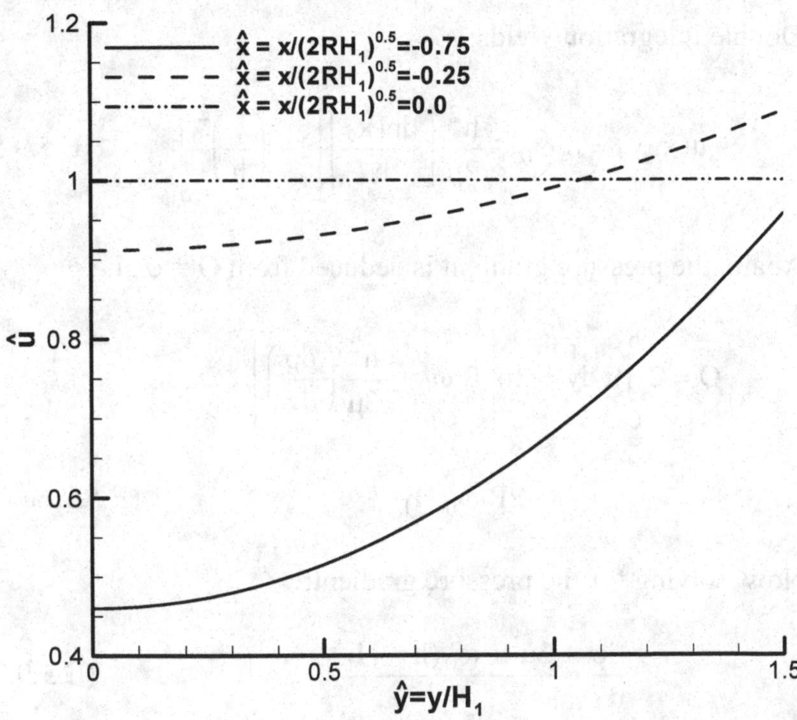

Comments:

- As the deformable sheet approaches the roller, i.e., $-x_{in} < x < 0$, the highly nonlinear material velocity profile flattens out and becomes uniform at x = 0.
- Because the axial coordinate appears as $\hat{x}^2$ in Eq. (E.5.9.9), the same velocity development occurs for $0 < x < x_{out}$.

5.5 Homework Assignments

Solutions to homework problems done individually or in, say, three-person groups should help to further illustrate fluid dynamics concepts as well as approaches to problem solving, and in conjunction with App. A, sharpen the reader's math skills (see Fig. 2.1). Unfortunately, there is no substantial correlation between good HSA results and fine test performances, just *vice versa*. Table 1.1 summarizes three suggestions for students to achieve a good grade in fluid dynamics – for that matter in any engineering subject. The key word is "independence", i.e., after studying the text and equipped with an equation sheet (see App. A), the student should be able: (i) to satisfactory answer all concept questions and (ii) to solve correctly all basic fluid dynamics problems.

The "Insight" questions emerged directly out of the Chap. 5 text, while some "Problems" were taken from lecture notes in modified, i.e., enhanced, form when using White (2006), Cimbala & Cengel (2008), and Incropera et al. (2007). Additional examples, concept questions and problems may be found in good UG fluid mechanics and heat transfer texts, or on the Web (see websites of MIT, Stanford, Cornell, Penn State, UM, etc.).

5.5.1 Text-Embedded Insight Questions and Problems

5.5.1 Derive Prandtl's boundary-layer equations (5.3 and 5.4a, b) based on scale analysis, in light of the conditions (5.2a, b). How do developing shear layers in a pipe's entrance region differ from, say, flat-plate boundary layers?

5.5.2 Derive Eq. (5.5) plus the associated boundary conditions (5.6a–c).

5.5.3 Why is Eq. (5.18) applicable to all types of flat-plate boundary layers, i.e., laminar and turbulent? What happens when the (external) wall is strongly curved, for example, for ship hulls?

5.5.4 List sources/causes of turbulence kinetic energy (see Eq. (5.20)).

5.5.5 Discuss turbulence, its origins, characteristics, and modeling approaches. Note: This is a Class Presentation topic.

5.5.6 Why is there a (linear) laminar sublayer at the wall in turbulent flows?

5.5.7 Produce $C_{Drag}(Re_D)$ -curve for a cylinder in cross flow and discuss the different flow mechanisms w.r.t. the drag curve and flow structures for $0.1 \leq Re_D \leq 10^{10}$.

5.5.8 As problem 5.5.7, but for a disk.

5.5.9 Derive Eq. (5.36) from first geometric principles.

5.5.10 The thin-film (or lubrication) analysis assume no rippling of the free surface, i.e., low Reynolds numbers. Discuss conditions, characteristics and potential solution methods for more realistic cases with $Re_h > 1$.

5.5.11 Revisiting Example 5.9: (a) Derive Eq. (E.5.9.1a) and Eq. (E.5.9.2b); (b) Determine the final sheet thickness H.

5.5.2 Problems

5.5.12 Blasius flow (i.e., $\frac{\partial p}{\partial x} \equiv 0$ and hence $U_{outer} = u_\infty = \not\subset$): (a) Why does the friction coefficient $c_f(x)$ decrease with downstream distance? (b) Determine the total drag force on a flat plate ($l = 5\,m$,

w = 1 m) subject to an oil stream ($u_\infty = 2 m/s$, $\nu = 1\times10^{-4} m^2/s$).
(c) Plot $c_f(x)$ for $0 < x \le 5m$, considering an air-stream ($u_\infty = 5m/s$, $\nu = 2\times10^{-5} m^2/s$), and comment.

5.5.13 Develop a relation $F_{Drag} = F_D(\chi\, u_\infty)$ for steady laminar flat-plate B-L flow, where χ is a positive integer.

5.5.14 Consider Blasius flow with $u_\infty = U_0$ and an assumed B-L velocity profile, i.e.,

$$u(x, y) = U_0 \tanh[\frac{y}{a(x)}]$$

(a) While the 1st B.C. is u(y=0)=0, the second one is $u[y = \chi a(x)] = U_0$. Determine the numerical value for χ.
(b) Solve for a(x) and hence find $\delta(x)$, $\tau_{wall}(x)$ and $c_f(x)$.

5.5.15 Consider steady incompressible *turbulent* B-L flow with $U = \not{c}$. Assuming a power-law velocity profile:

$$\frac{\bar{u}}{U} = (\frac{y}{\delta})^{1/n} \qquad (n \approx 7)$$

and the WSS-equation:

$$WSS \equiv \tau_w = \frac{f}{8}(\rho v^2)$$

where v is the mean velocity and the friction factor (Blasius)

$$f \approx 0.3164\,Re_\delta^{-1/4}$$

where $Re_\delta = \dfrac{v(2\delta)}{\nu} = O(10^7)$

(a) Calculate $v = v(n, U)$;
(b) Find an expression for $\tau_w = \tau_w(v; \delta; \rho; \nu)$;
(c) Develop an ODE for $\delta(x)$ and show that $\dfrac{\delta}{x} \approx \dfrac{0.38}{Re_x^{1/5}}$;

$$5\times10^5 < Re_x = \frac{Ux}{\nu} \le 10^8;$$

(d) Obtain an expression for the skin-friction coefficient and contrast laminar B-L and turbulent B-L flow results.

5.5.16 Discuss $F_{Drag}^{total} = F_{form} + F_{friction}$ for (a) blunt body and (b) a streamlined body, graphing flow field structures, and giving mathematical descriptions and physical explanations.

5.5.17 Spheres and cylinders in cross flow: (a) Show how the wakes in spheres and cylinders differ, considering $0.1 < Re_D < 10^6$; (b) for a cylinder in cross-flow, why does the drag coefficient suddenly drop when the flow becomes *turbulent*; although, c_D is supposed to increase with Re_D; (c) why is flow separation delayed in turbulent flow over a cylinder; (d) why do the $c_D(Re_D)$-curves for sphere and cylinder in App. B only hold for smooth submerged bodies?

5.5.18 Estimate the total force on a cylinder (e.g., a wire, pole, pipe, round tower, etc.) in high winds. Take $u_\infty = U = 50-100$ km/h, $0.1 \le D \le 10$ m, and $L = 1-100$ m. Plot the results.

5.5.19 Consider a ball ($d = 1$ cm, $\rho_b = 10^3$ kg/m^3) falling vertically on your head. Determine (a) the terminal velocity and (b) the force of impact.

5.5.20 What is the updraft (wind) velocity required to suspend a pollutant ($d = 0.1$ mm, $\rho_p = 2.1$ g/cm^3) in air (1 atm, 25 °C).

5.5.21 When computing the lift of airfoils (or on the pitched roof in Example 3.1) viscous effects are neglected – why? Why does the lift force suddenly decrease with an increase of the angle of attack?

5.5.22 "Induced drag" in airfoil theory is defined as the additional drag caused by the tip vortices. How should airplane and glider-plane wings be designed to reduce such drag. How did larger birds solve this problem?

5.5.23 A jet-aircarrier ($M_{max} = 4 \times 10^5$ kg with 500 passengers) has a take-off speed of 250 km/h. Assuming 150 kg for each passenger (plus luggage), plot the take-off speed vs. passenger seating, i.e., $v = v(N)$, $0 < N < 500$. Comment!

5.5.24 Consider a TT-ball ($d = 3.8$ cm, $m = 2.6$ g) suspended in an airstream (1atm, 25 °C). Determine v_{air}. What happens when the ball is pushed off-center, explain!

5.5.25 A glass ball falls with terminal velocity in an unknown fluid. Based on the experimental data $d = 3$ mm, $\rho_b = 2{,}500$ kg/m^3, $\rho_{fluid} = 875$ kg/m^3 and $v_t = 0.12$ m/s, determine the fluid viscosity.

5.5.26 Thin film on an incline: Consider gravity-driven flow of a thin layer of a Newtonian fluid on a flat slope with angle θ and fluid width W.

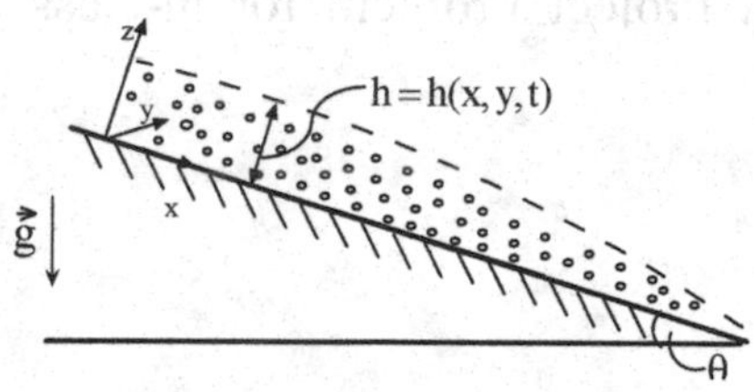

(a) Based on suitable assumptions/postulates, derive the governing equation for h(x,y,t), i.e.,

$$\frac{\partial h}{\partial t} + \frac{g \sin\theta}{3\nu}\frac{\partial h^3}{\partial x} = \frac{g\cos\theta}{3\nu}\left[\frac{\partial}{\partial x}\left(h^3\frac{\partial h}{\partial x}\right) + \frac{\partial}{\partial y}\left(h^3\frac{\partial h}{\partial y}\right)\right] \quad (1)$$

where $\nu \equiv \mu/\rho$ is the kinematic viscosity (see Leal 2007).

(b) Assuming steady state and a constant material supply $Q[L^3/T]$, consider a downstream test section where h is still h = h(x,y) but $\bar{h} \ll \bar{y}(x) \ll x$, where $\bar{h}$ and $\bar{y}$ are characteristic lengths, i.e., film thickness and cross-slope extent, respectively. Show that now

$$\frac{\partial h^3}{\partial x} = (\cot\theta)\frac{\partial}{\partial y}(h^3\frac{\partial h}{\partial y}) \quad (2)$$

where it is postulated that $h(x,y) = \bar{h}(x)\cdot H(\eta)$, $\eta = \frac{y}{\bar{y}(x)}$,

i.e.,

$$H(\eta) = \frac{3}{14}(\bar{\eta}^2 - \eta^2) \quad (3)$$

$$\bar{h}(x) \sim (\frac{\nu Q}{g\sin\theta})^{2/7}(x\cot\theta)^{-1/7} \quad (4)$$

and

$$\bar{y}(x) \sim (\frac{\nu Q}{g\sin\theta})^{1/7}(x\cot\theta)^{3/7} \quad (5)$$

(c) Confirm Eqs. (3)-(5), solve for Eq. (2), plot $h(x,y)$ for reasonable parameters θ and $\nu Q/(g\sin\theta)$, and comment.

Note: This is a Project Problem for in-class presentation and discussion.

References (Part B)

Bertin J.J., 2002, *Aerodynamics for the Engineer*, Prentice Hall, NJ.

Bird, R.B., Stewart, W.E., Lightfoot, E.N., 2002, *Transport Phenomena*, 2nd ed., Wiley, New York.

Blasius, H., 1908, Zeitschrift für Mathematik und Physik, Vol. 56, pp. 1–37.

Boussinesq, J., 1877, Mém. Acad. Sci. Inst. Nat. France, Vol. 23, pp. 1–680.

Chandran, K.B., Rittgers, S.E., Yoganathan, A.P., 2007, *Biofluid Mechanics: the Human Circulation*, CRC Press, Boca Raton, FL.

Cimbala, J.M., Cengel, Y.A., 2008, *Essentials of Fluid Mechanics: Fundamentals and Applications*, McGraw-Hill, New York, NY.

Cooney, D.O., 1976, *Biomedical Engineering Principles*, Marcel Dekker, New York.

Dittus, F.W., Boelter, L.M.K., 1930, *Heat Transfer in Automobile Radiators of the Tubular Type*, University of California Publications in Engineering 2, Berkeley, CA, Vol. (13), pp. 443–461.

Haaland, S.E., 1983, Journal of Fluid Engineering, Vol. 105(1), pp. 89–90.

Hoffmun, J.D., 2001, *Numerical Methods for Engineers and Scientists*, 2nd ed., CRC Press, Boca Raton, FL.

Incropera, F.P., DeWitt, D.P., Bergman, T.L., Lavine, A.S., 2007, *Introduction to Heat Transfer*, Wiley, New York.

Kleinstreuer, C., 1997, *Engineering Fluid Dynamics*, Cambridge University Press, New York, NY.

Kleinstreuer, C., 2006, *Biofluid Dynamics – Principles and Selected Applications*, Taylor & Francis, Boca Raton, FL/London/New York.

Kolmogorov, A.N., 1942, Izvestiya Academy of Sciences USSR, Phys. Vol. 6, pp. 56–58.

Middleman, S., 1972, *Transport Phenomena in the Cardiovascular System*, Wiley Interscience, New York.

Panton, R.L., 2005, *Incompressible Flow*, 3rd ed., Wiley, Hoboken, NJ.

Papanastasiou, T.C., 1994, *Applied Fluid Mechanics*, Prentice-Hall, Englewood Cliffs, NJ.

Polyanin, A.D., Zaitsev, V.F., 1995, *Handbook of Exact Solutions for Ordinary Differential Equations*, CRC Press, Boca Raton, FL.

Prandtl, L., 1904. *Verhandlungen des III. Internationalen Mathematiker-Kongress*, Heidelberg, pp. 484–491.

Sabersky, R.H., Acosta, A.J., Hauptmann, E.G., Gates, E.M., 1999, *Fluid Flow: A First Course in Fluid Dynamics*, Prentice-Hall, Englewood Cliffs, NJ.

Schlichting, H., Gersten, K., 2003, *Boundary-Layer Theory*, 2nd printing, McGraw Hill, New York.

Szeri, A., 1980, *Tribology: Friction, Lubrication and Wear*, 1st ed., McGraw-Hill, New York.

Truskey, G.A., Yuan, F., Katz, D.F., 2004, *Transport Phenomena in Biological Systems*, Pearson/Prentice Hall, Upper Saddle River, NJ.

von Kármán, T., 1921, Zeitschrift für Angewandte Mathematik und Mechanik, Vol. 1, pp. 233–252.

White, F.M., 2006, *Viscous Fluid Flow*, McGraw-Hill, New York.

White, R.A., 1974, *The Calculation of Supersonic Axisymmetric Afterbody Flow with Jet-Interference and Possible Flow Separation*, Aeronautical Research Institute of Sweden, Aerodynamics Dept, Stockholm.

Wilcox, D.C., 1998, *Turbulence Modeling for CFD*, 2nd ed., DCW industries, La Canada, CA.

Part C

Modern Fluid Mechanics Topics

PART C: Modern Fluid Dynamics Topics

Chapter 6

Dilute Particle Suspensions

6.1 Introduction

Natural and industrial two-phase flows, i.e., particle suspensions, are all around us, ranging from dust-storms in arid regions to bubbly flows in pipes or air-fuel injection in ICEs. Traditionally, two-phase flow theory/application was the domain of applied mathematicians, as well as chemical, environmental and nuclear engineers. However, for pipe-network design, pump sizing and applied force evaluation, mechanical engineers have to know basic two-phase flow modeling techniques. Furthermore, biomedical engineers encounter fluid-particle dynamics problems in both the cardiovascular and the pulmonary systems.

By definition, two-phase flow is the interactive flow of two distinct phases with common interfaces in, say, a conduit. Each phase, representing a volume fraction (or mass fraction) of solid, liquid or gaseous matter, has its own properties, velocity, and temperature. Typical dilute (or dense) particle suspension flows include droplets in gas flow, liquid–vapor, i.e., bubbly, flow as well as liquid or gas flow with solid particles. In addition to predicting the flow phases, it is also important to know the flow regimes, i.e., characteristic flow patterns based on the interfaces formed between the phases (see Fig. 6.1). Two-phase systems can be grouped into flows of separated phases, mixed phases, and dispersed phases. Examples of *separated* flows include liquid layers on a wall in gas flow, e.g., the mucus layer in lung airways, and liquid jets in gas flow (or vice versa). *Mixed-phase* flows are encountered in phase-change processes, such as boiling nuclear reactor channels and steam pipes with

C. Kleinstreuer, *Modern Fluid Dynamics: Basic Theory and Selected Applications in Macro- and Micro-Fluidics*, Fluid Mechanics and Its Applications 87,
DOI 10.1007/978-1-4020-8670-0_6, © Springer Science+Business Media B.V. 2010

vapor core and annular liquid wall film as well as heat pipes with large vapor bubbles and evaporating liquid layers on heated surfaces. Most frequently, two-phase flows appear as *dispersed* phases, such as dilute particle suspensions in gas or liquid flows, droplets in gas flow (e.g., sprays) or bubbles in liquid flows (e.g., chemical reactors). Clearly, alternative two-phase flow classifications exist. For example, solid particles, droplets or bubbles form the *dispersed (or particle) phase* while the carrier fluid is the *continuous (or fluid) phase*. The degree of phase coupling, i.e., from one-way for very dilute suspensions to four-way in dense suspensions. In the latter case, not only fluid flow affects particle motion and vice versa but particle–particle interactions due to collision are expected and particle-induced flow fields affect other particles, as in drafting.

Critical heat transfer may change the thermodynamic state of a phase or may generate two-phase flow in the first place, as discussed by Naterer (2002) and Faghri & Zhang (2006). Other recent two-phase or multiphase books include the texts by Crowe et al. (1998) and Kleinstreuer (2003), and the handbook edited by Crowe (2006).

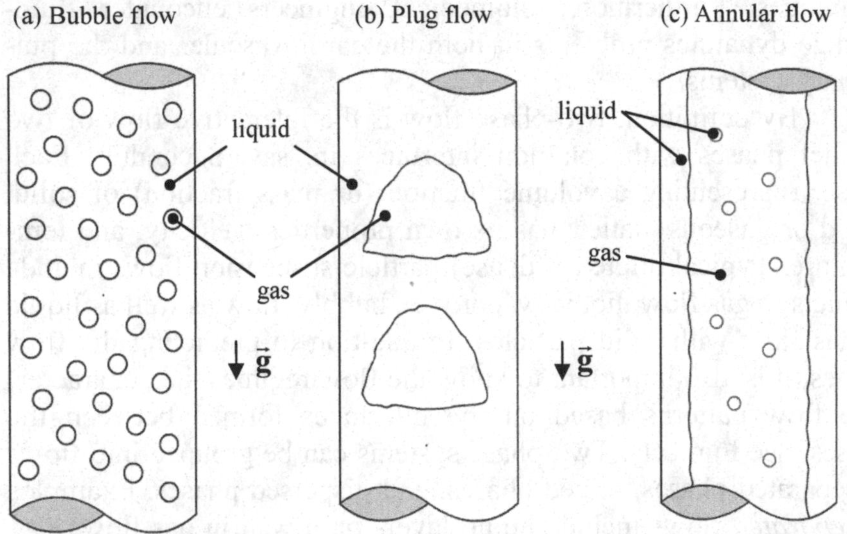

Fig. 6.1 Flow regimes in vertical pipes with co-current gas–liquid flows

6.2 Modeling Approaches

It should be evident from reading the Introduction (Sect. 6.1) that a very dilute suspension of uniformly distributed (micron or nano) particles which only slightly affect the mixture properties constitute the simplest two-phase flow example and hence the easiest case to solve. Such well-mixed suspensions, which are actually *pseudo* two-phase flows, can be described with homogeneous fluid mechanics equations and fall into the category of *flow mixture models* (see Fig. 6.2). In contrast, when a distinct particle phase interacts with the continuous phase, more complex *separate flow models* are needed to describe the two-phase dynamics.

6.2.1 Definitions

Focusing on dispersed flows, the particle phase is characterized by the particle volume fraction, mass concentration, and loading. In light of the continuum assumption, all quantities are defined as volume ratios with limits from the mixture volume sample $\delta\forall'$ which excludes the molecular range. It should be noted that *subscripts c, f or 1* indicates the continuous phase or carrier fluid, while *subscripts p, d or 2* refer to the particle or dispersed phase. For example, the volume (or void) fraction of the dispersed phase is:

$$\alpha_d \equiv \alpha = \lim_{\delta\forall \to \delta\forall'} \frac{\delta\forall_d}{\delta\forall} := \frac{\forall_{\text{particles}}}{\forall_{\text{mixture}}} \tag{6.1}$$

Then, the volume fraction of the continuous fluid phase is:

$$\alpha_c = \lim_{\delta\forall \to \delta\forall'} \frac{\delta\forall_c}{\delta\forall} := \frac{\forall_{\text{fluid}}}{\forall_{\text{mixture}}} \tag{6.2}$$

so that

$$\alpha_d + \alpha_c = 1 \tag{6.3}$$

(a) Hierarchy of multiphase flow models

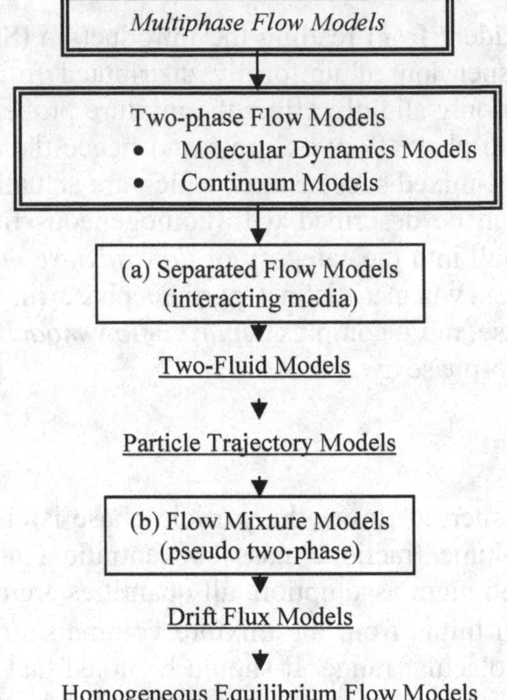

(b) Two-phase flow model applications

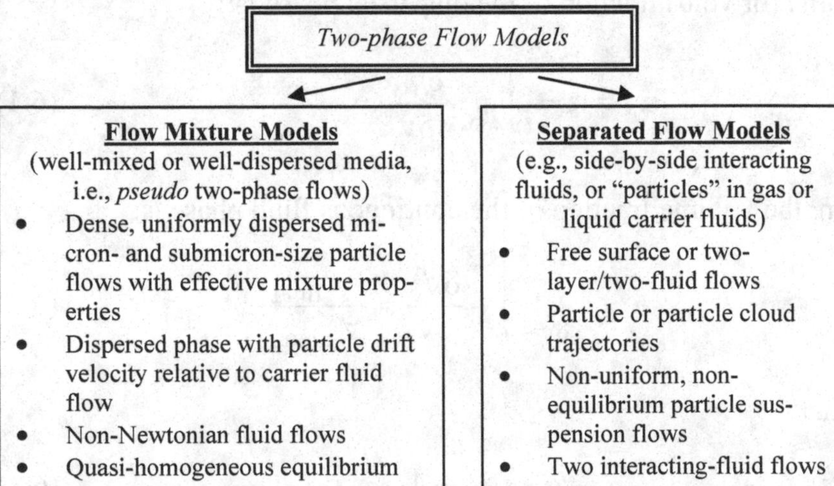

Fig. 6.2 Two-phase flow modeling categories

With the two-phase volume fractions defined, the mixture (or effective) density is:

$$\rho_m = \alpha_c \rho_c + \alpha_d \rho_d := \bar{\rho}_c + \bar{\rho}_d \qquad (6.4a, b)$$

where $\bar{\rho}_d = n\, m_p$ with n being the number of particles per unit volume and m_p being the particle's mass. The dispersed-phase concentration is given as:

$$c = \frac{\bar{\rho}_d}{\bar{\rho}_c} \qquad (6.5)$$

The local loading is the mass-flux ratio:

$$r = \frac{\bar{\rho}_d\, \bar{v}_p}{\bar{\rho}_c\, \bar{v}_c} \qquad (6.6a)$$

while the total loading is:

$$\kappa = \dot{m}_d / \dot{m}_c \approx \bar{\rho}_d / \bar{\rho}_c = c \qquad (6.6b)$$

implying that κ is approximately the concentration c.
Clearly, the mixture mass flow rate, important for internal flows, is:

$$\dot{m}_m = \dot{m}_c + \dot{m}_d = (\rho Q)_c + (\rho Q)_d \qquad (6.7a, b)$$

which leads to the "quality" (i.e., particle mass concentration):

$$x = \dot{m}_d / \dot{m}_m \qquad (6.8)$$

while the mixture mass flux is:

$$G_m = \frac{\dot{m}_m}{A} = G_c + G_d \qquad (6.9a, b)$$

and the volume flux is:

$$j_m = \frac{G_m}{\rho_m} = \frac{Q_c + Q_d}{A} = j_c + j_d = v_m \qquad (6.10a\text{–}d)$$

where v_m is the superficial velocity of the mixture, which is composed of the phase superficial velocities:

$$v_{c,s} = \alpha_c\, v_c \quad \text{and} \quad v_{d,s} = \alpha_d\, v_d \qquad (6.11a, b)$$

while the actual phase velocities are:

$$v_c = j_c / \alpha_c \quad \text{and} \quad v_d = j_d / \alpha_d \qquad (6.12a, b)$$

Finally, the relative (or slip) velocity between the two phases is:

$$v_r = v_c - v_d \qquad (6.13)$$

The drift velocity which indicates deviatory motion of the particle phase from the mixture flow is:

$$v^{drift} = v_d - v_m \qquad (6.14)$$

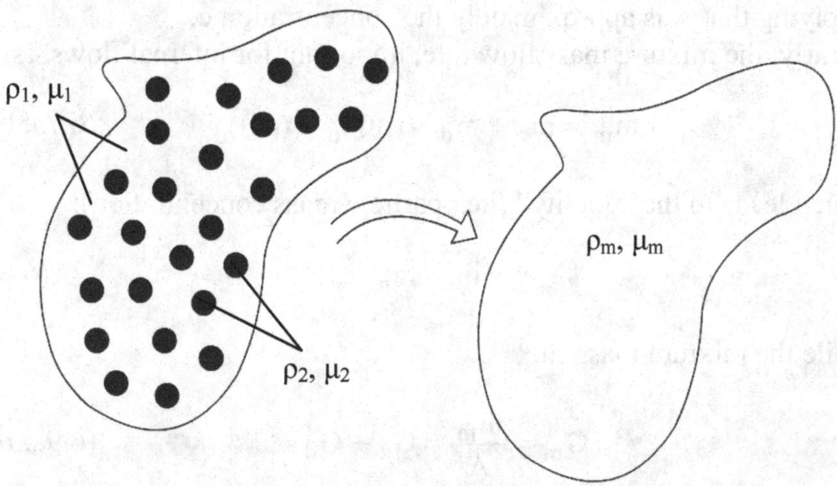

Fig. 6.3 Approximation of two-phase (dispersed) flow to a uniform mixture flow

With these basic phase definitions established, we can now consider mixture properties, such as density ρ_m and dynamic viscosity μ_m (see Fig. 6.3). Specifically,

$$\rho_m = \alpha\, \rho_2 + (1-\alpha)\, \rho_1 \qquad (6.15a)$$

or

$$\frac{1}{\rho_m} = \frac{x}{\rho_2} + \frac{1-x}{\rho_1} \qquad (6.15b)$$

where $\alpha \equiv \alpha_2 = \forall_2 / \forall = \forall_d / \forall_m$ and $x = \dot{m}_2 / \dot{m}$. At low volume fractions of spherical particles (say, $\alpha \leq 0.05$) the mixture viscosity is (Soo 1990; Zapryanov & Tabakova 1999):

$$\mu_m = \mu_1 \left(1 + 2.5\alpha \frac{\mu_2 + 0.4\mu_1}{\mu_2 + \mu_1} \right) \qquad (6.16a)$$

which for $\mu_2 >> \mu_1$ reduces to:

$$\mu_m \big|_{\text{solid spheres}} = \mu_1 (1 + 0.4\alpha) \qquad (6.16b)$$

or

$$\mu_m \big|_{gas\, bubbles} = \mu_1 (1 + \alpha) \qquad (6.16c)$$

Similar to Eq. (6.15b), for well-dispersed gas-liquid flows, e.g., droplets in air or bubbles in liquid flow, we have

$$\frac{1}{\mu_m} = \frac{x}{\mu_2} + \frac{1-x}{\mu_1} \qquad (6.17a)$$

or

$$\mu_m = x\, \mu_2 + (1-x)\, \mu_1 \qquad (6.17b)$$

Example 6.1: Poiseuille-Type Mixture Flow

Consider steady laminar fully-developed pipe flow (radius R, length L) where the void fraction of solid particles in air ranges from $\alpha = 0$ to $\alpha = 0.05$. Find u(r) and plot u / u_{max} vs. r/R.

Sketch	*Assumptions*	*Concepts*
p(x), r=R, ρ_m, μ_m, r, x, r=0, u(r), u_{max}	• Poiseuille flow • One-way coupling • Quasi-homogeneous equilibrium flow • Constant mixture properties	• Reduced N–S equations • Constant $-\frac{dp}{dx} = \frac{\Delta p}{L}$ • Mixture viscosity $\mu_m = \mu_1 (1 - 0.4\alpha)$ • $\alpha \equiv \frac{\forall_{particles}}{\forall_{mixture}}$

Solution: Based on the assumptions, continuity is fulfilled and the x-momentum equation reduces to (see Example 2.10):

$$0 = -\frac{dp}{dx} + \frac{\mu}{r}\frac{d}{dr}\left(r\frac{du}{dr}\right) \quad \text{(E.6.1.1)}$$

subject to $u(r = R) = 0$ and $du/dr|_{r=0} = 0$. Hence,

$$u(r) = \underbrace{\frac{R^2}{4\mu_m}\left(\frac{\Delta p}{L}\right)}_{u_{max}}\left[1 - \left(\frac{r}{R}\right)^2\right] \quad \text{(E.6.1.2)}$$

where according to Eq. (6.16b):

$$\mu_m = (1 + 0.4\alpha)\mu_{fluid} \text{ for } 0 \le \alpha \le 0.05. \qquad \text{(E.6.1.3)}$$

Graph:

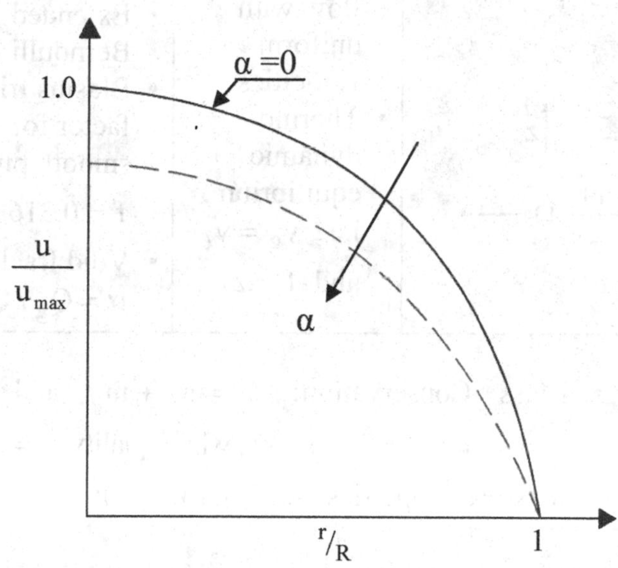

Comments:

- Clearly, for $\alpha = 0$ the original Poiseuille flow is recovered.
- When $\alpha > 0$, the flow rate decreases for the same (given) pressure drop because of the higher resistance, and hence the velocity profile is flatter.

Example 6.2: Bubbly Pipe Flow

Of interest are the total pressure drop and void fraction of steady turbulent bubbly flow in a vertical pipe (D = 25 mm, L = 45 cm). The upward flow is a mixture of water $m_\ell = 0.42\text{kg/s}, \rho_\ell = 10^3 \text{ kg/m}^3$) and air bubbles ($m_g = 0.01$ kg/s, $\rho_g = 1.1777$ kg/m^3.

Sketch	*Assumptions*	*Concepts*
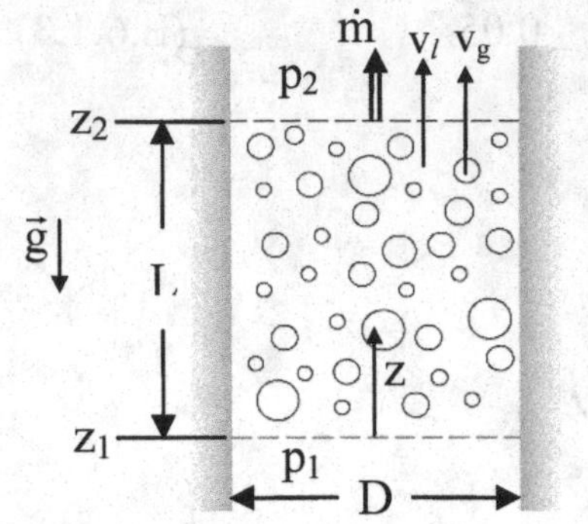	• Steady fully-developed turbulent flow with uniform velocities • Thermo-dynamic equilibrium, i.e., $v_g = v_\ell$ and $T = \not{c}$	• Constant mixture properties ρ_m and μ_m • Extended Bernoulli Eqs. • Blasius friction factor for smooth pipes $f = 0.316\, Re_D^{-1/4}$ • Void fraction $\alpha = Q_g / Q$

Solution:

- Mass Conservation: $\dot{m} = \dot{m}_\ell + \dot{m}_g$ and hence $v = v_\ell = v_g = \dot{m}/(\rho_m A)$, while quality $x = \dot{m}_g / \dot{m}$
- Mixture Properties (see Eq. (6.15b)):

$$\rho_m = \left(\frac{x}{\rho_g} + \frac{1-x}{\rho_\ell}\right)^{-1} \quad \text{and} \quad \mu_m = \left(\frac{x}{\mu_g} + \frac{1-x}{\mu_\ell}\right)^{-1}$$

Thus, the numerical property values are then with $x = 0.023$ and $v = 17.65$ m/s as follows:

$\rho_m \equiv \rho = 49.64\ \text{kg/m}^3$, $\mu_m \equiv \mu = 4.435 \times 10^{-4}\ \text{kg/(m}\cdot\text{s)}$, and $Re_D = \rho v D / \mu = 49{,}388$

- Wall Friction Factor:

After Blasius (see Sect. 5.2) the pipe friction factor is:

$$f = \frac{0.316}{Re_D^{0.25}} = 0.0212$$

- Extended Bernoulli Equation:

$$\frac{p_1}{\rho g}+\frac{v_1^2}{2g}+z_1=\frac{p_2}{\rho g}+\frac{v_2^2}{2g}+z_2+h_f$$

$$h_f=f\left(\frac{L}{D}\right)\frac{v^2}{2g}$$

Hence,

$$\Delta p=\Delta z+h_f:=31.7\ \text{kN}/\text{m}^2$$

- Volumetric Flow Rates:

$$Q_i=\left(\frac{\dot{m}}{\rho}\right)_i:=\begin{cases}0.00833\ \text{m}^3/\text{s} & \text{< bubbles >}\\ 0.00042\ \text{m}^3/\text{s} & \text{< water >}\end{cases}$$

- Void Fraction:

$$\alpha=\frac{\forall_g}{\forall}=\frac{Q_g}{Q}:=0.96$$

Comments:

Although the quality (x = 0.023) in rather low, the almost 1,000-fold density difference between the carrier fluid and the air-bubbles generates 20× the volumetric flow rate for air when compared to water flow. As a result, α = 0.96, i.e., the pipe is mainly filled with air (bubbles), under the assumption of homogeneous equilibrium flow.

6.2.2 Homogeneous Flow Equations

As indicated, certain two-phase flows can be treated as a single-phase flow if both phases are in quasi-thermodynamic equilibrium, i.e., their properties, velocities and temperatures do not deviate significantly. The flow regimes for gas-liquid flows are typically bubbly or misty flows. For solid-particle-suspension flows the void fraction is very low, say, less than 5% and the relative velocity $u_{fluid} - v_{particle}$ is minor. Area-averaging of both phases and the use of mixture properties (see Eqs. (6.15 and 6.16)) result in one-dimensional transport equations. Specifically, for a conduit of area A and inclined angle θ we obtain the following (see HWA Section 6.5).

Continuity:

$$A\frac{\partial\rho}{\partial t}+\frac{\partial}{\partial x}(\rho\, v\, A)=0 \tag{6.18}$$

Momentum:

$$A\frac{\partial(\rho v)}{\partial t}+\frac{\partial}{\partial x}(\rho\, v\, v\, A)=-\frac{\partial(pA)}{\partial x}+\tau_w\, P+\rho\, g\, A\sin\theta \tag{6.19}$$

Energy:

$$\frac{\partial(\rho e)}{\partial t}+\frac{1}{A}\frac{\partial}{\partial x}(v\, A\, \rho\, e)=\frac{P}{A}q_w+q_{gen}+\frac{\partial P}{\partial t} \tag{6.20}$$

Here, τ_w is the wall shear stress; P is the conduit perimeter; the specific energy is $e = h + v^2/2 + g_x \sin\theta$, with $h = u + p/\rho = h_c + x(h_\alpha - h_c)$ being the enthalpy; q_w is the wall heat flux; and q_{gen} is the internal heat generation, say, due to viscous effects or chemical reaction.

For steady homogeneous mixture flow in a rigid pipe of diameter D, the momentum equation (6.19) reduces to:

$$-\frac{dp}{dx}=\frac{\Delta p}{L}=\frac{\partial}{\partial x}(\rho\, v\, v)+\frac{4\tau_w}{D}+\rho\, g\sin\theta \tag{6.21}$$

while the energy equation (6.20) becomes with $dh = c_p\,dT$ the heat transfer equation:

$$\rho c_p \frac{dT}{dx} = -\frac{d}{dx}\left(\frac{\rho}{2}v^2\right) + \frac{4q_w}{vD} + \frac{q_{gen}}{v} - \rho g \sin\theta \tag{6.22}$$

Knowing the quality $x = \dot{m}_d / \dot{m}$, we can express the mixture density as:

$$\rho = \frac{\rho_d\,\rho_c}{\rho_c x + \rho_d(1-x)} \tag{6.23}$$

and the void fraction as:

$$\alpha = \frac{x}{x + (1-x)\,\rho_d/\rho_c} \tag{6.24}$$

while the momentum flux is:

$$\rho v = (\dot{m}_c + \dot{m}_d)/A = \dot{m}/A \tag{6.25}$$

For the wall shear stress, we recall: $\tau_w \sim \Delta p/L \sim h_f = h_f(f, \frac{L}{D}, v^2)$ as discussed in Sect. 4.2. Specifically,

$$-\left.\frac{dp}{dx}\right|_{friction} = \frac{4\tau_w}{D} = \frac{2f\rho v^2}{D} \tag{6.26a}$$

where Beattie & Whalley (1982) suggested for the annular and bubbly regimes (see Fig. 6.1):

$$f^{-1/2} = 3.48 - 4\log_{10}\left[2\left(\frac{\varepsilon}{D}\right) + \frac{9.35}{Re\sqrt{f}}\right] \tag{6.26b}$$

The Reynolds number $Re_D = \rho \, v \, D / \mu$ requires the evaluation of the dynamic viscosity, e.g.,

$$\mu = \left(\frac{x}{\mu_d} + \frac{1-x}{\mu_c} \right) \tag{6.27}$$

Summary The two Examples 6.1 and 6.2 as well as Eqs. (6.18) to (6.26) show that solving quasi-homogeneous flow problems is rather straight forward. It is required that the two phases form a uniform mixture with effective properties (see Fig. 6.3), and that the phases are in thermodynamic equilibrium, i.e., no velocity slip and equal temperatures. As indicated in Fig. 6.2, flow of non-Newtonian fluids (see Sect. 6.3) as well as flows with drift flux (Kleinstreuer 2003) fall also into the category of flow mixture models. More complicated are *separated flow* models (see Fig. 6.2). Examples include two-layer fluids flowing with a smooth interface and (dilute) spherical particle suspension flows (Sect. 6.4). More realistic aspects of two-phase flows can be described with two-way coupled *two-fluid models* where the phases interact, i.e., they influence each other.

6.3 Non-Newtonian Fluid Flow

We recall that gases and small-molecule liquids (e.g., water and basic oils) are Newtonian; because, the random thermally driven molecular motions (i.e., spin, vibration, collision) within such materials are sufficiently vigorous that they completely overcome any tendency of the fluid flow forces to produce a molecular configuration state (i.e., local molecular restructuring) that differs significantly from the isotropic, homogeneous state of statistical equilibrium. Clearly, subject to a shear stress ($F_{tang.} / A$) fluid-mass displaces, i.e., it flows continuously without changing the fluid configuration on the molecular level. In contrast, examples of *non-Newtonian* fluids are macro-molecule (MW > 400) fluids, such as paints, exotic oils, polymeric liquids, multi-fluid blends, and particle suspensions, such as blood (when $\dot{\gamma} < 200 \ s^{-1}$) slurries, etc.

As outlined in Sect. 2.4, for Newtonian fluids, such as air, water and basic oils, Stokes' hypothesis of a *linear* relationship between shear stress and shear rate holds (see Fig. 6.4). In contrast, some fluids, such as polymeric liquids, exotic lubricants, latex paints, food stuff, paste and certain particle suspensions, exhibit nonlinear viscous effects. Shear-rate dependence and/or memory of the viscosity of non-Newtonian fluids is due to their component make-up and/or molecular structure (Tanner 1998; Macosko 1994; Bird et al. 1987). Assuming steady incompressible isothermal fluid flow, only shear-rate (or shear-stress) dependent liquids are considered, i.e.,

$$\tau_{ij} = \mu\, \dot{\gamma}_{ij} \text{ is now replaced by } \tau_{ij} = \eta(\dot{\gamma}, \tau)\dot{\gamma}_{ij} \qquad (6.28a, b)$$

where η is the non-Newtonian (or apparent) viscosity.

Thus, non-Newtonian fluid flow phenomena such as rotating-rod climbing and jet-swelling after extrusion or the visco-elastic effect of fluid recoil, stress relaxation and overshoot are not discussed.

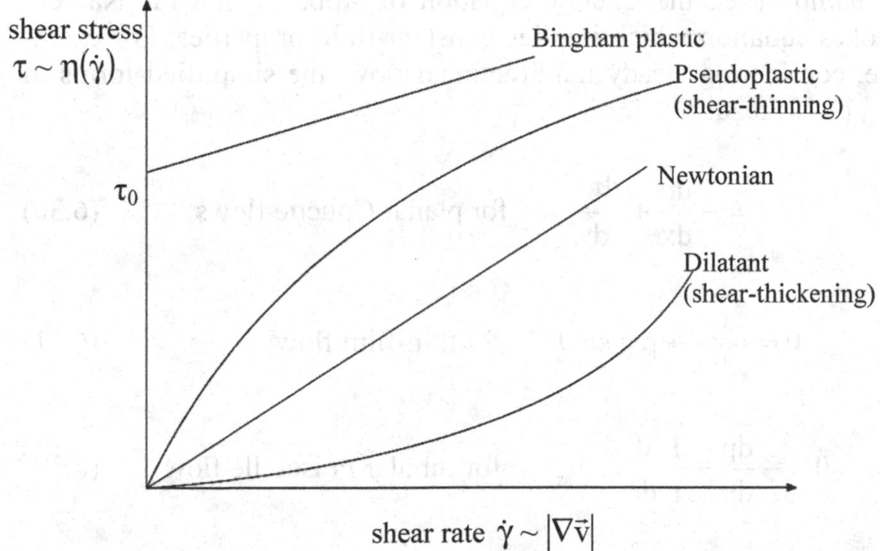

Fig. 6.4 Stress/shear-rate behavior of fluids

6.3.1 Generalized Newtonian Liquids

As indicated with Eqs. (6.28a, b), viscous inelastic liquids are also labeled generalized Newtonian liquids because of their similar constitutive equations. The simplest empiricism for $\eta(\dot{\gamma})$ is the two-parameter *power law* expression:

$$\eta = m\,\dot{\gamma}^{n-1} \tag{6.29}$$

where the constants m and n characterize the fluid. Clearly, when n = 1, m = μ and Eq. (6.28a) is recovered. If n < 1, the fluid exhibits shear-thinning (i.e., pseudoplastic) behavior, while for n > 1 the fluid is called dilatants (or shear thickening) as shown in Fig. 6.4. Although the power-law is being widely used, it cannot describe the viscosity at very low shear rates (e.g., blood) and the parameters m and n are *not* actual fluid properties.

Power-Law Modeling When using Eq. (6.29) to model viscous inelastic liquid flows, the point of departure is the reduced *equation of motion* (i.e., the Cauchy equation of App. A), not the Navier–Stokes equation which implies constant fluid properties. For example, considering steady unidirectional flow, the simplified forms of Eq. (2.18) read:

$$0 = -\frac{dp}{dx} + \frac{d\tau_{yx}}{dy} \quad \text{for planar Couette flows} \tag{6.30}$$

$$0 = \frac{d\tau_{yx}}{dy} + \rho\, g \sin\theta \quad \text{for thin-film flow} \tag{6.31}$$

$$0 = -\frac{dp}{dx} + \frac{1}{r}\frac{d}{dr}(r\,\tau_{rx}) \quad \text{for tubular Poiseuille flow} \tag{6.32}$$

Examples 6.3 to 6.5 illustrate the use of Eqs. (6.30–6.32).

Bingham Fluid For thick suspensions and pastes (e.g., ketchup and toothpaste), no flow occurs until a certain critical stress, called the *yield stress* τ_0, is reached as the result of an applied force. Then, the mixture flows like a Newtonian fluid (see Fig. 6.4). Thus, the two-parameter Bingham model:

$$\eta = \begin{cases} \infty & \text{for} \quad \tau < \tau_0 \\ \mu_0 + \dfrac{\tau_0}{\dot{\gamma}} & \text{for} \quad \tau \geq \tau_0 \end{cases} \tag{6.33a, b}$$

is an illustration of a constitutive equation for a "viscoplastic" material.

More accurate but also more complex non-Newtonian fluid models for specific applications are discussed in Bird et al. (1987), Macosko (1994) and Kleinstreuer (2006), among other texts. Convection heat transfer results in terms of Nusselt number correlations for pipe and slit flows with power-law fluids are summarized in Bird et al. (2002).

Example 6.3: Power-Law Fluid Flow in a Slightly-Tapered Tube

Sketch	*Assumptions*	*Concepts*
p(z); r=R₀; R(z); $\eta(\dot{\gamma})$; R_L; r; z; r=0; L	• Steady laminar unidirectional flow • Power-law fluid • $\vec{v} = [v_z(r, z); 0; 0]$ • $-\partial p/\partial z = \Delta p / L = \not{c}$	• Eq. (6.32) with $v_z(z)$ dependence via no-slip condition

Solution:

- Slightly tapered tube:

$$R(z) = R_0 - \frac{R_0 - R_L}{L} z \tag{E.6.3.1}$$

- Shear stress and power law:

$$\tau_{rz} = \eta \frac{dv_z}{dr} \text{ and } \eta = m\dot{\gamma}^{n-1} = m\left(\frac{dv_z}{dr}\right)^{n-1}$$

so that

$$\tau_{rz} = m\left(-\frac{dv_z}{dr}\right)^{n} \tag{E.6.3.2}$$

where the negative sign assures that $\dot{\gamma}$ stays a positive quantity.

- From a 1-D force balance for fully-developed flow:

$$\tau_{rz} = -\frac{\Delta p}{2L} r = \tau_w \frac{r}{R} \tag{E.6.3.3a, b}$$

which holds for turbulent flow as well.

- z-momentum equation (6.32):

$$\frac{1}{r}\frac{d}{dr}(r\,\tau_{rz}) = -\frac{p_0 - p_L}{L} \tag{E.6.3.4a}$$

or after integration

$$\tau_{rz} = -\frac{\Delta p}{2L} r + \frac{C_1}{r} \tag{E.6.3.4b}$$

Clearly, $C_1 \equiv 0$ because at the centerline r = 0 but τ_{rz} is finite, i.e., $\tau_{rz}(r=0)=0$ (see Eq. (E.6.3.3)). Combining (E.6.3.4b) with (E.6.3.3) and (E.6.3.2) yields:

$$m\left(-\frac{dv_z}{dr}\right)^{n} = \tau_w \frac{r}{R} \tag{E.6.3.5}$$

Taking the nth root of both sides and integrating results in

$$v_z = -\left(\frac{\tau_w}{mR}\right)^{1/n} \frac{n\, r^{\frac{n+1}{n}}}{n+1} + C_2$$

subject to $v_z[r = R(z) = R_0 - \frac{R_0 - R_L}{L} z] = 0$

Thus,

$$v_z(r, z) = \underbrace{\left[\left(\frac{\tau_w}{m R(z)}\right)^{1/n} \frac{n R(z)}{n+1}\right]}_{v_{max}} \left[1 - \left(\frac{r}{R(z)}\right)^{\frac{n+1}{n}}\right] \quad \text{(E.6.3.6)}$$

where $\tau_w = -\frac{\Delta p}{2L} R$ from Eq. (E.6.3.3).

- Pressure drop based on volumetric flow rate:

given a Q-value and with $Q = \int_A \vec{v} \cdot d\vec{A} = 2\pi \int_0^R v_z(r, z)\, r dr$, we obtain

$$\Delta p = \frac{2mL}{3n} \left[\frac{Q}{n\pi}(3n + 1)\right]^n \left(\frac{R_L^{-3n} - R_0^{-3n}}{R_0 - R_L}\right) \quad \text{(E.6.3.7)}$$

Graph:

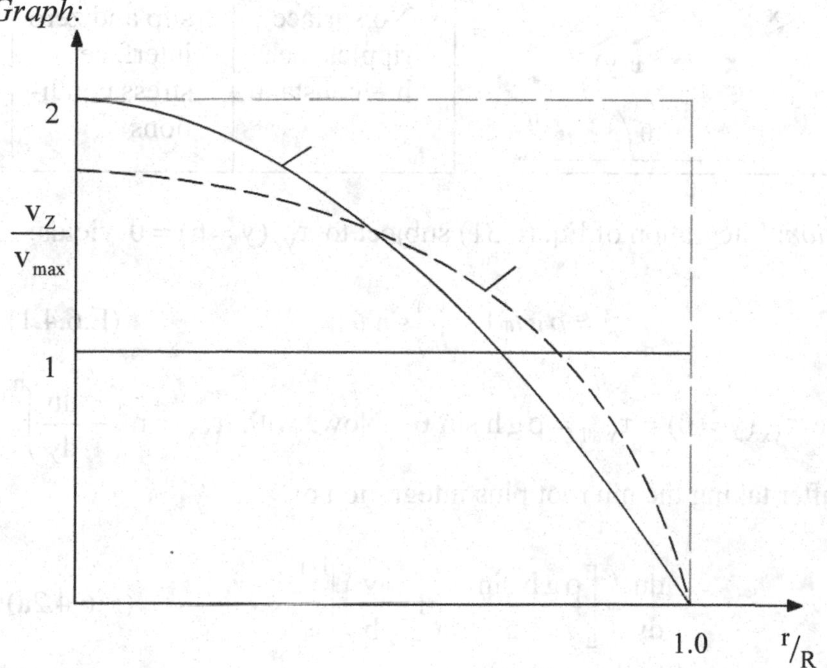

Comments:

- With n < 1.0, the axial velocity profile becomes flatter because of the shear-thinning effect.
- Compare these profiles to the results of Example 6.1 for dilute homogeneous particle suspension flows.

Example 6.4: Film Thickness of a Flowing Polymer

Consider a steady laminar thin layer ($h = \not\subset$) of a power-law fluid (m, n) moving down an incline (angle θ, width w) with a volumetric flow rate Q. Find the film thickness h.

Sketch	*Assumptions*	*Concepts*
y x u(y) h θ	• Steady laminar unidirectional flow • No surface ripples, i.e., h = constant	• Equation (6.31) subject to no-slip and zero interface stress conditions

Solution: Integration of Eq. (6.31) subject to $\tau_{yz}(y=h)=0$ yields:

$$\tau_{yx} = \rho g h\left(1-\frac{y}{h}\right)\sin\theta \tag{E.6.4.1}$$

where $\tau_{yx}(y=0)=\tau_{wall}=\rho g h \sin\theta$. Now, with $\tau_{yx}=m\left(-\frac{du}{dy}\right)^n$ and after taking the nth root plus integration of

$$-\frac{du}{dy}=\left[\frac{\rho g h \sin\theta}{m}\left(1-\frac{y}{h}\right)\right]^{1/n} \tag{E.6.4.2a}$$

subject to u(y = 0) = 0 we obtain:

$$u(y) = C\left[1 - \left(1 - \frac{y}{h}\right)^{\frac{n+1}{n}}\right] \tag{E.6.4.2b}$$

where

$$C \equiv \left(\frac{\rho g h \sin\theta}{m}\right)^{1/n} \frac{n h}{n+1} \tag{E.6.4.2c}$$

The volumetric flow rate is:

$$Q = w \int_0^h u(y)\,dy = w\,h\,C\left(1 - \frac{n}{2n+1}\right) \tag{E.6.4.3a}$$

$$Q = \frac{w h^2 n}{2n+1}\left(\frac{\rho g h \sin\theta}{m}\right)^{1/n} \tag{E.6.4.3b}$$

Given Q, we can solve for the film thickness, i.e.,

$$h = \left(\frac{m}{\rho g \sin\theta}\right)^{1/(2n+1)} \left[\frac{(2n+1)Q}{n\,w}\right]^{n/(2n+1)} \tag{E.6.4.4}$$

Graph:

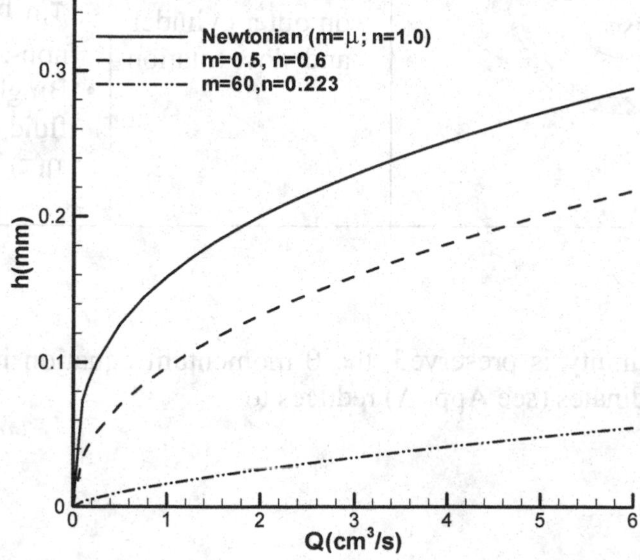

Comments:

- As expected, h(Q) is a nonlinearly increasing function.
- Recalling that power-law fluid with n < 1.0 are shear-thinning, generating a blunter velocity profile (see Example 6.3), the film thickness decrease with lower n-values.

Example 6.5: Cylindrical Couette Flow with a Bingham Plastic

Consider two *concentric* cylinders (length L, $R_{inner} = \kappa R$ and $R_{outer} = R$) where the outer one rotates at angular velocity ω_0 due to torque T, setting a Bingham fluid into steady laminar motion. Develop a relationship for $T = T(\omega_0; R, \kappa, L; \tau_0, \mu_0)$.

Sketch	*Assumptions*	*Concepts*
ω_0; $\kappa R \le r \le R$; $v_\theta(r)$; κR; R; fixed	• As stated • $\vec{v} = [0, v_\theta(r), 0]$ • $\nabla p = 0$ • Constant torque on outer cylinder and slow rotation	• Reduced θ-momentum equation with only $\tau_{r\theta}$ being non-zero • Bingham fluid model $\eta(\tau)$

Solution:

While continuity is preserved, the θ-momentum equation in cylindrical coordinates (see App. A) reduces to:

$$0 = \frac{1}{r^2}\frac{d}{dr}\left(r^2\,\tau_{r\theta}\right) \tag{E.6.5.1}$$

Integration yields:

$$\tau_{r\theta} = \frac{C}{r^2}$$

where

$$\tau_{r\theta}(r = R) = \tau_{wall} = \frac{T}{2\pi L R^2} = \frac{C}{R^2}$$

and hence

$$\tau_{r\theta} = \frac{T}{2\pi L}\,r^{-2} \tag{E.6.5.2}$$

As indicated with Eqs. (6.33a, b) there is a radial location r_0 where $\tau_{r\theta} = \tau_0$, the yield stress. Clearly, from Eq. (E.6.5.2):

$$\tau_0 = \left(\frac{T}{2\pi L\, r_0}\right)^{1/2} \tag{E.6.5.3}$$

Where r_0 has to be between κR and R to observe some form of fluid flow (see Eq. (6.33a)). Specifically, for $\kappa R < r < r_0$ there will be viscous flow and for $r_0 \le r \le R$ there will be uniform (or plug) flow.

Recalling that here with shear rate $\dot{\gamma}_{r\theta} = r\frac{d}{dr}\left(\frac{v_\theta}{r}\right)$ (see App. A), we rewrite Eq. (6.28b) as:

$$\tau_{r\theta} = \eta\left[r\frac{d}{dr}\left(\frac{v_\theta}{r}\right)\right] \qquad \text{(E.6.5.4)}$$

and with Eq. (6.33b) we have:

$$\eta = \mu_0 + \frac{\tau_0}{\dot{\gamma}} = \mu_0 + \frac{\tau_0}{r\frac{d}{dr}\left(\frac{v_\theta}{r}\right)} \qquad \text{(E.6.5.5)}$$

so that

$$\tau_{r\theta} = \tau_0 + \mu_0\, r\frac{d}{dr}\left(\frac{v_\theta}{r}\right) \qquad \text{(E.6.5.6)}$$

Combining (E.6.5.4 and (E.6.5.2) to solve for $v_\theta(r)$ when $\kappa R < r \le R$, we obtain:

$$\frac{d}{dr}\left(\frac{v_\theta}{r}\right) = \frac{T}{2\pi L\mu_0} r^{-3} - \frac{\tau_0}{\mu_0} r^{-1}$$

Integration and invoking the B. C. $v_\theta(r = r_0) = r_0\omega_0$ yields for $\kappa R \le r \le r_0$:

$$v_\theta(r) = \omega_0 r + \frac{T}{4\pi L\mu_0 r_0}\left(\frac{r}{r_0}\right)\left[1 - \left(\frac{r_0}{r}\right)^2\right] - \frac{\tau_0 r}{\mu_0}\ln\frac{r}{r_0} \qquad \text{(E.6.5.7a)}$$

while for $r_0 \le r \le R$:

$$v_\theta = \omega_0\, r_0 \qquad \text{(E.6.5.7b)}$$

The first graph shows schematically the impact regions of Eqs. (6.33a,b) in light of the $\tau_{r\theta}(r)$ function given by Eq. (E.6.5.2).

Graph I:

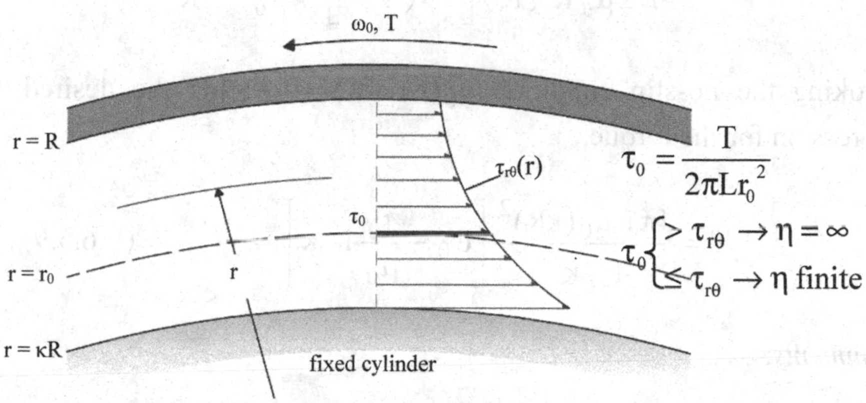

The second graph depicts $v_\theta(r)$ when r_0 is between κR and R as shown in Graph I.

Graph II:

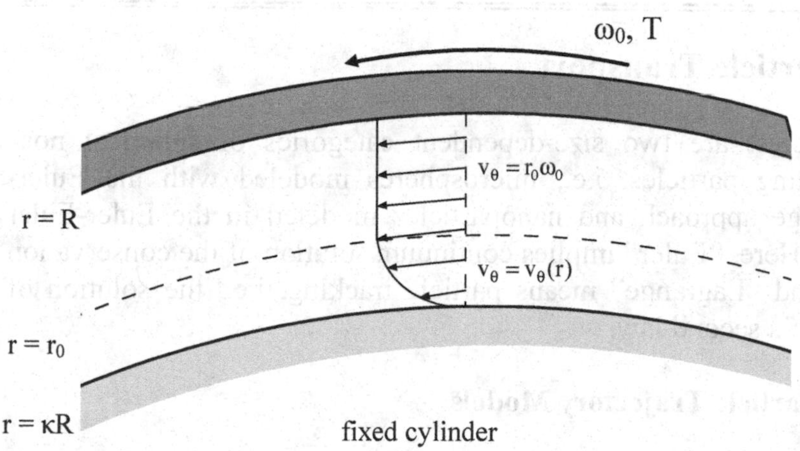

Now, if the yield stress τ_0 is exceeded in the entire gap, i.e., if $r_0 \geq R$, Eq. (E.6.5.7a) yields with the B.C. $v_\theta(r = R) = \omega_0 R$:

$$v_\theta(r) = \omega_0 r + \frac{T}{4\pi L \mu_0 R}\left(\frac{r}{R}\right)\left[1-\left(\frac{R}{r}\right)^2\right] - \frac{\tau_0 r}{\mu_0}\ln\frac{r}{R} \qquad \text{(E.6.5.8)}$$

Invoking the no-slip condition $v_\theta(r = \kappa R) = 0$ yields the desired expression for the torque.

$$T = \frac{4\pi L \mu_0 (\kappa R)^2}{1-\kappa^2}\left[\omega_0 - \frac{\tau_0}{\mu_0}\ln\kappa\right] \qquad \text{(E.6.5.9)}$$

Comments:

- Equation (E.6.5.9) is known as the (80-year old) Reiner–Rivlin equation. With the geometry of the "concentric-cylinder viscometer" given and T and ω_0 measured, the Bingham plastic parameters μ_0 and τ_0 can be determined.

6.4 Particle Transport

Of interest are two size-dependent categories of spherical non-interacting particles, i.e., microspheres modeled with the Euler-Lagrange approach and nanoparticles modeled in the Euler-Euler frame. Here, "Euler" implies continuum solution of the conservation laws and "Lagrange" means particle tracking, i.e., the solution of Newton's second law.

6.4.1 Particle Trajectory Models

As discussed in Sect. 6.1 and indicated in Fig. 6.2, suspensions of distinct particles with effective diameters, typically greater than $1\mu m$, fall into the category of *separated flows*. Consequently, two separate sets of equations are needed. One equation describes the particle dynamics and the other one describes the fluid flow; both may contain coupling terms which reflect possible two-phase interactions (see Michaelides 1997). Quite frequently the dispersed phase, i.e., solid particles, droplets or bubbles, is *uncoupled* from the continuous phase (or carrier fluid). Especially solid, non-rotating spheres with $d_p > 1\ \mu m$ and a high particle-to-fluid-density ratio simplify the trajectory equation significantly when the flow field is laminar. In any case, the combination of continuous fluid flow and discrete particle transport modeling is known as the Euler-Lagrange approach.

Considering relatively small quasi-spherical particles, as well as small particle and shear Reynolds numbers, i.e., $Re_p = d_p \mid v - v_p \mid / \nu << 1$ and $Re_s = v\, d_p^2 / (\nu L) << 1$, respectively, Newton's second law of motion is applicable in the form (Kleinstreuer 2003; among others):

$$m_p \frac{d\vec{\mathbf{v}}_p}{dt} = \vec{\mathbf{F}}_{drag} + \vec{\mathbf{F}}_{pressure} + \vec{\mathbf{F}}_{interactive} + \vec{\mathbf{F}}_{lift} + \vec{\mathbf{F}}_{Basset} + \underbrace{\vec{\mathbf{F}}_{gravity} + \vec{\mathbf{F}}_{virtual\ mass}}_{\Sigma \vec{\mathbf{F}}_{body}} \tag{6.34}$$

Here, m_p is the particle mass and $\vec{\mathbf{v}}_p = d\vec{\mathbf{x}}/dt$ is the particle velocity vector, while all external forces are *point forces* acting on the particle. For laminar flow with negligible particle lift $(\omega_p \approx 0)$, one-way coupling prevails, i.e., the particle presence does not influence the fluid flow and a high density ratio, i.e., $\rho_p / \rho_c >> 1$, assures that only $\vec{\mathbf{F}}_{drag}$ and perhaps $\vec{\mathbf{F}}_{gravity}$ are important (see Crowe et al. 1998; Buchanan et al. 2000; among others). Hence,

$$m_p \frac{d\vec{\mathbf{v}}_p}{dt} = \vec{\mathbf{F}}_D + \vec{\mathbf{F}}_G \tag{6.35a}$$

where

$$\vec{\mathbf{F}}_D = \frac{\pi}{8}\,\rho\, d_p^2\, C_{Dp}(\vec{\mathbf{v}}_p - \vec{\mathbf{v}})\,|\,\vec{\mathbf{v}}_p - \vec{\mathbf{v}}\,| \tag{6.35b}$$

which always keeps F_{drag} opposite to the flow direction. Furthermore,

$$\vec{\mathbf{F}}_G = m_p\,\vec{\mathbf{g}}\,; \quad \text{and} \quad m_p = \rho_p\,\pi\, d_p^3/6 \tag{6.35c, d}$$

and

$$C_{D_p} = C_D / C_{slip}; \qquad C_D = \frac{24}{Re_p}(1 + 0.15\,Re_p^{0.687}) \tag{6.35e, f}$$

As mentioned, the particle Reynolds number $Re_p = \rho d_p\,|\,v - v_p\,|\,/\,\mu$ is small and C_{slip} is the slip correction factor, O(1) after Clift et al. (1978).

For $Re_p \ll 1$, i.e., Stokes flow, $C_D = 24/Re_p$ and Eq. (6.35) reduces to (see Kleinstreuer et al. 2007 for turbulent flow):

$$\frac{U}{D}\,St\,\frac{d\vec{\mathbf{v}}_p}{dt} = (\vec{\mathbf{v}} - \vec{\mathbf{v}}_p) + v_{settling}\hat{g} \tag{6.36a}$$

where U is the mean fluid velocity, e.g., $0.5u_{max}$ in Poiseuille flow, D is the tube diameter, St is the Stokes number and $v_{settling}$ is the terminal velocity of a sphere. Specifically, for $C_{slip} = 1.0$:

$$St = \rho_p\, d_p^2 / (18\mu D) \tag{6.36b}$$

and after Stokes:

$$v_{settling} = \rho_p\, g\, d_p^2 / (18\mu) \tag{6.36c}$$

Knowing the flow field $\vec{v}(\vec{x}, t)$, the individual particle velocities $\vec{v}_p(t)$ can be obtained, subject to given initial conditions. A second integration, $\vec{x} = \int \vec{v}_p \, dt$, provides then the particle locations, i.e., trajectories. It is typically assumed that a particle has deposited on a surface when it approaches within one radius, i.e., the particle touches the wall. An application of Eq. (6.36) for micron-particle suspension flow in a horizontal pipe (D = 0.2 cm and L = 1 cm) is shown in Fig. 6.5 for U = 10 and 20 m/s. The deposition efficiency (DE) is defined as the number-ratio of particles deposited in a specific region to particles which have entered this region (Kleinstreuer et al. 2007).

For spherical, noninteracting *droplets*, alternative C_D- correlations apply (see Clift et al. 1978) because of the friction-induced internal circulation.

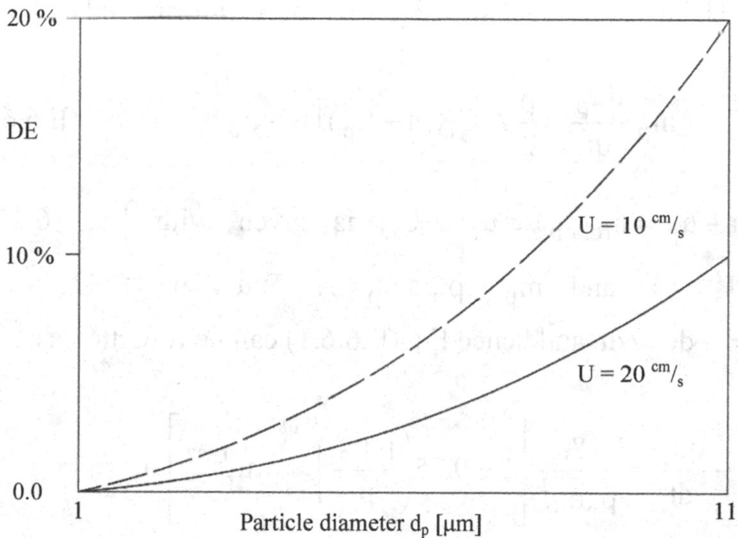

Fig. 6.5 Gravitational deposition of micron particle in a horizontal pipe

Example 6.6: Particle Being Accelerated from Rest by a Steady Uniform Air-Stream

Sketch:	*Assumptions*:	*Concepts*:
$u_\infty \equiv U$; a_p, u_p; m_p, d_p	• Spherical particle • Steady unidirectional (air)flow with $\rho/\rho_p \ll 1$ • Drag is the dominant point force	• Carrier fluid $\vec{v} = (u \equiv U, 0, 0)$ • Particle trajectory: $m_p\, du_p/dt = F_D$ $F_D \sim C_D\, A_{projected}$

Solution: Using Eq. (6.35a) with $\vec{F}_G = 0$, the 1-D form reads:

$$m_p \frac{du_p}{dt} = \frac{\rho}{2} A\, C_D (u - u_p) \,|\, u - u_p \,| \qquad \text{(E.6.6.1)}$$

where $u - u_p = u_{relative} \equiv u_r$, C_D is given with Eq. (6.35f), $A_{proj} = d_p^2\, \pi/4$ and $m_p = \rho_p\, \pi\, d_p^3/6$. With $u \equiv U = ⊄$, $du_r/dt = -\, du_p/dt$ and hence Eq. (E.6.6.1) can be rewritten as:

$$-\frac{du_r}{dt} = \frac{18\mu}{\rho_p {d_p}^2}\left[1 + 0.15\left(\frac{\rho\, d_p}{\mu}\right)^{0.687} u_r^{0.687}\right] u_r$$

or (E.6.6.2a,b)

$$\frac{du_r}{dt} = A\left[1 + B u_r^{0.687}\right] u_r$$

where $A = -\frac{18\mu}{\rho_p d_p^{\ 2}}$ and $B = 0.15\left(\frac{\rho\, d_p}{\mu}\right)^{0.687}$.

Subject to $u_p(t=0)=0$, i.e., $u_r(t=0)=u \equiv U$. Separation of variables and integration yields:

$$u_r = U - u_p = \left[\left(B + U^{-0.687}\right)\cdot \exp(-0.687At) - B\right]^{-1/0.687} \qquad \text{(E.6.6.3)}$$

Graph:

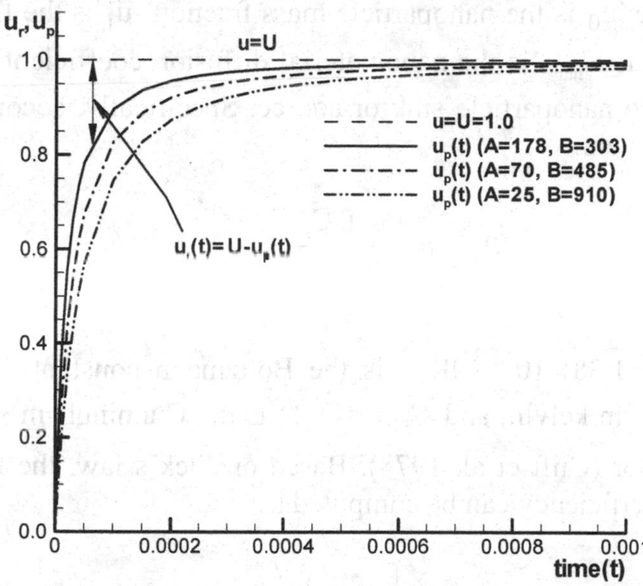

Comment: Pushed by the free stream, the particle accelerates from rest and reaches exponentially the fluid velocity at a point in time which depends on $A = 18\ \mu/(\rho d^2)_p$, i.e., $t_e = \ln 1/(0.687A)$.

6.4.2 Nanoparticle Transport

For submicron particles, Brownian motion becomes effective, i.e., random particle motion due to molecular bombardment by the surrounding fluid. That results in enhanced particle diffusion with decreasing nanoparticle diameter. Thus, rather than employing the Euler-Lagrange modeling approach (i.e., solving the fluid flow equations and then the particle trajectory equation), nanomaterial transport is best described in the Eulerian-Eulerian frame. Again, assuming laminar flow with low nanoparticle loading, say, below 6% by volume, and hence only one-way coupling, the momentum equation is solved first and then the mass transfer equation, viz:

$$\frac{\partial y}{\partial t}+\frac{\partial}{x_i}(u_i\, y)=\frac{\partial}{\partial x_i}\left[\mathcal{D}_{nano}\frac{\partial y}{\partial x_i}\right]\pm S_y \tag{6.37}$$

where $y \equiv c/c_0$ is the nanoparticle mass fraction, u_i is the fluid velocity vector, $\mathcal{D}_{nano}$ is the nanomaterial diffusion coefficient, and S_y is a possible nanoparticle sink or source. Specifically, according to Stokes–Einstein:

$$\mathcal{D}_{nano}=\frac{k_B\, T\, C_{slip}}{3\pi\, \mu\, d_p} \tag{6.38}$$

where $k_B = 1.38\times10^{-23}\ \mathrm{JK}^{-1}$ is the Boltzmann constant, T is the temperature in kelvin, and $C_{slip}=O(1)$ is the Cunningham slip correction factor (Clift et al. 1978). Based on Fick's law, the regional deposition efficiency can be computed as:

$$DE=\left(\mathcal{D}\left.\frac{\partial y}{\partial n}\right|_{n=0} A\right)\Big/(Q_{in}\; y_{in}) \tag{6.39}$$

where n is the surface normal, A is the surface area, Q_{in} is the inlet volumetric flow rate, and y_{in} is the inlet mass fraction.

Applications of nanoparticle transport and deposition related to biomedical engineering is discussed in Kleinstreuer (2006), while nanofluid flow in microchannels is given in Sect. 7.6.

Example 6.7: Nanoparticle Convection, Diffusion and Uptake from a Planar Source

Consider steady 1-D flow of a liquid through a porous plug which releases a low constant concentration of nanoparticles which disperse and dissolve/vanish according to a first-order reaction.

Sketch	*Assumptions*	*Concepts*:
x=0 x→∞ Q=vA c(x) x $c(x=0) = c_0$	• Steady 1-D isothermal plug flow • Constant properties • Low-volume nanoparticle release • Idealized nanoparticle sink	• Uniform flow $v = \frac{Q}{A} = ¢$ • Reduced form of Eq. (6.37) • Axial diffusion only

Solution:

In light of the assumptions, Eq. (6.37) can be reduced, with y →c, u→v and S_c→–kc, to:

$$v\frac{dc}{dx} = \mathcal{D}\frac{d^2c}{dx^2} - kc \qquad \text{(E.6.7.1)}$$

where v = Q/A, while the binary diffusion coefficient $\mathcal{D}$ is given with Eq. (6.38), and k is a constant reaction coefficient.

Equation (E.6.7.1) can be recast as:

$$c'' - \frac{v}{\mathcal{D}}c' - \frac{k}{\mathcal{D}}c = 0 \qquad \text{(E.6.7.2)}$$

subject to the B.C.s:

$$c(x=0) = c_0 \text{ and } c(x \to \infty) \to 0 \qquad \text{(E.6.7.3a, b)}$$

The trial solution $c(x)=e^{ax}$ satisfies Eq.(E.6.7.2) where

$$a = \frac{v}{2\mathcal{D}}\left[1 - \sqrt{1 + \frac{4k\mathcal{D}}{v^2}}\right]$$

to match the B.C.s. Hence,

$$c(x) = c_0 \exp\left[-\frac{vx}{2\mathcal{D}}\left(\sqrt{1+\frac{4k\mathcal{D}}{v^2}}-1\right)\right] \quad \text{(E.6.7.4)}$$

Parameter values:

- The nanoparticle diffusion coefficient for T = 300 K, $C_{slip} \approx 1.0$, $\mu_{water} = 0.9\times10^{-3}\ \frac{kg}{ms}$ and d_p = 10 nm is

$$\mathcal{D} = 4.88\times10^{-10}\ m^2/s$$

- Assuming the ratio $4k\mathcal{D}/v^2$ in Eq. (E.6.7.4) to be in the range of

$$0.1 \le \frac{4k\mathcal{D}}{v^2} \le 10$$

we can now graph a family of curves:

$$\frac{c(x)}{c_0} \quad \text{vs.} \quad \frac{v}{2\mathcal{D}}x$$

Graph:

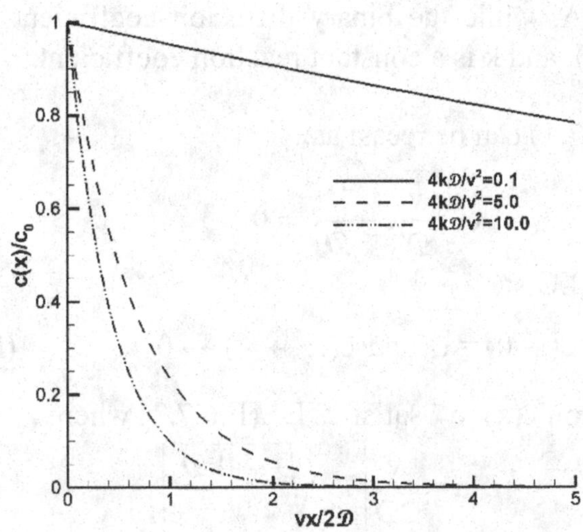

Comment:

As expected, c(x) decays swiftly for $\frac{4k\mathcal{D}}{v^2} > 1.0$, i.e., due to the first-order uptake.

6.5 Homework Assignments and Course Projects

Solutions to homework problems done individually or in, say, three-person groups should help to further illustrate fluid dynamics concepts as well as approaches to problem solving, and in conjunction with App. A, sharpen the reader's math skills (see Fig. 2.1). Unfortunately, there is no substantial correlation between good HSA results and fine test performances, just vice versa. Table 1.1 summarizes three suggestions for students to achieve a good grade in fluid dynamics – for that matter in any engineering subject. The key word is "independence", i.e., equipped with an equation sheet (see App. A), the student should be able: (i) to satisfactory answer all concept questions and (ii) to solve correctly all basic fluid dynamics problems.

The "Insight" questions emerged directly out of the Chap. 6 text, while some "Problems" were taken from lecture notes in modified, i.e., enhanced, form when using White (2006), Cimbala & Cengel (2008), Incropera et al. (2007), and Bird et al. (2002). Additional examples, concept questions and problems may be found in any UG fluid mechanics and heat transfer text, or on the Web (see websites of MIT, Stanford, Cornell, UM, etc.).

6.5.1 Guideline for Project Report Writing

A project is more than a homework problem. As a starter, it typically requires a brief literature review to document the state-of-the-art for the given task. The antagonistic interplay between a rather accurate mathematical description of an assigned problem and finding a tractable solution method puts a premium on making the right assumptions and selecting a suitable equation solver, often in terms of available software and hardware.

The resulting work has to be documented in an appropriate format, i.e., *project reports* should have the following features:

- **ToC:** Table of Contents with page numbers (optional for smaller reports)
- **Abstract:** Summary of the fluid dynamics problem analyzed, solution method employed, and novel results with physical insight presented
- **Nomenclature:** Optional but recommended for complex problem solutions
- **Introduction:** Project objectives and task justification; System sketch with state-of-the-art description, basic approach or concept, and references of the *literature review* in terms of "author (year)," e.g., White (1998)
- **Theory:** Basic assumptions, governing equations, boundary conditions, property values, ranges of dimensionless groups, closure models, and references
- **Model Validations:** (a) Numerical accuracy checks, i.e., mesh-independence of the results as well as mass & momentum residuals less than, say, 10^{-4}; (b) Quantitative proof, e.g., comparison with an exact solution of (typically a much) simpler system, and/or experimental data comparisons, or (at least) qualitative justification that the solution is correct
- **Results and Discussion:** Graphs with interpretations, i.e., explain physical insight (note: number the figures and discuss them concisely); draw conclusions and provide applications
- **Future Work**
- **References:** List all literature cited in alphabetical order
- **Appendices:** Lengthy derivations, computer programs, etc.

6.5.2 Text-Embedded Insight Questions and Problems

6.5.1 List examples of natural and man-made (i.e., industrial) flow: (i) quasi-homogeneous mixture flows; and (ii) separated (or truly two-phase) flows.

6.5.2 Definitions: (a) What are the underlying assumptions for $\kappa \approx c$ to hold (see Eq. (6.6b)); (b) is Eq. (6.8) related to the definition of "steam quality" in thermodynamics? (c) give an example with sketch and explanation for the particle drift velocity (see Eq. (6.14)); (d) derive and discuss Eq. (6.16b) vs. Eq. (6.17b).

6.5.3 Derive Eqs. (6.18) - (6.20) from first principles and set up a typical application.

6.5.4 Set up the governing equations and boundary conditions for power-law fluid flow in an annulus formed by a stationary (cylindrical) housing and a rotating and translating shaft.

6.5.5 In Graph I of Example 6.5 the shear stress $\tau_{r\theta}$ varies inside the gap; however, $v_\theta = r_0\omega_0 = \not\subset$ (see Graph II). Explain that mathematically and physically.

6.5.6 Devise an experiment based on Eq. (E.6.5.9) with which both unknown Bingham plastic parameters τ_0 and μ_0 can be determined.

6.5.7 Derive and discuss the different Reynolds numbers in fluid-particle flows.

6.5.8 The forces in Eq. (6.34) are so-called "point forces"; what does that mean? Discuss alternative (numerical) methods to solve for (spherical) particle trajectories.

6.5.9 What does the correction factor C_{slip} accomplish?

6.5.10 Derive Eq. (6.36a), interpret Eq. (6.36b), and prove Eq. (6.36c).

6.5.11 Solve the falling sphere problem and plot $v_p(t)$ for different fluids, e.g., oil, syrup, and a power-law fluid.

6.5.12 Discuss Eq. (6.38), i.e., research its derivation and state its limits of application. This is Project Assignment (d_P<100nm)

6.5.3 Problems

6.5.13 Consider steady fully-developed flow of a power-law fluid (m,n) in a slit of small spacing 2h formed by two plates (L × W).

(a) Compare your velocity profile with the Poiseuille solution ($m = \mu$, $n = 1$):

$$u(y) = \frac{h^2}{2\mu}(\frac{\Delta p}{L})[1-(\frac{x}{h})^2]$$

(b) Show that the mass flow rate is:

$$\dot{m} = \frac{2\rho W h^2}{(1/n)+2}\left[\frac{h}{m}\left(\frac{\Delta p}{L}\right)\right]^{1/n}$$

(c) How would your results change if the plates are vertical, i.e., 100% downward flow?

6.5.14 Commercial capillary viscometers (see Bird et al., 2002) can be modeled as cylindrical Couette flow. Assuming the inner cylinder (radius κR, $0 < \kappa < 1$) be stationary and the outer one (radius R, forming a small gap of $R - \kappa R$ filled with an unknown power-law fluid) rotating at low angular velocity ω_0 due to a measured torque, how would you obtain the power-law fluid parameters m and n?

6.5.15 Consider flow of a power-law fluid (m, n; ρ) in a horizontal annulus (κR and R). The fluid is kept in motion due to the translatory motion (v_0) of the inner cylinder (κR).

(a) Derive the velocity profile $v_z(r; v_0, \kappa, R; n)$ and show that it simplifies to $\frac{v_z}{v_0} = \frac{\ln(r/R)}{\ln \kappa}$ for $n \to 1$.

(b) Find an expression for the mass flow rate $\dot{m}$ for $n \neq 1/3$; what is $\dot{m}\,(n \neq 1/3)$?

6.5.16 Consider flow of two immiscible fluids (viscosities μ_1 and μ_2 with different densities) between parallel horizontal plates a small distance 2h apart, driven by $\Delta p / L = \text{const.}$ Assuming that the interface stays planar and each fluid takes up half the channel, find the shear stress τ_{yx} and velocities $u_1(y)$ for $-h \le y \le 0$ and $u_2(y)$ for $0 \le y \le h$.

6.5.17 Consider steady laminar flow in a network of 12 tubes forming a cube. Driven by an effective pressure drop $\Delta p = p_{(1)} - p_{(8)}$, where (1) is the (corner) inlet and (8) is the diagonally opposite outlet.

(a) Find an expression for the mass flow rate $\dot{m} = \dot{m}(\Delta p; \rho, \mu, R, L)$, where R and L are the same radius and length of each of the 12 tubes.
(b) Convert this system to a simple "flow through porous media" model by extracting Darcy's velocity. Comment!

6.5.18 Provide definitions, characteristics, differences and applications of *constitutive* equations vs. *closure* models in two-phase flow analysis.

6.5.19 Red blood cells (RBCs) in tubular flow appear to congregate in the tube's central core, resulting in a cell-free plasma layer along the vessel wall. Thus, there are two regions, say, a constant plasma layer of thickness δ with viscosity μ_p and a homogeneous mixture region, $0 \le r \le (R-\delta)$ with μ_m. Typical parameter values are $R = 10\ \mu m$, $\delta = 1\ \mu m$, and $\mu_p / \mu_m = 0.25$.

(a) Find u(r) for each of the two domains;
(b) Find $Q = Q_P + Q_M$ and compare the result with the Hagen-Poiseuille expression:

$$Q = \frac{\pi R^4 \Delta p}{8 \mu L}$$

and relate μ to μ_p / μ_m for $\delta / R << 1$.
(c) Plot for the given numerical data $u_P(r) + u_M(r)$ as well as Poiseuille's u(r).

6.5.20 A spherical particle of mass m is released at the upper plate corner $(x = 0; y = h)$ in Poiseuille flow between two horizontal parallel plates, a small distance 2h apart. For $Re \le 1.0$, find the minimum channel length when the particle just touches the lower plate (i.e., $x = L_{min}, y = -h$).

6.5.4 Projects (see Sect. 6.5.1 for proper report writing)

6.5.21 Consider steady 2-D laminar flow ($Re_p = \rho v d_p / \mu << 1$) in a 90°-curved channel (width 2h and centerline radius R). A micron particle (d_p, ρ_p) is released on the centerline at the fully-developed flow entrance. Assuming a suitable (Dean's flow) velocity profile, determine the particle trajectory and hence the particle drift δ from the centerline at the channel outlet. Is your result for δ a conservative estimate in light of possible particle deposition?

6.5.22 Actual, i.e., natural or man-made, particles are typically non-spherical. However, most correlations and analyses assume spherical particles by introducing the sphericity $\psi = \frac{\pi^{1/3}(\forall_p)^{2/3}}{A_p}$ and shape factor $n = \frac{3}{\psi}$ as well as various effective (spherical) particle diameters, d_p.

(a) Derive the ψ-equation based on the assumption that sphericity is defined as ψ = (surface area of a sphere with the same particle volume/surface area of the particle).
(b) Tabulate the various "effective spherical particle" diameter definitions and comment on applications and accuracy.
(c) Use Eq. (10) in Holzer & Sommerfeld (2008); (in: Powder Technology, Vol. 184, pp. 361–365) and compare non-spherical particle trajectories with those for equivalent spheres of Problem 6.5.20.

6.5.23 Survey modeling approaches as well as associated analytic and numerical solution methods for fluid-particle flows. Distinguish between particle size ($d_p \gtrless 1\,\mu m$), particle shape (spherical vs.

non-spherical), and solid vs. liquid particles. Demonstrate an application with a solved example.

6.5.24 Survey and discuss "porous media flow" as a suitable model for porous plugs, catalytic converters, filters, soft/hard tissues, etc. Solve the problem of flow through: (a) a tube filled with porous material; and (b) a tube with partially porous wall.

6.5.25 Survey natural and modern engineering lubrication problems and their solutions. Solve the slider-bearing problem for a non-Newtonian fluid. What happens when a particle is entrained, say, due to wear?

Chapter 7

Microsystems and Microfluidics

7.1 Introduction

Overview The main focus of micro-scale research and development is on device fabrication and expansion of microsystem applications. This implies innovative advances in the material sciences, manufacturing technology/methodology as well as creation of supportive design software. However, in this chapter we consider only the fluid flow, heat/mass transfer and particle transport aspects of microsystems, where phase changes such as boiling and condensation as well as thermal radiation are excluded (for boiling and radiation processes see Tong & Tang 1997 and Zhang 2007; among others).

Although *microfluidics* deals with fluid behavior in systems with "small" length scales, conventional (i.e., macro-scale) flow theory is typically applied, at least for liquid flows in microchannels with $D_{hydraulic} \geq 10\ \mu m$ and standard gas flows when $D_h \geq 100\ \mu m$. However, for microchannel gas flows in the slip regime, i.e., $0.01 \leq Kn \leq 0.1$, where the Knudsen number is the ratio of the molecular mean free path over a system length scale, modification to the velocity and temperature boundary conditions have to be made. Clearly, when the Knudsen number is 1.0 or above, alternative system equations and numerical solution techniques have to be considered.

In any case, for proper communication and relative size appreciation, prefixes for parameter-scale variations and a few length-scale examples are given below:

C. Kleinstreuer, *Modern Fluid Dynamics: Basic Theory and Selected Applications in Macro- and Micro-Fluidics*, Fluid Mechanics and Its Applications 87,
DOI 10.1007/978-1-4020-8670-0_7,

Unit Prefixes

Femto	→	10^{-15}	33	Kilo	→	10^{3}
Pico	→	10^{-12}	34	Mega	→	10^{6}
Nano	→	10^{-9}	35	Giga	→	10^{9}
Micro	→	10^{-6}	36	Tera	→	10^{12}
Milli	→	10^{-3}	37	Peta	→	10^{15}

Length Scales of Microsystems

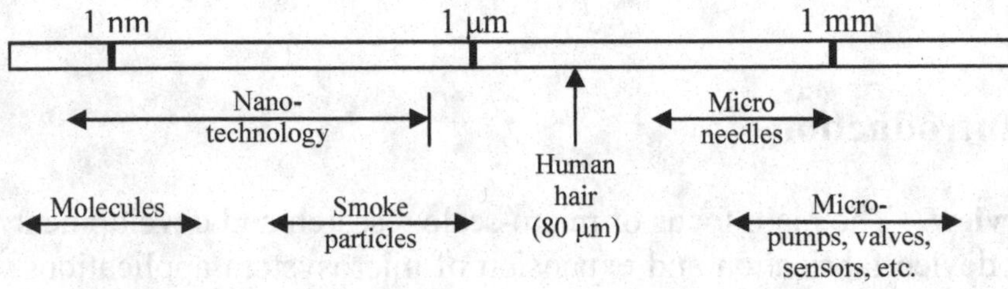

In general, the surface-to-volume ratio varies as the inverse of the system's length scale, i.e., 1/L, which implies that for *microsystems* large surface areas cause significant viscous resistance. In turn, it requires relatively powerful actuators and pumps to drive a microfluidic device, at least at Reynolds numbers $\mathrm{Re} \geq \mathbf{O}(10)$. In order to have such pumps/actuators/valves as integral parts of the microfluidic device, new challenges had to be overcome. However, when $\mathrm{Re} \leq 1$, instead of employing mechanical actuators with moving parts, surface-modulated phenomena were used, such as electrokinetic pumping (e.g., electro-osmosis) and capillary surface-tension effects, electro-magnetic force fields, and acoustic streaming.

For conventional, i.e., macro-scale and limited micro-scale analyses, fluid flow is described by velocity and pressure fields and by its properties. They are characterized as interacting groups, such as kinematic (i.e., velocity and strain rate), thermodynamic (i.e., pressure and temperature), transport (i.e., viscosity, conductivity, diffusivity), and miscellaneous (i.e., surface tension, vapor pressure, etc.). In contrast, on the *submicron-scale*, solid, liquid, or gaseous matter is more realistically described in terms of interacting atoms/molecules. For example, molecules in a solid are densely packed and arranged in a lattice, where each molecule is held in place by large repulsive forces according to the Lennard–Jones (L–J) potential (see Bird 1994; among others). Nevertheless, when solving problems of fluid flow in microchannels, the continuum mechanics assumption is preferable over any molecular approach. In the latter case the state of each molecule in terms of position and velocity has to be known, and then one has to evolve that state forward in time for each molecule. That implies the solution of Newton's second law of motion with the L–J force (i.e., the spatial derivative of the L–J potential) for billions of molecules. In contrast, when continuous fluid flow behavior can be assumed, i.e., system length scales $L_{liquid} \geq 10$ μm and $L_{gas} \geq 100$ μm (or Kn < 0.01 for gas flow), we just solve numerically the conservation laws subject to key assumptions and appropriate boundary conditions, as discussed in Sect. 7.2.5.

Microfluidics Devices Clearly, *microfluidics is the study of transport processes in microchannels*. Of interest are methods and devices for controlling and manipulating fluid flow, finite liquid-volume delivery and particle transport on a nano- and micro-scale. Microfluidics devices consist typically of reservoirs, channels, pumps, valves, mixers, actuators, sensors, filters and/or heat exchangers (see Fig.7.1 as an example of a biomedical micro-electro-mechanical system). Applications include hydraulic machine parts, heat sinks and combustors in mechanical engineering, lab-on-a-chip systems and reactors in chemical engineering, and bio-MEMS for drug delivery in biomedical engineering. In this chapter we focus on transport phenomena in microchannels. Specifically, we consider pure fluid and nanofluid flows in microchannels as applied to micro-heat sinks and bio-MEMS. Additional information, including micro-scale device manufacturing methods may be found in Tabeling (2005) and Nguyen & Wereley (2006). A review of engineering flows in small devices has been provided by Stone et al. (2004).

Microsystem Modeling Assumptions As mentioned, electromechanical components of consumer goods, vehicles and machinery as well as entire devices, especially medical implants and laboratory test equipment, are being built on a *micro-scale. Clearly, it is the low production cost, controlled multi-functionality, compactness, and low operating cost which make micro-scale fluid devices attractive alternatives to conventional flow systems.*

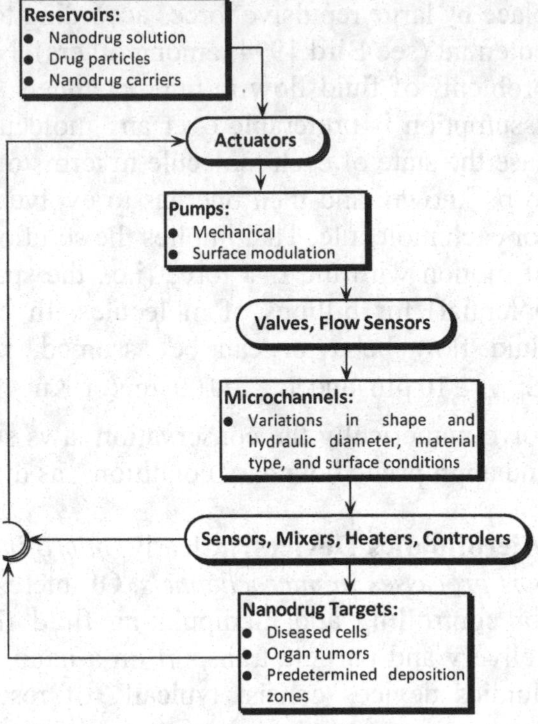

Fig. 7.1 Bio-MEMS components and flow chart

One of the key elements of all microsystems is the microchannel (soon becoming a nanochannel) with hydraulic diameter, $D_h = 4A/P$, ranging typically from 10 to 500 μm. This is rather small in light of the fact that the diameter of the human hair is about 80 μm. When considering fluid flow in such tiny conduits, we should recall that the underlying macro-scale assumptions for the equations/solutions in Chaps. 1–6 are only valid when:

(i) The fluid is "infinitely divisible" (see definition of a continuum in Sect. 1.2)
(ii) All flow quantities are in local thermodynamic equilibrium (see mean-free path of molecules for gases or mean molecular distance for liquids)

Concerning the *continuum mechanics assumption*, the two main classes of fluids, i.e., gases and liquids differ primarily by their densities and by the degrees of interaction between the constituent molecules. Considering gases, say, air at 20°C and 1 atm, its density is $\rho_{air} \approx 1\ \text{kg/m}^3$ with a mean-free path $\lambda \approx 65$ nm and an intermolecular distance $\delta = 3\ \text{nm}$. In contrast, $\rho_{liquid} \approx 10^3\ \text{kg/m}^3$ with an intermolecular distance $\delta = 0.3\ \text{nm}$. Now, if the key system length scale, e.g., the microchannel D_h, is of the order of 10 μm or more, fluids with those characteristics appear continuous and hence the Navier–Stokes equations hold. Recall, *local thermodynamic equilibrium* implies that all macroscopic quantities within the fluid have sufficient time to adjust to their surroundings. That process depends on the time between molecular collisions and hence the magnitude of the mean-free path travelled. However, for certain "dilute" gases in channels with relatively high Knudsen numbers, i.e., $0.01 \leq \text{Kn} = \lambda/\text{L} \leq 0.1$ velocity slip and temperature jump at the wall as well as wall roughness and/or viscous heating may have to be taken into account (see Sects. 7.2 and 7.3). Furthermore, when considering rarefied gases ($\lambda > 10$ μm), microscopic modeling (e.g., molecular dynamics simulations) are required, because the Knudsen number is $\text{Kn} \geq 1.0$.

The conventional driving force for flow in microchannels is still the net pressure force, using micro-pumps, when substantial flow rates, i.e., $\text{Re} = \text{v}D_h/\upsilon > 1.0$ are desired, as for micro-heat exchangers. However as mentioned, certain microfluidics devices for biomedical, chemical and pharmaceutical applications employ more esoteric driving forces, such as surface tension (i.e., capillary or Marangoni effects) and electrokinetic phenomena (i.e., electrophoresis or electro-osmosis). This works very well in microdevices where the fluid volumes are in the pico-to micro-liter range and the distance to be travelled are often only on a micrometer scale.

In summary, it is not surprising that *fluid-flow in microchannels* may differ from macrochannel flow behavior in terms of entrance, wall and thermal flow effects (see Koo & Kleinstreuer 2003; among others). As alluded to, because of the typically short microchannel length, entrance effects (i.e., developing 2-D or 3-D flows) may be dominant. At the micro-channel wall, the "no slip" conditions may not hold, electrokinetic forces may come into play, and surface roughness effects may be substantial. Early onset of laminar-to-turbulent flow transition may occur and viscous dissipation of heavy liquids in high shear-rate fields may increase the fluid temperature measurably.

7.2 Microfluidics Modeling Aspects

Although we still assume the continuum assumption to be valid for flow in microchannels, say, $D_h \geq \mathbf{O}(10\ \mu m)$, one should be aware of certain short-range forces and micro-scale transport phenomena which can be neglected in the macro-world. Specifically, because of the flip-side to the advantageous high *surface-to-volume* ratios for microdevices, analyses have to deal with all kinds of "surface effects," such as wall roughness, surface tension and electrokinetics. Those $A/\forall$-ratios, and hence surface effects, scale as 1/L, where L is a linear system dimension, e.g., the effective microchannel diameter. Some resulting microscale phenomena are discussed in Sects. 7.3 and 7.4, illustrating gas flow with wall slip, charged fluid-particle flow and nanofluid flow with enhanced heat transfer.

To begin with, a few familiar concepts and flow variables are revisited and discussed from a molecular perspective. An understanding of the behavior and action of molecules leads to more fundamental explanations. For example, what are actually the causes of stresses and pressures in a fluid flow field, or how does heat conduct? Then some background material in electrohydrodynamics and nanofluid flow is reviewed to set the stage for Sects. 7.3–7.5.

7.2.1 Molecular Movement and Impaction

When considering fluid flow in conduits with hydraulic (or effective) diameter $D_h = 4A/P$ in the micrometer range, we first have to recall what any material is actually made of. Specifically, solids and fluids consist of molecules which are differently arranged and excited based on the thermodynamic state of the matter. For example, H_2O-molecules in ice are arranged in a regular lattice, vibrating about equilibrium positions. As water and steam, they move around randomly, where the average speed depends on the temperature. More specifically, liquid molecules move around randomly in a limited collision range, while gas molecules are much further apart and fluctuate much more vigorously, hardly colliding with each other. Clearly, the average molecular distance increases when going from the solid to the gaseous phase.

Gas Flow Characteristics Based on *kinetic gas theory*, i.e., assuming a gas molecule moving with constant velocity on a straight line until (infrequent binary) collision, molecule and hence gas characteristics can be computed. For example, a basic result relates the gas pressure to its temperature (Probstein 1994):

$$p = n\, k_B\, T \tag{7.1}$$

where n is the number density $[\frac{\#}{m^3}]$, $k_B = 1.3805 \times 10^{-23}\ J/K$ is the Boltzmann constant, and T[K] is the gas temperature. When compared to the ideal gas law $p = \rho\, R\, T$, which a number of real gases obey at standard conditions, we can express the density as:

$$\rho = \frac{n\, k_B}{R} \tag{7.2}$$

where R is the specific gas constant, $\rho \approx 1.204\,\text{kg/m}^3$ for air at $20\,°\text{C}$. From Eq. (7.1) the number density of the gas can be obtained. For example, air at standard conditions yields,

$$n_{air} = 2.7 \times 10^{25}\ m^{-3} \tag{7.3a}$$

from which the average molecular spacing in air can be deduced as:

$$\delta = n^{-1/3} := 3.3\,\text{nm} \tag{7.3b}$$

Clearly, the ratio of molecular spacing to diameter provides an indication if a given gas is dilute or dense. For practical purposes, $\delta/d \geq 7$ implies a "*dilute*" (in the sense of rarefied) gas.

As mentioned, an even more important molecular length scale is the mean free path λ, i.e., the average distance traveled by a molecule before collision. Bird (1994) showed that for a simple gas represented by hard spheres at thermodynamic equilibrium:

$$\lambda = \frac{1}{\sqrt{2}\pi n d^2} \tag{7.3c}$$

or

$$\lambda = \frac{\mu}{p}\sqrt{\frac{\pi RT}{2}} \tag{7.3d}$$

$\lambda_{air} \approx 65\,\text{nm}$ for air at STP conditions and as expected $\lambda_{air} >> d$, the effective diameter of the gas molecule. Bird (1994) also derived the mean-square speed of molecules as:

$$\bar{c} = \sqrt{3R\,T} \tag{7.4}$$

and the speed at which sound propagates through a gas:

$$c_s = \sqrt{\kappa\,R\,T} \tag{7.5}$$

where $\kappa = c_p / c_v$ is the ratio of specific heats. Now three important dimensionless groups can be formed:

- Mach number $Ma = \frac{v}{c_s}$ (7.6)
  where v is the mean or maximum fluid velocity; hence, Ma < 1 implies subsonic and Ma > 1 supersonic flow
- Reynolds number $Re = \frac{vL}{\upsilon}$ (7.7)
  where L is a system length scale (e.g. tube diameter) and kinematic viscosity $\upsilon \equiv \mu/\rho = \frac{1}{2}\bar{c}\lambda$
- Knudsen number (for gas flow) $Kn = \frac{\lambda}{L} = \sqrt{\frac{\kappa\,\pi}{2}}\frac{Ma}{L}$ (7.8)

  where for $Kn < 10^{-3}$ the Navier–Stokes equations are valid (some argue $Kn < 10^{-2}$ is sufficient), while for the range 0.01 < Kn < 0.1 the Navier–Stokes solutions are subject to velocity slip and temperature jump conditions. Clearly, 0.1 < Kn < 1.0 is the transition regime, while for $Kn = \mathcal{O}(1)$ molecular dynamics (MD) simulation is necessary.

Liquid Flow Because of the rather complex behavior of molecules in liquids, a "kinetic liquid theory" does not exist yet. Thus, basic properties (ρ, μ) and molecular length scales (δ, λ) cannot be derived but have to be measured or indirectly deduced. Compared to gases, liquid molecules are much closer together and collide frequently (e.g., $n_{water} = 3.34 \times 10^{28} / m^3$ so that $\delta_{mean} \approx 0.3$ for water). They continuously vibrate while driven by the plus/minus forces given by the Lennard–Jones model (see Bird 1994). Hence, the concepts of a mean-free-path and a Knudsen number, indicating distinct flow regimes with associated equations, are not applicable. Nevertheless, the ratio of mean-molecular-spacing-to-system-characteristic-length-scale, δ/L, may be used to determine if the continuum assumption holds for a given microchannel flow problem. Specifically, for the following ratio we have two basic modeling approaches assuming water at STP conditions:

$$\frac{\delta}{L}\begin{cases} \leq 3\times10^{-5} & \text{continuum mechanics equations} \\ >10^{-3} & \text{molecular dynamics simulations} \end{cases}$$

which implies a large transition zone.

Brownian Motion Robert Brown (1827) observed under the microscope some "animated motion of pollen grains in still water." The random collision with water molecules pushed the small particles of low inertia around. This bombardment effect is not that noticeable for large (i.e., $d_p > 1\,\mu m$) particles. The Stokes–Einstein equation, based on *kinetic theory*, describes the diffusivity of spherical (nano) particles in the range $1 < d_p \leq 100\,nm$, as a result of Brownian motion:

$$D = k_B T/(3\pi\mu d_p) \tag{7.9}$$

where $k_B = 1.4 \times 10^{-23}$ J/K is the Boltzmann constant, and T is the temperature in kelvin. On a *macro-scale*, the Brownian motion effect for submicron particles may negate gravity as seen in colloidal suspensions (e.g., milk), suspensions of iron particles in water (i.e., ferrofluids), and inhaled toxic/therapeutic particles in air or metallic particles in liquids (i.e., nanofluids discussed in Sect. 7.5).

Thermal Conduction Recalling Fourier's law $\vec{\mathbf{q}} = -k\nabla T$, we have for 1-D heat transfer the conductivity $k_x = -\frac{q_x}{\partial T / \partial x}$, where for isotropic material $k_x = k_y = k_z = k$. In general, $k_{metal} > k_{fluid} > k_{gas}$ largely due to the differences in intermolecular spacings, $\delta = n^{-1/3}$. To explain what causes heat to conduct in solids vs. liquids/gases, we consider the effectiveness of thermal energy transport in these two material groups.

As alluded to, a *solid* is comprised of free electrons and atoms bounded in a periodic arrangement, called the *lattice*. The *thermal energy transport* is due to two energy carriers:

(i) Migration of free electrons as in pure metals

or

(ii) Lattice-vibrational waves as in non- and semi-conductors

Note, as part of the wave-corpuscle duality, the ΔT -induced lattice-vibration quanta are termed *phonons*. In case of *fluids*, the thermal energy transport is less effective because of the much larger intermolecular spacings and the randomness of molecular motion. Thus, the impact of temperature and pressure, with the resulting interplay of the material properties $c_v, \rho, \bar{c}$, δ and λ, determine the thermal conductivity, i.e., k = k(fluid, T, p, fluid properties).

All this is under the assumption of thermal conduction in *bulk* materials. In contrast, when the system's characteristic dimension, e.g., L or D_h, is very small, the molecular motion (and hence the energy carrier) is restricted due to solid boundaries or interfaces. Examples include micro-devices and ultra-thin films. As a result, when a length-scale $L_{critical} \sim \lambda$ is in the y-direction $k_y < k_x < k_{bulk}$. For example, in a thin solid alumina film, when, say, $L < y < L_{critical} \leq 30\text{ nm}$, $k_y^{film} < k_{bulk}$ for the same material, because $\lambda \approx 5$ nm for Al_2O_3.

Note, quite a different story are nanofluids, i.e., dilute suspensions of (metallic) nanoparticles ($1<d_p<100$ nm) in liquids, which may generate effective thermal conductivities measurably higher than the pure liquid values (see Sect. 7.5).

Forces, Stresses and Fluid Motion In contrast to the continuum approach taken in Sect. 2.4, here the equation of motion is derived on a molecular basis. On the microscale, only molecules directly surrounding a fluid element $\delta\forall$ exert a force on the fluid molecules (or particles) inside. Such interactive molecular forces are of very short range, i.e., they drop to zero beyond a molecular diameter or so. Hence, the resulting fluid force is proportional to the *surface area of* $\delta\forall$, not to its volume. Such molecular bombardment (or impaction) causes net surface forces which we classify as pressure, $p = F/A$, or stress, $\tau = F/A$. So, the force on molecules inside of fluid element $\delta\forall$ due to the surrounding layer of molecules can be expressed as:

$$F_i = \oiint_S \tau_{ij} \cdot n_j \cdot dS \tag{7.10a}$$

where n_j is the surface normal unit vector, τ_{ij} is the 9-component total stress tensor which encompasses pressure, normal stress and shear stress, as discussed in Sect. 1.4. Recall, it consists of three diagonal entries, i.e., pressure plus normal stresses and 2×3 symmetric shear stresses. Employing the Divergence Theorem (see App. A) to Eq. (7.10a) yields:

$$F_i = \iiint_\forall \frac{\partial \tau_{ij}}{\partial x_j} d\forall \tag{7.10b}$$

or for a small $\Delta\forall$ in which the expression $\partial\tau_{ij} / \partial x_j \, \Delta\forall$ hardly varies,

$$F_i \approx \frac{\partial \tau_{ij}}{\partial x_j} \Delta\forall \tag{7.10c}$$

Now, using Eq. (7.10c) in Newton's second law of motion as one of the external forces causing fluid element acceleration, we have:

$$(\rho\Delta\forall)a_i = \frac{\partial \tau_{ij}}{\partial x_j} \Delta\forall + \sum f_i^{ext} \Delta\forall \tag{7.11}$$

where gravity is typically the major external force, i.e., $\sum f_i^{ext} = \rho g_i + ...$ and $\tau_{ij} \equiv \tau_{ij}^{total} = p\delta_{ij} + \tau_{ij}^{viscous}$. Clearly, p is the thermodynamic pressure, which is caused by normal forces (as τ_{ii}), while $\tau_{ij}^{viscous}$ is due to viscous forces, i.e., normal and mainly tangential. As alluded to, obtaining τ_{ij} from molecular properties requires a kinetic theory for gases (as developed by Boltzmann) and for liquids, has not established yet. In any case, the fluid particle acceleration is:

$$a_i = \lim_{\Delta t \to 0}[v_i(\vec{x} + \Delta\vec{x}, t + \Delta t) - v_i(\vec{x}, t)]/\Delta t$$

where $\vec{x} = (x_1, x_2, x_3)$ and $\Delta\vec{x} = (\Delta x_1, \Delta x_2, \Delta x_3)$.

With Taylor's Theorem (see App. A) we have:

$$a_i = \lim_{\Delta t \to 0}\left[\frac{\partial v_i}{\partial t}\frac{\Delta t}{\Delta t} + \frac{\Delta x_1}{\Delta t}\frac{\partial v_i}{\partial x_i} + \frac{\Delta x_2}{\Delta t}\frac{\partial v_i}{\partial x_2}\frac{\Delta x_3}{\Delta t}\frac{\partial v_i}{\partial x_3}\right]$$

Taking the limit, where $\lim_{\Delta t \to 0}\frac{\Delta x_j}{\Delta t} = v_j$, yields (see Sect. 1.3):

$$a_i = \frac{\partial v_i}{\partial t} + v_1\frac{\partial v_i}{\partial x_1} + v_2\frac{\partial v_i}{\partial v_2} + v_3\frac{\partial v_i}{\partial x_3} \equiv \frac{D\vec{v}}{Dt} = a_{local} + a_{conv.} \quad (7.12)$$

Equation (7.12) can now be written on a per-unit-volume basis as (see Sect. 2.4):

$$\rho\left(\frac{\partial v_i}{\partial t} + v_k\frac{\partial v_i}{\partial x_k}\right) = \frac{\partial \tau_{ij}^{total}}{\partial x_j} + \rho g_i = -\frac{\partial p}{\partial x_j} + \frac{\partial \tau_{ij}^{visc}}{\partial x_j} + \rho g_i \quad (7.13a)$$

or in vector notation (see Equation of Motion in Sect. 2.4):

$$\rho\frac{D\vec{v}}{Dt} = -\nabla p + \nabla\cdot\vec{\vec{\tau}} + \rho\vec{g} \quad (7.13b)$$

As mentioned and implied with Eq. (7.13), the total stress tensor can be decomposed into two parts, i.e., one for the static body of fluid and the other one for fluid in motion:

$$\tau_{ij} = -p\delta_{ij} + \tau_{ij}^{visc} \tag{7.14}$$

where p is the (scalar) thermodynamic pressure; δ_{ij} is the Kronecker delta which is 1 when i = j and 0 when i ≠ j. It should be noted that $p + \tau_{ii}$ are often lumped together as the "normal stresses," while τ_{ij} with i ≠ j are the shear stresses. As indicated in Fig. 7.2, the pressure force will be always perpendicular to the exposed surface.

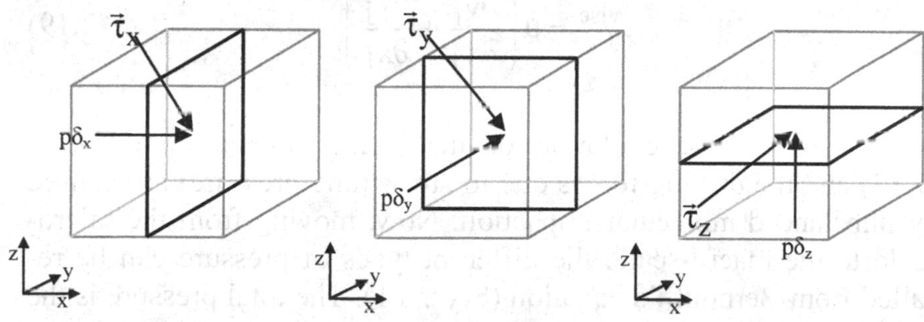

Fig. 7.2 Stress vectors and scalar pressure acting in/on the x-, y- and z-planes

On the microscopic scale, it can be shown that the *dynamic pressure* stems from the *translational* part of the kinetic energy (KE) of molecules. Such a molecular impact can be visualized/modeled as boundary work (W_b) when pushing a fluid in a piston-tube assembly. Thus, on a differential basis:

$$\delta W_b = dKE \tag{7.15}$$

or

$$p d\forall = F ds = m \frac{dv}{dt} ds = m v dv \tag{7.16a–c}$$

Integration yields:

$$p\forall = \frac{m}{2} v^2 \tag{7.17}$$

or

$$p_{dynamic} = \frac{\rho}{2} v^2 \tag{7.18}$$

Returning to Eq. (7.14), the viscous part of the total stress tensor relates for Newtonian constant-property fluids linearly to the velocity gradient (see Stokes' hypothesis in Sects. 1.4 and 2.4):

$$\tau_{ij}^{visc} = \mu\left(\frac{\partial v_i}{\partial x_j} + \frac{\partial v_j}{\partial x_i}\right) \tag{7.19}$$

Clearly, the evolution of the linear momentum equation (7.13) and the driving forces due to stress (and pressure) are caused by unbalanced molecular impaction. Now, moving from the micro-scale to the macro-scale, the different types of pressure can be recalled from Bernoulli's equation (Sect. 3.1). The total pressure is the sum of the thermodynamics pressure (measured via a static pressure tap), the dynamics pressure (measured by a Pitot-static probe), and the hydrostatic pressure which depends on the system's elevation:

$$p_{total} = p + \frac{\rho}{2} v^2 + \rho g z = p_{stag} + \rho g z \tag{7.20}$$

The stagnation pressure, $p_{stag} = p + \frac{\rho}{2} v^2$, measured with a Pitot tube, represents the pressure at a point where the fluid is brought to a complete stop isentropically. The thermodynamic pressure p is the actual pressure in the fluid flow field, as felt by an observer moving with the flow.

7.2.2 Movement and Impaction of Spherical Micron Particles

Many microfluidics applications include the dynamics of micron particles where the drag as well as electrokinetic and surface tension forces are the most important point forces (see also Sect. 6.4). Focusing first on the *total drag* on a sphere of radius R exposed to uniform flow U_0 where $Re_D \leq 1$, Stokes' law states:

$$F_{total} = F_{pressure} + F_{friction}$$

or

$$= F_{form} + F_{viscous}$$

$$= 2\pi\mu U_0 R + 4\pi\mu U_0 R = 6\pi\mu_0 R \qquad (7.21a–d)$$

It is insightful to examine:

- One-third of the drag is contributed by the pressure changes in front and behind the sphere, also known as the *form drag*.
- However, it is also the viscosity, i.e., frictional effects, which determines local flow pattern, hence causes F_{form} based on the pressure drop.
- For inviscid fluids ($\mu \equiv 0$), $F_{drag} = 0$ (known as "D'Alambert's Paradox") which again indicates that $F_{pressure}$ is also of viscous origin.

Applications The key assumptions for Stokes' law, $U_0 = ¢$ and $Re_D \leq 1$ are not that restrictive, when recalling that in fluid particle dynamics Re_D is actually $Re_{relative} = | v_{fluid} - v_{particle} | d_p/\upsilon$ and in the vicinity of small particles, say, of micro-diameter d_p, the flow field is quasi-uniform. Here are two applications of Stokes' law of interest in microfluidics and beyond.

Problem (i): When there are small particles of *irregular shape*, one can postulate that the actual drag is bounded by Stokes' law for two concentric spheres submerged in the same flow. One inscribed into the non-spherical body and the other fully surrounding the actual particle (see Fig. 7.3).

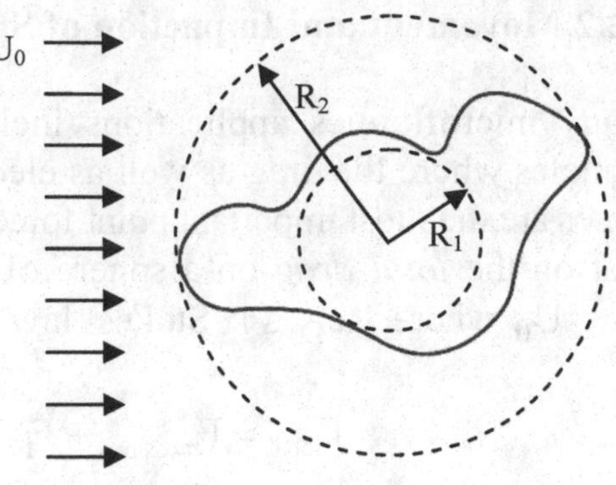

Fig. 7.3 Schematic to Hill and Powell theorem

Thus, we can state that the actual drag is bounded by F_{drag} of the inner and outer spheres:

$$F_{drag}^{inner} < F_{drag}^{actual} < F_{drag}^{outer} \tag{7.22a}$$

or

$$6\mu\pi U_0 R_1 < F_{actual} < 6\mu\pi U_0 R_2 \tag{7.22b}$$

or

$$1 < \frac{F}{6\mu\pi\, U_0 R_1} < \frac{R_2}{R_1} \tag{7.22c}$$

Problem (ii): Another (indirect) application of Stokes' law appears (courtesy of Prof. Sandip Ghosal) when considering pressure-force damping (or load lubrication) of a large/heavy object approaching a solid stationary surface. Examples include the cerebro-spinal fluid when an accelerated brain-mass hits the inside of a skull, air-cushioning of falling assembly-line parts, including glass-sheets descending on planar surfaces, as well as suspension and lubrication systems. Of interest is the fluid-resistive force on the approaching body, say, a large sphere of radius R as discussed in the next example.

Example 7.1: Squeeze Film Force vs. Sudden Impact Force

Sketch:	*Assumptions*:	*Concepts*:
	• Quasi-steady 2-D flow • Very small gap in height and extent • $\vec{\mathbf{v}}[u(r, z), v \approx 0, 0]$ Constant properties	• Reynolds lubrication assumption • Geometric approximations • Global mass balance

Solution:

- Sphere surface line equation h(r):

  For the right triangle,

$$R^2 = r^2 + [R - (h - h_0)]^2$$

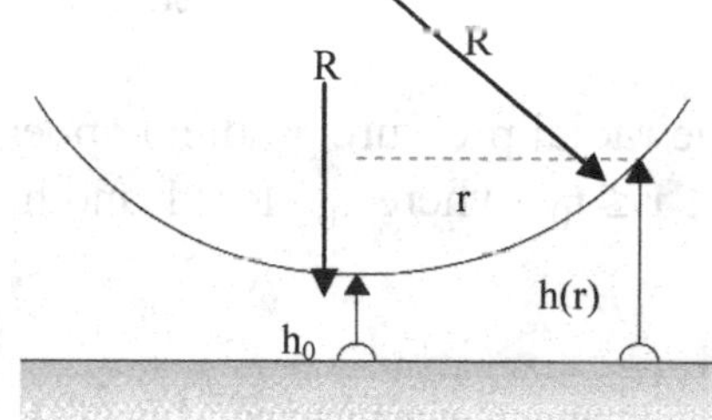

or

$$h - h_0 = R\left[1 - \sqrt{1 - \left(\frac{r}{R}\right)^2}\right] \qquad \text{(E.7.1.1a)}$$

For $\frac{r}{R} < 1$ (small gap region):

$$\sqrt{1 - \left(\frac{r}{R}\right)^2} \approx 1 - \frac{1}{2}\left(\frac{r}{R}\right)^2 \qquad \text{(E.7.1.1b)}$$

so that

$$h(r) = h_0 + \frac{r^2}{2R} \qquad \text{(E.7.1.1c)}$$

- Reduced N–S equations based on lubrication assumptions (see Sect. 4.3 with $h_0/R \ll 1$ and $Re_{h_0}(\frac{h_0}{R}) \ll 1$, where $Re_{h_0} = v_0\, h_0/\nu$) so that:

$$\text{(r-momentum)} \qquad 0 = -\frac{\partial p}{\partial r} + \mu \frac{\partial^2 u}{\partial z^2} \tag{E.7.1.2}$$

and

$$\text{(z-momentum)} \qquad 0 = -\frac{\partial p}{\partial z} \tag{E.7.1.3}$$

subject to no-slip conditions u(z = 0) = u[z = h(r)] = 0. Double integration yields:

$$u(r, z) = \frac{1}{2\mu}\left(\frac{\partial p}{\partial r}\right) z^2 \left(1 - \frac{h}{z}\right) \tag{E.7.1.4}$$

The radial pressure gradient appears in a global mass balance for gap $0 \le r \le r_0$, where $r_0/R < 1$ and $h/R \ll 1$:

$$\rho v_0(\pi r^2) = 2\pi r\rho \int_0^h u(r, z)\, dz \tag{E.7.1.5}$$

or

$$\int_0^h u(r, z)\, dt = \frac{v_0}{2} r$$

Using u(r, z) yields:

$$\frac{1}{2\mu}\left(\frac{dp}{dr}\right)\int_0^h z^2\left(1 - \frac{u}{z}\right) dz = \frac{v_0}{2} r$$

So that after integration:

$$\frac{dp}{dr} = -\frac{6\mu v_0}{h^3} r \qquad \text{(E.7.1.6)}$$

where $h - h_0 + \frac{r^2}{2R}$ and $\int_r^\infty dp = p(r) - p_{atm}$. Hence,

$$p(r) = p_{atm} - 6\mu v_0 \int_r^\infty \frac{r}{[h(r)]^3} dr$$

or

$$p(r) - p_{atm} = \frac{6\mu v_0 R^2}{h_0\left(1 + \frac{r^2}{2h_0R}\right)} \qquad \text{(E.7.1.7)}$$

Thus,

$$p(r=0) = p_{max} = p_{atm} + 6\mu v_0\left(\frac{R^2}{h_0}\right), \text{ where } \frac{R^2}{h_0} >> 1$$

The pressure force is:

$$F = \int_0^\infty \Delta p(2\pi r)dr = \frac{12\pi\mu v_0 R^2}{h_0} \int_0^\infty \frac{rdr}{1 + \frac{r^2}{2h_0R}}$$

or (E.7.1.8a, b)

$$F = 6\pi\mu v_0 R\left(\frac{R}{h_0}\right) = F_{stokes} \cdot \left(\frac{R}{h_0}\right)$$

1-D Force balance ($Re_D \le 1$), using Eq. (E.7.1.8b):

$$m\frac{dv}{dt} = F_{weight} - F_{fluid} = mg - 6\pi\mu v_0 \frac{R^2}{h_0}$$

(E.7.1.9 a–e)

$$\frac{dv}{g - Kv} = dt\,;\;\; K = \frac{6\pi\mu R^2}{mh_0}\,;\;\; m = \frac{4\pi}{3}\rho R^3\,;$$

subject to $v(t = 0) = v_0$. Thus,

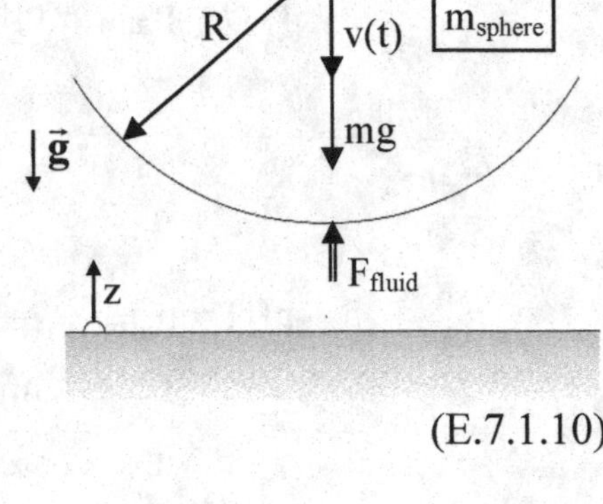

$$\ln\frac{g - Kv}{g - Kv_0} = -Kt$$

or

$$v(t) = \frac{g}{K} - \left(\frac{g}{K} - v_0\right)e^{-Kt} \qquad \text{(E.7.1.10)}$$

Acutally $\frac{g}{K} << 1$, so that:

$$v(t) = v_0 e^{-Kt}$$

where $K \equiv \dfrac{9\upsilon}{2Rh_0}\left[s^{-1}\right]$ can be interpreted as the ratio $F_{fluid}/(mv)$ and hence

$$F_{fluid}(t) = \frac{mv}{t}\ln\left(\frac{v_0}{v}\right) \qquad \text{(E.7.1.11)}$$

Note: At t = 0, $\ln(\frac{v_0}{v}) = 0$ because $v = v_0$, and according to L'Hospital's rule $F_{fluid} = 0$.

Graphs:

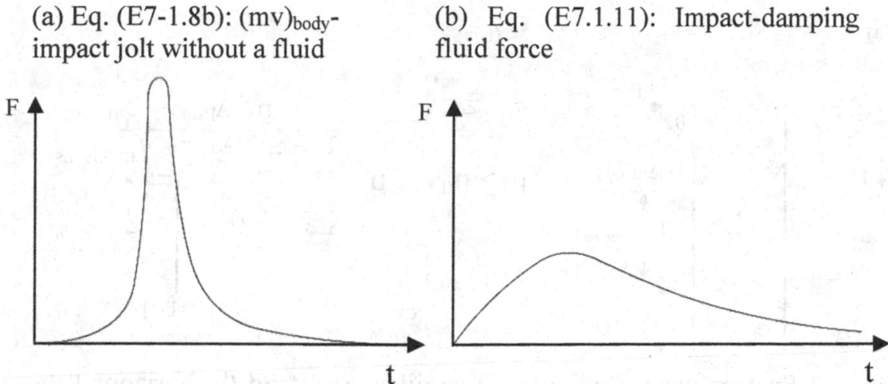

Comments:

Clearly, as p(r) peaks when $r \to 0$, F shoots up as the body/sphere approaches, i.e., as $h_0(t)$ is reduced. As a result, v_0 decreases rapidly and the momentum impact, $(mv_0)_{body}$ is greatly damped by the fluid by: (a) significantly reducing the peak force and (b) distributing the impact over time.

7.2.3 Pumps Based on Microscale Surface Effects

In addition to mechanical fluid-volume movers, such as micro-scale pumps and syringes, surface-modulated "pumps" are very suitable for microsystems (say, MEMS) when small Reynolds number, i.e., $10^{-4} < Re < 1$, suffice. Surface tension, Marangoni effect and electrokinetic effects are the underlying principles for low-pressure pumps.

Surface Tension The most familiar example of unusual behavior of a liquid in a small-diameter tube is the "capillary rise" where a liquid column moves upwards against gravity or sideways due to differences in surface tension, which is a directional force divided by the length of the contact interface (Fig. 7.4). Derivations of the upcoming section equations are left as homework assignments (see Sect. 7.6).

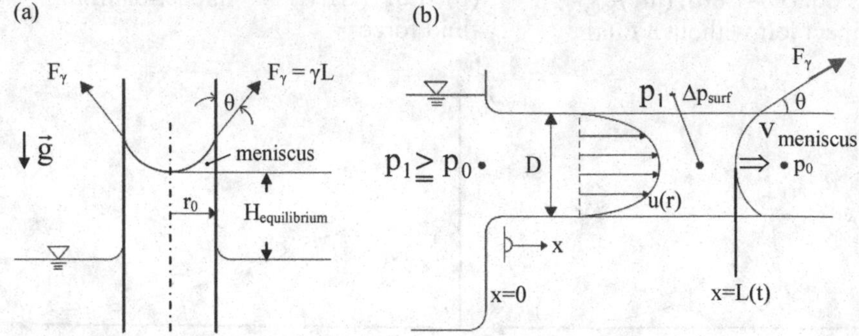

Fig. 7.4 Surface tension effects: (a) capillary rise; and (b) horizontal displacement

At a liquid–gas interface the liquid molecules exposed to the gas have a higher energy when compared to all other liquid molecules, because of the missing chemical bonds in the direction of the surface. That creates the *surface tension* which depends on the two materials forming the interface, e.g., combinations of liquid–gas, solid–liquid or solid–gas surfaces. Another basic concept is the *contact angle*. It appears at the contact line between, say, the solid capillary wall and the two immiscible fluids, typically liquid/gas, inside the tube (see Fig. 7.4a). Assuming equilibrium, the contact angle Θ is determined according to Young's force-balance equation, involving the three surface tensions γ_{sg}, γ_{sl} and γ_{lg} at the three interfaces solid–gas, solid–liquid, and liquid–gas:

$$\gamma_{sg} - \gamma_{sl} = \gamma_{lg} \cos \Theta \tag{7.23}$$

A solid surface is hydrophilic for $0° < \Theta < 90°$ and hydrophobic if Θ is in the range of 90° to 180°. For example, measuring the capillary rise H and contact angle Θ, the liquid–gas surface tension can be computed from a vertical force balance (see Fig. 7.4a):

$$2\pi r_0 (\gamma_{sg} - \gamma_{sl}) = \rho g H \pi r_0^2 \tag{7.24a}$$

With Young's equation (7.23) we have:

$$\gamma_{lg} \equiv \gamma = \frac{\rho g r_0 H}{2 \cos \Theta} \qquad (7.24b)$$

As an example, for water at 20°C (i.e., $\gamma = 73$ mN/m) in a plastic tube of radius $r_0 = 100$ μm and contact angle $\Theta = 74°$, we obtain with Eq. (7.24b) the significant rise of H = 42 mm. Across curved interfaces, characterized by the radii of curvature R_1 and R_2, a pressure drop Δp_{surf} occurs according to the Young–Laplace equation:

$$\Delta p_{surf} = \gamma \left(\frac{1}{R_1} + \frac{1}{R_2} \right) \qquad (7.25)$$

Equation (7.25) reduces for a spherical bubble where $R_1 = R_2 = R$ to:

$$\Delta p_{surf} = \frac{2\gamma}{R} \qquad (7.26a)$$

For a liquid column with meniscus radius R (see Fig. 7.4a), the radius of curvature is $R = r_0/\cos \Theta$ and hence

$$\Delta p_{surf} = \frac{2\gamma \cos \Theta}{r_0} \qquad (7.26b)$$

The interplay of surface tension, gravitational, viscous and inertial forces is captured in three dimensionless groups, i.e., the

Bond number: $$Bo = \frac{F_{gravity}}{F_{surf.\,tens.}} = \frac{\pi g r_0^2}{\gamma} \qquad (7.27)$$

Capillary number: $$Ca = \frac{F_{viscous}}{F_{surf.\,tens.}} = \frac{\mu v}{\gamma} \qquad (7.28)$$

and the

$$\text{Weber number:} \qquad We = \frac{F_{inertia}}{F_{surf.\ tens.}} = \frac{\rho v^2 r_0}{\gamma} \tag{7.29}$$

Example 7.2: The Capillary Pump: A Microfluidics Application

Of interest is the time it takes for a liquid column/meniscus to advance in a long *horizontal* microchannel of cross sectional area $A = h \cdot w$, where surface tension overcomes frictional effects (see Fig. 7.4b). Given h, γ, μ and θ, find $v_{meniscus}$ and extent travelled L(t).

Sketch	*Assumptions*	*Concepts*
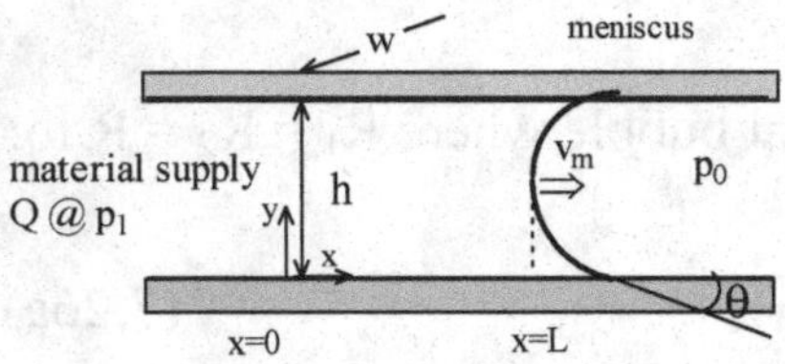	• Poiseuille flow between parallel plates (w >> h) • Constant properties • Uniform meniscus motion	• Mass conservation • $\Delta p_{surf.}$ is the driving force • Hagen-Poiseuille flow rate • $v_{meniscus} = dL/dt$

Solution:

- Mass conservation, $Q = \bar{u}A = ¢$, and the use of the Hagen–Poiseuille flow solution yields:

$$Q = \left(\frac{\Delta p_{surf}}{L}\right)\frac{h^2}{12\mu} A \tag{E.7.2.1}$$

- The Young–Laplace equation states (see Eq. (7.26b)):

$$\Delta p_{surf} = \frac{2\gamma \cos \Theta}{h} \tag{E.7.2.2}$$

Note: $$v_{meniscus} \equiv \frac{dL}{dt} = \frac{Q}{hw} = \frac{\Delta p}{L}\frac{h^2}{12\mu} \qquad \text{(E.7.2.3a, b)}$$

Hence

$$LdL = \frac{\Delta p h^2}{12\mu} dt; \text{ with } L(t=0) = 0 \qquad \text{(E.7.2.4a, b)}$$

so that

$$\frac{L(t)}{h} = \sqrt{\frac{t}{t^*}}; \text{ where } t^* \equiv \frac{3\mu h}{\gamma \cos\theta} \qquad \text{(E.7.2.5a, b)}$$

and the advancing meniscus speed is:

$$v_m \equiv \frac{dL}{dt} = \frac{h}{2\sqrt{t^*}} t^{-1/2} \qquad \text{(E.7.2.6)}$$

Graphs:

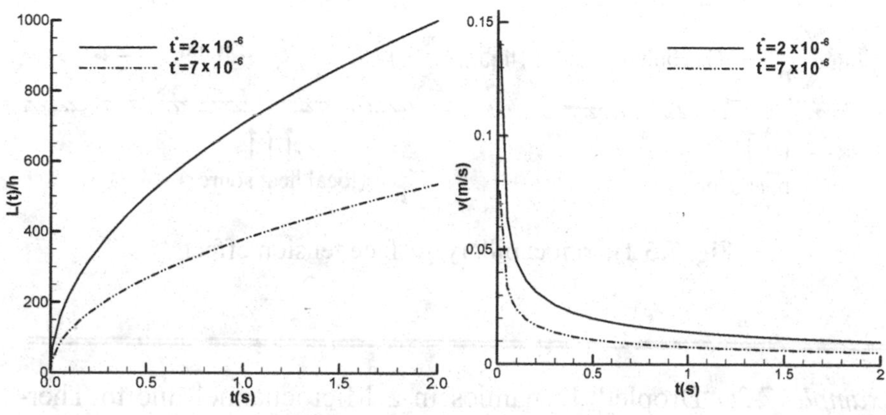

Comments: In light of the key assumptions, the liquid micro-column covers a distance following a parabolic function while the meniscus speed is essentially hyperbolic, i.e., a sudden maximum velocity which quickly diminishes.

Marangoni Effect So far, it was assumed that the surface tension γ is constant; however, spatial changes in surfactant concentrations and/or temperature cause gradients in the surface tension as well. Specifically, while $\gamma = F/L$, $\nabla\gamma$ implies a force per unit area, called the Marangoni or thermocapillary effect, which can move gas bubbles or liquid droplets in partially heated microchannels (see Fig. 7.5a). The one-sided heat flux generates a temperature gradient across the gas bubble, where the local surface tension is lower at the higher temperature. As a result the Marangoni force reads:

$$f_M = \nabla\gamma \tag{7.30}$$

which pushes the bubble towards the heat source. The inverse happens for the liquid plug [see (Fig. 7.5b)]. For example, γ_{water} at 20°C is 72.75 × 10^{-3} N/m but 62.6 × 10^{-3} N/m at 80°C.

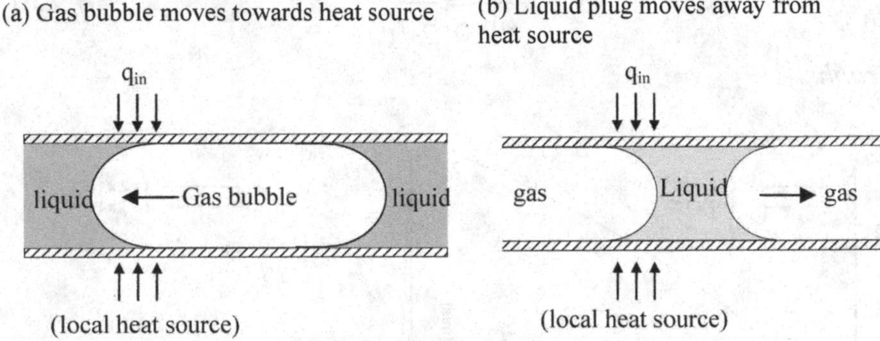

Fig. 7.5 Thermocapillary surface tension effect

Example 7.3: "Droplet" Dynamics in a Microchannel due to Thermocapillary Effect

Consider a horizontal microtube of radius r_0 with a liquid plug (i.e., a droplet) of length L. A locally applied heat flux q_s pushes the liquid away from the heat source because of the Marangoni effect. Assume a constant axial temperature gradient and that all system parameter values are given.

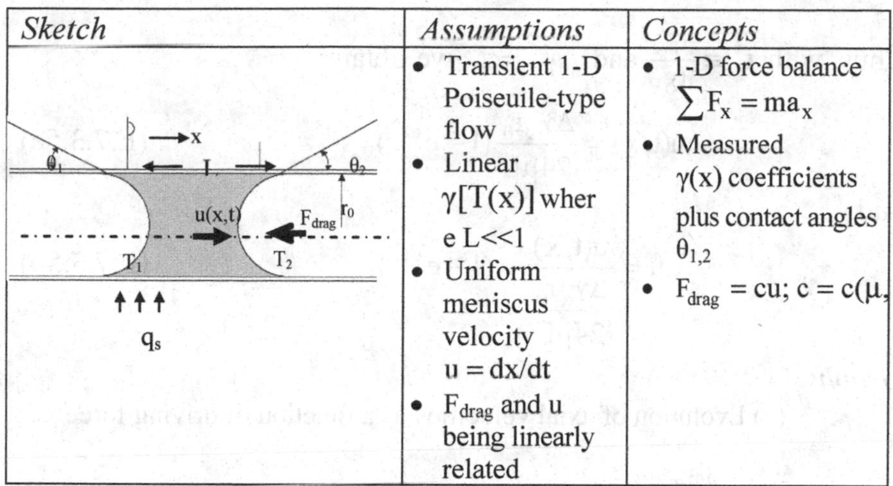

Sketch	*Assumptions*	*Concepts*
	• Transient 1-D Poiseuile-type flow • Linear $\gamma[T(x)]$ where $L<<1$ • Uniform meniscus velocity $u = dx/dt$ • F_{drag} and u being linearly related	• 1-D force balance $\sum F_x = ma_x$ • Measured $\gamma(x)$ coefficients plus contact angles $\theta_{1,2}$ • $F_{drag} = cu;\ c = c(\mu,$

Solution:

$$ma_x = F_{net\ pressure} + F_{Marangoni} - F_{drag} \tag{E.7.3.1}$$

$$ma_x = \frac{\Delta p}{L}\forall + \frac{\Delta\gamma}{L}\Lambda_{surf} - c\cdot u \tag{E.7.3.2}$$

Here, we assume:

$$F_{drag} = c\cdot u = 56\pi\mu L\cdot u \tag{E.7.3.3 a}$$

where 56 is a balancing number. With $\frac{\Delta p}{L} = \frac{8\mu}{r_0^2}u$ (see Sect. 2.4), Eq. (E.7.3.2) now reads:

$$m\frac{du}{dt} = 8\pi\mu L u + 2\pi r_0[\gamma(x = L)\cos\theta_2 - \gamma(x = 0)\cos\theta_1] \tag{E.7.3.3b}$$

With the linear approximation for $\gamma(T)$ (while $L << 1$ and $\Theta_1 \approx \Theta_2$), we have:

$$\gamma(L)\cos\theta_2 - \gamma(0)\cos\theta_1 = \gamma_2\cos\theta - \gamma_1\cos\theta = -\Delta\gamma\cos\theta \tag{E.7.3.4}$$

Hence, with $m = \rho\pi r_0^2 L$ and $\upsilon = \mu/\rho$ we have an ODE for u(x,t):

$$\frac{du}{dt} - 8\frac{\nu}{r_0^2}u - \frac{2}{\rho r_0 L}\Delta\gamma_x + 56\frac{\nu}{r_0^2}u = 0 \tag{E.7.3.5}$$

where $\Delta\gamma_x = \frac{T_1 - T_2}{T_1 - T}(\gamma - \gamma_1)$ with T(x) and $\gamma(x)$ being linear functions. The initial condition is $u(t=0) = 0$.

Thus, with $t^* \equiv \dfrac{r_0{}^2}{48\nu}$ and $\Delta\gamma_x = ⊄$, we obtain:

$$u(t,x) = \frac{\Delta\gamma_x r_0}{24\mu L}(1 - e^{t/t^*}) \tag{E.7.3.5a}$$

and

$$\hat{u} = \frac{u(t,x)}{\dfrac{\Delta\gamma_x r_0}{24\mu L}} = (1 - e^{t/t^*}) \tag{E.7.3.5b}$$

Graphs:

(a) Evolution of axial velocities as a function of driving force

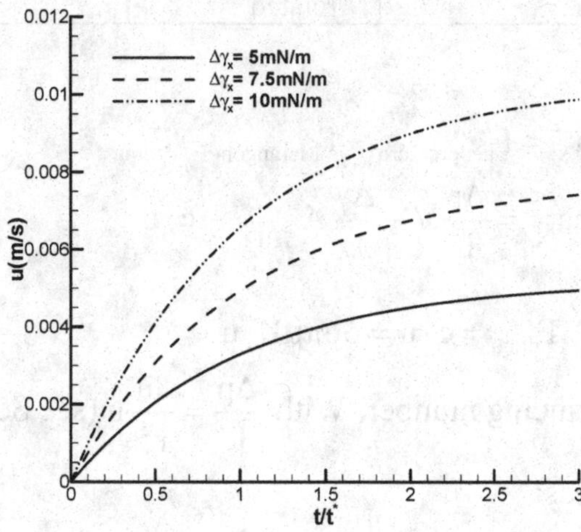

(b) Dimensionless axial velocity with time

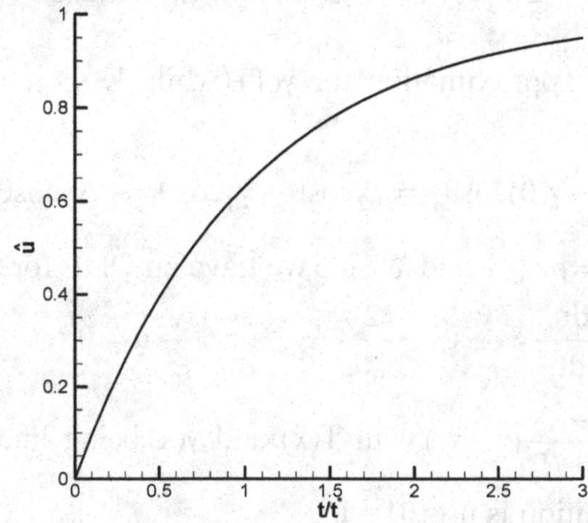

Comments:

- The meniscus velocity reaches asymptotically the maximum velocity if $\Delta\gamma_x(T)$ can be maintained.
- For a more realistic analysis, temperature dependence T(x) (i.e., also $\gamma(x)$) should be considered.

7.2.4 Microchannel Flow Effects

As discussed in Sect. 7.2.1, for gas flow the Knudsen number has to be checked. For liquid flow in microchannels with $D_h > 10$ μm the continuum assumption holds, while on hydrophobic surfaces liquid flow velocity-slip at the wall should be considered (see Sect. 7.2.5). In any case, certain size-related phenomena may have to be accounted for. They include:

- *Entrance effects* when the channel is short, the aspect ratio (i.e., height vs. width) is high, and/or the Reynolds number is relatively large.
- *Viscous dissipation* when liquids have a high viscosity and/or the flow field exhibits steep velocity gradients, which maybe partly due to small channels, say, $D_h \leq 50$ μm.
- *Surface roughness effects* when the channel walls are not smooth, especially in tubes, and when the Reynolds number is high.

The friction factor, $f = f(\text{Re}, \frac{e}{D_h}, \frac{L_e}{D_h})$, as well as the Nusselt number, $\text{Nu} = \text{Nu}(\text{Re}, \text{Pr}, f)$, encapsulate these effects, where e is the roughness height, L_e the entrance length, and $\text{Pr} = \upsilon/\alpha$ is the Prandtl number. It is not surprising that the macroscale correlations for f and Nu (see Sects. 3.6 and 4.2) do not apply to many microchannel flows (see Koo & Kleinstreuer 2003).

Microchannel Entrance Flow For microchannels, the semi-empirical entrance length is for $\text{Re}_{D_h} < 10^3$:

$$\frac{L_e}{D_h} = a\, Re_{D_h}\,;\ a = 0.08 - 0.09 \tag{7.31}$$

For example, when Re = 100 and D_h = 50 μm, L_e = 4 mm which may exceed the actual length of a MEMS device. When compared to hydrodynamic entrance lengths of macro-conduits, the coefficient a in Eq. (7.31) ranges from 0.056 to 0.085 for laminar flow in circular to rectangular channels.

Thus, depending on channel geometry, surface roughness and Reynolds number range, we could state:

$$\frac{L_e|\,\text{micro}}{L_e|\,\text{macro}} \geq 10 \tag{7.32}$$

Microchannel Surface Roughness Depending on the type of channel material (e.g., silicon vs. steel) and manufacturing technique, the relative surface roughness $\varepsilon = e/D_h$ may fall in the range $1 \leq \varepsilon \leq 4\%$. Clearly, significant ε-values affect the friction factor, may influence the onset of turbulence at elevated Reynolds numbers, and could perturb classical flow patterns, e.g., axisymmetric tubular flow turns into 3-D flow when a circumferential velocity component emerges. Rough wall surfaces have been depicted as geometrically regular/random protrusions, sine- or cosine-functions, or in form of a porous-medium layer (see Kleinstreuer & Koo 2004; among others). Especially for rough microtubes, the effective friction factor increases with higher Reynolds numbers and smaller diameters (Li & Kleinstreuer 2008).

Microchannel Cross Section In contrast to microscale conduits which come in all shapes and forms, microchannels are presently either tubular, rectangular or trapezoidal, where the latter is popular because it allows for a more convenient manufacturing technique. The aspect ratio of trapezoidal channel widths, $\kappa = \overline{W}_{bottom}/\overline{W}_{top}$, greatly influences fully-developed laminar flow in smooth trapezoidal microchannels. In terms of the Poiseuille number $Po \equiv f\,Re$, we recall that for macrotubes $Po = 16$ (see Sect. 3.2), while for trapezoidal microchannels (Wu & Cheng 2003 as well as Li & Kleinstreuer 2008) found:

$$\text{Po} \equiv f\,\text{Re} = \frac{\Delta p D_h^{\ 2}}{2\,\rho\,\mu\,v\,L} = 11.43 + 0.8e^{2.67\kappa} \qquad (7.33)$$

Viscous Dissipation in Microchannels With decreasing micro-channel size as well as a high Reynolds number and large aspect ratio, viscous dissipation, $\mu\,\Phi \sim [\nabla v]^2$, becomes significant and hence the fluid temperature may increase, influencing fluid properties and flow parameters. For example, when $D_h < 50\ \mu m$ and the fluid is water the viscous dissipation effect should be considered (see Koo & Kleinstreuer 2004; among others). More complicated aspects of microfluidics, such as near-wall electrostatic effects as well as velocity-slip and temperature-jump boundary conditions are discussed in the remaining sections.

7.2.5 Wall Boundary Conditions

As emphasized several times, the Navier–Stokes equations are based on the conditions that:

(i) The fluid is infinitely divisible (i.e. the continuum assumption holds).
(ii) All flow quantities are in local thermodynamic equilibrium, i.e. macroscopic quantities within the fluid have sufficient time to adjust to their surroundings, a process which depends on the time between molecular collisions and hence the mean-free path travelled before molecules collide.

Clearly, when the time scale or length scale, e.g., the mean-free path $\lambda = \frac{\mu}{p}\sqrt{\frac{\pi R T}{2}}$ of the molecular collisions are small in comparison to the macroscopic flow variations, then the fluid will be able to adjust quickly to the new conditions. Furthermore, if the condition $\lambda \ll L_{system}$ holds, where the flow systems' length scale is typically the hydraulic diameter $D_h = \frac{4A}{P}$, then the macroscopic flow quantities will basically vary linearly in space; as for example shear stress versus strain rate (Stokes) or heat flux versus temperature gradient (Fourier). Based on kinetic theory, i.e. the hard-sphere ideal gas model, the fluid–wall interactions are rather well defined for *gas flow* in terms of the Knudsen number, $Kn = \lambda / D_h$. As mentioned in Sect. 7.2.1, when $Kn < 10^{-2}$, the no-slip condition is valid. However, for the range $0.01 < Kn < 0.1$ solutions to the Navier–Stokes equations have to rely on suitable first-order slip-flow conditions, i.e., $u_{slip} = u_{gas} - u_{wall}$, in the so-called transition region when $0.1 < Kn < 1$, a second-order slip-velocity boundary condition may have to be invoked.

For *liquid flows* the breakdown of the thermodynamic equilibrium condition is not well defined. However, because liquid molecules are much closer together and do not move in such random fashion, the Navier–Stokes equations with no-slip wall condition can be employed for most microchannel flows, say, when $D_h \geq O(10\ \mu m)$ for water at STP level.

7.2.5.1 Review of Boundary Conditions for Macro-Scale Flows

As mentioned, the underlying assumptions for the use of the Navier–Stokes equations to solve macro-scale flows subject to the no-slip condition are that the fluid is a continuum and thermodynamic equilibrium exists. The no-slip velocity at the wall requires that:

$$\vec{v}_{fluid} = \vec{v}_{wall} \tag{7.34}$$

For example, for unidirectional flow where $\vec{v} = (u,0,0)$, Eq. (7.34) implies that:

- $u_{fluid} = \begin{cases} u_{wall} = 0 \text{ in axial direction because the wall is stationary} \\ u_{wall} \neq 0 \text{ because the wall or plate moves} \end{cases}$

- $v_{fluid} = v_{wall} = 0 \begin{cases} \text{because the wall is impermeable and does not} \\ \text{move in the normal direction} \end{cases}$

One-Dimensional Flow Applications Standard 1-D reductions of the Navier–Stokes equations describe Couette flow, Poiseuille flow and thin-film flow. The implications are that such flows are steady, laminar, fully-developed with constant fluid properties and zero or constant pressure gradient. Based on a 1-D balance of driving force and resistance force, the second-order ODE for the axial velocity is of the form (see Sect. 2.4 and App. A):

$$\frac{d^2u}{dy^2} = K = \text{constant} \tag{7.35}$$

The necessary two boundary conditions (B.C.'s) are illustrated for several cases in Fig. 7.6.

Couette Flow:

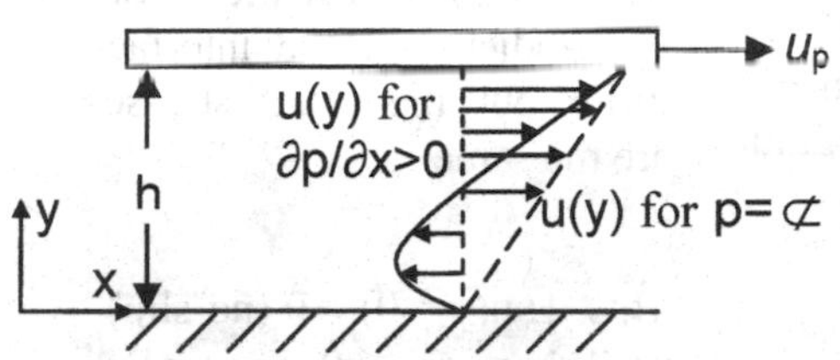

B.C. 1: $u(y=0)=0$
B.C. 2: $u(y=h)=u_p=$ ¢

Note: When the constant pressure gradient is adverse, u(y) is non-linear and backflow occurs

Poiseuille Flow:

(a) Tube

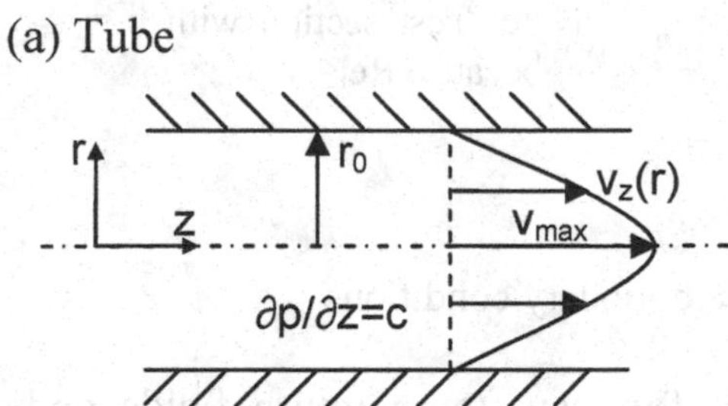

B.C. 1: $v_z(r = r_0) = 0$ (no-slip)
B.C. 2: $dv_z/dr(r = 0) = 0$ (symmetry)

Note: When $v_z(r = r_0) = v_{max}$ or $u(y = 0) = 0 = u_{max}$ is not helpful because the centerline velocity is unknown.

(b) Parallel-Plate Flow:

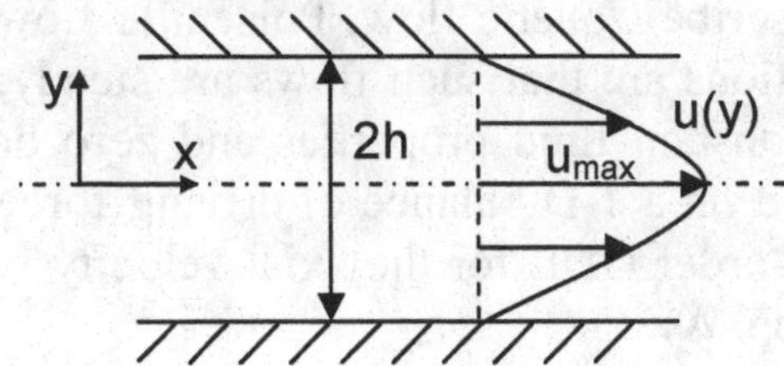

B.C. 1: $u(y = h) = 0$
B.C. 2: $du/dy(y = 0) = 0$

(c) Two-Fluid Flow:

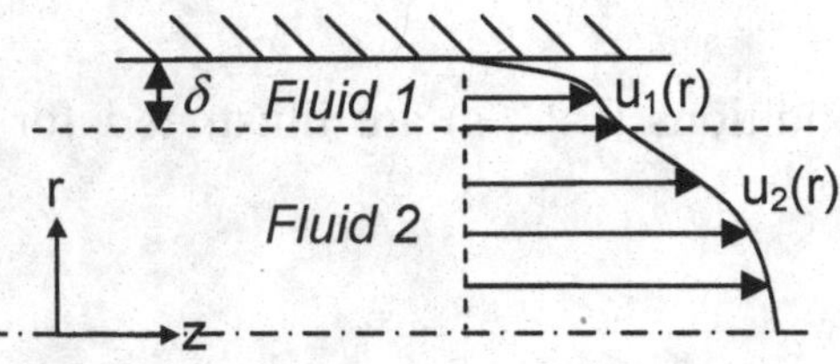

B.C. 1: $u_1(r = r_0) = 0$ (no-slip)
B.C. 2: $du_2/dr(r = 0) = 0$ (symmetry) as before.

At interface $r = r_0 - \delta$:
B.C. 3: $u_1 = u_2$
B.C. 4: $\mu_1 du_1/dr = \mu_2 du_2/dr$ which implies that the velocity gradients at the interface differ but the shear stresses are the same

Note: Scalar transport equations for, say, temperature fields and concentration distributions require very similar "magnitude and flux" boundary conditions.

(d) Thin-Film Flow:

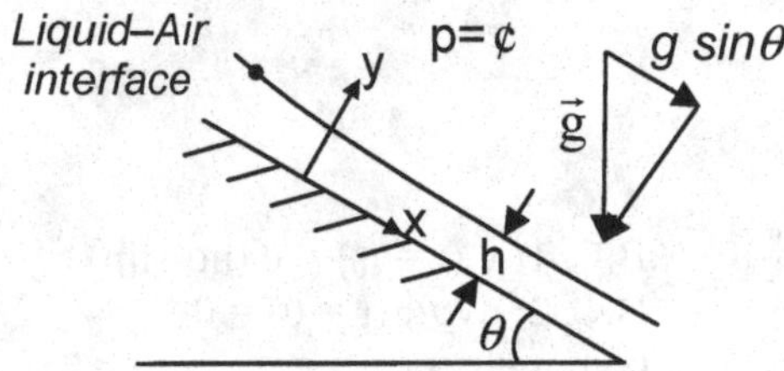

B.C. 1: $u(y = 0) = 0$ (no-slip)
B.C. 2: $\tau_{interface} \approx 0$, i.e. negligible frictional impact, so that $du/dy(y = h) = 0$

Note: Test section with $h = c$, because $Re \leq 1$.

Fig. 7.6 Macro-scale boundary conditions

Note: Scalar transport equations for, say, temperature fields and concentration distributions require very similar "magnitude and flux" boundary conditions.

7.2.5.2 Wall-Boundary Conditions for Micro-conduit Flows

Thermodynamic non-equilibrium conditions may occur at a solid (heated) surface in case the fluid is a "dilute" (i.e., near rarefied) gas or a liquid on a *hydrophobic* surface with micro-scale roughness.

Gas Flow For dilute to rarefied gas flows in microchannels, the form of interaction between gas molecules and the wall plays a crucial role in possible velocity slip and temperature jump. Neglecting adsorption, as molecules impinge on the wall they will be reflected after collision with other molecules near the wall and the wall surface. The two extreme types of reflection are:

(i) Specular, i.e., mirrored incidence and reflection on ideal surfaces so that the tangential momentum (m v) remains the same and perfect slip results.
(ii) Diffuse, on rough surfaces when the molecules randomly reflected, acquiring equilibrium with the wall and losing on average their entire tangential momentum.

For real surfaces, molecular reflection is a random mixture of both, resulting in a possible net tangential momentum in the axial flow direction. Concerning Eq. (7.37), $\sigma \rightarrow 0$ for ideal specular reflection (i.e., effectively representing inviscid flow). In contrast, $\sigma \approx 1.0$ for 100% diffuse reflection, a σ-value assumed for most microscale engineering systems. In terms of the Knudsen number, for $\mathrm{Kn} < 0.01$ (or $< 10^{-3}$) the gas is a continuum obeying the Navier–Stokes equations with no-slip and no temperature-jump boundary conditions. As Kn increases, collision of gas molecules with the wall becomes more frequent, because fewer molecules are around to collide with as the mean free path λ increases.

Focusing on simple conduction (i.e., thermal diffusion) of a stationary gas with variable Kn # between two heated walls, Fig. 7.7 illustrates the various temperature profiles for $\mathrm{Kn} \leq 0.01$ (continuum thermal diffusion), $0.01 < \mathrm{Kn} \leq 0.1$ (slip-flow and temperature-jump regime), and $Kn = \mathrm{O}(1)$, i.e., molecular flow.

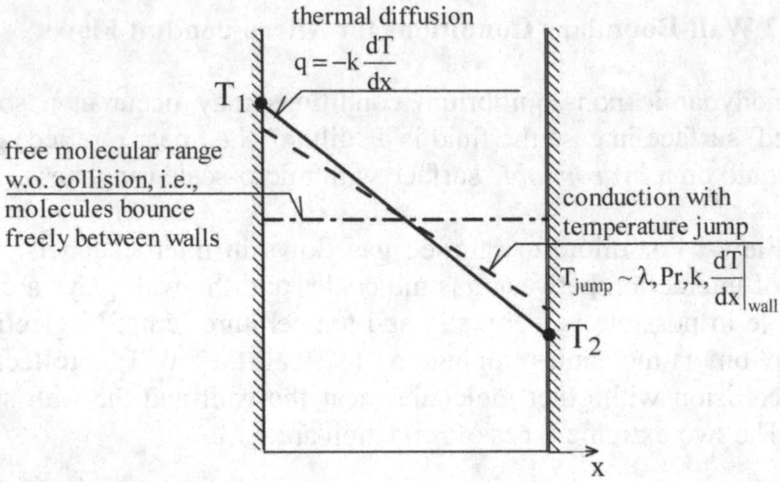

Fig. 7.7 Kundsen number effects on linear temperature profiles in heated gas flow

Liquid Flow Possible first-order velocity slip can be accommodated with Navier's "slip-length model". Liquid flow slip may occur in microchannels with hydrophobic and rough-wall surfaces. The degree of hydrophobicity is measured by the angle that a droplet's free surface makes with the solid on which it rests. For example, clean glass is hydrophilic where water forms a contact angle θ close to zero. In contrast, water droplets on silanized surfaces, i.e., silicon hydride (SiH) coated hydrophobic surfaces, are somewhat rounder than hemispherical, exhibiting $\theta \geq 100°$. In addition to the right liquid-and-surface pairing to generate hydrophobic effects leading to velocity slip, the degree of *surface roughness* in microchannels is an important factor as well. Specifically, it can be envisioned that a liquid, say water, does not fill the micro-valleys between roughness peaks and hence forms an unstable water-air-pocket-solid-surface interface. The resulting surface tension pushes the water elements forwards gliding on the air–water interfaces from peak to peak over the rough surface. Experiments showed that by reducing the roughness height but increasing the peak spacing, laminar flow drag reduction could be achieved (see Ou et al. 2004; among others).

7.2.5.3 Slip Flows

The concept of slip boundary conditions can be illustrated in two ways, following Navier's model based on the "slip-length" δ, and Maxwell's model relying on the mean free path λ. In both cases the slip velocity is proportional to the wall velocity gradient. For simple shear flows, these models are illustrated in Fig. 7.8.

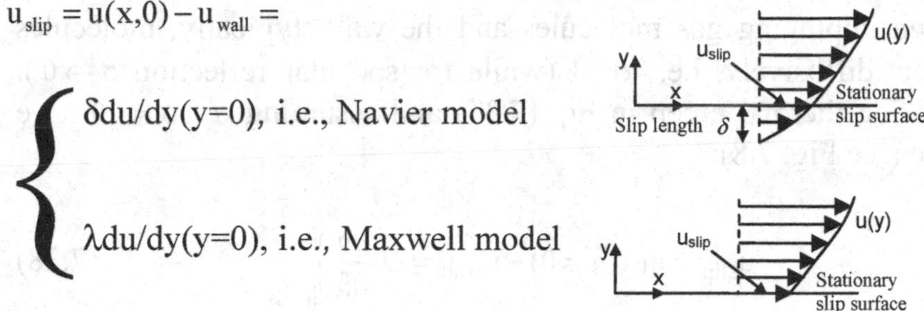

Fig. 7.8 Velocity slip models

Navier's Slip-Length Model Very large slip lengths, say $\delta \geq 100\ \mu\text{m}$, have been observed for certain polymeric liquids while $\delta < 100$ nm for basic fluids flowing over smooth hydrophobic surfaces. As discussed by Ou et al. (2004), δ can be estimated for parallel-plate flow (Fig. 7.8) from measuring the volumetric flowrate per unit width:

$$\hat{Q} = \frac{H^3}{4\mu}\left(\frac{\Delta p}{L}\right)\left(\frac{1}{3} + \frac{\delta}{H}\right) = \hat{Q}_{\text{Poiseuille}} + \hat{Q}_{\text{slip}} \tag{7.36}$$

where $\hat{Q} = Hu_{\text{slip}}$ and $u_{\text{slip}} = \delta \left.{}^{du}\!/_{dy}\right|(y=0)$. Equation (7.36) implies that a 20% drag reduction requires a slip length of $\delta = {}^{H}\!/_{15}$, or $H = 15\delta$. Thus, for, say, $\delta = 33$ nm it implies a channel height of half a micrometer. Clearly, employing nanochannels and fluids with both great hydrophobicity and ordered surface roughness will significantly decrease drag and increase throughput (see Eqs. (7.36), (7.40), or reduce power requirements.

Maxwell's First-Order Slip-Velocity Model For *gas flow* on planar surfaces:

$$u_{slip} = u_{gas} - u_{wall} = \alpha \frac{2-\sigma}{\sigma} \lambda \left(\frac{\partial u}{\partial \hat{n}} \right) \Bigg|_{\hat{n}=0} + \frac{3}{4} \frac{\mu}{\rho T} \left(\frac{\partial T}{\partial x} \right) \Bigg|_{wall} \tag{7.37}$$

where $\alpha = 1.1466$, based on kinetic analysis of the Boltzmann equation, σ is the tangential momentum accommodation coefficient which represents the average streamwise momentum exchange between impinging gas molecules and the wall; typically, molecules reflect diffusively, i.e. $\sigma = 1$ (while for specular reflection $\sigma \to 0$). Ignoring thermal creep in Eq. (7.37) and assuming $\alpha \approx \sigma \approx 1.0$, we have (see Fig. 7.8):

$$u_{slip} = u(x, \hat{n} = 0) - u_{wall} = \lambda \left(\frac{\partial u}{\partial \hat{n}} \right) \Bigg|_{\hat{n}=0} \tag{7.38}$$

where the mean free path:

$$\lambda = \frac{\mu}{p} \sqrt{\frac{\pi R T}{2}} \tag{7.39}$$

and $\hat{n}$ is the surface normal vector. For example, for Couette flow (Fig. 7.9):

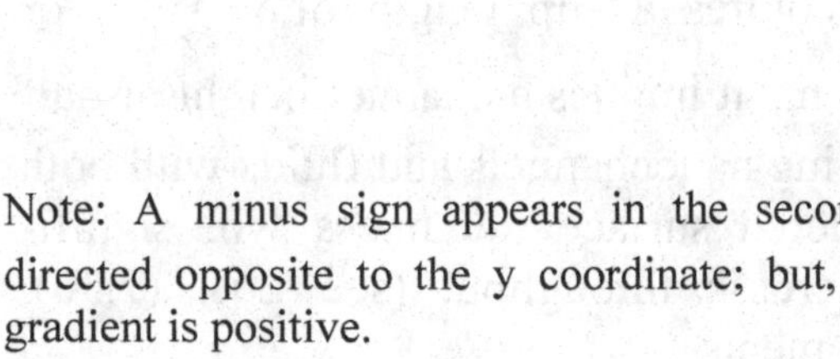

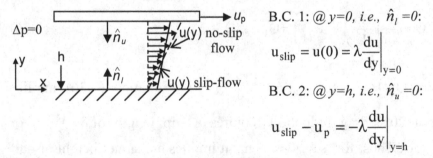

B.C. 1: @ *y=0, i.e.,* $\hat{n}_l = 0$:

$$u_{slip} = u(0) = \lambda \frac{du}{dy} \Bigg|_{y=0}$$

B.C. 2: @ *y=h, i.e.,* $\hat{n}_u = 0$:

$$u_{slip} - u_p = -\lambda \frac{du}{dy} \Bigg|_{y=h}$$

Note: A minus sign appears in the second slip-equation because $\hat{n}_u$ is directed opposite to the y coordinate; but, in both cases the wall-velocity gradient is positive.

Fig. 7.9 Couette flow comparing linear no-slip to slip flow profiles

Using Eqs (7.38) and (7.39) for gas flow in a long rectangular microchannel (H × W × L) subject to a constant pressure gradient, the volumetric flowrate is:

$$Q = \frac{H^3 W}{24\mu}\left(\frac{\Delta p}{L}\right)\left(\frac{p_1 - p_0}{p_0} + 12 Kn_0\right) \tag{7.40}$$

where $\Delta p = p_1 - p_0$ and Kn_0 is the outlet Knudsen number.

7.2.5.4 Temperature Jump Condition

In case thermodynamic equilibrium cannot be maintained, for example for gas-flow along the walls of microconduits when the Knudsen number is in the range of $0.01 < Kn < 0.1$ and the flow is non-isothermal, Smoluchowski's boundary condition is appropriate:

$$T_{gas} = T(x, n = 0) = T_{wall} + T_{jump} = T_w + \frac{2-\sigma_T}{\sigma_T}\frac{2\gamma}{\gamma+1}\frac{\lambda}{Pr}\left.\frac{\partial T}{\partial \hat{n}}\right|_{\hat{n}=0} \tag{7.41}$$

Clearly, there is a temperature discontinuity at the gas–solid interface where the thermal accommodation coefficient $\sigma_T = O(1)$, $\gamma = {c_p}/{c_v}$ and $Pr = {\nu}/{\alpha}$ is the Prandtl number.

Example 7.4: Poiseuille Flow of a Hot Gas in a Microchannel

Consider steady laminar fully-developed airflow in a parallel-plate microchannel of height 2h (see property data given by S. Colin in "Heat Transfer and Fluid Flow in Mini and Microchannels", Elsevier (2005)).

Given: $p_{in} = 0.25 MPa$, $p_{out} = 15 kPa$, $h = 7\mu m$, $L = 10 mm$, $T_{fluid} = 325 K$, $\lambda = k_2 \, {\mu}/{p} \sqrt{RT}$ where $k_2 = \frac{2(7-\omega)(5-2\omega)}{(15\sqrt{2\pi})}$ and for air the "temperature exponent" $\omega = 0.77$, and the gas constant

$R = 287 J/kg \cdot K$. The temperature-dependent air viscosity is $\mu(T) = \mu_0 \left(T/T_0 \right)^{\omega}$, where $T_0 = 273$ K and $\mu_0 = 171.9 \times 10^{-7}$

Find:

- $Kn_m = \lambda_m / D_h$, where $\lambda_m = \frac{1}{2}(\lambda_{in} + \lambda_{out})$ and $D_h = 4A/P$
- Velocity profile $u(y, Kn_m)$ and mean velocity $u_m(Kn_m)$
- Poiseuille number $Po = f \times Re$, where $f = 2\tau_{wall} / \rho u^2$,

  $\tau_{wall} = \frac{D_h}{4} \frac{dp}{dx}$ and $Re = D_h u_m / \nu$

Graph:

- Velocity profiles u(y) for different mean Kn-numbers
- Po(Kn)

Sketch	*Assumptions*	*Concepts /Approach*
p_{in}, p(x), p_{out}, y, h, u(y), x, u_{max}	• As stated • Slip-flow regime $10^{-3} < Kn < 0$	• Differential approach • First-order slip condition

Solution:

Reduced Navier–Stokes equations based on the postulates $\vec{\mathbf{v}} = (u,0,0)$, u=u(y) only, and $dp/dx = -\Delta p / L = \not{c}$:

$$\mu \frac{d^2u}{dy^2} = \frac{dp}{dx} = -\frac{\Delta p}{L} \tag{E.7.4.1}$$

subject to $u(y=h) = u_{slip} = -\lambda_m \left.\frac{du}{dy}\right|_{y=h}$ and $\left.\frac{du}{dy}\right|_{y=0} = 0$.

Double integration and invoking the BCs yields:

$$u(y) = \frac{h^2}{2\mu(T)}\left(\frac{\Delta p}{L}\right)\left(1-\left(\frac{y}{h}\right)^2 + 2\frac{\lambda_m}{h}\right) = K\left(1-\left(\frac{y}{h}\right)^2 + 2\frac{\lambda_m}{h}\right) \quad (E.7.4.2)$$

with $Kn_m = \lambda_m / D_h$, $D_h = 4A/P = 4h$:

$$u(y, Kn_m) = K\left(1-\left(\frac{y}{h}\right)^2 + 8Kn_m\right) \quad (E.7.4.3)$$

and hence:

$$u_m(Kn_m) = \frac{h^2}{3\mu(T)}\left(\frac{\Delta p}{L}\right)(1+12Kn_m) \quad (E.7.4.4)$$

The Poiseuille number $Po = f(Re)$ can be expressed as:

$$Po = \frac{D_h^2}{2\mu(T)u_m}\left(\frac{\Delta p}{L}\right) = \frac{24}{1+12Kn_m} \quad (E.7.4.5)$$

Numerical values:

$$\mu(T) = \mu_0\left(\frac{T}{T_0}\right)^{\omega} = 1.966\times10^{-5}\,\text{Pa.s}$$

$$k_2 = \frac{2(7-\omega)(5-2\omega)}{(15\sqrt{2\pi})} = 1.005$$

$$\lambda_i = \frac{k_2\mu(T\sqrt{R_{air}T})}{p_i}, \text{ i=in, out}$$

$$\therefore \qquad \lambda_m = \frac{\lambda_{in} + \lambda_{out}}{2} = \frac{24.1375 \text{ nm} + 402.29 \text{ nm}}{2} = 213.2 \text{ nm}$$

and hence:

$$Kn_m = \frac{\lambda_m}{D_h} = \frac{\lambda_m}{4h} = 7.616 \times 10^{-3} \quad \text{(slip-flow regime!)}.$$

Finally:

$$\boxed{u_m = 21.3 \text{ ms}^{-1}}$$

Graphs:

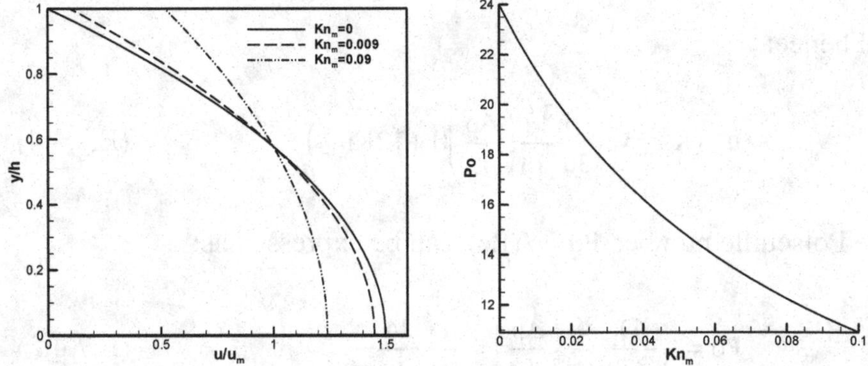

Comments: As the Knudsen number jumps from near zero (i.e., 100% continuous) to Kn = 0.1, the friction factor (or Po#) drops significantly and major wall-slip ($u/u_m \approx 1.1$) increases the volumetric flow rate.

Example 7.5: Thermal Couette Gas Flow in Slip-flow Regime

Consider thermal Couette flow with viscous heating in the slip-flow regime. Specifically, assume simple Couette flow, i.e. plate spacing H, upper plate velocity u_p, zero pressure gradient, insulated stationary wall, conduction–convection heat flux at the moving plate. Find T(y), T_m and Nu all based on Kn. The following data are given: $H = 10\,\mu m$, $u_p = 10\,ms^{-1}$, $T_p = 325\,K$, $T_\infty = 303K$, $k = 0.243 W/mK$, $h = 4.0 W/m^2K$, $\kappa = 1.4$, $Pr = 0.7$, $\mu = 1.966\times10^{-5}\,Pa\,s$.

Sketch	*Assumptions*	*Concepts/ Approach*
T_∞, T_p, u_p, H, y, x, u(y), T(y), u_{slip}	• Steady laminar unidirectional flow • Constant pressure and fluid properties • Zero plate thickness • u(y) is linear	• Reduced heat-transfer equation • Temperature jump condition • Example 7.4 analysis with slip conditions given in Fig.7.9

Solution:

Based on the postulates that u = u(y) and T = T(y) only, the heat transfer equation reduces to (see App. A):

$$k\frac{d^2T}{dy^2} + \mu\Phi = 0 \tag{E.7.5.1}$$

where k is the thermal conductivity and the dissipation function for the present case is just:

$$\Phi = \left(\frac{du}{dy}\right)^2 \tag{E.7.5.2}$$

From the previous analysis (see Fig. 7.9 and Example 7.4):

$$u(y) = \frac{u_p}{1+2Kn}\left(\frac{y}{H} + Kn\right) \tag{E.7.5.3}$$

and hence $\frac{du}{dy} = \frac{u_p}{H(1+2Kn)}$, where $Kn = \lambda/D_h$ with $D_h = 2H$ and λ being the mean free path, assumed to be constant here.
Hence:

$$\frac{d^2T}{dy^2} = -\frac{\mu}{k}\left(\frac{u_p}{H(1+2Kn)}\right)^2 \equiv K = \text{const.} \tag{E.7.5.4a}$$

The two necessary B.C.s are:

(i) Insulated wall

$$dT/dy\Big|_{y\approx 0} = 0 \tag{E.7.5.4b}$$

and

(ii) Heat flux balance:

$$\underbrace{-k\,dT/dy\Big|_{y=H}}_{\text{Conduction through plate}} = \underbrace{h(T_p - T_\infty)}_{\text{Convection to ambient}} \tag{E.7.5.4c}$$

Alternatively to the BC (E7.5.4c), the fluid temperature at the moving plate T_{wall} is not equal to the plate temperature T_p. Specifically, due to the thermodynamic non-equilibrium at walls when the Knudsen number is in the range $0.01 < Kn < 0.1$ (slip-flow regime), the fluid temperature at the moving plate is not equal to the plate temperature, i.e. the discontinuity (or temperature jump) is approximately (with $\sigma_T \approx 1.0$):

$$T(x, y = H) - T_p = -\frac{2\kappa}{\kappa+1}\frac{\lambda}{Pr}\frac{\partial T}{\partial y}\Big|_{y=H} \tag{E.7.5.5}$$

where $\kappa = c_p/c_v$ and $Pr = \nu/\alpha$.

Double integration yields:

$$T(y) = \frac{K}{2} y^2 + Ay + B \qquad \text{(E.7.5.6)}$$

Application of the B.C.s yields:

$$A = 0 \text{ and } B = -\frac{H^2 K}{2} - \frac{4\kappa}{1+\kappa} \frac{Kn}{Pr} H^2 K + T_p \qquad \text{(E.7.5.7a, b)}$$

due to the temperature jump at the moving plate. Now the mean temperature is defined based on an enthalpy balance, i.e.,

$$\dot{m} c_p T_m = W \int_0^H \rho c_p u T dy \qquad \text{(E.7.5.8)}$$

so that with c_p =constant:

$$T_m = \frac{2}{u_p H} \int_0^H u(y) T(y) dy \qquad \text{(E.7.5.9)}$$

where u(y) is given in Equation (E 7.5.3).
Thus,

$$T_m = \frac{2}{H(1+2Kn)} \int_0^H \left(\frac{y}{H} + Kn \right) \left(\frac{K}{2} y^2 + B \right) dy \qquad \text{(E.7.5.10)}$$

so that:

$$T_m = T_p - KH^2 [\frac{1}{2} + \frac{4\kappa Kn}{(1+\kappa) Pr}] \qquad \text{(E.7.5.11)}$$

The Nusselt number is defined as:

$$Nu = \frac{h D_h}{k} \qquad \text{(E.7.5.12)}$$

where $D_h = 4A/P = 2H$ and $h = \dfrac{-k \partial T/\partial y \big|_{y=H}}{T_m - T_p}$, so that:

$$Nu = -2H \frac{\partial T/\partial y \big|_{y=H}}{T_m - T_p} \qquad \text{(E.7.5.13)}$$

Inserting $\left.\frac{dT}{dy}\right|_{y=H} = -HK$ as well as the derived T_m expression (Eq. (E. 7.5.11)), the Nusselt number can be simplified after some algebra to:

$$\mathrm{Nu} = \frac{4}{[1 + \frac{8\kappa \mathrm{Kn}}{(1+\kappa)\mathrm{Pr}}]} \tag{E.7.5.14}$$

Graphs:

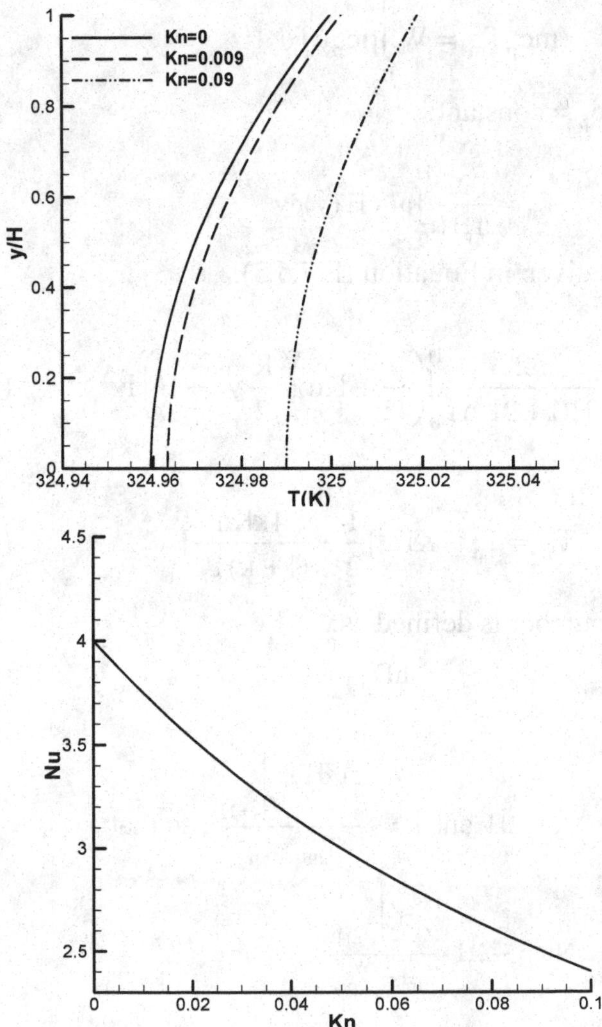

Comments:

- The Nu# is directly proportional to the dimensionless temperature gradient (see Eg. (E.7.5.13), which is diminished due to the temperature jump at the plate (see Eq. (E.7.5.5) and T(y)-graph).
- As a result, with increasing Kn#, T(y) becomes more uniform and the Nu# decreases rapidly.

7.3 Electro-Hydrodynamics in Microchannels

As alluded to in Sect. 7.1, for liquid flows in microchannels (not to mention nanochannels) wall forces due to an electric charge may scale with the hydraulic diameter or even the conduit length. Such wall electric forces can be effective in pushing bulk liquid, or moving particles relatively small distances at very low Reynolds numbers. Specifically, phenomena in *electrokinetics*, i.e., movement induced by an externally applied electric field, include electro-osmosis and electro-phoresis. In an *electrolyte liquid*, e.g., salt water, the fluid moves relative to a charged microchannel surface (*electro-osmosis:* EO), while in *electrophoresis* charged particles migrate relative to a liquid body.

For example, in electro-osmotic flow caused by an EO-based "micropump", two metallic electrodes are situated at each end of a channel, in which charge separation occurs only in a very thin wall layer, whose extent is known as the Debye length λ_D. Typically, $\lambda_D = 1$ to $100\,\text{nm}$ thin, with "equilibrium charge density" $\rho_{el}^{eq.}(y) \sim Q_{el} / \forall [C/m^3]$. The applied DC-potential difference $\Delta V = \Delta\phi_{ext}$ results in an electric field, E_{ext} [V/m], where

$$\vec{\mathbf{E}}_{ext} \equiv -\nabla\phi_{ext} \tag{7.42}$$

Clearly, the applied electric potential is only effective at the wall, i.e., it ranges from $\phi_{ext}(y=0)=\phi_{wall}$ to $\phi_{ext}(y=\lambda_D)\approx 0$. The product $\rho_{el}^{eq}\vec{\mathbf{E}}_{ext}=\vec{\mathbf{f}}_{el}$ is an electric force per unit volume acting inside a diffuse layer of thickness λ_D, the Debye length. This body force moves the liquid (diffuse) wall layer opposite to the (axial) direction of E_{ext} and in turn by viscous effects drags along the charge-neutral adjacent liquid body (see Figs. 7.10 and 7.11). Mathematically, the new (conservative) body force, $\vec{\mathbf{f}}_{el}=-\rho_{el}\nabla\phi$, can be added to the Navier–Stokes equation, i.e.,

$$\rho\left[\frac{\partial\vec{\mathbf{v}}}{\partial t}+(\vec{\mathbf{v}}\cdot\nabla)\vec{\mathbf{v}}\right]=-\nabla p+\mu\nabla^2\vec{\mathbf{v}}+\rho\vec{\mathbf{g}}-\rho_{el}\nabla\phi \tag{7.43}$$

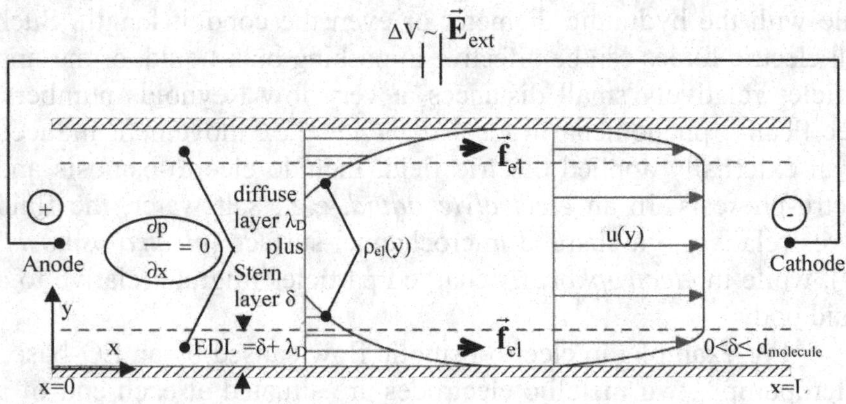

Fig. 7.10 Idealized electro-osmotic flow in a microchannel with zero pressure gradient and electric double layer (EDL)

The applied electric field, however, has limits because of possible Joule heating, i.e., when an associated strong current, I~E, may cause, in the worst case, boiling of the solution. This new body force $\vec{\mathbf{f}}_{el}\equiv\rho_{el}\vec{\mathbf{E}}=-\rho_{el}\nabla\phi$ is discussed in the next section and applied in Example 7.6.

7.3.1 Electro-Osmosis

Ignoring any magnetic and radiative effects, the fundamental electro-static equations for continuous media after Maxwell are:

$$\nabla \times \vec{\mathbf{E}} = 0 \tag{7.44}$$

$$\nabla \cdot (\varepsilon \vec{\mathbf{E}}) = \rho_{el} \tag{7.45}$$

and

$$\varepsilon \vec{\mathbf{E}} = \varepsilon_0 \vec{\mathbf{E}} + \vec{\mathbf{P}} = \vec{\mathbf{D}} \tag{7.46}$$

where the electric field $\vec{\mathbf{E}}$, polarization field $\vec{\mathbf{P}}$ and displacement field $\vec{\mathbf{D}}$ are locally averaged vectors, and ε is the dielectric constant. As indicated with Eq. (7.42), the conservative electric field $\vec{\mathbf{E}}$ can be expressed as the gradient of an electric potential $\phi(\vec{\mathbf{r}})$, i.e.,

$$\vec{\mathbf{E}} = -\nabla\phi(\vec{\mathbf{r}}) \tag{7.47}$$

which leads with Eq. (7.45) and a constant ε to a Poisson equation:

$$\nabla^2\phi(\vec{\mathbf{r}}) = -\frac{1}{\varepsilon}\rho_{el}(\vec{\mathbf{r}}) \tag{7.48a}$$

In terms of an electric force per unit volume:

$$\vec{\mathbf{f}}_{el} \equiv \rho_{el}\vec{\mathbf{E}} = -\varepsilon\vec{\mathbf{E}}\nabla^2\phi \tag{7.48b}$$

Specifically, after an external electric field $\vec{\mathbf{E}}$ is applied (see Fig. 7.10), the charge density ρ_{el} along the wall is mainly confined to an electric double layer (EDL), following a Boltzmann distribution. In the Debye–Hückel limit, $\rho_{el}(y) = -\frac{\varepsilon}{\lambda_D^2}\phi$ can be approximated (see Fig. 7.11), and Eq. (7.48a) then reads:

$$\nabla^2\phi = \frac{\phi}{\lambda_D^2} \tag{7.49}$$

where the nanometer Debye length λ_D represents the decay distance for the potential. Integration of Eq. (7.49) for the 1-D one-wall case depicted in Fig. 7.11 yields:

$$\phi(y) = \phi_w e^{-y/\lambda_D} \tag{7.50a}$$

Note, Eq. (7.50) is based on the BCs:

$$\phi(y=0) = \phi_{wall} \approx \zeta \tag{7.50b}$$

where ζ is the zeta potential and

$$\phi(y=\lambda_D) = \frac{\phi_{wall}}{e} << 1 \tag{7.50c}$$

However, when applying $\left.\frac{d\phi}{dy}\right|_{y=\mathcal{C}} = 0$ (channel symmetry), the result is $\phi(y) \sim \zeta \cosh(y/\lambda_D)$ as shown in Example 7.6.

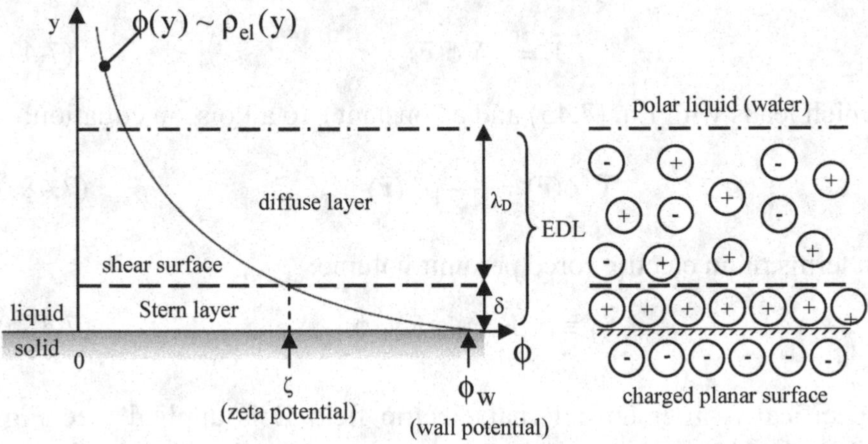

Fig. 7.11 Decay of the electric potential in an electric double layer (EDL)

Concerning Fig.7. 11, when a polar liquid such as water is in contact with a charged solid surface, ions in the liquid accumulate on the wall. They are forming a very thin layer, i.e., the *Stern layer*, containing immobilized "counter ions". That in turn influences an adjacent liquid region, the thicker *diffuse layer* of Debye length λ_D, with a net charge that can be moved with an applied electric field. Hence, a shear surface with an electric potential, $\phi(y = \delta) = \zeta$, called the zeta potential, is created. Clearly, the maximum potential appears at the wall, i.e., $\phi(y = 0) = \phi_w$, where $\phi_w \approx \zeta$ because $\delta <<< 1$.

Hence, the charge density $\rho_{el}(y) = -\varepsilon d^2\phi/dy^2$ is now with Eq. (7.50) for a single wall:

$$\rho_{el}(y) = -\frac{\varepsilon\zeta}{\lambda_D^2} e^{-y/\lambda_D} \tag{7.51}$$

Example 7.6: Electric-Force-Driven Flow in a Parallel-Plate Micro-channel

Consider a parallel-plate microchannel with positively charged walls, where the external electric field is applied in the negative x-direction, i.e., $\vec{E} = -E_0\hat{i}$. Assuming a constant ζ-potential along the walls, a constant E_0, zero pressure gradient and a Debye length much much smaller than the plate spacing, derive $\phi_{el}(y)$ and then u(y) and find the volumetric flow rate Q per unit width.

Sketch	*Assumption*	*Concepts*
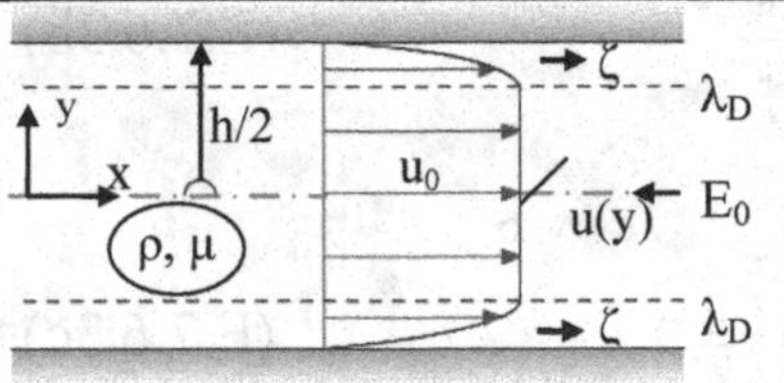 y, x, h/2, u_0, u(y), ζ, λ_D, E_0, ρ, μ	• Steady 1-D flow • $\nabla p = 0$; $\vec{v} = [u(y), 0, 0]$ • Constant zeta potential, electric field, and fluid properties	• Reduced N-S equations with new electric force term • Mass conservation

Typical system parameter values: Zeta potential $\zeta = 100$ mV, Debye lenth $\lambda_D \approx 20$ nm, electro-osmotic velocity $u_{EO} = 1$ mm/s and electro-osmotic mobility $\eta_{EO} = 10^{-7}\ \frac{m^2}{s\ V}$, to be defined below.

Solution:

- Applied electric potential (see Eq. (7.49)) in 1-D form reads:

$$\frac{d^2\phi}{dy^2} - \frac{\phi}{\lambda_D^2} = 0\,; \quad \phi(y = \{h/2) \approx \zeta = \not{c} \text{ and } \left.\frac{d\phi}{dy}\right|_{y=0} = 0 \quad \text{(E.7.6.1a,b)}$$

Its general solution (see p. 130 in Polyamin & Zartsev 1995) has the form:

$$\phi(y) = \frac{\zeta \cosh(y/\lambda_D)}{\cosh(h/2\lambda_D)} \quad \text{(E.7.6.2)}$$

- Based on the listed assumptions, the x-momentum equation of Eq. (7.43) reduces to:

$$0 = \mu\frac{d^2u}{dy^2} + f_x^{el} \quad \text{(E.7.6.3a)}$$

where, according to Eq. (7.48b),

$$f_x^{el} = \varepsilon E_0 \frac{d^2\phi}{dy^2} \quad \text{(E.7.6.3b)}$$

Thus, we have to solve

$$\frac{d^2}{dy^2}\left[u(y) + \frac{\varepsilon E_0}{\mu}\phi(y)\right] = 0 \quad \text{(E.7.6.3c)}$$

subject to:

$$u(y = \{h/2) = 0 \text{ and } \left.\frac{du}{dy}\right|_{y=0} = 0 \quad \text{(E.7.6.3d,e)}$$

The result is:

$$u(y) = u_{EO}\left[1 - \frac{\cosh\left(\frac{y}{\lambda_D}\right)}{\cosh\left(\frac{h}{2\lambda_D}\right)}\right] \quad \text{(E.7.6.4a)}$$

where u_{EO} is the electro-osmosis (EO) velocity, i.e., u_{EO} is given as:

$$u_{EO} = \frac{\varepsilon\zeta}{\mu} E_0 \quad \text{(E.7.6.4b)}$$

Another EDL parameter is the "EO mobility" η_{EO} which is defined as:

$$\eta_{EO} \equiv \frac{u_{EO}}{E_0} = \frac{\varepsilon\zeta}{\mu} \quad \text{(E.7.6.5)}$$

Clearly, for $\lambda_D \lll h/2$ and no additional driving force:

$$u \approx u_{EO} = -\eta_{EO}\, E_0 \text{ and hence } Q \approx u_{EO}(wh) \quad \text{(E.7.6.6,7)}$$

Graphs:

(a) Electric potential $\phi(2y/h)$

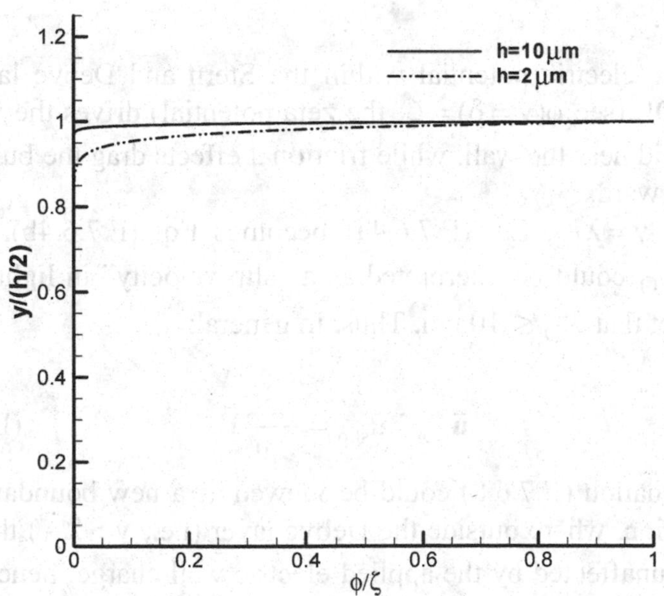

(b) Dimensionless velocity $\frac{u}{u_{EO}}(2y/h)$

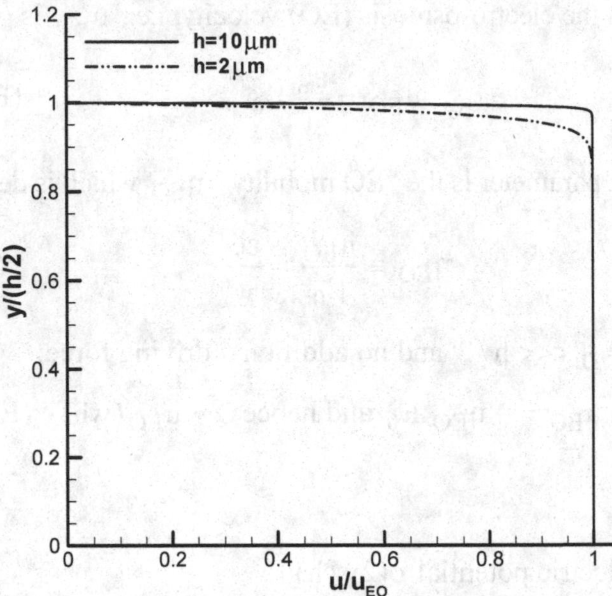

Comments:

- The electric potential within the Stern and Debye layer, or EDL (see $\phi(y=\delta)=\zeta$, the zeta potential) drives the viscous fluid near the wall, while frictional effects drag the bulk fluid forward.
- At $y=\lambda_D$, Eq. (E.7.6.4a) becomes Eq. (E.7.6.4b), where u_{EO} could be interpreted as a "slip velocity" in light of the fact that $\lambda_D \leq 10$ nm. Thus, in general:

$$\vec{\mathbf{u}}_{EO} \triangleq \vec{\mathbf{u}}_{slip} = -\frac{\varepsilon\zeta}{\mu}\vec{\mathbf{E}} \tag{E.7.6.8}$$

- Equation (E.7.6.8) could be viewed as a new boundary condition, where outside the Debye layer (i.e., $y > \lambda_D$) the fluid is unaffected by the applied electric wall charge; hence, generally only subject to a pressure gradient or gravity.

Example 7.7: Pressure-Driven and Electro-Osmotic Flow in a Microchannel

Consider a microchannel of height h containing an ionized aqueous solution (with zeta potential ζ_0, Debye length λ_D, and permittivity, or dielectric constant ε, subject to a constant pressure gradient and electric field E. Derive expressions for u(y) and u_{mean} for $\pm E_x$, i.e., the electric force acting in the positive or negative x-direction.

Sketch	*Assumptions*	*Concepts*
ΔV; $\frac{\partial p}{\partial x}$; +; -; h; u(y); y; x; L	• Superposition of Poiseuille and electro-osmotic flows • $\nabla p \rightarrow \frac{dp}{dx} = \not{c}$ • $\vec{E} \rightarrow E_x = \not{c}$	• Reduced Eq.(7.43) • Electric charge: $\rho_{el} = -\frac{\varepsilon \zeta_0}{\lambda_D^2} e^{-y/\lambda}$ (see Eq. (7.51))

Solution:

Based on the given system and stated assumptions, Eq. (7.43) reduces to:

$$0 = -\frac{\partial p}{\partial x} + \mu \frac{\partial^2 u}{\partial y^2} \pm \rho_{el} E_x \qquad \text{(E.7.7.1a,b)}$$

where,

$$-\frac{\partial p}{\partial x} = \frac{\Delta p}{L} = \not{c}, \text{ u = u(y) only, and } \rho_{el}(y) \text{ as given.}$$

Hence,

$$\frac{d^2u}{dy^2} = -\frac{1}{\mu}\frac{\Delta p}{L} \mp \frac{\varepsilon\,\zeta_0}{\mu\,\lambda_D^2}\,E_x e^{-y/\lambda_D} \qquad \text{(E.7.7.1c,d)}$$

has to be integrated twice subject to the no-slip velocity condition at the wall as well as $\left.\frac{du}{dy}\right|_{y=h/2} = 0$. The results are for $\pm E_x$ applied:

$$u(y) = \frac{h^2}{2\mu}\left(\frac{\Delta p}{L}\right)\left[\frac{y}{h} - (\frac{y}{h})^2\right] + \frac{\varepsilon\zeta_0 E_x}{\mu}(1 - e^{-y/\lambda_D})$$

$$-\frac{\varepsilon\zeta_0 E_x y}{\mu\lambda_D}e^{-h/2\lambda_D} \qquad \text{(E.7.7.2a)}$$

and

$$u(y) = \frac{h^2}{2\mu}\left(\frac{\Delta p}{L}\right)\left[\frac{y}{h} - (\frac{y}{h})^2\right] - \frac{\varepsilon\zeta_0 E_x}{\mu}(1 - e^{-y/\lambda_D})$$

$$+\frac{\varepsilon\zeta_0 E_x y}{\mu\lambda_D}e^{-h/2\lambda_D} \qquad \text{(E.7.7.2b)}$$

The flow rates per unit width are:

$$Q = 2\int_0^{h/2} u(y)dy := \frac{h^3\,\Delta p}{12\mu\,L} + \frac{2\varepsilon\,\zeta_0 E_x}{\mu}[\lambda_D(e^{-h/2\lambda_D} - 1)]$$

$$-\frac{h^2\varepsilon\,\zeta_0 E_x}{4\mu\lambda_D}e^{-h/2\lambda_D} + \frac{\varepsilon\,\zeta_0 E_x h}{\mu} \qquad \text{(E.7.7.3a)}$$

and

$$Q = 2\int_0^{h/2} u(y)dy := \frac{h^3\,\Delta p}{12\mu\,L} - \frac{2\varepsilon\,\zeta_0 E_x}{\mu}[\lambda_D(e^{-h/2\lambda_D} - 1)]$$

$$+\frac{h^2\varepsilon\,\zeta_0 E_x}{4\mu\lambda_D}e^{-h/2\lambda_D} - \frac{\varepsilon\,\zeta_0 E_x h}{\mu} \qquad \text{(E.7.7.3b)}$$

Therefore, the mean velocity is:

$$u_{mean} = \frac{Q}{h \cdot w} \qquad \text{(E.7.7.4)}$$

Graphs:

(a) Negative pressure gradient and electrokinetic flow (dimensional velocity)

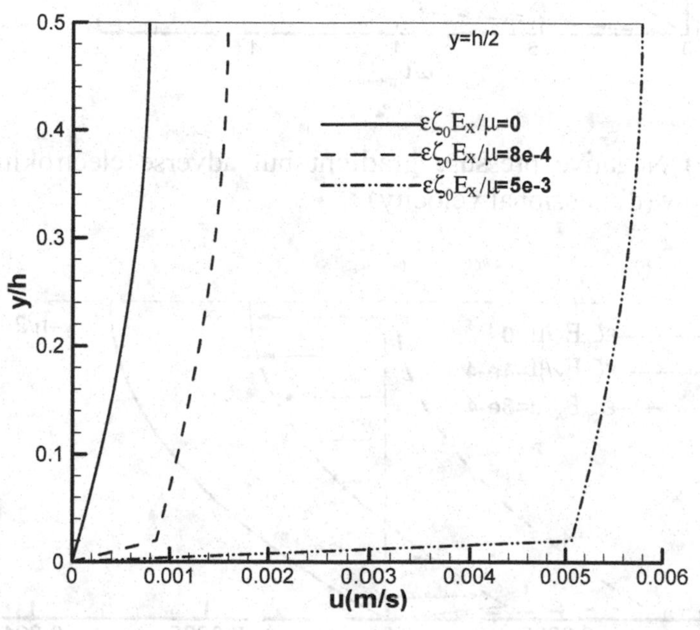

(b) Negative pressure gradient and electrokinetic flow (dimensionless velocity)

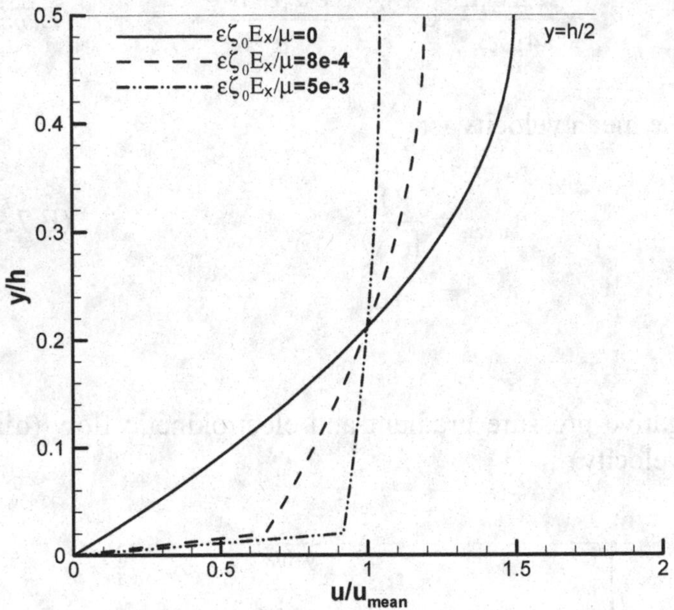

(c) Negative pressure gradient but adverse electrokinetic flow (dimensional velocity)

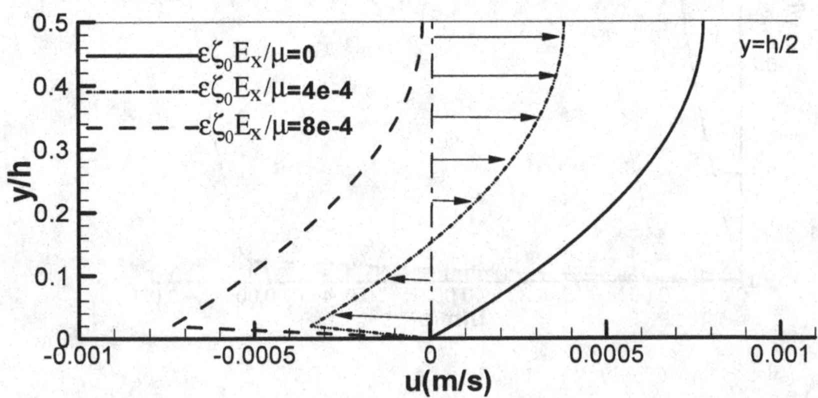

Comments:

- Graph (a) shows the strong effect of electrokinetics in microchannel flow, i.e., $f_{electr.} = \rho_{el} E_x$ is aiding the conventional pressure force.
- Graph (b) indicates the usefulness of nondimensionalization (of Graph (a)), i.e., figure compactness and the appearance of a unique channel location $y/h \approx 0.21$ where all velocity profiles intersect at $u = u_{mean}$, recalling that $u_{mean} = \frac{Q}{h \cdot w}$ is changing with Q due to E_x.
- Graph (c) demonstrates the impact of reversing the axial direction of the electro-osmotic force, generating mixed forward/backward flows as well as all backward flow, i.e, against the negative pressure gradient.

7.3.2 Electrophoresis

An applied electric field $\vec{E}$ can move a charged spherical particle through a (stagnant) liquid of low electric conductivity, e.g., de-ionized water. Electrophoresis is often combined with capillary effects to separate charged particles/molecules according to their sizes in small-diameter tubes, say, $D \leq 50\,\mu m$.

Similar to the electro-osmotic phenomenon, an electric double layer (EDL) of approximate Debye thickness λ_D (see Sect. 7.3.1) is formed, surrounding each micron particle. Again, for steady 1-D particle movement, the electric force, F_{el}, is balanced by Stokes' drag:

$$F_{el} = F_{drag} \tag{7.52a}$$

$$Q_{el}\, E_{el} = 6\pi\, \mu\, u_{EP}\, r_0 \tag{7.52b}$$

where the total surface charge is $Q_{el} = \rho_{el}\forall = -\varepsilon\nabla^2\phi\cdot\forall$, u_{EP} is the electrophoretic (EP) particle velocity relative to the liquid, and $r_0 = d_p/2$ is the radius of the sphere of surface S.
In spherical coordinates (see Eq. (7.48a)):

$$Q_{el} = -\frac{\varepsilon}{r^2}\frac{d}{dr}(r^2\frac{d\phi}{dr})\bigg|_{r=r_0}\cdot(\frac{4}{3}\pi r_0^{\ 3}) \tag{7.53}$$

And hence with Eq. (7.52b) an expression for the EP-velocity can be deduced:

$$\vec{u}_{EP} = \frac{(Q\,\vec{E})}{6\pi\,\mu\,r_0} = \eta_{ion}\,\vec{E} \tag{7.54a,b}$$

where η_{ion} is the "ionic mobility". As mentioned, the EP-velocity can be employed as a "drift velocity" to sort out bio-particles such as proteins, DNA, cells, etc.

Example 7.8: Micron-Particle Dynamics in Water for Particle Mass of m = 1 pg

Consider a 1 μm sphere (e.g., a self-propelled bacterium) floating in water ($\mu = 0.001\ N\cdot s/m^2$). How long would it take the single cell to reach from zero a speed of, say, 30 μm/s, and what distance does it travel?

Sketch	*Assumptions*	*Concepts*
propelling flagella; cell ≙ ; 1μm; F_D; v; x	• 1-D Stokes flow • Cell ≙ sphere • $F_{Drag} = F_{Thrust} \approx$ co[nst.] Constant acceleration	• Stokes' law with terminal velocity • F = ma; a = dv/dt • Distance traveled $x = vt - \frac{a}{2}t^2$ for a = ¢

Solution:

- Stokes' law states (see Eq. (7.21d): $F_D = 6\pi\mu r_p v$
  $\therefore F_D = 2.8\times10^{-13}$ N $= F_{Thrust}$
- Acceleration: $a = \frac{F_{Thrust}}{m} := 280\ m/s^2$ for m = 10^{-12} g Time required: $a = \frac{dv}{dt} \approx \frac{\Delta v}{\Delta t} \succ t = \frac{v}{a} = 10^{-7}\,s$
- Distance travelled (a = constant):

$$a = \frac{d^2x}{dt^2} \Rightarrow \frac{dx}{dt} = v = at + c_1 \text{ and hence } x = \frac{a}{2}t^2 + c_1 t + c_2$$

Here, $x(t = 0) = 0 \succ c_2 \equiv 0$ and $\left.\frac{dx}{dt}\right|_{t=t_{final}} = v$, i.e., $v = at + c_1$ or $c_1 = v - at$. Finally,

$$x = vt - \frac{a}{2}t^2 := 2\times10^{-12}\ m$$

Comments:

The tiny, light-weight "bug" accelerates tremendously and hence reaches top speed in no time after a very short distance traveled.

7.4 Entropy Generation in Microfluidic Systems

Inevitable entropy generation is associated with waste of energy. Hence minimizing entropy production is a key design objective, especially in Microsystems. We recall that the Second Law of Thermodynamics stipulates that the *rate of entropy generation* must be positive for all thermophysical processes/devices, i.e., $\dot{S}_{gen} > 0$, known as the "increase-of-entropy principle". Specifically, over observation (or process) time Δt:

$$S_{gen} \equiv \Delta S_{total} = \Delta S_{system} + \Delta S_{surrounding} > 0 \tag{7.55}$$

By including all irreversibilities, mainly due to friction, fluid flow and heat transfer, an "entropy balance" can be formed. On a rate bases we can formulate the following equations.

(i) Open System:

$$\sum\left(\frac{\dot{Q}}{T_{amb}}\right) + \sum \dot{m}S\Big|_{in} - \sum \dot{m}S\Big|_{out} + \dot{S}_{gen} = \frac{\Delta S}{\Delta t}\Big|_{C.\forall.} \tag{7.56a}$$

(ii) Closed System:

$$\sum\left(\frac{\dot{Q}}{T_{amb}}\right) + \dot{S}_{gen} = \frac{\Delta S}{\Delta t}\Big|_{system} \tag{7.56b}$$

where T_{amb} is the ambient (or surrounding) temperature and $\sum \dot{Q}$ are the exchange heat flow rates.

Clearly, for a process to proceed or a device to work, both the first law (see Sect. 2.5.1) and the second law have to be fulfilled. It should be noted that the direction of $\dot{Q}$ and the point of reference for T_{amb} are important considerations.

Example 7.9: Entropy Generation due to Heat Transfer

Consider a frictionless piston-cylinder device with a saturated liquid-vapor H_2O-mixture at $100\,°C$. During an isobaric process $\dot{Q} = 600\ kW/s$ is transferred to the surrounding air at $25\,°C$. Determine the total rate-of-entropy generated.

Sketch	*Assumptions*	*Concepts/ Approach*
piston; $\dot{Q}$; $T_{H_2O} = ¢$; $\rho = ¢$; $T_{amb.}$	• Closed system • $\dot{S}_{internal} \equiv \dot{S}_{system} = 0$ • Constant T_{H_2O} and $T_{ambient}$	• Eq. (7.56b), where $\frac{\dot{Q}}{T_{H_2O}} = \frac{\dot{Q}}{T_{system}}$ is negative and $\frac{\dot{Q}}{T_{amb}} = \frac{\dot{Q}}{T_{surr}}$ is positive

Solution:

- For a closed system

$$\dot{S}_{gen} = \left.\frac{\Delta S}{\Delta t}\right|_{system} - \sum \frac{\dot{Q}}{T_{ref}} = \dot{S}_{system} - \frac{\dot{Q}}{T_{system}} + \frac{\dot{Q}}{T_{surr}}$$

$$= 0 - \frac{600kJ}{373K} + \frac{600kJ}{298K}$$

$$\dot{S}_{gen} = 0.40 kW/s \cdot K > 0$$

Comment:

With respect to Eq. (7.55), $\dot{S}_{gen}$ can be interpreted as:

$$\dot{S}_{gen} = \underbrace{\dot{S}_{system} - \frac{\dot{Q}}{T_{system}}}_{\Delta\dot{S}_{system}} + \underbrace{\frac{\dot{Q}}{T_{amb}}}_{\Delta\dot{S}_{surrounding}}$$

7.4.1 Entropy Minimization

It is obvious that an engineering device performs the better the lower the (irreversible) losses are, i.e., the closer it operates at an isentropic efficiency. That directly implies that $\dot{S}_{gen}$ should be minimized as part of any device/process design or improvement. This task is most vital for microsystems.

For example, considering simple heat transfer from an ambient reservoir at $T_0 = ¢$, Eq. (7.56a) can be rewritten as:

$$\dot{S}_{gen} = \frac{\partial S}{\partial t} - \sum \frac{\dot{Q}}{T_0} - \sum \dot{m}S\Big|_{in} + \sum \dot{m}S\Big|_{out} > 0 \qquad (7.57)$$

Effectively, $\dot{S}_{gen} \sim \dot{W}_{loss}$, i.e., in general, power loss is due to system, heat transfer and fluid flow irreversibilities:

$$P_{loss} \equiv \dot{W}_{loss} = T_0 \dot{S}_{gen}$$
$$= T_0 \left(\frac{\partial S}{\partial t} - \frac{\dot{Q}}{T_0} - \sum \dot{m}S\Big|_{in} - \sum \dot{m}S\Big|_{out} \right) \qquad (7.58)$$

Focusing on friction (or viscous effects) as the main cause of irreversibilities, and hence entropy generation, the rate of irreversible conversion from flow energy to heat can be expressed as (see Eq. (2.52)):

$$\tau_{ij} \frac{\partial v_i}{\partial x_j} \equiv \mu \Phi = \frac{P_{loss}}{\forall} = T_0 \frac{\dot{S}_{gen}}{\forall} \qquad (7.59)$$

Clearly, in order to minimize entropy production in pressure-driven microchannels, we have to reduce the viscous dissipation function Φ, i.e., achieve minimization of:

$$\frac{\dot{S}_{gen}}{\forall} = \frac{\mu}{T} \Phi = \frac{\mu}{T} \left[\left(\frac{\partial u}{\partial y} + \frac{\partial v}{\partial x} \right)^2 + 2 \left(\frac{\partial u}{\partial x} \right)^2 + 2 \left(\frac{\partial v}{\partial y} \right)^2 \right] \qquad (7.60)$$

In general, and for fully-developed flow exclusively, the term $\frac{\mu}{T}(\frac{\partial u}{\partial y})^2$ is most important. If wall slip is significant, that u(y)-velocity profile is greatly affected and hence the channel pressure drop.

Gas Flow in a Microchannel Naterer & Camberos (2008) depicted a nice example of entropy production due to developing nitrogen gas flow ($\mu = 1.6 \times 10^{-5} \mathrm{Ns/m^2}$) in a microchannel (height h = 1 μm, varying length L) at very low Reynolds numbers, i.e., $0.001 \le \mathrm{Re} \le 0.003$. Different slip coefficient values were generated by changing the momentum accommodation coefficient σ. The re-computed results are shown in Figs. 7.12 and 7.13, where the slip boundary condition with $\lambda = \mathrm{hKn}$ reads (see Sect. 7.2.5):

$$u\big|_{slip} = \frac{2-\sigma}{\sigma}\lambda\frac{\partial u}{\partial y}\bigg|_{wall} \tag{7.61}$$

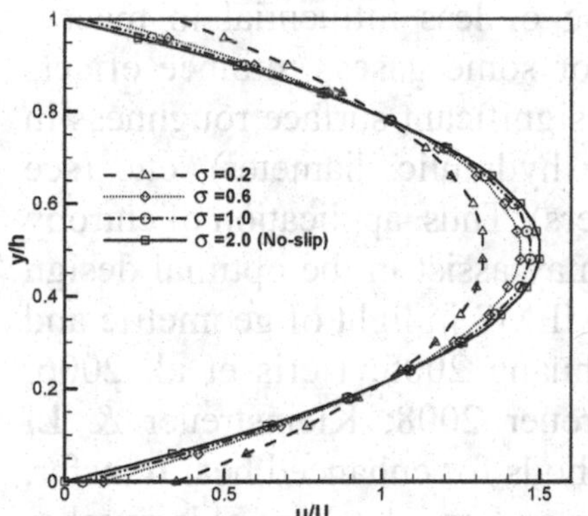

Fig. 7.12 Impact of wall-velocity slip on the fully-developed velocity profile

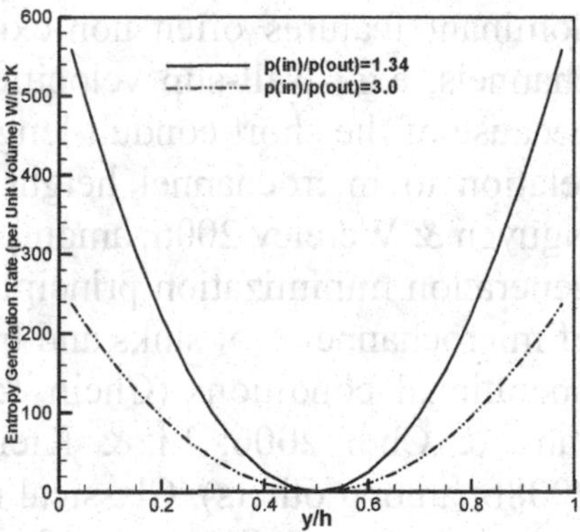

Fig. 7.13 Cross-sectional entropy production rate for two pressure ratios

For $\sigma = 2.0$, i.e., no velocity slip at the wall, the Poiseuille flow solution is recovered. As the slip coefficient decreases (e.g., $\sigma = 0.2$) the axial velocity profile is more blunt (Fig. 7.12), less kinetic energy dissipates (see Eq. (7.60) and the reduced contribution of the $\frac{\mu}{T}(\frac{\partial u}{\partial y})^2$ -term), and hence less power (or Δp) is required to achieve the same flow rate.

As expected, $\frac{\dot{S}_{gen}}{\forall}$ (see Eq. (7.59)) is zero at mid channel as shown in Fig. 7.13. With lower pressure ratios $\frac{p_{in}}{p_{out}} \leq 3.0$, the energy production rate decreases because of milder velocity gradients.

Liquid Flow in a Microchannel Minimization of entropy generation as a design tool to determine best device geometry and operation, especially for heat exchangers, has been established for macro-scale configurations (see Bejan 1996, 2002; Sahin 1998, 2000; Mahmud & Fraser 2003; Mansour et al. 2006; Khan et al. 2007; Ko & Wu 2009; among others). However, fluid flow in microchannels exhibits dominant features often non-existing or less influential in macro-channels, e.g., wall-slip velocities for some gases, entrance effects because of the short conduit length, significant surface roughness in relation to microchannel height (or hydraulic diameter), etc. (see Nguyen & Wereley 2006; among others). Thus, application of entropy generation minimization principles may assist in the optimal design of microchannel heat sinks and bio-MEMS in light of geometric and operational conditions (Chein & Chuang 2005; Heris et al. 2006; Jang & Choi 2006; Li & Kleinstreuer 2008; Kleinstreuer & Li 2008b; among others). Classical methods for enhanced heat transfer, e.g., an increase of heat transfer area and/or inlet Reynolds number are limited options for microchannel flow. Thus, the use of nanofluids as coolants, e.g., CuO or Al_2O_3 nanospheres with diameters in the range of $5\,\text{nm} < d_p < 150\,\text{nm}$ in water, oil or ethylene glycol, is a third option (see Sect. 7.5). In case of nanomedicine delivery with bio-MEMS, nanofluid flow is a given and measurable reduction of entropy generation is desirable as well.

In this section, entropy generation is minimized for steady laminar pure-water and nanofluid flows in a representative trapezoidal microchannel in terms of most suitable channel aspect ratio and Reynolds- number range.

One effective operational parameter is the inlet Reynolds number, where Fig. 7.14 indicates a desirable range of $425 \leq \mathrm{Re} \leq 1100$ for all fluids and aspect ratios considered, when ignoring "slit flow" for $AR = 0.9337$. Due to slightly enhanced frictional effects (the increase of viscosity), $\hat{S}^{nanofluid}_{G,total} > \hat{S}^{water}_{G,total}$. Here, the overall entropy generated in the entire flow field is used in integral form for different scenarios:

$$\hat{S}_{G,total} = \frac{1}{\dot{m}c_p} \iiint_V \frac{\dot{S}_{gen}}{\forall} dV \tag{7.62}$$

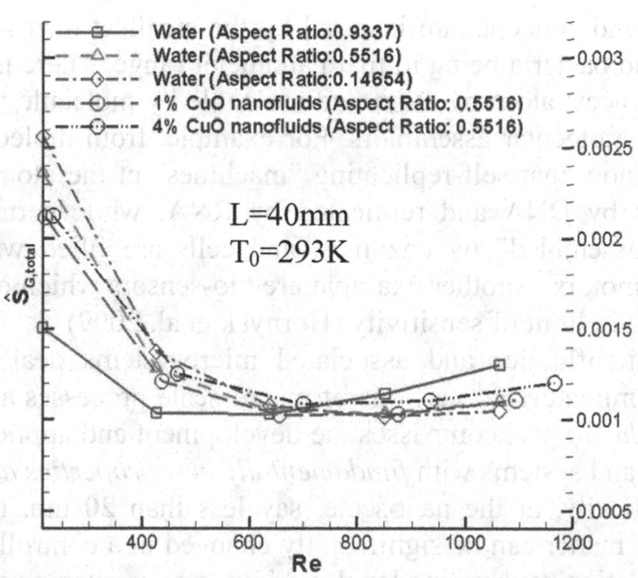

Fig. 7.14 System entropy generation vs. Reynolds number (after Li & Kleinstreuer 2010)

7.5 Nanotechnology and Nanofluid Flow in Microchannels

Nanotechnology "Nano" (Greek) means "dwarf" and, indeed, nanoparticles are very small as discussed in previous sections. For example, if a nanosphere is compared to a soccer ball, that ball can be compared to our planet:

$$\frac{d_p^{nano}}{D_{ball}} = \frac{D_{ball}}{D_{Earth}} \tag{7.63}$$

Nanotechnology is a science & engineering branch which focuses on constructing material, objects and devices smaller than 100 nm in dimension. Thus, nanomaterials, microchannels and micro-actuators are integral parts of nanotechnology. It also deals with nanofluid flow and nanoparticle dynamics in micro- and nano-scale devices. Some examples and concepts are inspired by the smallest forms of life, e.g., cells and bacteria being in that nanometer range. There is a vision to build objects atom by atom and molecule by molecule via self-assembly or molecular assemblers. For example, from molecular biology we know that self-replicating "machines" at the atomic level are guided by DNA and replicated by RNA, while certain molecules are "assembled" by enzymes, and cells are filled with molecular-scale motors. Another example are bio-sensors which operate at the quantum limit of sensitivity (Hornyak et al. 2009)

While microfluidics and associated microsystems deal in most cases with miniaturized versions of macro-scale processes and devices, *nanotechnology* encompasses the development and application of materials and systems with *fundamentally new properties and functions*. Specifically, at the nanoscale, say less than 20 nm, the characteristics of matter can be significantly changed in a controlled manner. Manipulation at that size-level, various effects such as the dominance of the interfaces, size confinement and quantum mechanics can change a material's mechanical, electrical, optical, biological and/or chemical properties. Thus, progress in nanoscale science and engineering requires the collaboration of physicists, chemists and biologists as well as engineers from most disciplines. It implies that the interdisciplinary basics of nanotechnology have to be well integrated into the curriculum, in order to educate a new generation of experts in nanoscience and nanoengineering.

In any case, this section provides only a small aspect of nanotechnology, i.e., nanofluid flow in microchannels for improved device design and operation.

Nanofluids Nanofluids, already employed insect. 6.4, are dilute suspensions of nanoparticles ($5\,\text{nm} \leq d_p < 150\,\text{nm}$ at volume fractions of 0.1% to 5% or more) in a liquid, say, water, ethylene glycol (EG), oil, etc. Combinations of metal, metal-oxide or carbon-based nanoparticles in EG or machine oil can be used as coolants in micro-heat sinks because of their elevated thermal conductivity without the specter of filter clogging and/or excessive pressure drops if particle coagulation can be avoided. In contrast, nano-size drugs in porous microspheres or in aqueous solutions are part of bio-MEMS for controlled internal drug delivery.

Clearly, the presence of nanoparticles alters the properties of the quasi-homogeneous mixture (see Sect. 6.2), such as the density, viscosity, and thermal conductivity. The next two sections provide applications to nanofluid flow in microchannels, employing CFD techniques outlined in Chap.10.

7.5.1 Microscale Heat-sinks with Nano-coolants

Microscale cooling devices, such as micro-channel heat sinks (MCHS), are increasingly important in current and future heat removal applications. Specifically, a coolant flows through a large number of parallel, micro-machined or etched conduits with the purpose to remove heat from and assure quasi-uniform temperature distributions in small devices. Microsystems with high transient loads requiring effective heat removal include MEMS, integrated circuit boards, laser-diode arrays, high-energy mirrors, and compact electronic consumer products. The microchannels in those devices are typically silicon-, metal-, or plastic-based, featuring often rectangular or trapezoidal cross sections with hydraulic diameters ranging from 100 nm to 1 mm. their very large heat transfer surface-to-volume ratio in conjunction with exotic coolants, e.g., nanofluids, allow for controlled cooling and thereby avoiding hot spots and ultimately device failure (Kleinstreuer & Li 2008a).

Key to the cooling performance of a nanofluid is its *effective thermal conductivity*, k_{eff}, or convective heat transfer coefficient h (see Sect. 3.6). As first demonstrated at Argonne National Laboratory, those liquid-nanoparticle mixtures exhibited k_{eff}-values 10%–150% higher than the ones of the base fluids (Choi 1995; Chopkar et al. 2007; among others). The effective thermal conductivity of any nanofluid depends mainly on the nanoparticle volume fraction, conductivity and diameter, as well as the carrier-fluid temperature and conductivity (see Fig. 7.15). For example, Koo & Kleinstreuer (2004) proposed a k_{eff}-model based on *micromixing* induced by Brownian nanoparticle motion (see Sect.7.2). However, controversies arose when more precise, i.e., optical methods vs. transient hot-wire techniques, were employed and the validity of the origins for the unusual thermal effect was questioned (see Kleinstreuer & Li 2008b and cited papers therein). The underlying k_{eff}-model hypothesis is that k_{eff} is composed of a static part and a kinetic part:

$$k_{eff} = k_{static} + k_{motion} \tag{7.64}$$

The static part, k_{static}, is based on the effective medium (Hamilton-Crosser) theory. For spheres,

$$\frac{k_{static}}{k_f} = \frac{2 + k_p / k_f + 2(k_p / k_f - 1)\varphi}{2 + k_p / k_f - 2(k_p / k_f - 1)\varphi} \approx 1 + 3\varphi + O(\varphi^2) \quad (7.65a, b)$$

where k_f is the fluid conductivity and k_p the particle conductivity, while φ is the volume fraction. When $\varphi << 1$ and $k_p / k_f \geq 20$, the approximation Eq. (7.65b) holds.

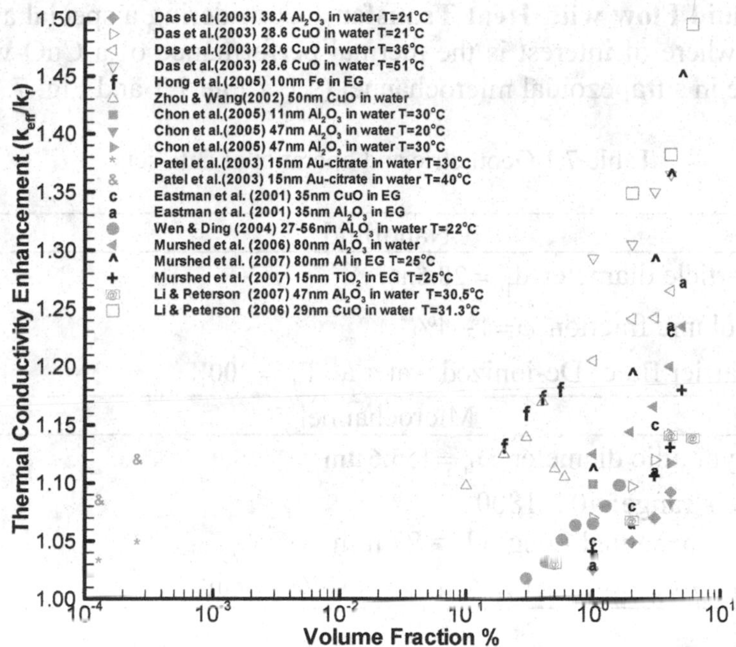

Fig. 7.15 Sample of experimental evidence for elevated k_{eff} of nanofluids

The motion part, k_{motion}, is based on kinetic theory together with Stokes' flow of micro-scale convective heat transfer (Koo & Kleinstreuer 2004 and Kleinstreuer & Li 2008b).

For dilute nanoparticle suspensions, the other effective mixture properties are:

$$\rho_{eff} = \varphi\rho_p + (1-\varphi)\rho_f \tag{7.66}$$

$$\mu_{eff} = \mu_f(1-\varphi)^{-2.5} \tag{7.67}$$

and

$$(\rho c_p)_{eff} = \varphi(\rho c_p)_p + (1-\varphi)(\rho c_p)_f \tag{7.68}$$

Nanofluid Flow with Heat Transfer Considering a special application where of interest is the thermal performance of a CuO-water mixture in a trapezoidal microchannel (see Table 7.1 and Fig. 7.16).

Table 7.1 Geometric and operational data sets

Nanofluid
• Particle diameter $d_p = 28.6\text{nm}$ • Volume fraction $\varphi = 1-4\%$ • Carrier fluid: De-ionized water at $T_{in} = 300\text{K}$
Microchannel
• Hydraulic diameter $D_h = 155.6\,\mu\text{m}$ • Re # range: 400–1800 • Microchannel Length L = 27 mm • Imposed wall heat flux $q_{wall} = 431466\,\text{W}/\text{m}^2$

(a) Representative Single-channel Model

(b) Finite-volume Mesh

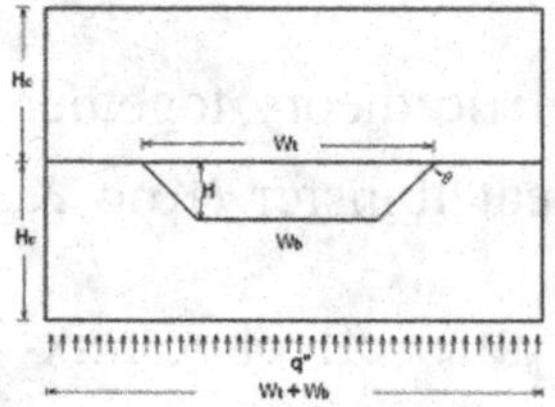

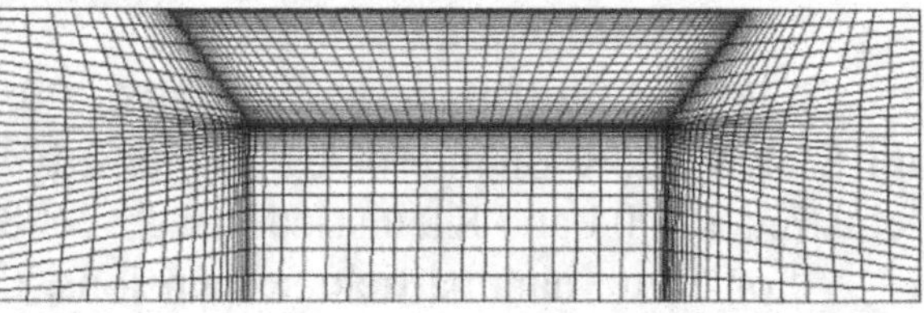

Fig. 7.16 Representative microchannel heat transfer model: System sketch and finite-volume mesh

The equations for thermal mixture flow are given in Li & Kleinstreuer (2008). As a first step, a steady global energy balance based on the water temperature difference between channel inlet and outlet was performed, i.e.,

$$\rho u_m c_p A_c (T_{out} - T_{in}) = q_{wall} A_w \tag{7.69}$$

Figure 7.17 compares for a fixed q_{wall}-value, the Reynolds-number dependence of $\Delta T = T_{out} - T_{in}$, assuming constant and variable properties of a *pure fluid*, say, water.

For relatively small Reynolds numbers, say, $Re \leq 800$, , the fluid residence time is long enough to produce a significant temperature increase caused by the constant heat flux from the microchannel bottom. For $Re \leq 1200$, the effect of temperature on the water properties k, ρ and c_p becomes measurable. For example, at elevated water temperatures the thermal capacity is reduced.

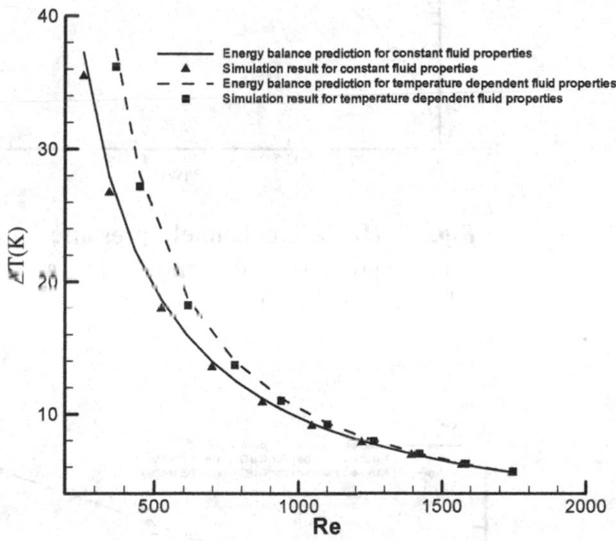

Fig. 7.17 Temperature rise as a function of inlet Reynolds number (after Li & Kleinstreuer 2008)

The necessary pumping power, P, to maintain a given volumetric flow rate, Q, is directly proportional to the pressure drop over the microchannel length:

$$P = \Delta p Q \tag{7.70}$$

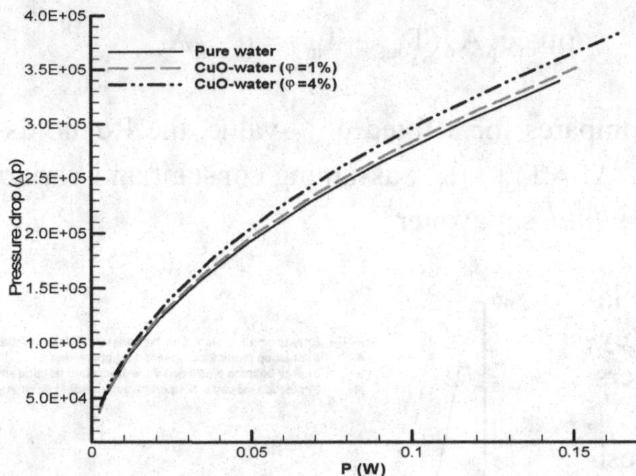

Fig. 7.18 Microchannel pressure drop for water and different nanofluids (after Li & Kleinstreuer 2008)

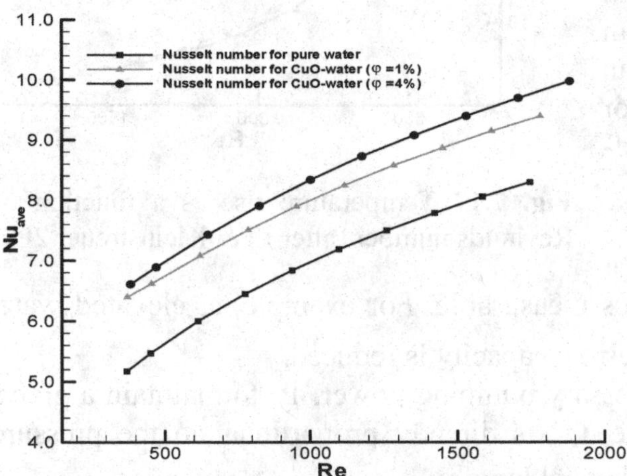

Fig. 7.19 Average Nusselt number for water and nanofluids as a function of Reynolds numbers (after Li & Kleinstreuer 2008)

Compared to water, Fig. 7.18 indicates only an average 2% increase for CuO-water with a 1% volume fraction and about 8% when $\varphi = 4\%$. Still, it was assumed that the nanoparticles were uniformly distributed, which could be approximately achieved via surface coating/charging and/or the addition of surfactants.

Of major interest is the heat transfer performance, in terms of the Nusselt number, of nanofluids in microchannels. The average Nusselt number can be defined as:

$$\overline{Nu} = \frac{q_w D_h}{(T_{w,av} - T_{f,av})k} \tag{7.71}$$

where q_w is the wall heat flux, D_h is the hydraulic diameter, $T_{w,av}$ is the average wall (or channel surface) temperature, $T_{f,av}$ is the average bulk-fluid temperature, and k is the pure water thermal conductivity. Thus, Fig. 7.19 demonstrates that nanofluids can improve the thermal performance. Specifically, compared to water an average 15% $\overline{Nu}$ -enhancement was achieved for a 1% CuO-water pairing and a 20% $\overline{Nu}$ -increase when 4% of CuO-nanoparticles were added to the base fluid. Clearly, the change in $\overline{Nu}$ -values for nanofluids (see Eq. (7.71)) stems from an elevated $T_{f,av}$ due to the more effective heat conduction in nanofluid flow.

7.5.2 Nanofluid Flow in Bio-MEMS

Microfluidics deals with methods and devices for manipulating and controlling fluid-particle flow in microchannels (Kleinstreuer et al. 2008). A recent application area is nanomedicine with the goal of controlled nanodrug delivery to specified target areas (see Saliterman 2006; Labhasetwar & Leslie-Pelecky 2007; among others). A key aspect of this goal is the development of integrated drug delivery systems to monitor and control target-cell responses to pharmaceutical stimuli, to understand biological cell activities, or to facilitate drug development processes. An important part of such drug- delivery systems, which belong to the family of bio-MEMS, are active or passive micro-mixers (Nguyen & Wu 2005; Nguyen 2008) to assure near-uniform nanodrug concentrations. Static micro-mixers, not requiring any external energy source, rely on chaotic advection and/or enhanced diffusion, typically to mix two fluids (see Stroock et al. 2002; Kim et al. 2004; Munson & Yager 2004; Hardt et al. 2005; Floyed-Smith et al. 2006; Chang & Yang 2007; Chung & Shih 2008; Kang et al. 2008; Brotherton et al. 2008; among others). For example, Hardt et al. (2005) reviewed recent developments in micromixing technology, focusing on liquid mixing with passive micro-mixers. Four kinds of mixers which employed different hydrodynamic principles are discussed: *hydrodynamic focusing*, *flow separation*, *chaotic advection*, and *split-and-recombine flows*. Diffusive mixing can be improved by increasing the interfacial contact area between the different fluids and reducing the diffusion length scale. Thus, selecting the right type of micro-mixer for a specific application is very important.

Li and Kleinstreuer (2009) analyzed rapid *nanoparticle mixing* in a carrier fluid, employing low-cost micro-mixers and heat transfer to achieve two system design goals, i.e., uniform exit particle concentration and minimum required channel length. Specifically, a microfluidics device for controlled nanofluid flow in microchannels (Fig. 7.20) is investigated for basic nanomedicine applications. Presently planned for laboratory-scale testing, a predetermined uniform concentrations of a stimulus (e.g., cocaine particles) should be delivered via multiple microchannels to an array of wells containing brain cells to measure cell responses (e.g., dopamine production levels). Their study focuses on device miniaturization in light of the ultimate goal of bio-MEMS implantation into the diseased brain region of, say, Parkinson's patients. Most importantly, the impact of two types of static micro-mixers (Fig. 7.20c) is analyzed to achieve uniform nanoparticle concentrations at the exit of a representative micro-channel of minimum length.

(a) Laboratory-scale nanodrug supply device

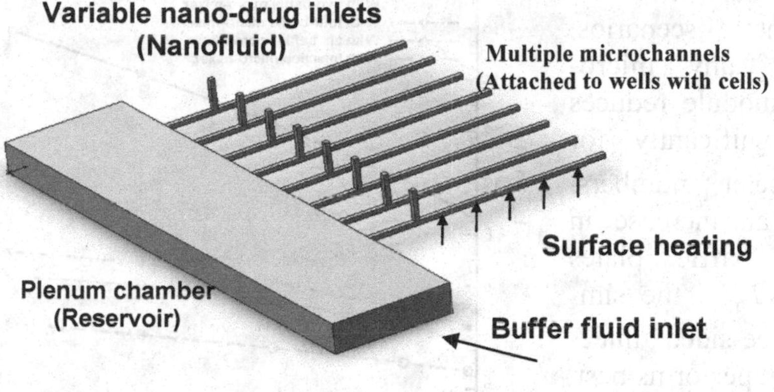

(b) Representative microchannels

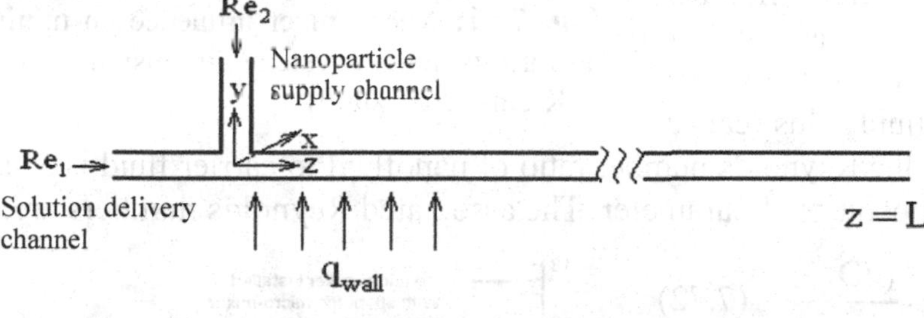

(c) Static mixers

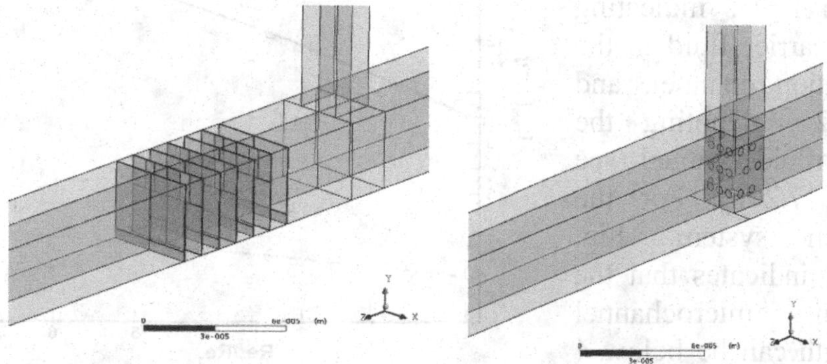

(i) Six-baffle unit with slits (ii) Injection unit with holes on three sides

Fig. 7.20 Microfluidics system: (a) Laboratory-scale nanodrug supply device, (b) Representative microchannels, (c) Static mixers

Figure 7.21 shows L_{min} (Pe) for the different scenarios. Clearly, any micro-mixer module reduces L_{min} significantly for all Peclet numbers. While an increase in slotted baffle plates reduces L_{min}, the simple three-sided injection unit performs best.

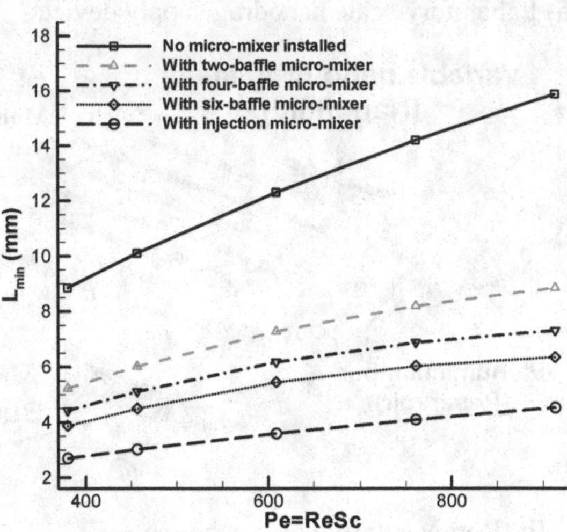

Fig. 7.21: Micro-mixer influence on minimal uniformity length/system dimension (after Li & Kleinstreuer 2009)

Alternative to the Peclet number, which is based on the average velocity of nanofluid plus carrier fluid, the Reynolds number ratio of nanofluid to carrier fluid is a suitable operational parameter. The associated Reynolds numbers are:

$$\mathrm{Re}_i = \frac{(uD_h)_i}{\nu_i} \quad (7.72)$$

with $i=1$, indicating the carrier fluid in the solution channel, and $i=2$, denoting the nanofluid channel (see Fig. 7.20b). For the given system, Fig. 7.22 indicates that the main microchannel length can be below 4 mm when employing an injection micro-mixer, i.e., a 70% reduction in channel length.

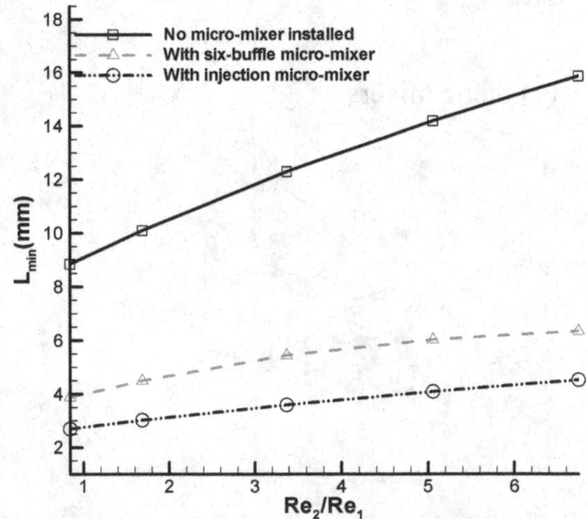

Fig. 7.22 Minimal uniformity length vs. Reynolds number ratio (after Li & Kleinstreuer 2009)

The addition of nanoparticles and certainly the installment of micro-mixers increases the pressure drop in both channels and hence the pumping-power requirements. Pumping power is defined as the product of the pressure drop across the channel (Δp) and the volumetric flow rate (Q), i.e.,

$$P = \Delta p \cdot Q \tag{7.73}$$

The pressure differences occur between the nanofluid inlet or carrier fluid inlet and the system outlet, i.e., required minimal length. The volumetric flow rate is the sum of the volumetric flow rates of both nanofluid and carrier fluid.

Figures 7.23a,b depict the relationship of pressure drop and pumping power for the two cases. Clearly, the added micro-mixer increases the local pressure drop, but the decreased system length may reduce any negative effect caused by the micro-mixer. As shown in Figs. 7.23a,b, the power requirement even decreases in some case, i.e., when employing the injection micro-mixer. The employment of baffle-slit micro-mixers slightly increases the pressure drop; however, when the pumping power/volumetric flow rate gets larger and larger, the negative effect appears to be less and less. For example, for the two or four-baffle micro-mixer, the pressure drop is even smaller than that without any micro-mixer when the nanofluid

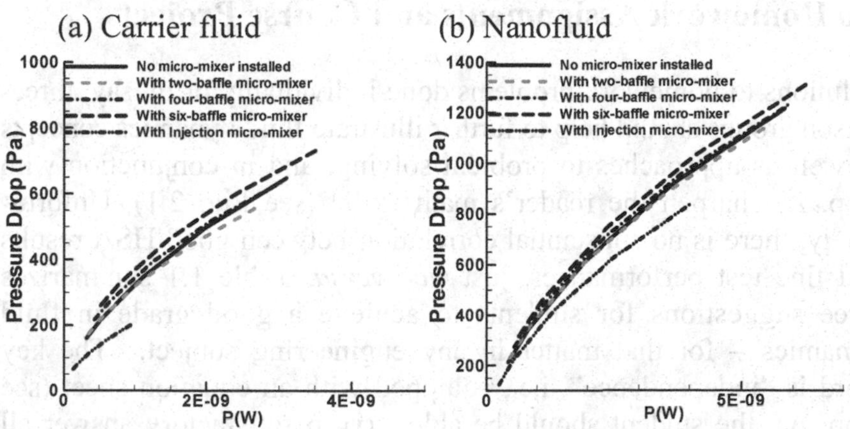

Fig. 7.23 Pressure drop vs. pumping power: (a) Carrier fluid inlet (b) Nanofluid inlet (after Li & Kleinstreuer 2009)

supply rate is increased to 8mm/s. In summary, employing an appropriate micro-mixer decreases the system dimension as well as the associated power requirement.

A heat flux was used to ensure that mixture delivery to the living cells occurs at a required temperature of 37°C. The change of fluid properties and nanoparticle diffusivity, caused by the added heat flux, also benefits system miniaturization. As shown in Fig. 7.24, the added heat flux greatly decreases the system dimension, i.e., an average 35% reduction is observed.

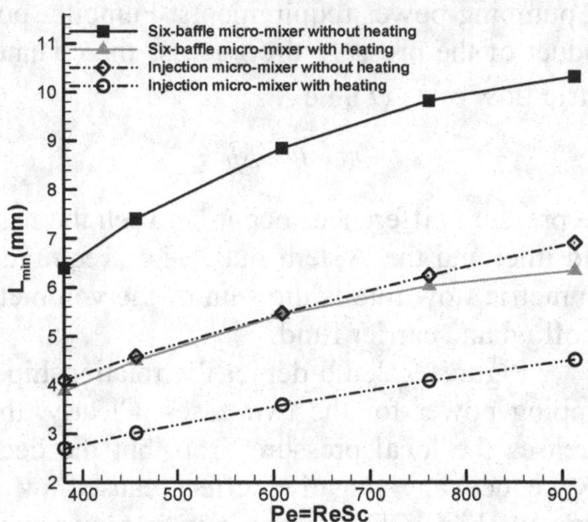

Fig. 7.24 Heat flux influence on minimal uniformity length (after Li & Kleinstreuer 2009)

7.6 Homework Assignments and Course Projects

Solutions to homework problems done individually or in, say, three-person groups should help to further illustrate fluid dynamics concepts as well as approaches to problem solving, and in conjunction with App. A, sharpen the reader's math skills (see Fig. 2.1). Unfortunately, there is no substantial correlation between good HSA results and fine test performances, just *vice versa*. Table 1.1 summarizes three suggestions for students to achieve a good grade in fluid dynamics – for that matter in any engineering subject. The key word is "independence", i.e., equipped with an equation sheet (see App. A), the student should be able: (i) to satisfactory answer all concept questions and (ii) to solve correctly all basic fluid dynamics problems.

The “Insight” questions emerged directly out of the Chap.7 text, while some “Problems” were taken from lecture notes in modified, i.e., enhanced, form when using White (2006), Cimbala & Cengel (2008), and Incropera et al. (2007). Additional examples, concept questions and problems may be found in any UG fluid mechanics and heat transfer text, or on the Web (see websites of MIT, Stanford, Cornell, UM, etc.).

End-of-chapter projects are “more involved: HW problems, which should be followed-up with in-class presentations.

7.6.1 Guideline for Project Report Writing

A project is more than a homework problem. As a starter, it typically requires a brief literature review to document the state-of-the-art for the given task. The antagonistic interplay between a rather accurate mathematical description of an assigned problem and finding a tractable solution method puts a premium on making the right assumptions and selecting a suitable equation solver, often in terms of available software and hardware.

The resulting work has to be documented in an appropriate format, i.e., *project reports* should have the following features:

- **ToC:** Table of contents with page numbers (optional for smaller reports)
- **Abstract:** Summary of the fluid dynamics problem analyzed, solution method employed, and novel results with physical insight presented
- **NOMENCLATURE:** Optional but recommended for complex problem solutions
- **Introduction:** Project objectives and task justification; System sketch with state-of-the-art description, basic approach or concept, and references of the *literature review* in terms of “author (year),” e.g., White (1998)
- **Theory:** Basic assumptions, governing equations, boundary conditions, property values, ranges of dimensionless groups, closure models, and references
- **Model Validations:** (a) Numerical accuracy checks, i.e., mesh-independence of the results as well as mass & momentum

residuals less than, say, 10^{-4} ; (b) quantitative proof, e.g., comparison with an exact solution of (typically a much) simpler system, and/or experimental data comparisons, or (at-least) qualitative justification that the solution is correct

- **Results and Discussion:** Graphs with interpretations, i.e., explain physical insight (Note: number the figures and discuss them concisely); draw conclusions and provide applications
- **Future Work**
- **REFERENCES:** List all literature cited in alphabetical order
- **APPENDICES:** Lengthy derivations, computer programs, etc.

7.6.2 Homework Problems and Mini-Projects

7.6.1 In light of the condition $\delta / d \geq 7$, check if air at STP is a dilute gas or not.

7.6.2 Derive Eq. (7.8) and discuss that for continuum gas flow the N-S equations are applicable only if $Kn \leq 10^{-2}$. (Mini Project)

7.6.3 Contrast gas flow vs. liquid flow on a molecular level and discuss the criteria when the continuum mechanics assumption is valid. (Mini Project)

7.6.4 Similar to Problem 7.7.3 for convection heat transfer. (Mini Project)

7.6.5 Provide the physical insight gained from deriving Eq. (7.13) on a micro-scale rather than from a macro-scale viewpoint given in Sect. 2.4.

7.6.6 Derive Eq. (7.21d) from first principles.

7.6.7 Examples 5.10 and 7.1 indicate that a smooth sphere in a viscous fluid cannot touch a solid surface, contrary to experimental evidence. Comment!

7.6.8 Derive Eq. (7.23) from first principles and prove Eqs. (7.24) – (7.26).

7.6.9 Revisiting Example 7.2 and Fig. 7.4b, explain how an up-stream parabolic velocity profile turns into a (uniform) meniscus speed.

7.6.10 With the continuum assumption to be valid, list the differences in the considerations and hence solutions of problems for flows in macrochannels vs. microchannels.

7.6.11 Explain in detail for entrance lengths of micro- vs. macro-conduits the condition given as Eq. (7.32).

7.6.12 Discuss Eq. (7.33), i.e., graph $\mathrm{Po}(\kappa)$ and comment!

7.6.13 List and explain the conditions/assumptions for electrokinetic and electrohydrodynamic effects in microchannel flows.

7.6.14 Revisit the examples in Sect. 7.3, collect some numerical values for the key system parameters and show that, indeed, the resulting Re-numbers are very small.

7.6.15 Concerning slip-flow (see Fig. 7.8), how exactly do the models by Navier and Maxwell differ? Are there cases where δ and λ are the same?

7.6.16 Is the net through-put for simple Couette flow (see Fig. 7.9) the same for the two velocity wall conditions.

7.6.17 Do the analyses in Sects. 7.2.5.3 and 7.2.5.4 only hold for gas flows (at $0.01 < \mathrm{Kn} < 0.1$)? If so, what do you suggest to in cases of liquid nonequilibrium flows?

7.6.18 Using Example 7.5 as a guide, solve the problem of thermal gas flow in the slip-flow regime for Poiseuille flow.

7.6.19 Comparing Eqs. (7.55) and (7.56), an *inequality* (representing the Second Law of Thermodynamics) turned into an *equation* for entropy generation. Explain that mathematically and physically.

7.6.20 In nanofluids, even *metallic* particles, say, in water, do not settle. Explain with math and physics arguments.

7.6.21 Discuss the differences between k_{static} and k_{motion} in Eq. (7.64).

7.6.22 Correlations for effective properties of nanofluids are very important. Based on a literature review, tabulate and discuss the applications of reliable correlations for basic nanofluids, i.e., common nanoparticle-and-liquid pairings. Do the nanoparticles for increased heat transfer have to be metallic? (Class Project)

7.6.23 Section 7.5 extolls the advantages of nanofluids and their applications in microchannel heat exchangers and bio-MEMS. Search the literature for updated applications, but also critically review drawbacks.

7.6.3 Course Projects

7.6.24 When the effective diameter of a microchannel is comparable with the mean-free-path of the flowing gas molecules, the fluid is not any more in thermodynamics equilibrium.

(a) Perform a recent literature review of experimental/computational articles discussing (rarefied) micro-to-nano gas flows in the slip-flow regime

(b) Provide a basic (computer) simulation of such gas flow as described in an *experimental* paper

(c) Write a project report

7.6.25 A combination of electro- and hydro-dynamic driving forces allow for a wide range of Re# operations of liquid flow in micro-channels.

(a) Perform a recent literature review of experimental/computational articles discussing this topic

(b) Provide a basic (computer) simulation of such a topic as described in an *experimental* paper

(c) Write a project report

7.6.26 Minimization of entropy generation is a key criterion in the design and operation of microfluidics devices.

(a) Perform a recent literature review of experimental/computational articles discussing this topic

(b) Provide a basic (computer) simulation of such a topic as described in an *experimental* paper

(c) Write a project report

7.6.27 The use of nanofluids in micro-cooling devices has definite advantages.

(a) Perform a recent literature review of experimental/computational articles discussing this topic

(b) Provide a basic (computer) simulation of such a topic as described in an *experimental* paper

(c) Write a project report

Chapter 8

Fluid–Structure Interaction

8.1 Introduction

Almost all problems solved in the previous chapters dealt with *static* fluid–structure scenarios where just the configuration of a solid structure at rest (or moving in the flow direction) determined the fluid flow field. Examples included internal flow through a conduit or external flow past submerged bodies. Still, not only the static structure via the boundary conditions determined the flow field; but, in turn fluid stresses and forces on the structure walls could be evaluated as well.

In contrast to the static cases, dynamic fluid–structure interactions are characterized by the following two-way coupled mechanisms (see Fig. 8.1):

- Transient, typically fluctuating, displacements of a structure wall induces fluid motion, which stays in contact with the wall.
- The unsteady flow field imposes varying shear stresses and pressure loads onto a fluid–solid interface and hence initiates or modifies the motion of the structure.

Clearly, most often such dynamic coupling occurs when transient fluid flow causes structure movement (or even structure reconfiguration). Examples include blood flow past heart valves and in elastic arteries, turbulent flow causing conduit or machine-part vibration, container or structure motion due to surface waves, vibro-acoustic

C. Kleinstreuer, *Modern Fluid Dynamics: Basic Theory and Selected Applications in Macro- and Micro-Fluidics*, Fluid Mechanics and Its Applications 87,
DOI 10.1007/978-1-4020-8670-0_8, © Springer Science+Business Media B.V. 2010

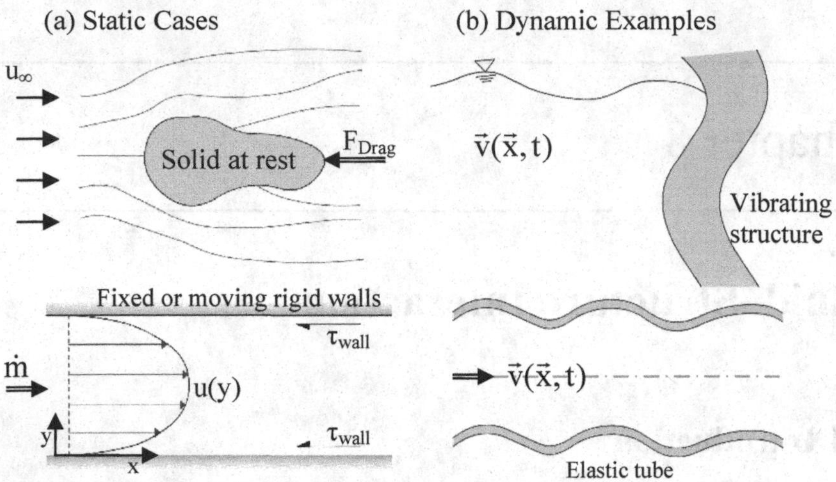

Fig. 8.1 Fluid–structure interactions: (a) Static cases; and (b) dynamic examples

coupling in speakers, wind effects on wires, towers and bridges, or ocean waves affecting off-shore structures. Specifically, high winds, strong river currents, major ocean waves and atmospheric explosions can cause severe damage of all kinds of structures, while on a micro-scale fluid-flow induced vibration may lead to material fatigue, internal/external cracks, and ultimately structure collapse.

Fluid–structure interactions are modeled based on the equation of motion (Sect. 2.4) and the dynamic equations of deformable solids (Sect. 8.2). Suitable coupling conditions at the fluid–structure interface have to be enforced. Specifically, the fluid and solid wall have the same motion (i.e., wetted surface), and the fluid and structural stresses exerted on the wall are balanced, i.e., they are locally in dynamic equilibrium.

Section 8.2 deals briefly with stress fields in solids, conditions of equilibrium, stress–strain relations, and compatibility conditions. Subsequently, applications to slender-body dynamics and flow-induced vibrations in terms of mass-spring-damper systems are discussed in Sects. 8.3 and 8.4, respectively.

8.2 Solid Mechanics Review

As mentioned in Sect. 8.1, fluid–structure interactions (FSI) are an integral part of most transport phenomena in biological, environmental and industrial systems.

In this section, the basic solid mechanics equations are reviewed which, in conjunction with the conservation laws (Chap. 2), prepare to solve FSI problems. Typically, a pressure due to fluid flow generates a load on a structure which results in wall stress, strain and deformation.

8.2.1 Stresses in Solid Structures

As outlined in Sect. 2.4, the state of material interaction at any point in a body is specified by the stress tensor:

$$\sigma_{ij} = \begin{bmatrix} \sigma_{11} & \sigma_{12} & \sigma_{13} \\ \sigma_{21} & \sigma_{22} & \sigma_{23} \\ \sigma_{31} & \sigma_{32} & \sigma_{33} \end{bmatrix} = \begin{bmatrix} \sigma_{xx} & \tau_{xy} & \tau_{xz} \\ \tau_{yx} & \sigma_{yy} & \tau_{yz} \\ \tau_{zx} & \tau_{zy} & \sigma_{zz} \end{bmatrix} \tag{8.1a}$$

where σ_{ii} (no sum on i) are the normal stresses, i.e., generated by forces F_i perpendicular to surfaces A_i, and the shearing stresses $\sigma_{ij} \hat{=} \tau_{ij}$ $(i \neq j)$ are caused by tangential forces F_j acting on surfaces A_i $(i \neq j)$ with normal vector $\hat{n}_i$ (see Fig. 8.2).

As eluded to in Sect. 2.4, the stress tensor is symmetric, i.e., $\sigma_{ij} = \sigma_{ji}$. This equality of shearing stresses can be readily shown in applying the static conditions $\sum \vec{\mathbf{M}} = \mathbf{I}\vec{\boldsymbol{\alpha}} \equiv 0$ to the solid cube of Fig. 8.2a, where $\vec{\mathbf{M}} = \vec{\mathbf{r}} \times \vec{\mathbf{F}}$ is the applied moment or torque, i.e., force times lever distance, $\mathbf{I}$ is the moment of inertia, and $\vec{\boldsymbol{\alpha}}$ is the angular acceleration vector. The normal and shearing stresses relate to the following forces and moments (Fig. 8.2b):

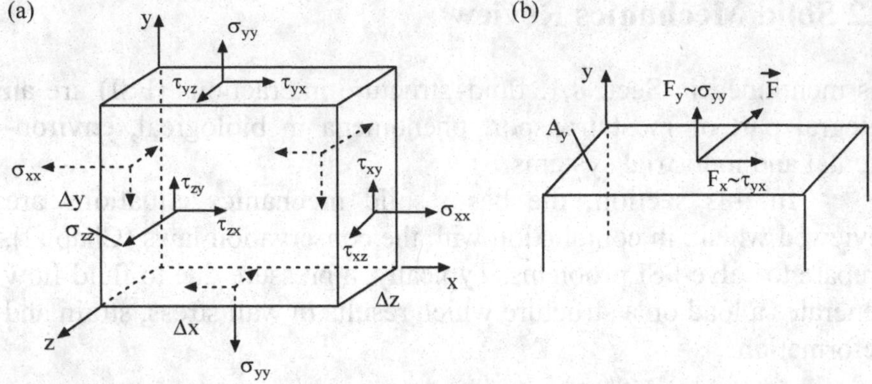

Fig. 8.2 Stresses and forces: (a) 3-D solid element with positive stresses; (b) 2-D force components and stresses in the A_y-plane

- *Axial force* ~ $\int \sigma_{ii} dA$, where σ_{ii} (no sum on i) are either tensile or compressive stresses
- *Shear force* ~ $\int \tau_{ij} dA$ $(i \neq j)$, which try to move (shear) adjacent parts of a solid
- *Moments* (or torques), which may bend a beam, e.g., $M_y = \int \sigma_{xx} z dA$, $M_z = -\int \sigma_{xx} y dA$, etc., or may twist a body, e.g., $T = \int (\tau_{xz} y - \tau_{xy} z) z dA$

Clearly, the symmetric stress tensor has six independent components. These can be further reduced to three principal stresses which act in mutually perpendicular, principal directions. They act normal to three principal planes on which all shearing stresses are zero. Thus, Eq. (8.1a) now reads after the coordinate transformation from σ_{ij} to σ_{ii}:

$$\sigma_{ij} = \begin{bmatrix} \sigma_1 & & \emptyset \\ & \sigma_2 & \\ \emptyset & & \sigma_3 \end{bmatrix} \tag{8.1b}$$

where $\sigma_1, \sigma_2, \sigma_3$ are the principal stresses. The principal stresses can be combined according to Von Mises' deformational theory in terms of an effective stress σ_e, known as the *Von Mises stress*.

$$\sigma_e = \frac{1}{\sqrt{2}}[(\sigma_1 - \sigma_2)^2 + (\sigma_2 - \sigma_3)^2 + (\sigma_3 - \sigma_1)^2]^{1/2} \tag{8.2}$$

Equation (8.2) is typically used to determine the locations and magnitudes of maximum stresses in human hard or soft tissues as well as medical devices and implants.

The next two examples illustrate how internal stresses are computed for 2-D plate bending and cylinder twisting (Example 8.1) as well as internal stress distributions, including principal stresses, for a bar under axial tension (Example 8.2).

Example 8.1: Simple Bending and Torsion Formulas for 2-D Solids of Linearly Elastic Material

Consider a 2-D beam subject to pure bending and then a cylinder subject to simple torsion. Show that $\sigma_{xx} = -M_z y / I$, where I is the moment of inertia, and $\tau_{\rho\theta} = Tr / J$, where J is the polar moment of inertia.

Sketches:

(a) Bending

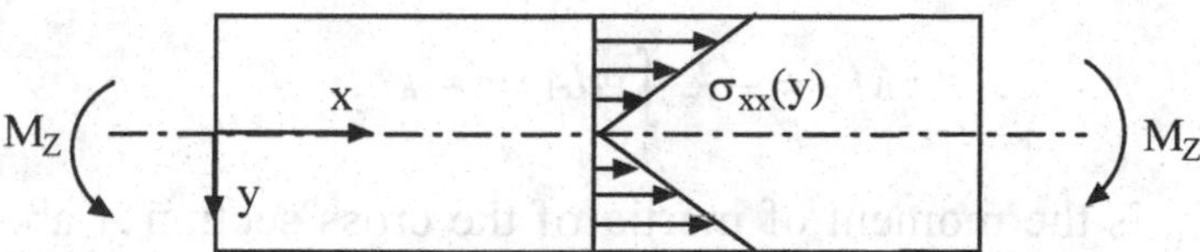

(b) Torsion

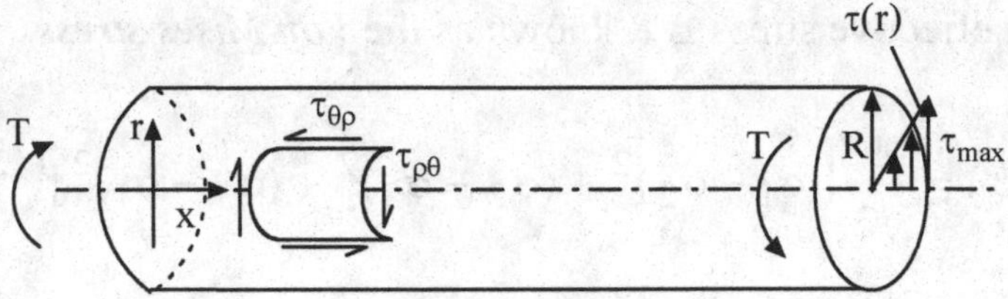

Assumptions:

- The normal stress $\sigma_{xx}(y)$ is linear in y, and all other stresses in the beam are zero (pure bending).
- The resultant of the internal force of the beam is zero, i.e., $\int_A \sigma_{xx} dA = 0$.
- The torsional loading does not reshape or re-orientate the cross-sectional planes of the cylinder.

Solution:

(a) Bending

The moments of the internal forces about the beam's centerline (i.e., normal axis) are equal to the applied moment around the *z*-axis (see Sketch (a)):

$$M_z = -\int_A y\sigma_{xx} dA$$

With $\sigma_{xx} = ky$, where *k* is a constant, we have

$$M_z = -k\int_A ydA := -kI_z$$

where I_x is the moment of inertia of the cross section *A* about the *z*-axis, e.g., $I_x = bh^3/12$ for a rectangular beam with $A = bh$ and $I_z = \pi R^4/4$ for a cylinder of radius R, i.e., $A_0 = \pi R^2$.

Thus, with $k = -M_z / I_z$

$$\sigma_{xx} = -M_z y / I_z$$

(b) Torsion

The resultant of the shearing stress distribution being equal to the applied torque yields (see Sketch (b)):

$$T = r\tau dA; \qquad \tau(r) = \frac{r}{R}\tau_{max}$$

Thus,

$$T = \frac{\tau_{max}}{R}\int_A r^2 dA := \frac{\tau_{max}}{R} J$$

where $J = \pi R^4 / 2$ is the polar moment of inertia of the cylinder's cross section. Finally, with $\tau_{max} / R = \tau / r$, we obtain:

$$\tau = \tau_{\rho\theta} = Tr / J$$

Example 8.2: Stresses on an Inclined Plane of a Bar Under Uniaxial Tension

Consider a bar ($A = 10^3$ mm^2) subject to a tensile load of $N = 100$ kN. Determine the stress in a plane slanted by $\alpha = 35°$ and verify stress symmetry and the existence of a principal plane.

Sketch	*Assumption*
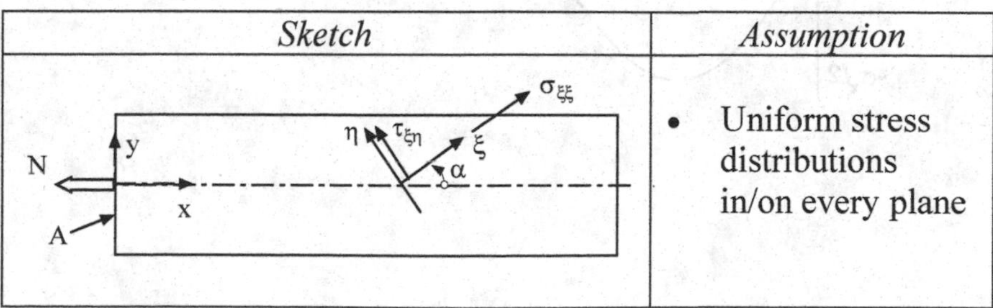	• Uniform stress distributions in/on every plane

Solution: With respect to the incline, the loading force N generates a tensile stress $\sigma_{\xi\xi}$ because of the normal $N\cos\alpha$ and a shearing stress $\tau_{\xi\eta}$ because of the tangential $N\sin\alpha$. Thus, with $\sigma_{xx} = N/A$ we have:

$$\sigma_{\xi\xi} = \frac{N\cos\alpha}{A/\cos\alpha} = \sigma_{xx}\cos^2\alpha$$

and

$$\tau_{\xi\eta} = -\frac{N\sin\alpha}{A/\cos\alpha} = -\sigma_{xx}\sin\alpha\cos\alpha$$

In letting the "cutting angle" vary, i.e., $0° \le \alpha \le 180°$, $\sigma_{\xi\xi}$ and $\tau_{\xi\eta}$ undergo maxima and minima. For example,

$$\sigma_{\xi\xi}(\alpha = 0°, 180°) = \sigma_{xx} = \sigma_{max} \quad \text{when} \quad \tau_{\xi\eta} = 0$$

and

$$\tau_{\xi\eta}(\alpha = 45°, 135°) = \tau_{max} \quad \text{when} \quad \sigma_{\xi\xi} = \pm\frac{1}{2}\sigma_{xx}$$

Graph:

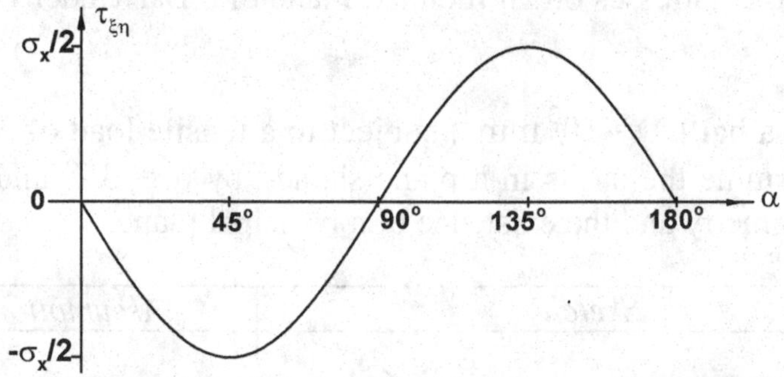

Comments:

When plotting $\tau_{\xi\eta}$ it is revealed that both symmetry of the stress tensor (see Eq. (8.1a)) and principal planes or axes (see Eq. (8.1b)) actually exist. Specifically,

$$\left|\tau_{\xi\eta}(\theta)\right| = \left|\tau_{\xi\eta}(\theta + 90°)\right|$$

and at $\alpha = 0°$ and $180°$

$$\sigma_{\xi\xi} = \sigma_{\max} \equiv \sigma_{xx} \quad \text{when} \quad \tau_{\xi\eta} = 0$$

With the given data,

$$\sigma_{xx} = \frac{N}{A} = 100 \text{ MPa}, \quad \sigma_{\xi\xi} = \sigma_{xx} \cos^2 \alpha = 67.11 \text{ MPa}, \text{ and}$$
$$\tau_{\xi\eta} = -\sigma_{xx} \sin\theta \cos\theta = -47 \text{ MPa}$$

8.2.2 Equilibrium Conditions

So far we assumed uniformly distributed stresses across each surface caused by an external load. More likely, stress components vary from point to point in a body; nevertheless, all external forces and internal stresses have to be in static equilibrium, i.e.,

$$\sum \vec{\mathbf{F}} = 0 \quad \text{and} \quad \sum \vec{\mathbf{M}} = 0 \tag{8.3.4}$$

For example, considering a thin 2-D solid element of differential area $dA_z = dxdy$ (see Fig. 8.3), we take the moments about the origin [see Eq. (8.4)] and can show that $\tau_{xy} = \tau_{yx}$. Now, employing $\sum F_x = 0$ [see Eq. (8.3)], we obtain:

$$\left(\sigma_{xx} + \frac{\partial \sigma_{xx}}{\partial x} dx\right) dy - \sigma_{xx} dy + \left(\tau_{xy} + \frac{\partial \tau_{xy}}{\partial y} dy\right) dx - \tau_{xy} dx + f_s dx dy = 0$$

or

$$\frac{\partial \sigma_{xx}}{\partial x} + \frac{\partial \tau_{xy}}{\partial y} + f_x = 0$$

Similarly,

$$\frac{\partial \sigma_{yy}}{\partial y} + \frac{\partial \tau_{xy}}{\partial x} + f_y = 0$$

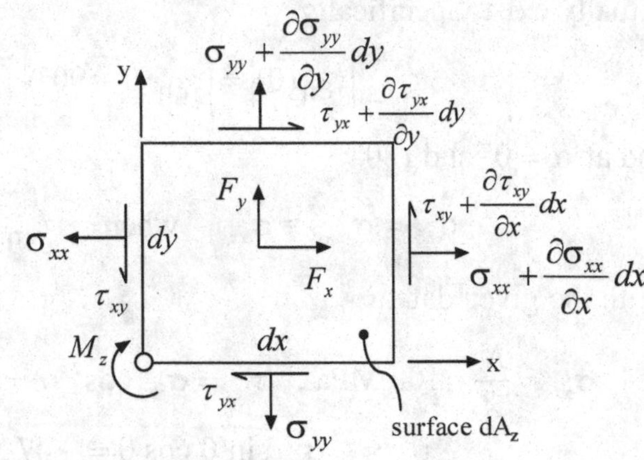

Fig. 8-3 Differential 2-D solid element with varying forces and stresses

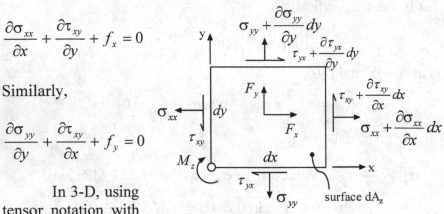

In 3-D, using tensor notation with the summation convection of repeated indices, we have the equilibrium condition:

$$\frac{\partial \sigma_{ij}}{\partial x_j} + f_i = 0 \tag{8.5a}$$

or for the general 3-D case:

$$\sigma_{ij,j} + f_i = 0; \quad i, j = x, y, z, \qquad \sigma_{ij} = \sigma_{ji} \ (i \neq j) \tag{8.5b}$$

Clearly, given the external forces per unit volume f_i, Eq. (8.5) contains six unknown stress components, i.e., three additional equations have to be found. This brings us to the three basic principles for solving solid mechanics problems:

(i) *Conditions of Equilibrium* [see Eq. (8.5)]
(ii) *Stress–Strain Relations* where material properties in terms of constitutive equations correlate forces, causing stresses in a given solid with body, displacements.
(iii) *Conditions of Compatibility,* where body continuity is everywhere preserved, consistent with local strain distributions and deformations.

Equation (8.5) can be readily extended to *dynamic structures*, where in general (see Newton's second law of motion):

$$\rho a_i = \sigma_{ij,j} + f_i \quad \text{in} \quad \Omega(t) \tag{8.6}$$

with $a_i = dv_i / dt$ being the acceleration of a material point in the domain Ω at time t, f_i are the external forces per unit volume, and ρ equals the mass density. Equation (8.6) is subject to appropriate boundary conditions and necessary stress–strain relations as discussed in the next section.

8.2.3 Stress–Strain Relationships

When a force and/or moment is applied to an object, material deformation may occur in form of body elongation/contraction, beam bending, rod twisting, and/or simple shearing. In all these cases, body elements undergo shape changes. In contrast, during rigid-body motion, e.g., translatory displacement due to an axial force as for any flying/moving, non-deforming object such as balls, cars, planes, ships, etc., no stress is induced. Thus, deformation relates directly to stress when body element distortions, i.e., shape changes occur, say, due to stretching, compression, and/or twisting.

Deformation Analysis As illustrated in Fung (1994) and other texts, a line element in a 3-D body, $\overline{AA'} = ds_A$, translates, stretches, and rotates, because of body deformation, to $\overline{BB'} = ds_B$ (see Fig. 8.4).

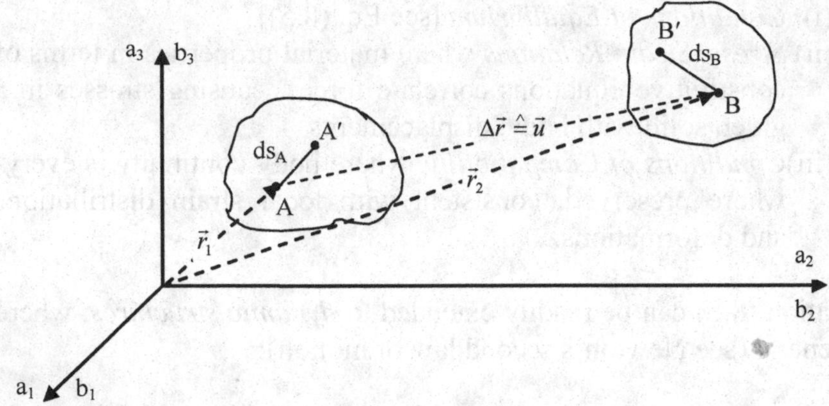

Fig. 8.4 Line element changes during body deformation (after Fung 1994)

Clearly, with point A at (a_1, a_2, a_3) and point B at (b_1, b_2, b_3) the distance to the neighboring point $A'(a_1 + da_1, a_2 + da_2, a_3 + da_3)$ is after Pythagoras:

$$(ds_A)^2 = (da_1)^2 + (da_2)^2 + (da_3)^2$$

Similarly,

$$(ds_B)^2 = (db_1)^2 + (db_2)^2 + (db_3)^2$$

Now, determination of any continuous body deformation requires known mapping functions

$$b_i = b_i(a_1, a_2\ a_3) \ \text{ or } \ a_i = a_i(b_1, b_2\ b_3)$$

for every point in the body. The components of the displacement vector $\vec{\mathbf{u}} = \Delta\vec{\mathbf{r}}$ (see Fig. 8.4) are then

$$u_i = b_i - a_i$$

Given the mapping functions, we can write

$$db_i = \frac{\partial b_i}{\partial a_j} da_j \text{ and } da_i = \frac{\partial a_i}{\partial b_j} db_j$$

so that the displacements can be expressed, using the Kronecker delta, as

$$ds_A^2 = \delta_{ij} da_i da_i = \delta_{ij} \frac{\partial a_i}{\partial b_l} \frac{\partial a_j}{\partial b_m} db_l \, db_m$$

and

$$ds_B^2 = \delta_{ij} db_i db_i = \delta_{ij} \frac{\partial b_i}{\partial a_l} \frac{\partial b_j}{\partial a_m} da_l \, da_m$$

Forming the difference, i.e., *a measure of body deformation*, we obtain

$$ds_B^2 - ds_A^2 = \begin{cases} 2E_{ij} da_i da_j \\ \text{or} \\ 2\varepsilon_{ij} db_i db_j \end{cases}$$

where the symmetric tensor E_{ij} is Lagrangian (or Green's) strain tensor and the tensor ε_{ij} is the Eulerian (or Cauchy's) strain tensor. Clearly, if the length of each line element stays the same, $ds_B^2 - ds_A^2 = 0$, implying that $E_{ij} = \varepsilon_{ij} = 0$, and the body is at rest or undergoing rigid-body motion.

Using the displacement vector $u_i = b_i - a_i$, we can form:

$$\frac{\partial b_i}{\partial a_j} = \frac{\partial u_i}{\partial a_j} + \delta_{ij} \text{ and } \frac{\partial a_i}{\partial b_j} = \delta_{ij} - \frac{\partial u_i}{\partial b_j}$$

where $u_i = (u_1, u_2, u_3)$ and $b_i = (x_1, x_2, x_3)$. Thus, inserting these relations into the displacement expression and neglecting squares and products of the derivatives of the displacement components u_i, the Cauchy infinitesimal strain terms reduces to:

$$\varepsilon_{ij} = \frac{1}{2}\left(\frac{\partial u_j}{\partial x_i} + \frac{\partial u_i}{\partial x_j}\right) \tag{8.7}$$

or in terms of the engineering shearing strain (i.e., the shear-rate tensor of Sect. 2.4):

$$\gamma_{ij} = \frac{\partial u_i}{\partial x_j} + \frac{\partial u_j}{\partial x_i} \quad (i \neq j) \tag{8.8}$$

Equation (8.7) can be written as:

$$\varepsilon_{ij} = \begin{bmatrix} \varepsilon_{xx} & \frac{1}{2}\gamma_{xy} & \frac{1}{2}\gamma_{xz} \\ \frac{1}{2}\gamma_{yx} & \varepsilon_{yy} & \frac{1}{2}\gamma_{yz} \\ \frac{1}{2}\gamma_{zx} & \frac{1}{2}\gamma_{zy} & \varepsilon_{zz} \end{bmatrix} \tag{8.9}$$

Clearly, for infinitesimal displacements $E_{ij} \approx \varepsilon_{ij}$ because it is immaterial if the derivatives $u_{i,j}$ are calculated at the position of a point before or after deformation.

Now the cause of displacements, i.e., stresses, have to be correlated to strain. Specifically, a stress–strain relationship describes the mechanical property of a material and is therefore a *constitutive equation*. For linear elastic materials, e.g., hard tissue (bones) and basic metals, for which *Hooke's law* holds,

$$\sigma_{ij} = C_{ijkl}\varepsilon_{kl} \tag{8.10}$$

where the tensor of rank four, C_{ijkl}, is symmetric, i.e., the $3^4 = 81$ elastic constants (or moduli) can be reduced to 36 (Fung 1994). For an *isotropic* elastic solid, i.e., when the properties are identical in all directions, Eq. (8.10) reduces to:

$$\sigma_{ij} = \lambda \varepsilon_{kk} \delta_{ij} + 2\mu \varepsilon_{ij} \tag{8.11}$$

Clearly, instead of 36 material values, only two, i.e., the *Lamé constants* λ and μ, are needed. Writing Eq. (8.11) out and solving for the six ε-components, we obtain:

$$\varepsilon_{ij} = \frac{1+\nu}{E}\sigma_{ij} - \frac{\nu}{E}\sigma_{kk}\delta_{ij} = \begin{cases} \frac{1}{2}\left(\frac{\partial u_i}{\partial x_j} + \frac{\partial u_j}{\partial x_i}\right) \\ \frac{1}{2}\gamma_{ij} \end{cases} \tag{8.12a–c}$$

where the "shearing strain" $\gamma_{ij} = \tau_{ij}/G$, E is *Young's modulus*, $\nu = -\varepsilon_{yy}/\varepsilon_{xx} = -\varepsilon_{zz}/\varepsilon_{xx}$ is the *Poisson ratio* expressing (lateral strain)/(axial strain), and $G \equiv \mu$ is the *shear modulus*. Because only two independent constants are needed for homogeneous isotropic elastic materials [see Eq. (8.11)], E, ν, and G are related, i.e.,

$$G = \frac{E}{2(1+\nu)} \tag{8.13}$$

Actual materials, such as biological soft and most hard tissues, rubber, shape memory alloys (SMAs), and non-Newtonian fluids, do not follow Hooke's law. Clearly, compared to metals, these other materials exhibit nonlinear relationships and hysteresis, i.e., the non-linear loading and unloading curves differ. Mechanical models, a combination of (linear) springs and viscous dashpots, mimic some viscoelastic behavior. For example, as outlined in Humphrey and Delange (2004) and other solid mechanics books, the *Kelvin model* relates the load F to the deflection (or displacement) $u(t)$ as:

$$F + a\dot{F} = b(u + c\dot{u}) \tag{8.14a}$$

subject to

$$aF(t = 0) = bcu(t = 0) \tag{8.14b}$$

where a to c are constant system parameters.

Simplifications Hooke's law, for a linear elastic, isotropic material, i.e., Eq. (8.12), can be rewritten as:

$$\varepsilon_{xx} = \frac{1}{E}[\sigma_{xx} - v(\sigma_{yy} + \sigma_{zz})] \tag{8.15a}$$

$$\varepsilon_{yy} = \frac{1}{E}[\sigma_{yy} - v(\sigma_{xx} + \sigma_{zz})] \tag{8.15b}$$

$$\varepsilon_{zz} = \frac{1}{E}[\sigma_{zz} - v(\sigma_{xx} + \sigma_{yy})] \tag{8.15c}$$

$$\varepsilon_{xy} = \frac{1+v}{E}\tau_{xy} = \frac{1}{2G}\tau_{xy} := \frac{1}{2}\gamma_{xy} \tag{8.15d}$$

$$\varepsilon_{yz} = \frac{1+v}{E}\tau_{yz} = \frac{1}{2G}\tau_{yz} := \frac{1}{2}\gamma_{yz} \tag{8.15e}$$

$$\varepsilon_{zx} = \frac{1+v}{E}\tau_{zx} = \frac{1}{2G}\tau_{zx} := \frac{1}{2}\gamma_{zx} \tag{8.15f}$$

Clearly, for plane stress analysis, $\sigma_z = 0$, and the remaining normal stresses are:

Rectangular Coordinates	Polar Coordinates
$\sigma_{xx} = \frac{E}{1-v^2}(\varepsilon_{xx} + v\varepsilon_{yy})$	$\sigma_{rr} = \frac{E}{1-v^2}(\varepsilon_{rr} + v\varepsilon_{\theta\theta})$
$\sigma_{yy} = \frac{E}{1-v^2}(\varepsilon_{yy} + v\varepsilon_{xx})$	$\sigma_{\theta\theta} = \frac{E}{1-v^2}(\varepsilon_{\theta\theta} + v\varepsilon_{rr})$

(8.16a–d)

Provided that Young's modulus E and Poisson's ratio ν are known, each of these equations contain extra unknowns; thus, equilibrium equations have to be established first which relate forces to stresses, as illustrated next.

Example 8.3: Displacement of a Non-uniform Rod under Axial Stress

Consider a vertical, axially loaded rod with suddenly changing cross section and material property as shown (cf. bone with implanted metal segment). Find the total axial displacement *u*.

Sketch	*Assumptions*	*Concepts*
L_1 A_1 E_1 x L_2 A_2 E_2 N	• Segmentally constant A_i and E_i • Only normal, i.e., axial force N applied • Hooke's law holds	Use of Eq. (8.15a), where $\sigma_{yy} = \sigma_{zz} = 0$

Solution:

The reduced governing equations read:

- $\varepsilon_{xx} = \frac{\sigma_{xx}}{E}$ and $\Delta u = \int_0^L \varepsilon_{xx} dx$, where $\sigma_{xx} = \frac{N}{A}$

Specifically, with axial displacement $u(x = 0) = 0$:

- $u_{total} = \int_0^{L_1} \frac{N}{A_1 E_1} dx + \int_{L_1}^{L} \frac{N}{A_2 E_2} dx\,;\;\; L = L_1 + L_2$

and hence,

- $u_{total} = N\left(\frac{L_1}{A_1 E_1} + \frac{L_2}{A_2 E_2}\right) = \frac{N L_1}{A_1 E_1}\left(1 + \frac{L_2}{L_1}\frac{A_1}{A_2}\frac{E_1}{E_2}\right)$

Plane Stress Analysis In case the body force $F_z = 0$, then $\sigma_{zz} = \tau_{xz} = \tau_{yz} = 0$ (cf. "thin plate problem"). Thus, three equations are necessary to calculate σ_{xx}, σ_{yy}, and τ_{xy}, subject to *surface tractions*, i.e., surface forces per unit area:

$$p_x = \sigma_{xx} l + \tau_{xy} m \tag{8.17a}$$

and

$$p_y = \tau_{xy} l + \sigma_{yy} m \tag{8.17b}$$

where $l = \cos(\hat{n}, x)$ and $m = \cos(\hat{n}, y)$ are the direction cosines for the normal vector $\hat{n}$. The first two equations for σ_{xx}, σ_{yy}, and τ_{xy} are obtained from Eq. (8.5):

$$\frac{\partial \sigma_{xx}}{\partial x} + \frac{\partial \tau_{xy}}{\partial y} + f_x = 0 \tag{8.18a}$$

$$\frac{\partial \sigma_{yy}}{\partial y} + \frac{\partial \tau_{xy}}{\partial x} + f_y = 0 \tag{8.18b}$$

The third one is the equation of compatibility in terms of stress, i.e.,

$$\left(\frac{\partial^2}{\partial x^2} + \frac{\partial^2}{\partial y^2}\right)(\sigma_{xx} + \sigma_{yy}) = -\frac{1}{1-\nu}\left(\frac{\partial f_x}{\partial x} + \frac{\partial f_y}{\partial y}\right) \tag{8.19a}$$

also

$$\frac{\partial^2 \varepsilon_{xx}}{\partial y^2} + \frac{\partial^2 \varepsilon_{yy}}{\partial x^2} = \frac{\partial^2 \gamma_{xy}}{\partial x \partial y} \tag{8.19b}$$

Equations (8.19a, b) can be derived from the compatibility condition for strain components which directly relate to the displacements u and v.

When the body forces f_x and f_y are zero, the task of evaluating planar stress, strain, and displacement subject to Eqs. (8.17a, b) is less challenging. Similar to the stream function $\Psi(x, y)$ satisfying the continuity condition, $\nabla \cdot \vec{v} = 0$, automatically, a *stress function* $\Phi(x, y)$ is introduced where

$$\sigma_{xx} \equiv \frac{\partial^2 \Phi}{\partial y^2}, \; \sigma_{yy} \equiv \frac{\partial^2 \Phi}{\partial x^2}, \; \text{and} \; \tau_{xy} \equiv -\frac{\partial^2 \Phi}{\partial x \partial y}$$

so that Eqs. (8.18a, b) balance, and Eq. (8.19a) yields:

$$\frac{\partial^2 \Phi}{\partial x^4} + \frac{2\partial^4 \Phi}{\partial x^2 \partial y^2} + \frac{\partial^4 \Phi}{\partial y^4} = 0 \tag{8.20}$$

Traditionally this biharmonic equation (8.20) has been approximately solved, subject to (8.17a, b), with polynomial functions of various degrees (see Ugural & Fenster 2003).

8.3 Slender-Body Dynamics

Structure vibrations caused by fluid flow may lead to noise, material fatigue, system damage or even total destruction. However, controlled fluid-induced vibration can be desirable as with music using wind instruments, as well as for enhanced mixing or particle sorting.

A body, subject to a flow field, experiences surface pressures and stresses, resulting in surface forces such as drag and lift (see Sect. 5.3). In turn, the body may not only moves, but also deforms under these exerted forces. Such body deflections change the flow boundary conditions and hence the flow field, which again alters the surface forces. Thus, fluid-dynamic forces and structural motions may interact significantly. In any case, of interest are ***flow instabilities*** such as vortex-sheding from a fixed body (see Fig. 8.5a) and elastic instabilities, i.e., oscillating structures, as well as the coupled fluid–structure interactions of both (see Fig. 8.5b). The fundamental aspects of fluid-flow and elastic-body phenomena are important for the safe design of tall buildings, suspension bridges, towers, offshore structures, piping systems and power lines on a large scale, as well as medical implants, such as stent-grafts, and MEMS for cooling or drug delivery on a small scale.

Classification of flow-induced vibration has been done in terms of the sources of excitation, according to the nature of the vibration, or via the type of fluid flow, i.e., steady or unsteady (see Païdoussis 1998). Specifically, steady flows may cause instabilities in terms of self-excited vibrations and vortex-induced vibrations. An example of self-excited oscillations (or flutter) is the thrashing, shaking motion of a garden-hose when dropped on the ground. Transient flows causing structure vibrations include random/chaotic (i.e., turbulent) flows, as well as wavy and oscillatory flows. A commonly observed phenomenon of internal oscillatory flow is the water hammer, a quite noisy occurrence of pressure waves in pipe networks after sudden valve opening.

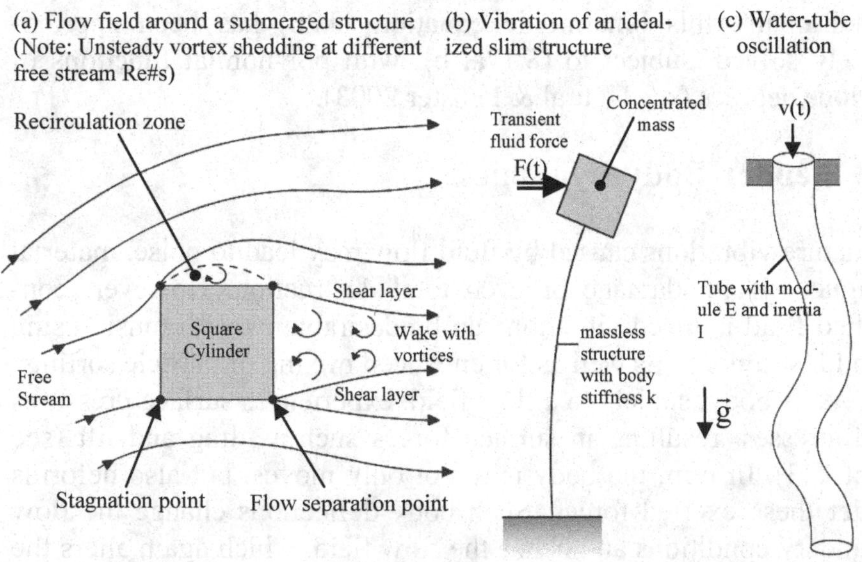

Fig. 8.5 Example of fluid–structure transients: (a) Flow field around a square cylinder submerged in an angled free stream; (b) 1-D cantilever vibration; and (c) oscillation of an elastic tube caused by fluid flow

Flow-Induced Slender-Body Oscillations Consider planar isothermal fluid flow ($u_{mean} \equiv U$) in a flexible pipe (elasticity-module E and mass-moment of inertia I) with lateral deflection $w(x, t)$ of small amplitude and long wavelength compared to the pipe diameter (see Fig. 8.5c). For a representative elementary volume $A\Delta s$ (Fig. 8.6) with accelerating fluid mass m_f, the following forces have to be taken into account: net pressure, gravity and fluid–pipe interaction forces, i.e., tangential (wall shear stress) and normal (reaction) forces. Thus, applying Newton's second law in the x- and z-direction where $\partial s \approx \partial x$ and the deflections w are small, we have per unit length Δs:

$$-A\frac{\partial p}{\partial x} - \tau_w S + m_f g + R\frac{\partial w}{\partial x} = m_f a_{f,x} \tag{8.21a}$$

and

$$-R - A\frac{\partial}{\partial x}\left(p\frac{\partial w}{\partial x}\right) - \tau_w S\frac{\partial w}{\partial x} = m_f a_{f,z} \tag{8.21b}$$

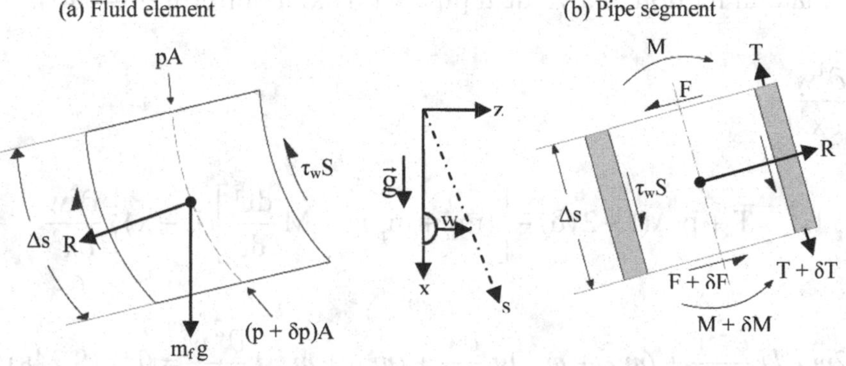

Fig. 8.6 Forces on accelerating fluid element and coupled pipe segment

Here, A is the cross sectional area, p is the gage pressure, τ_w is the wall shear stress, S is the wall surface, R is the reaction force and m is the mass, where subscript f denotes fluid and p indicates pipe. The pipe segment is subject to longitudinal tension T and a transverse shear force F, i.e.,

$$F = \partial M/\partial x \approx EI \frac{\partial^3 w}{\partial x^3}. \quad (8.22)$$

Thus, following Païdoussis (1998):

$$\frac{\partial T}{\partial x} + \tau_w S - m_p g - R\frac{\partial w}{\partial x} = m_p a_{p,x} \approx 0 \quad (8.23a)$$

and

$$\frac{\partial F}{\partial x} + R + \frac{\partial}{\partial x}\left(T\frac{\partial w}{\partial x}\right) + \tau_w S\frac{\partial w}{\partial x} = m_p a_{p,z} \quad (8.23b)$$

Recalling that $a_p = \frac{\partial^2 w}{\partial t^2}$, while $a_{f,x} = \frac{dU}{dt}$, and $a_{f,z} \approx \left[\frac{\partial}{\partial t} + U\frac{\partial}{\partial x}\right]^2 w$, substituting Eq. (8.22), $F \approx EI\frac{\partial^3 w}{\partial x^3}$, using Eqs. (8.21a, b) to eliminate *R*, and combining terms yields a PDE for small lateral motion of a vertical pipe with axial uniform flow *U*(*t*):

$$EI\frac{\partial^4 w}{\partial x^4} +$$

$$\left\{m_f U^2 - \overline{T} + \overline{p}A(1-2\nu\delta) - \left[(m_f + m_p)g - M\frac{dU}{dt}\right](L-x)\right\}\frac{\partial^2 w}{\partial x^2}$$

$$+ 2m_f U\frac{\partial^2 w}{\partial x \partial t} + (m_f + m_p)g\frac{\partial w}{\partial x} + (m_f + m_p)\frac{\partial^2 w}{\partial t^2} = 0 \quad (8.24a)$$

For Eq. (8.24a) averaged tension and pressure over pipe length L were assumed, while ν is the Poisson ratio and δ = 0 implies no pipe constraint to axial motion at *x* = *L*. Neglecting gravity, pressure, and wall tension effects and assuming U = constant, Eq. (8.24a) reduces to:

$$\underbrace{EI\frac{\partial^4 w}{\partial x^4}}_{\substack{\text{Flexural}\\ \text{restoring force}}} + \underbrace{m_f U^2 \frac{\partial^2 w}{\partial x^2}}_{\text{Centrifugal force}} + \underbrace{2m_f U \frac{\partial^2 w}{\partial x \partial t}}_{\text{Coriolis force}}$$

$$+ \underbrace{(m_f + m_p)\frac{\partial^2 w}{\partial t^2}}_{\text{Inertial force}} = 0 \qquad (8.24b)$$

Clearly, Eq. (8.24b) subject to appropriate initial/boundary conditions has to be solved numerically for $w(x, t)$, representing "free vibration" (see Sect. 8.4).

Flow-Induced Concentrated-Mass Oscillations In the previous section we considered the lateral small-scale deflection, $w = w(x, t)$, of a flexible pipe, induced by transient or steady axial flow. The describing fourth-order PDE for $w(x, t)$ is essentially the same as for an oscillating cantilever. In fact, a cantilevered tube conveying fluid at sufficiently high velocities may exhibit flutter of the single-mode type (see Païdoussis 1998; among others), as seen with wings and beams.

In this section we focus on a (vertical) elastic cantilever, e.g., a tower or high-rise, where the mass of the structure is concentrated at the end point, and the displacement $x(t)$ is caused by a transient load $F(t)$, say, a gusty wind. Figure 8.7 depicts this single-degree-of-freedom system (see also Fig. 8.5b). Specifically, in response to the excitation force $F(t)$, the body mass m accelerates, $\ddot{x}$, and moves, $x(t)$, while it is opposed to a restoring force $R = kx$ and slowed down by a damping force $D = c\dot{x}$. Thus, Newton's second law of motion, $\sum \vec{F}_{ext} = m\vec{a}$, describes the basic mass-spring-damper system in the form:

$$F(t) - kx - c\dot{x} = m\ddot{x} \qquad (8.25)$$

Here for linear systems k (the "spring" constant or body stiffness) and c (the damping coefficient combining internal body friction and external fluid resistance) are independent of the displacement.

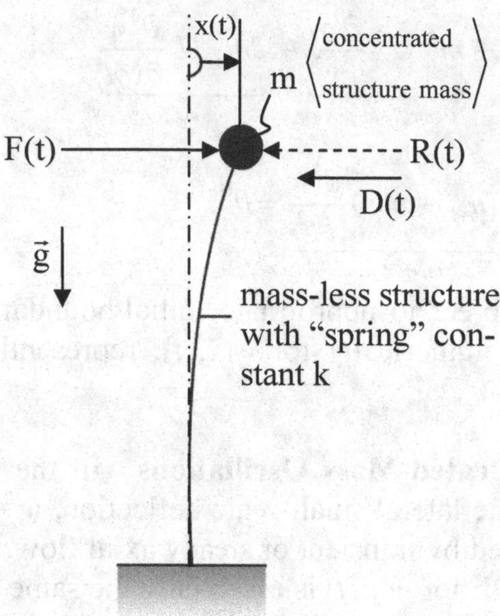

Fig. 8.7 Mass-spring-damper model representing slender elastic body dynamics

As shown in the next section, Eq. (8.25) is commonly written in more "vibrational terms" as:

$$\ddot{x} + 2\kappa\, \omega_N\, \dot{x} + \omega_N^2\, x = \frac{F(t)}{m} \tag{8.26}$$

where $\kappa \equiv \frac{c}{2\sqrt{km}}$ is the damping ratio and $\omega_N \equiv 2\pi f_N = \sqrt{\frac{k}{m}}$ is the system's natural angular velocity (or circular frequency). For $\kappa = 1$, free motion of the system is non-oscillatory; but, when $\kappa \ll 1$ and $\omega_{forced} = \omega_N$ (or $F_{forced} = F_{natural}$) resonance oscillation may occur, leading potentially to system destruction.

Example 8.4 summarizes a vibrating slender-body application, illustrating a solution to Eq. (8.26). Section 8.4 then provides in more detail the theoretical background for both translator and rotational vibration, induced by unbalanced forces.

Example 8.4: Slender-Body Oscillations Caused by a Harmonic Load

Consider an excitation function of the form $F(t) = F_0 \cos \omega t$; although, in reality $F(t)$ is non-harmonic when structures are subject to, say, turbulent flow or vortex shedding.

Sketch	*Model*	*Assumption*	*Concept*
F(t), m, R, D, k, c	F(t), m, k, c, x(t)	• Linear, single-degree of freedom system	Newton's second law of motion

Solution: Equation (8.26) with $F(t) = F_0 \cos \omega t$, $\omega = 2\pi f$, has the solution (see Polyanin & Zaitsex 1995; p. 130):

$$x(t) = F_0\, H(\omega) \cos(\omega t - \varphi) \tag{E. 8.4}$$

where

$$H(\omega) = \left\{ m^2\, \omega_N^4 \left[\left(1 - \left(\frac{\omega}{\omega_N} \right)^2 \right)^2 + 4\kappa^2 \left(\frac{\omega}{\omega_N} \right)^2 \right] \right\}^{-1/2} \tag{E. 8.5}$$

where the damping ratio

$$\kappa \equiv \frac{c}{2\sqrt{mk}} \tag{E. 8.6}$$

and

$$\varphi = \tan^{-1} \left[\frac{2\kappa \left(\frac{\omega}{\omega_N} \right)}{1 - \left(\frac{\omega}{\omega_N} \right)^2} \right] \tag{E. 8.7}$$

Graphs

(a) Resonance example with damping (b) Forced oscillation with damping effects

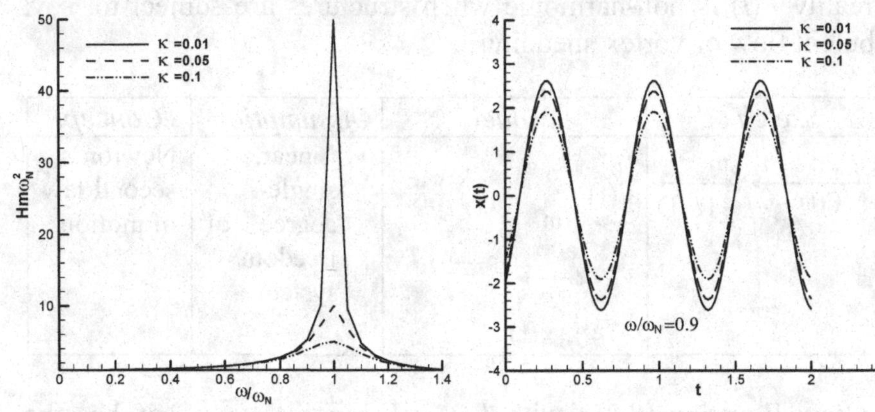

Comments

- As expected, the amplitude of the resulting forced vibration spikes at $\omega = \omega_{natural}$ with values greatly dependent on $\kappa \sim c$, the damping coefficient (Graph (a)).
- Forced oscillation are strong when ω is near $\omega_{natural}$. Again, $\kappa \sim c$ has a measurable effect on the amplitude F_0H (see Eq. (E8.4) and Graph (b)).

8.4 Flow-Induced Vibration

Dynamics, i.e., body (or fluid) motion caused by forces, encompasses: (i) kinematics, i.e., the study of motion disregarding the forces; and (ii) kinetics, i.e., the relationship between forces and the resulting motion. As indicated in the last part of Sect. 8.3, the dynamic response to forced vibration (or oscillation) of a linear, single-degree-of-freedom system is the solution to:

$$m\ddot{x} + c\dot{x} + kx = F(t) \quad \text{or} \quad I\ddot{\theta} + c\dot{\theta} + k\theta = T(t) \qquad (8.27a,\ b)$$

where *F*(*t*) in Eq. (8.27a) can be any transient input, such as a harmonic forcing function $F(t) = F_0 \sin \omega t$ or displacement *x*(*t*), while *T*(*t*) is a moment or torque. As mentioned, such excitations may be caused by gusty winds, unbalanced rotary parts, earth-quakes, or uneven surfaces. The homogeneous solutions of Eqs. (8.27a, b), i.e., $F(t) = T(t) = 0$, is known as the free response (or free vibration), triggered solely by the initial conditions. The particular solutions of Eqs. (8.27 a, b) have the form of the forcing function plus its derivative to match the second-order ODE.

Numerous applications of Eqs. (8.27a, b) and their solutions are discussed in mechanical-vibration texts (e.g., Palm 2007). In the framework of fluid–structure interactions, of special interest here are the different forms of the damping coefficient *c*, as related to viscous flow. The (linear) spring constants, *k*, are given in strength-of-material books or deduced from experiments (see Roark 2001). In general,

$$k = \begin{cases} dF/dx & \text{for force – deflection elements} \\ dT/d\theta & \text{for torque – angular – twist elements} \end{cases} \qquad (8.28a,\ b)$$

Damping is the fluid resistance force or torque, i.e.,

$$F = c\,\bar{v} \quad \text{or} \quad T = c\,\omega \qquad (8.29a,\ b)$$

where *c* is the damping coefficient, $\bar{v}$ is the mean velocity of the moving part, and $\omega = 2\pi f$ is the angular velocity in radians per unit time, while the frequency *f* is in cycles per unit time, i.e., in hertz [Hz].

Figure 8.8 summarizes the two basic damping elements, i.e., translational and rotational (or torsional) dampers. Oscillating machine-part lubrication, shock absorbers and door closers are just a few sample applications.

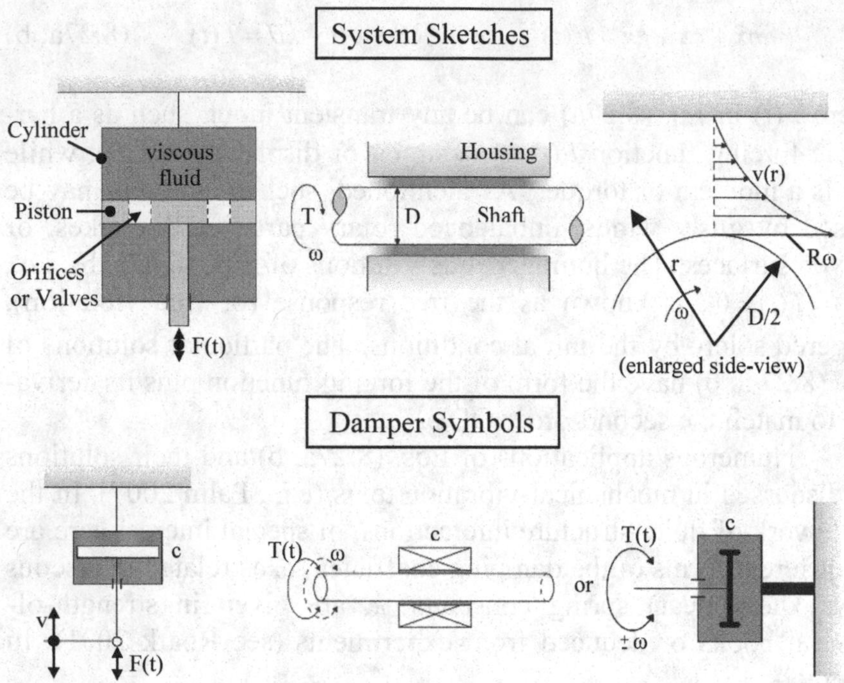

Fig. 8.8 Basic viscous flow systems and associated damper/dashpot symbols

---

Example 8.5: Derivations of Damping Coefficients

Correlations for $c = F/\bar{v}$ are derived, considering translatory and rotational Couette flows with approximately linear velocity profiles, as well as a basic piston-orifice-cylinder system

(A) Simple Couette Flow (planar plate lubrication):

Sketch	*Concepts*
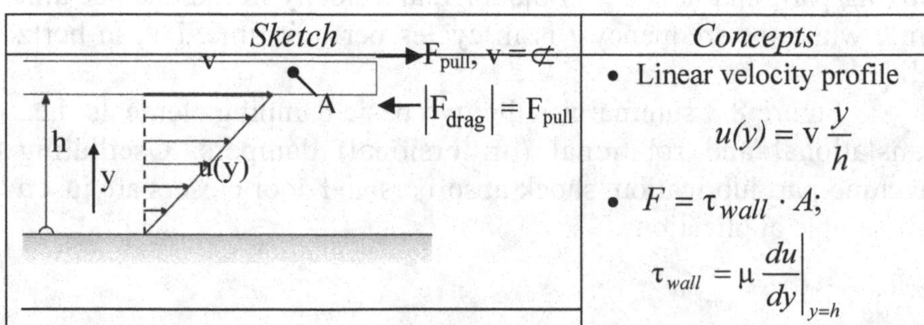	• Linear velocity profile $u(y) = v\,\dfrac{y}{h}$ • $F = \tau_{wall} \cdot A$; $\tau_{wall} = \mu \left.\dfrac{du}{dy}\right\rvert_{y=h}$

Solution:

$$c = \frac{F}{\overline{\mathrm{v}}}$$

$$F = \tau_w A = \mu \left.\frac{du}{dy}\right|_{y=h} \cdot A = \mu A \frac{\mathrm{v}}{h}$$

Hence, with $\mathrm{v} \equiv \overline{\mathrm{v}}$:

$$\boxed{c = \frac{F}{\overline{\mathrm{v}}} = \frac{\mu A}{h}}$$

(B) Cylindrical Couette Flow (Journal Bearing)

Sketch	*Concepts*
v_r, ω, r, R, h	• Shaft (R, L) and housing (R + h) forming a small gap, so that $\mathrm{v}(r) \approx \mathrm{v}_R \frac{r}{h}$ • Torque $T = F_{drag} R = A \tau_w R$

Solution:

$$\mathrm{c} = \frac{\mathrm{T}}{\omega}$$

$$\mathrm{T} = \mathrm{A}\,\tau_w\,\mathrm{R} = (2\pi\,\mathrm{L}\,\mathrm{R})\left(\mu \left.\frac{d\mathrm{v}_\theta}{dr}\right|_{r=R}\right)\mathrm{R}$$

$$\tau_w = \mu \left.\frac{du}{dr}\right|_{r=R} = \mu \frac{\mathrm{v}_R}{h}; \quad \mathrm{v}_R = \omega R$$

Thus,

$$\mathrm{T} = \frac{2\pi\,\mu L\,R^3}{h}\,\omega$$

and hence

$$c = \frac{T}{\omega} = 2\pi\,\mu\,R^3\,\frac{L}{h}$$

(C) Piston-orifice-cylinder damper (part of suspension system)

Sketch	*Concepts*
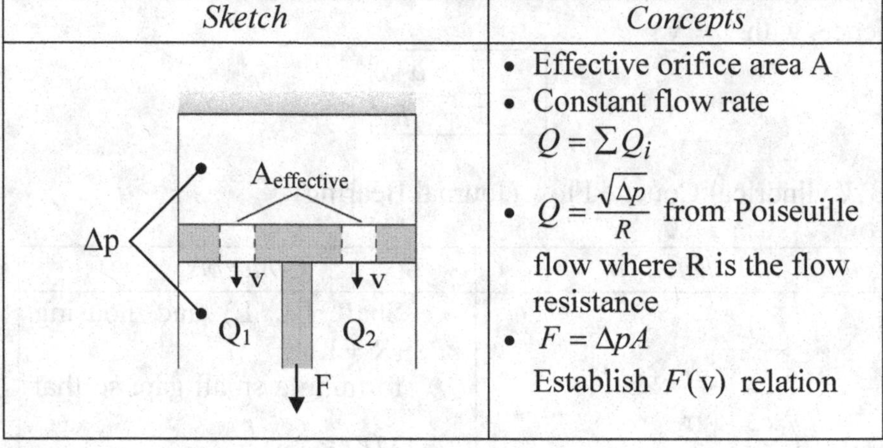	• Effective orifice area A • Constant flow rate $Q = \Sigma Q_i$ • $Q = \frac{\sqrt{\Delta p}}{R}$ from Poiseuille flow where R is the flow resistance • $F = \Delta p A$ Establish $F(\mathrm{v})$ relation

Solution:

$$F = \Delta\,p\,A = (QR)^2\,A = (v\,A\,R)^2\,A$$

$$F = R^2\,A^3\,v^2$$

Hence,

$$\boxed{c = \frac{F}{v} = R^2\,A^3\,v}$$

Now, with the fluid–structure damping coefficient illuminated and "spring constant" k given based on structure properties (see Roak 2001; among others), Eqs. (8.27a, b) can be solved step-by-step, starting with the free-response case, i.e., when $F(t)$ or $T(t)$ are zero.

8.4.1 Harmonic Response to Free Vibration

Undamped Vibration Case Clearly, when there is zero (or negligible) damping the system oscillates at its natural frequency, i.e., the case of *free undamped vibration*. Equations (8.27a, b) reduce to:

$$m\ddot{x} + kx = 0 \quad \text{or} \quad I\ddot{\theta} + k\theta = 0 \tag{8.30a, b}$$

where the translatory mass m is equivalent to the mass moment of inertia I. With the trial solution x = A sin ωt, $\ddot{x} = -\omega^2 A \sin \omega t$ so that Eq. (8.30a) reads:

$$(-m\,\omega^2 + k)\left(A \sin \omega t\right) = 0$$

This implies,

$$-m\,\omega^2 + k = 0$$

Hence, the frequency of the free response is the natural frequency:

$$\omega = \omega_N = \begin{cases} \sqrt{k/m} & \text{for linear oscillations} \\ \sqrt{k/I} & \text{for rotational oscillations} \end{cases}$$

Damped Vibration Case 1-D free vibration with damping is described by:

$$m\ddot{x} + c\dot{x} + kx = 0 \quad \text{or} \quad I\ddot{\theta} + c\dot{\theta} + k\theta = 0 \tag{8.31a, b}$$

With the trial solution to, say, Eq. (8.31a) $x = A\,e^{st}$, the characteristic equation reads:

$$m\,s^2 + cs + k = 0 \tag{8.32a}$$

and hence,

$$s_{1,2} = \frac{-c \pm \sqrt{c^2 - 4mk}}{2m} \tag{8.32b}$$

Equation (8.32b) implies three distinct cases depending upon the character of the roots $s_{1,2}$, i.e., $s_1 = s_2$ real, $s_1 \neq s_2$ real, or $s_{1,2}$ complex.

Specifically, Eq. (8.32b) implies:

(i) Critical damping when $c^2 - 4km = 0$, i.e.,

$$c_{crit} = 2\sqrt{km} \tag{8.33}$$

(ii) Overdamping when $c > c_{crit}$ < real roots >

(iii) Underdamping when $c < c_{crit}$ < complex roots >

The three cases are summarized via the damping ratio which is defined as:

$$\kappa \equiv \frac{c}{c_{crit}} \begin{cases} = 1 \text{..........critically damped system} \\ > 1 \text{...........exponential behavior occurs} \\ < 1 \text{.......... oscillation occurs} \end{cases} \tag{8.34a–c}$$

Now, with $\kappa = \frac{c}{2\sqrt{km}}$ and $\omega_N = \sqrt{\frac{k}{m}}$ (see Eq. (8.26)), $c = 2\kappa\sqrt{km} = 2\kappa\, m\, \omega_N$ and Eq. (8.32b) can be rewritten as ($i \equiv \sqrt{-1}$):

$$s = \frac{-2\kappa\, m\, \omega_N \pm i\, 2\sqrt{km}\sqrt{1 - \left(\frac{c}{2\sqrt{km}}\right)^2}}{2m} \tag{8.35}$$

Hence,

$$s = -\kappa\, \omega_N \pm i\, \omega_N \sqrt{1 - \kappa^2}$$

or

$$s_{1,2} = a \pm ib$$

The inverse of the first term is a time constant, i.e., $\tau = a^{-1}$ (see the trial solution $x = Ae^{st}$) and the second term is the system's damped (natural) frequency:

$$\omega_D = \omega_N \sqrt{1-\kappa^2} \tag{8.36}$$

Clearly, we always have $\omega_D < \omega_N$, which implies that only the underdamped case with $\kappa < 1$ lends physical meaning to these frequencies ω_D and ω_N. With the two roots of the characteristic equation, the general solution is a linear combination of the postulate $x = Ae^{st}$, i.e.,

$$x = A_1\, e^{s_1 t} + A_2\, e^{s_2 t}$$

With the initial conditions $x(t=0) = x_0$ and $\dot{x}(t=0) = \mathrm{v}_0$, the coefficients are:

$$A_1 = \frac{\mathrm{v}_0 - s_2\, x_0}{s_1 - s_2} \quad \text{and} \quad A_2 = x_0 - A_1$$

For the underdamped case, $\kappa < 1$, $s_1 = a + ib$ and $s_2 = a - ib$, so that

$$x = A_1\, e^{(a+ib)t} + A_2\, e^{(a-ib)t} = e^{at}\,(A_1 e^{ibt} + A_2 e^{-ibt})$$

or in equivalent form, employing Euler's identities (see Sect. A.2.4 of App. A):

$$x = B\, e^{at}\, \sin(bt + \phi) \tag{8.37}$$

where B is the amplitude and ϕ is the phase angle. Applying the ICs, $x(t=0) \equiv x_0 = B \sin \phi$ and $\dot{x}(t=0) \equiv v_0 = a\, B \sin \phi + b\, B \cos \phi$, yields:

$$\sin \phi = \frac{x_0}{B} \quad \text{and} \quad \cos \phi = \frac{v_0 - a\, x_0}{b\, B}$$

so that the phase angle

$$\phi = \tan^{-1}\left(\frac{b\, x_0}{v_0 - a\, x_0}\right) \tag{8.38}$$

while the amplitude B can be deduced via the identity $\sin^2 \phi + \cos^2 \phi = 1$. The free response of the mass-spring-damper system to the ICs $x = x_0$ and $\dot{x} = v_0$ at $t = 0$ is shown in Fig. 8.9. Here, $a \equiv -\kappa\, \omega_N < 0$ and $b \equiv \omega_N$ in radians per unit time is the natural frequency. Clearly, due to the damping effect ($\kappa \neq 0$), the amplitude of the oscillations decay exponentially to zero and the system mass gains a stable equilibrium.

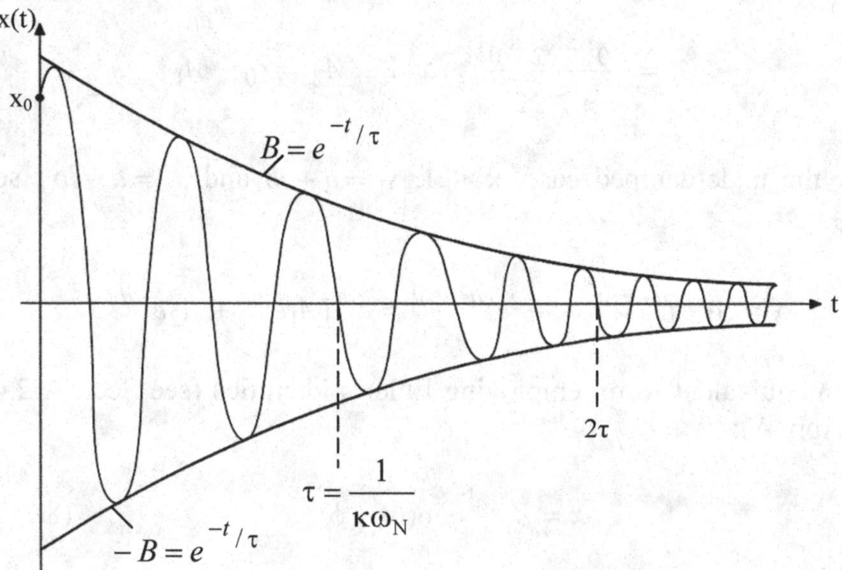

Fig. 8.9 Free response oscillations of an underdamped case

Example 8.6: Torsion Viscometer

Consider a thin disk of diameter D and inertia I, suspended by a torsion wire of "spring constant" k, oscillating in a viscous fluid layer of height h and damping coefficient c. Develop an equation for the fluid viscosity $\mu = \mu(D, h, I, \omega_N, \text{and } \omega)$.

Sketch	*Assumptions*	*Concepts*
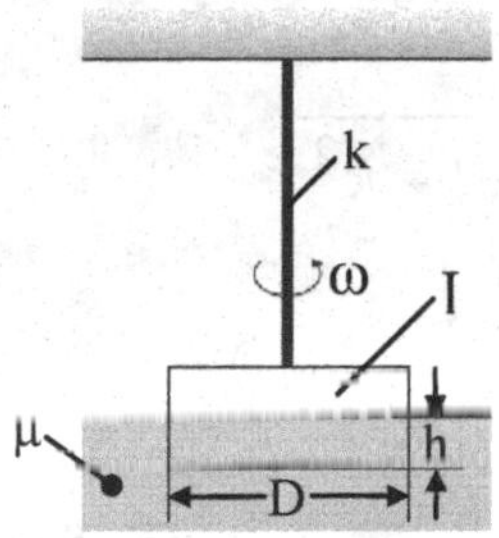 k; ω; I; h; μ; D	Free response, single-degree-of-freedom mass-damper-spring system for rotating disk	• Newton's second law of motion • Rotational damper $T = c\omega$

Solution: The equation of motion of the "freely" oscillating disk is (see Eq. (8.27b)):

$$I\ddot{\theta} + c\dot{\theta} + k\theta = 0 \tag{E. 8.6.1}$$

where I is the disk inertia, c is the damping coefficient, and k is the wire (spring) constant.

Similar to Example 8.1B, the damping constant can be derived as (see also Von Karman's "rotating disk in an infinite fluid reservoir" in, for example, Kleinstreuer 1997):

$$c = \frac{\mu \pi D^4}{32h} \tag{E. 8.6.2a}$$

or

$$\mu = 32hc/(\pi D^4) \tag{E. 8.6.2b}$$

With the postulate $\theta = A\,e^{st}$, the characteristic equation of (E. 8.6.1) reads:

$$Is^2 + cs + k = 0 \qquad \text{(E. 8.6.3a)}$$

and hence

$$s_{1,2} = \frac{-c \pm \sqrt{c^2 - 4kI}}{2I} \qquad \text{(E. 8.6.3b)}$$

As shown (see Eqs. (8.33) and (8.34)):

$$\kappa = \frac{c}{2\sqrt{kI}} \quad \text{and} \quad \omega_D = \omega_N \sqrt{1 - \kappa^2}$$

or

$$\kappa = \sqrt{1 - \left(\frac{\omega_D}{\omega_N}\right)^2} = \frac{c}{2\sqrt{kI}} \qquad \text{(E. 8.6.4a)}$$

so that

$$c = 2\sqrt{kI}\,\sqrt{1 - \left(\frac{\omega_D}{\omega_N}\right)^2} \qquad \text{(E. 8.6.4b)}$$

With $k = I\,\omega_N^2$,

$$c = 2\,I\,\omega_N \sqrt{1 - \left(\frac{\omega_D}{\omega_N}\right)^2} \qquad \text{(E. 8.6.5)}$$

and finally with Eq. (E8.6.2b):

$$\mu = \frac{64hI}{\pi D^4}\,\omega_N \sqrt{1 - \left(\frac{\omega_D}{\omega_N}\right)^2} \qquad \text{(E. 8.6.6)}$$

Example 8.7: Free Response to a Linear Mass-Damper-Spring System $2\ddot{x} + 6\dot{x} + 18x = 0$ Based on Initial Body Displacement $x(0) = 2$

Sketch	*Assumptions*	*Concepts*
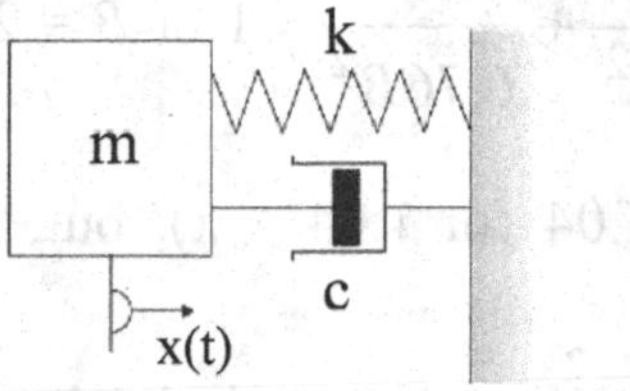 k, m, c, x(t)	As stated 1-D, under-amped system	• Newton's second law of motion parameter identification via equation comparison

Solution:

- Compare given ODE with Eq. (8.27a), i.e., $m = 2$, $c = 6$, and $k = 18$. Hence, $\kappa = \frac{c}{2\sqrt{km}} = 0.5$, i.e., underdamped, $\omega_N = \sqrt{k/m} = 3$ rad/s, and $\omega_D = \omega_N\sqrt{1-\kappa^2} = 2.6$ rad/s

- The roots of the characteristic equation

$$2s^2 + 6s + 18 = 0$$

are

$$s_{1,2} = -1.5 \pm 2.6i = a \pm bi$$

which implies: $\tau = a^{-1} = 2/3s$; $\omega = b = 2.6\ rad/\sec$; and period $p = \frac{2\pi}{\omega_N} = 2.42s$

- Hence, in comparison with Eq. (8.37):

$$x(t) = Be^{-1.5t}\sin(2.6t + \phi)$$

where the parameters B and ϕ depend on the initial conditions.

- ICs: $x(t=0)=2$ and $\dot{x}(t=0)=0$, so that $\sin\phi = \dfrac{x_0}{B} = \dfrac{2}{B}$ and $\cos\ \phi = \dfrac{v_0 - a\ x_0}{b\ B} = \dfrac{3}{2.6\ B}$. Employing the identity $\sin^2\phi + \cos^2\phi = 1$, we have $\dfrac{4}{B^2} + \dfrac{9}{6.76B^2} = 1$ or $B = 2.31$. Hence, $\phi = \tan^{-1}\left(\dfrac{bx_0}{v_0 - ax_0}\right) = 1.04$ (or $1.04 + \pi$); but, with $\sin\phi = \dfrac{2}{2.31} > 0$ and $\cos\phi = \dfrac{3}{2.6 - 2.31} > 0$, ϕ is in the first quadrant.

- Finally,

$$\boxed{x(t) = 2.31e^{-1.5t}\ \sin(2.6t + 1.04)}$$

Graph:

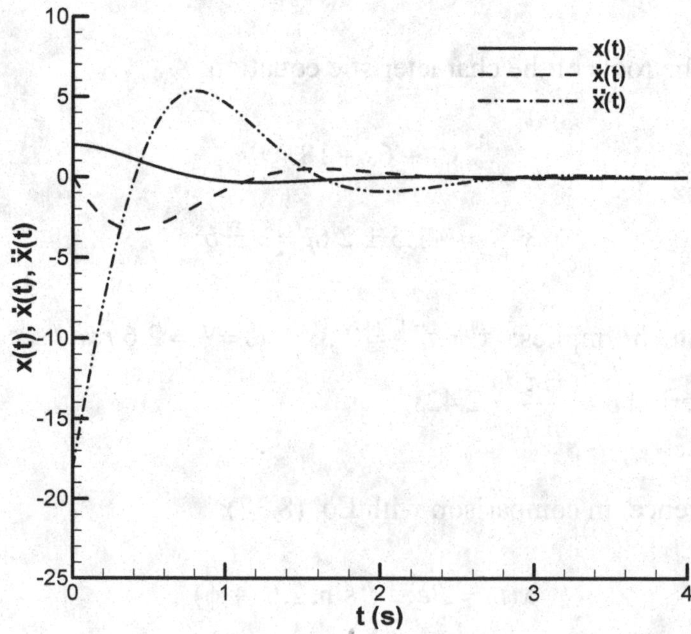

Comments:

- For the given system values, say $m = 2\,kg$, $k = 18\,kg/s^2$ and $c = 6 \times 10^2\, N \cdot s / cm$, the initial displacement ($x_0 = x(t = 0) = 2\,cm$) is quickly damped out, i.e., after $t \approx 2\,s$. That causes within $0 \le t \le 1.4\,s$ rather large mass velocities $\dot{x}(t)$ and accelerations $\ddot{x}(t)$.

8.4.2 Harmonic Response to Forced Vibration

Free vibration is a one-way fluid–structure coupling where a mass-spring system, set into oscillatory motion via initial conditions, is damped due to the interaction with a viscous fluid. If in addition to the initial conditions $x(t = 0) = x_0$ and $\dot{x}(t = 0) = v_0$, the system is subjected to a forcing function, say $F(t) = F_0 \sin \omega t$ caused by a periodic stream, we have a *two-way coupled* fluid–structure interaction case:

$$m\ddot{x} + c\dot{x} + kx = F_0 \sin \omega t \tag{8.39}$$

The trial solution reads:

$$x(t) = \underbrace{Ae^{at} \sin (bt + \phi)}_{\text{free response}} + \underbrace{B \sin \omega t + C \cos \omega t}_{\text{forced response plus derivation}} \tag{8.40}$$

For the underdamped case, we have two complex conjugate roots $s_1 = a + ib$ and $s_2 = a - ib$, where

$$a = -\frac{c}{2m} \quad \text{and} \quad b = \frac{\sqrt{4mk - c^2}}{2m} \tag{8.41a, b}$$

implying that $4mk - c^2 > 0$. As before, the phase angle ϕ depends on the ICs.

With the ICs $x(0) = \dot{x}(0) = 0$, we obtain the forced response as:

$$x(t) = \frac{F_0}{(k - m\omega^2)^2 + (c\omega)^2}\left\{e^{at}\left[-\frac{\omega}{b}(ac + k - m\omega^2)\sin bt + c\omega\cos bt\right] + \left[(k - m\omega^2)\ \sin\omega t - c\omega\cos\omega t\right]\right\} \quad (8.42)$$

With $a \equiv \tau^{-1} = -\frac{c}{2m} = -\kappa\,\omega_N < 0$, the terms multiplied by e^{at} vanish with time, i.e., the first bracket is the disappearing transient response. What remains with time are the sin ωt and cos ωt terms, known as the "steady-state" response:

$$x_{ss}(t) = \frac{F_0}{(k - m\omega^2)^2 + (c\omega)^2}[(k - m\omega^2)\sin\omega t - c\omega\cos\omega t] \quad (8.43)$$

Example 8.8: Harmonic Response of an Underdamped System

Consider the forced response to the mass-damper-spring system described by

$$\ddot{x} + \dot{x} + x = \sin\omega t \quad (E.8.8.1)$$

for the two imposed frequencies $\omega_1 = 1$ and $\omega_2 = 2$.

Sketch	*Assumptions*	*Concepts*
m c k	• Linear, single DoF, second-order system • Zero initial conditions • One-way coupled FSI	• Identification of coefficients m, c, k, and F_0 by comparison with Eqs. (8.39) and (8.42) • Solution $x(t)$ given by Eq. (8.42)

Solution:

- Comparison of the given ODE with Eq. (8.39) yields:

  $m = c = k = F_0 = 1$, so that for both cases:

  $\kappa = \dfrac{c}{2\sqrt{km}} = 0.5$, i.e., we have an underdamped system.

- The roots are $s_{1,2} = a \pm bi$, where

  $$a = -\frac{c}{2m} = -0.5\ \text{s}^{-1} \equiv \tau^{-1}$$

  $$\text{and } b = \frac{\sqrt{4mk - c^2}}{2m} = \frac{\sqrt{3}}{2} = 0.866\ \text{rad/s} \equiv \omega_N$$

- Case I ($\omega_1 = 1$):

  Thus, in comparison with Eq. (8.42):

$$x(t) = e^{-0.5t}[0.5774 \sin(0.866t) + \cos(0.866t)] - \cos t \quad \text{(E.8.8.2)}$$

Note: With $\tau = \frac{1}{a} = -2$, $\left. e^{-\frac{t}{2}} \right|_{t=8} \approx 0$ and hence the "steady-state" response is:

$$x_{ss}(t) = -\cos t \quad \text{(E.8.8.3)}$$

- Case II ($\omega_2 = 2$):

  Again in comparison with Eq. (8.42):

$$x(t) = 0.0769\{e^{-0.5t}[8.0829\sin(0.866t) + 2\cos(0.866t)]$$

$$-[3\sin 2t + 2\cos 2t]\} \qquad \text{(E.8.8.4)}$$

Graph:

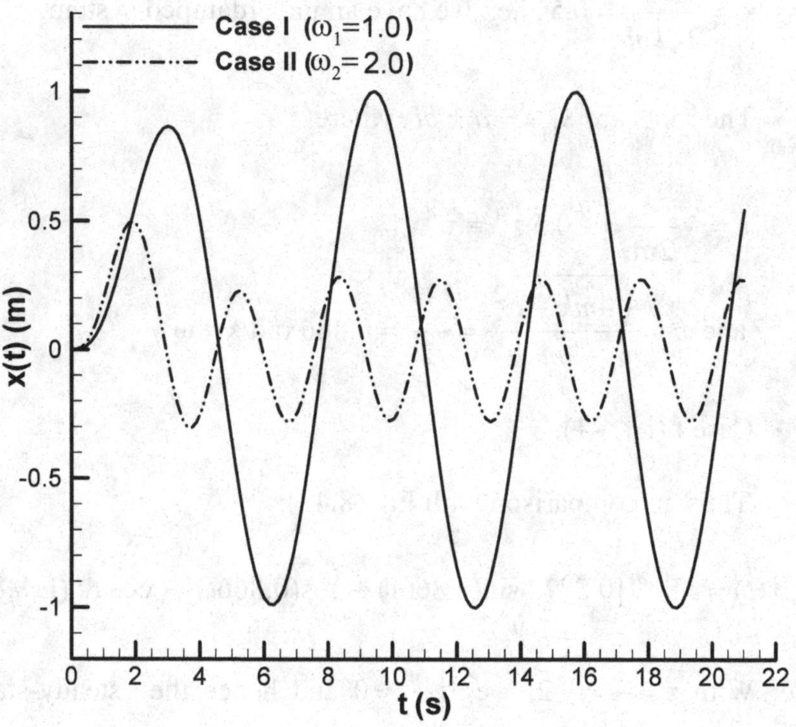

Comments:

- Case I ($\omega_1 = 1$) with forcing function sin 1t: The forced oscillation is damped during $0 \le t \le 8$ s., and then Eq. (E8.8.3), i.e., the "steady-state" oscillations take over.
- Case II ($\omega_2 = 2$) with forcing function sin 2t: At the higher frequency, the initial "overshoot" is damped and then at $t \approx 6\,s.$, $x_{ss}(t) = 0.0769(-3\sin 2t - 2\cos 2t)$ prevails.

8.5 Homework Assignments and Course Projects

Solutions to homework problems done individually or in, say, three-person groups should help to further illustrate fluid dynamics concepts as well as approaches to problem solving, and in conjunction with App. A, sharpen the reader's math skills (see Fig. 2.1). Unfortunately, there is no substantial correlation between good HSA results and fine test performances, just *vice versa*. Table 1.1 summarizes three suggestions for students to achieve a good grade in fluid dynamics – for that matter in any engineering subject. The key word is "independence", i.e., equipped with an equation sheet (see App. A), the student should be able: (i) to satisfactory answer all concept questions and (ii) to solve correctly all basic fluid dynamics problems.

8.5.1 Guideline for Project Report Writing

A project is more than a homework problem. As a starter, it typically requires a brief literature review to document the state-of-the-art for the given task. The antagonistic interplay between a rather accurate mathematical description of an assigned problem and finding a tractable solution method puts a premium on making the right assumptions and selecting a suitable equation solver, often in terms of available software and hardware.

The resulting work has to be documented in an appropriate format, i.e., *project reports* should have the following features:

- **ToC**: Table of contents with page numbers (optional for smaller reports).
- **Abstract**: Summary of the fluid dynamics problem analyzed, solution method employed, and novel results with physical insight presented.
- **Nomenclature**: Optional but recommended for complex problem solutions.
- **Introduction**: Project objectives and task justification; System sketch with state-of-the-art description, basic approach or concept, and references of the *literature review* in terms of "author (year)," e.g., White (1998).

- **Theory**: Basic assumptions, governing equations, boundary conditions, property values, ranges of dimensionless groups, closure models, and references.
- **Model Validations**: (a) Numerical accuracy checks, i.e., mesh-independence of the results as well as mass and momentum residuals less than, say, 10^{-4}; (b) Quantitative proof, e.g., comparison with an exact solution of (typically a much) simpler system, and/or experimental data comparisons, or (at least) qualitative justification that the solution is correct.
- **Results and Discussion**: Graphs with interpretations, i.e., explain physical insight (Note: number the figures and discuss them concisely); draw conclusions and provide applications.
- **Future Work.**
- **References**: List all literature cited in alphabetical order.
- **Appendices**: Lengthy derivations, computer programs, etc.

8.5.2 Text-Embedded Insight Questions and Problems

8.5.1 Why is Couette flow, although having a moving plate, is categorized as a static FSI case?

8.5.2 Prove that $\sigma_{ij} = \sigma_{ji}$, i.e., $\vec{\vec{\sigma}}$ being a symmetric tensor.

8.5.3 Show how to transform Eq. (8.1a) for $\sigma_{ij} = \sigma_{ji}$ to Eq. (8.1b) and express σ_1 to σ_3.

8.5.4 Derive Eq. (8.2) and discuss its usefulness.

8.5.5 Relate Eqs. (8.3 and 8.4) as well as Eq. (8.6) to fluid mechanics.

8.5.6 Derive Eqs. (8.12a–c) and discuss how they relate to (similar) fluid flow equations.

8.5.7 Sketch, derive and discuss the Kelvin model for viscoelastic material behavior. Graph F(t), u(t) and F(u).

8.5.8 Describe the causes of elastic tube oscillation as indicated in Fig. 8.5c. What happens when the inlet velocity v = constant?

8.5.9 Derive Eq. (8.22) as well as Eqs. (8.23a, b).

8.5.10 Derive Eq. (8.26) and discuss the system parameters κ and ω_N. Provide an FSI example which illustrates c, κ and ω_N.

8.5.11 Discuss Eqs. (8.28a, b and 8.29a, b) with FSI examples.

8.5.12 Redo Example 8.8 with higher damping, i.e., c = 1.5.

8.5.3 Projects

8.5.13 Analyze and simulate material dynamics models (e.g., the Kelvin model), representing non-Hookean material behavior of:
(i) A power-law fluid
(ii) A viscoelastic tubular wall

8.5.14 Compare computationally FSI results for flow-induced slender-body oscillation with the (simple) 1-D forced mass-spring-damper (MSD) system:
(i) Select a (nice) case study from the open literature
(ii) Duplicate the system numerically (or transfer the results)
(iii)Develop and solve the equivalent 1-D MSD system

8.5.15 Considering the "viscous flow systems" described in Fig. 8.8 as a start, develop correlations for parameters k and c (Eqs. (8.27a, b)) for a family of:
(i) Mechanical or (ii) biomedical linear and nonlinear single DoF systems. Provide a couple of illustrative examples.

8.5.16 Using Eq. (8.39), develop and solve three illustrative applications in modern fluid dynamics.

Chapter 9

Biofluid Flow and Heat Transfer

9.1 Introduction

Fluid mechanics, dissolved species transport and particle dynamics phenomena play major roles in normal and pathological processes occurring in the human body. Most evident on the macro-scale are the transport phenomena associated with blood flow supplying oxygen and nutrients to organs, and airflow in the lung enabling the $O_2 - CO_2$ gas exchange. On the micro-scale, it appears that complex particle-hemodynamics can trigger biochemical responses at the cellular level that could lead to stenosed arteries, aortic heart-valve failure, or aneurysm rupture.

Clearly, the study of biofluid mechanics relies greatly on the traditional and modern topics presented in Chaps. 1–8. It also benefits from advanced computational fluid-particle dynamics and computational fluid–structure interaction (FSI) simulations (see Chap. 10). Such (validated) results can be used to gain physical insight into complex flow phenomena to a depth simply not attainable with experiments alone. However, the ultimate goals on a patient-specific basis, i.e., an understanding of the biofluid transport processes and subsequently the development of therapeutic techniques or medical devices, pose major challenges. For example, on a micro-scale most biochemical processes area not well understood, i.e., comprehensive equations/models and accurate data sets are not established. Best numerical techniques for multi-scale problems with a broad range of

C. Kleinstreuer, *Modern Fluid Dynamics: Basic Theory and Selected Applications in Macro- and Micro-Fluidics*, Fluid Mechanics and Its Applications 87,
DOI 10.1007/978-1-4020-8670-0_9, © Springer Science+Business Media B.V. 2010

Reynolds and Stokes numbers as well as FSI phenomena are still under development. Simulating transport phenomena in complex organs, e.g., patient-specific lung airways, requires peta-scale computing which is presently even taxing for the world's fastest and largest supercomputer.

Figure 9.1 illustrates the key steps in medical device design for individual patients, e.g., stent-grafts to form new (synthetic) blood vessels or a smart inhaler for optimal drug-aerosol targeting of a lung tumor. In any case, stent-graft improvement for endovascular aneurysm repair may serve as an example. Based on tests and scans the recommended treatment for a patient is the insertion of a stent-graft, say, to repair an abdominal aortic aneurysm (see Sect. 3.2 in Kleinstreuer 2006). The image file of the patient's aorta is sent from the doctor's office to the computer lab for geometry-file conversion. In addition, mesh generation, mathematical modeling and computer simulation are accomplished. The results are validated data sets in terms of the blood velocity and pressure fields plus their derivatives, i.e., wall shear stress, wall stress and strain as well as axial forces. Such information leads to best possible implant configuration and *in vivo* placement for the particular patient by either adjusting/ improving a suitable, off-the-shelf stent-graft or starting with a new design.

When developing a *new* medical device or implant, the underlying concept/methodology as well as the design and operation of the new system can be quantitatively assessed and illustrated in virtual reality, i.e., via computational analysis and simulation. While this is an important first step, an experimental proof-of-concept has to be provided to convince people (e.g., your boss, funding agencies, etc.) that the new device actually works. In case of a costly, high-stake biomedical system, prototypes have to be built, laboratory tested FDA approved, and clinical trials conducted before the invention can be considered for manufacturing and large-scale marketing.

Biomedical system development is a very suitable research area to demonstrate the need for a modern two-phase research approach in order to be successful.

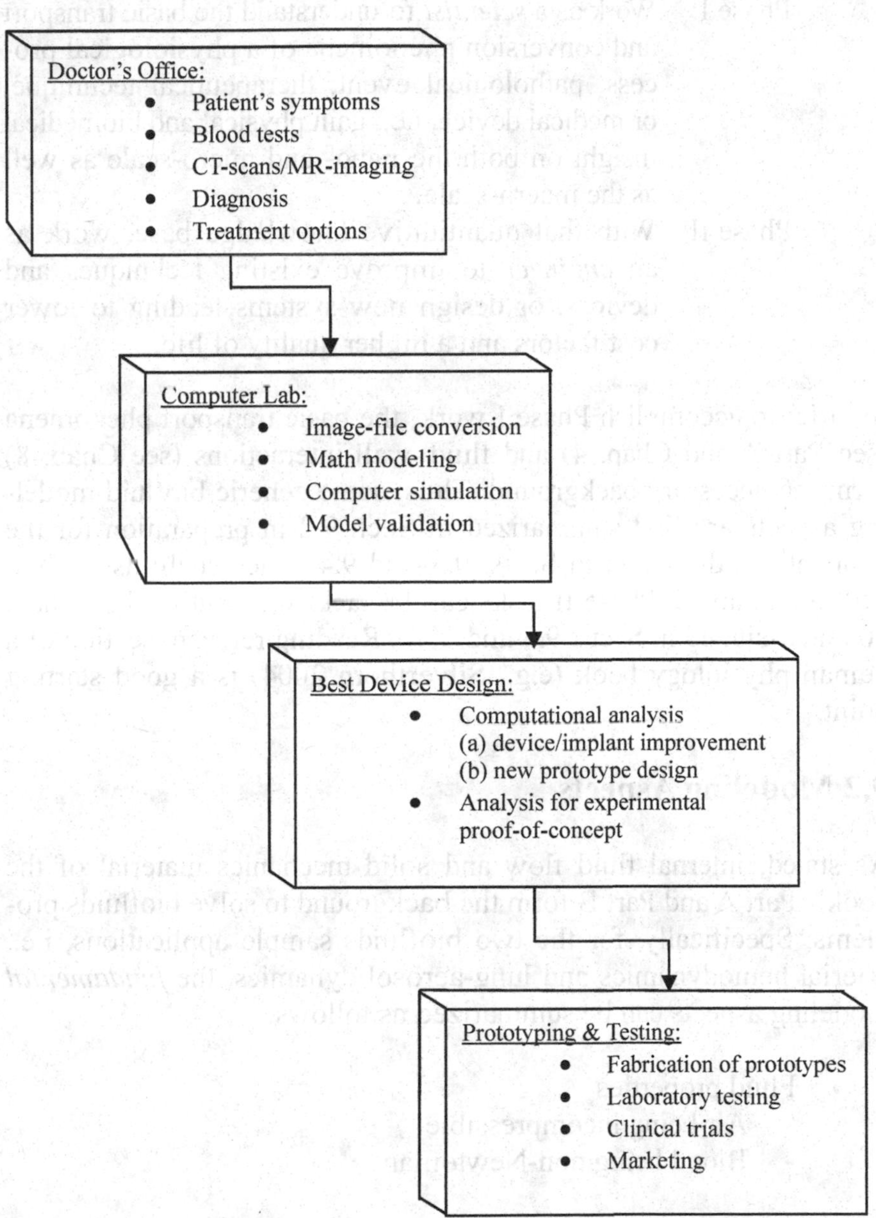

Fig. 9.1 Flow chart for new patient-specific paradigm

Phase I: Work as a *scientist* to understand the basic transport and conversion phenomena of a physiological process, pathological event, therapeutical technique, or medical device, i.e., gain physical and biomedical insight on both the nano- and micro-scale as well as the macro-scale.

Phase II: With that quantitative knowledge base, work as an *engineer* to improve existing techniques and devices, or design new systems leading to lower cost factors and a higher quality of life.

In order to accomplish Phase I work, the basic transport phenomena (see Part A and Chap. 4) and fluid–wall interactions (see Chap. 8) form the necessary background. Thus, some generic biofluid modeling aspects are first summarized in Sect. 9.2 in preparation for the applications discussed in Sects. 9.3 and 9.4. Once such insight and skill are attained, Phase II tasks can be tackled as outlined in Chap. 10 and assigned in Sects. 9.5 and 10.6. Reading relevant section of a human physiology book (e.g., Silverthorn 2004) is a good starting point.

9.2 Modeling Aspects

As stated, internal fluid flow and solid mechanics material of the book's Part A and Part B form the background to solve biofluids problems. Specifically, for the two biofluids sample applications, i.e., arterial hemodynamics and lung-aerosol dynamics, the *fundamental* modeling aspects can be summarized as follows:

- Fluid properties
  - Air being incompressible
  - Blood being non-Newtonian

- Geometric Configurations
  - Single or triple bifurcation

- Out-of-plane multiple bifurcations with curved non-circular tubes
- Large segments of the respiratory tract or vascular tree

• Flow Dynamics (or Fluid Rheology)
  - Quasi-steady vs. transient (or pulsatile) flow
  - Laminar vs. turbulent flow

• Fluid-Particle Dynamics
  - Dilute (i.e., non-interacting) suspensions
  - Spherical vs. non-spherical (or deformable) particles
  - Micron particles vs. nanomaterial vs. vapor

• Fluid–Structure Interactions
  - Rigid-wall (i.e., one-way) coupling
  - Linearly elastic or anisotropic nonlinear wall behavior with two-way coupling

Blood property and blood rheology modeling are briefly discussed in order to set the stage for Sects. 9.3 and 9.4.

Blood Properties Whole blood is a suspension of red and white blood cells as well as platelets in an aqueous solution of three proteins, i.e., fibrinogen, globulin and albumin. Red blood cells (RBCs), consisting of hemoglobin carrying oxygen from the lungs to the body's tissue/organs, make up about 45% by volume, referred to as the hematocrit (Ht). Unstressed, an RBC has a bi-concave disk-like shape with an 8.5 μm diameter and a varying thickness from 1 μm at the center to 2.5 μm at the outer edge. However, by assuming a bent shape they can migrate through capillaries only 5 μm in diameter.

At shear rates $\dot{\gamma}(\sim \partial u / \partial y) > 200\,\mathrm{s}^{-1}$, whole blood behaves like a Newtonian fluid with an effective viscosity of $\mu \approx 0.0348$ dyne·s/cm^2 (or $\nu = \mu/\rho \approx 0.033$ cm^2/s) at Ht = 45% and a body temperature of 37°C. However, near centerlines and in recirculation regions $\dot{\gamma}_{actual} < 200\,s^{-1}$ and hence non-Newtonian (shear-thinning) effects come into play (see Figs. 9.2a, b and Sect. 6.3).

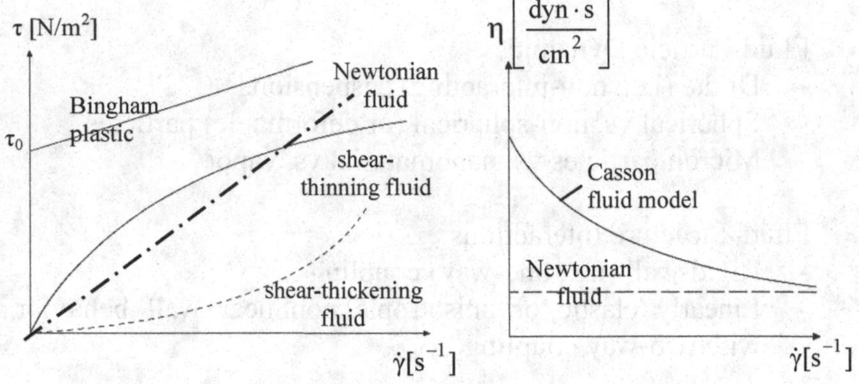

Fig. 9.2 Shear-rate dependent fluids

Clearly, a shear-rate dependent blood viscosity model is needed to capture actual blood rheology, where the Newtonian μ is now replaced by the apparent viscosity $\eta(\dot{\gamma})$. Numerous such models are available in the literature (see Sect. 9.6 assignment), where the Casson fluid and Quemada fluid models are often employed. Of course, the simplest $\eta(\dot{\gamma})$-correlation is provided by the power law, discussed in Sect. 6.3. The Casson model, similar to the Bingham plastic (see Sect. 6.3 and Fig. 9.2a), exhibits a yield stress, i.e., $\tau_{blood}(\dot{\gamma}=0)=\tau_0$. Specifically, a curve-fitted correlation (Chandran et al. 2007) yielded:

$$\sqrt{\frac{\tau_{blood}}{\mu_{plasma}}} = 2.0 + 1.53\sqrt{\dot{\gamma}} \tag{9.1}$$

where $\mu_p \approx 0.012\,\frac{\text{dyn}\cdot\text{s}}{\text{cm}^2}$ and $\dot{\gamma}[\text{s}^{-1}]$ is the shear rate. Equation (9.1) is a variation of the original Casson model of the form:

$$\sqrt{\tau_{ij}} = \begin{cases} 0 & \text{for } \tau \le \tau_0 \\ \sqrt{\tau_0} + K_c\sqrt{\dot{\gamma}_{ij}} & \text{for } \tau > \tau_0 \end{cases} \tag{9.2}$$

where τ_0 is the yield stress and we are assuming $\dot{\gamma}_{ij} > 1\text{s}^{-1}$.

The Quemada model, being more flexible and accurate, carries additional "constants":

$$\eta(\dot{\gamma}) = \eta_p \left[1 - 0.5\left(\frac{k_0 + k_\infty\sqrt{\dot{\gamma}_{rel}}}{1 + \sqrt{\dot{\gamma}_{rel}}}\right)\text{Ht}\right]^{-2} \tag{9.3}$$

where $\eta_p = 0.012\,\frac{\text{dyn}\cdot\text{s}}{\text{cm}^2}$ is the plasma viscosity; $k_0 = 7.6$ and $k_\infty = 2.8$ are empirical dimensionless terms called the intrinsic viscosities at zero and infinite shear rates, respectively; and $\dot{\gamma}_{rel} \equiv \dot{\gamma}/\dot{\gamma}_c$ with $\dot{\gamma}_c = 0.58\text{s}^{-1}$ being the critical shear rate. Clearly, all three coefficients depend on the hematocrit (Ht) value, which may be as low as 25% for dialysis patients.

Use of the Casson model is illustrated in Sect. 6.3 and an application of the Quemada model in modified form is discussed by Buchanan & Kleinstreuer (1998).

Blood Rheology Rheology is descriptive of continuous material deformation, i.e., flow of blood in the present case. In order to assess the fluid-dynamic impact on arterial diseases, such as atherosclerosis, or on posthetic devices, such as heart valves or stents, the local blood rheology has to be well understood (see Kleinstreuer 2006; Chandran et al. 2007; among others for sample applications). The simplest application is steady laminar fully-developed blood flow through an artery, i.e., a rigid circular tube (see Example 9.1).

Example 9.1: Poiseuille-Type Blood Flow using a Casson Model

Sketch	*Assumptions*	*Concepts*
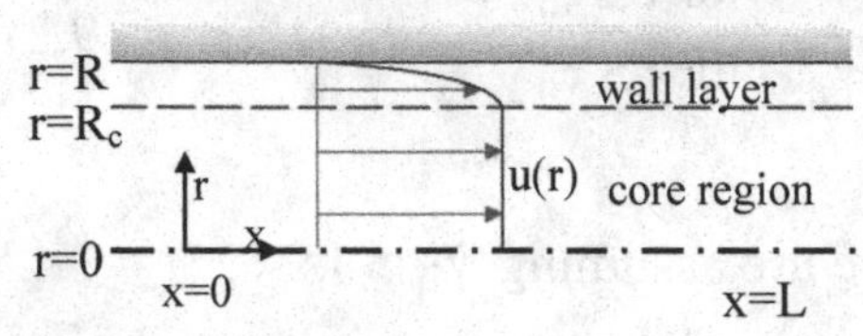 r=R, r=R_c, wall layer, u(r), core region, r=0, x=0, x=L	• Poiseuille-type flow • Non-Newtonian fluid	• Reduced Eq. of Motion • Casson-type Eq. (9.1)

Solution:

• Momentum transfer

Based on the postulates:

$$\vec{v} = [u(r); 0;0], \quad -\frac{\partial p}{\partial x} = \frac{\Delta p}{L} = \not{c} \text{ and } \dot{\gamma}_{rx} = -\frac{du}{dr}$$

the reduced equation of motion (App. A) reads:

$$0 = \frac{\Delta p}{L} + \frac{1}{r}\frac{d}{dr}(r\,\tau_{rx}) \quad \text{with} \quad \tau_{rx}(r=0) = 0 \qquad \text{(E.9.1.1a, b)}$$

Integration yields:

$$\tau_{rx} = \frac{\Delta p}{2L} r \quad \text{with} \quad \tau_{rx}\big|_{r=R_c} = \tau_0 = \frac{\Delta p}{2L} R_c \qquad \text{(E.9.1.2a, b)}$$

and

$$\tau_{rx}\big|_{r=R} = \tau_w = \frac{\Delta p}{2L} R \qquad \text{(E.9.1.3)}$$

• Casson model

Equation (9.1) squared yields for $\tau > \tau_0$:

$$\dot{\gamma}_{rx} = \frac{1}{1.53^2}\left[\sqrt{\frac{\tau_{rx}}{\mu_p}} - 2.0\right]^2 \quad \text{(E.9.1.4)}$$

or

$$u = -0.4272 \int \left[\sqrt{\frac{\Delta p}{2\mu_p L}}\sqrt{r} - 2.0\right]^2 dr + C \quad \text{(E.9.1.5)}$$

where u(r = R) = 0. Thus, for $r \geq R_c$,

$$u(r) = a\left[\frac{b}{2}(R^2 - r^2) - \frac{8}{3}\sqrt{b}\left(R^{3.2} - r^{3/2}\right) + 4(R - r)\right] \quad \text{(E.9.1.6)}$$

where a = 0.4272 and $b = \frac{\Delta p}{2\mu_p L}$. The radius of the core region is:

$$R_c = \frac{2\tau_0 L}{\Delta p} \quad \text{(E.9.1.7)}$$

and the core velocity is:

$$u_c = u(r = R_c) \quad \text{(E.9.1.8)}$$

Hence, the volumetric flow rate $Q = \int_A u dA = 2\pi \int_0^R u(r) \cdot r dr$ is:

$$Q = u_c(2\pi R_c^2) + 2\pi \int_{R_c}^{R} u(r) r dr \quad \text{(E.9.1.9)}$$

which should collapse for $\tau_0 = 0$ (i.e., zero yield stress) to the Hagen–Poiseuille solution

$$Q_{H-P} = \frac{\pi R^4}{8\mu_p}\left(\frac{\Delta p}{L}\right) \quad \text{(E.9.1.10)}$$

9.3 Arterial Hemodynamics

Several books (e.g., Chandran et al. 2007; Nichols & O'Rourke 2005; Kleinstreuer 2006) discuss the fluid flow and wall mechanics of arteries. Here, only some effects of geometric features, fluid-particle dynamics and fluid–structure interaction are outlined.

Geometric Configurations and Flow Fields The geometric features of blood vessels greatly determine the local flow structures. Figure 9.3 depicts some basic conduits and their associated laminar flow patterns. More complex flow fields occur, for example, in pathological cases of stenosed arteries which may also feature transitional and turbulent flows (see Kleinstreuer 2006) as well as fluid–particle and fluid–structure interactions.

(a) Flow in a curved pipe (Dean's Flow)

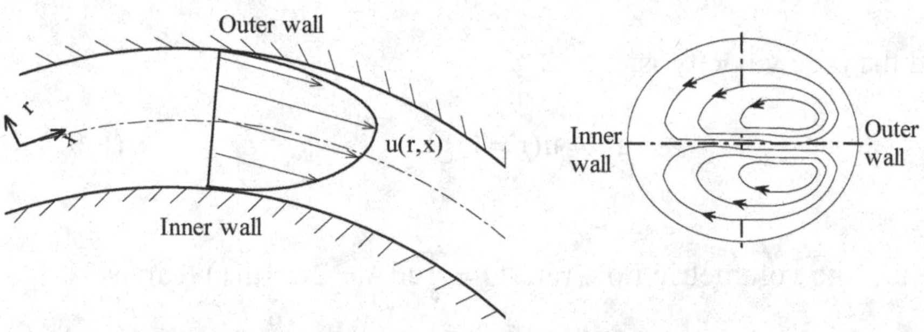

Sheared axial velocity profile due to opposing radial pressure force and centrifugal force

Dean vortices due to viscous effects generating strong secondary motion

(b) Flow in a symmetric bifurcation

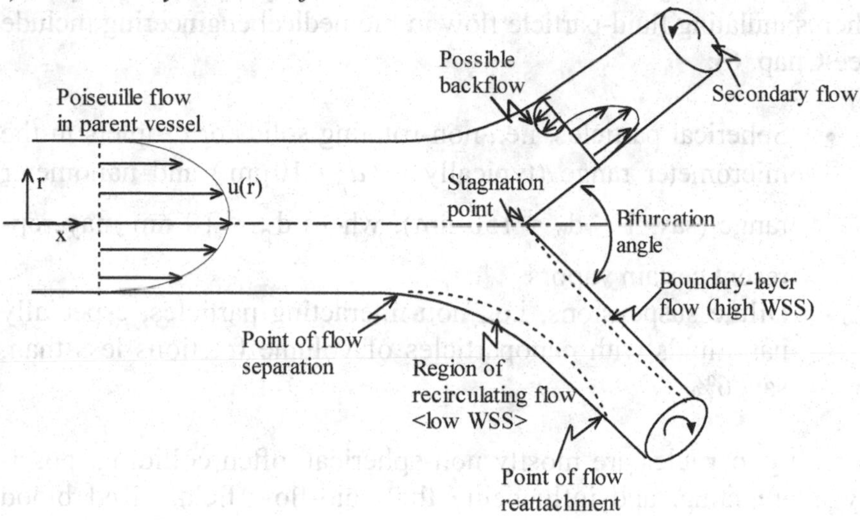

(c) Transient (or pulsatile) flow in parent tube with minor daughter branches

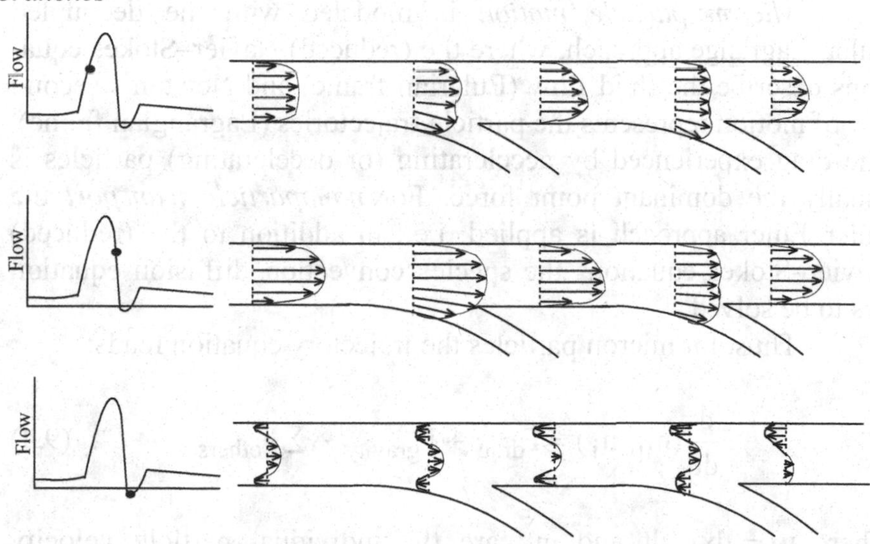

Fig. 9.3 Steady and transient velocity profiles and flow structures in a bent tube and in different bifurcations

Fluid-Particle Dynamics Presently, the underlying key assumptions when simulating fluid-particle flow in biomedical engineering include (see Chap. 6):

- Spherical particles, i.e., non-rotating solids or droplets in the micrometer range (typically $1 < d_p < 10\,\mu m$) and nanometer range (say, $1 \leq d_p < 150$ nm), where $d_p < 10$ nm may represent certain vapors
- Dilute suspensions, i.e., non-interacting particles, especially nanofluids with nanoparticles of volume fractions less than, say, 6%

In reality, particles are mostly non-spherical, often colliding, possibly aggregating, and influencing the fluid-flow field. Red blood cells with volume fractions up to 45% (hematocrit) as well as dense clouds of inhaled toxic or therapeutic aerosols, typically in fiberous or amorphous forms, are extreme examples.

Micron particle motion is modeled with the decoupled Euler–Lagrange approach, where the (reduced) Navier–Stokes equations describe the fluid flow (Eulerian frame) and Newton's second law of motion represents the particle trajectories (Lagrangian frame). The drag experienced by accelerating (or decelerating) particles is usually the dominant point force. For *nanoparticle transport* the Euler–Euler approach is applied, i.e., in addition to the (reduced) Navier–Stokes equations the species convection–diffusion equation has to be solved.

Thus, for micron particles the trajectory equation reads:

$$\frac{d}{dt}(m_p\ u_i) = F_{drag} + F_{gravity} + \sum F_{others} \tag{9.4}$$

where $u_i = dx_i / dt$ and m_p are the individual particle velocity vector and mass, respectively, while $\sum F_{others}$ signifies near-wall forces, interaction forces, etc. In case of large particle-to-fluid density ratios, dilute suspensions and negligible particle rotation, drag is the dominant force away from walls, i.e., $F_{drag} = \frac{\rho}{2} A_{proj}\, C_D$. Omitting the last term, ΣF_{others}, Eq. (9.4) can be written as:

$$\frac{d}{dt}(m_p\, u_i) = \frac{\pi}{8}\rho d_p^2 C_D (v_i - u_i)\,|\, v_j - u_j \,| + m_p\, g_i \qquad (9.5)$$

where v_i is the fluid velocity vector associated with each particle and C_D is the drag coefficient. Specifically,

$$C_D = \begin{cases} 24(1 + 0.15\mathrm{Re}_p^{0.687})/\mathrm{Re}_p & \text{for droplets with } 0 < \mathrm{Re}_p < 10^3 \\ \left.\begin{array}{ll} 24/\mathrm{Re}_p & \text{for } 0 < \mathrm{Re}_p \le 1.0 \\ 24/\mathrm{Re}_p^{0.646} & \text{for } 1 < \mathrm{Re}_p \le 400 \end{array}\right\} & \text{for solid spheres} \end{cases} \qquad (9.6a\text{–}c)$$

Two important dimensionless groups are the particle Reynolds number

$$\mathrm{Re}_p = \frac{|\, v_j - u_j \,|\, d_p}{\nu} \qquad (9.7)$$

and the Stokes number

$$\mathrm{St} = \frac{\rho_p\, d_p\, U}{18\mu\, D} \qquad (9.8)$$

where ν is the kinematic fluid viscosity, U is a characteristic fluid velocity and D a length scale; for example, for tubular flow U is the mean inlet velocity and D is the tube diameter.

The Stokes number, which can be interpreted as the ratio of particle relaxation time and flow characteristic time, is important in particle deposition studies (see Sect. 9.4 and Kleinstreuer 2006). In general, particle impaction/interception and sedimentation are the main deposition mechanisms for micronparticles, whereas nanomaterial convects and/or diffuses towards solid boundaries.

Nanoparticle (or vapor) transport is described by the species convection–diffusion equation (see Sect. 2.5):

$$\frac{\partial c}{\partial t} + \frac{\partial}{\partial x_j}(v_j c) = \frac{\partial}{\partial x_j}\left[\mathcal{D}\frac{\partial c}{\partial x_j}\right] \tag{9.9}$$

where c is the nanoparticle concentration, v_j is the fluid velocity vector and D is the Stokes–Einstein diffusion coefficient, i.e.,

$$\mathcal{D} = \frac{k_B T}{3\pi\mu d_p} \tag{9.10}$$

Here, $k_B = 1.38 \times 10^{-23}$ J/K is the Boltzmann constant, T[K] is the temperature, μ is the fluid viscosity and d_p is the nanoparticle diameter. Assuming the wall to be a perfect sink, we set:

$$c(x_{normal} = 0) = c_{wall} = 0 \tag{9.11}$$

Example 9.2: Nanomaterial Transport and Deposition

For viscous flow the distribution of the axial velocity u in a tube is parabolic and is give by the Poiseuille flow result $u = 2U(1 - \frac{r^2}{r_0^2})$.

Find the distribution of the concentration of nanoparticles in the circular tube, assuming that the tube is cylindrically symmetry and the diffusion coefficient D is constant.

Sketch	*Assumptions*	*Concepts*
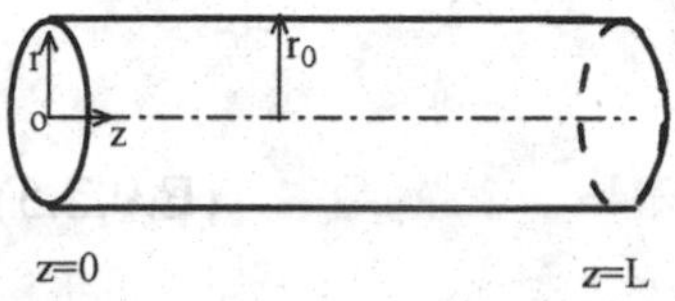	• At the inlet, the nanoparticle concentration is c_0 • The deviation of the velocity profile from the Pouseuille formula at the entry is of negligible importance. • The particle concentration c = 0 at the tube wall $r = r_0$ for $z > 0$ • The fluid flow is laminar steady and fully developed	• Reduced Eq. (9.9) • Nanoparticle diameter $d_p \leq 100\text{nm}$ in order to use Eq. (9.10)

Solution:

With these assumptions, the Poiseuille velocity distribution is substituted into Eq. (9.9). Note that the term $\frac{\partial^2 c}{\partial z^2}$ may be omitted if it is assumed that the axial change in concentration is much less than the radial change; thus, the steady-state mass diffusion equation can be written as:

$$2U[1-(\frac{r}{r_0})^2]\frac{\partial c}{\partial z} = \mathcal{D}[\frac{1}{r}\frac{\partial}{\partial r}(r\frac{\partial c}{\partial r})] \tag{E.9.2.1}$$

The corresponding boundary conditions are:

$$c = c_0 \ \text{ at } z = 0 \text{ for all } r$$

$$c = 0 \ \text{ at } r = r_0 \text{ for } z > 0 \tag{E.9.2.2a, b}$$

Introducing non-dimensional variables

$$r' = \frac{r}{r_0},\ c' = \frac{c}{c_0},\ z' = \frac{z}{L} \tag{E.9.2.3a–c}$$

Eq. (E9.2.1) reduces to

$$\frac{1}{2\Lambda}(1-r'^2)\frac{\partial c'}{\partial z'}=\frac{\partial^2 c'}{\partial r'^2}+\frac{1}{r'}\frac{\partial c'}{\partial r'} \qquad \text{(E.9.2.4)}$$

where

$$\Lambda=\frac{\mathcal{D}L}{4Ur_0^{\,2}} \qquad \text{(E.9.2.5)}$$

Then, the boundary conditions, Eq. (E,9.2.2a, b), become

$$c'=1 \text{ at } z'=0 \text{ for all } r'$$

$$\text{(E.9.2.6a, b)}$$

$$c'=0 \text{ at } r'=1 \text{ for } z'>0$$

The above Eq. (E9.2.4) can easily be solved numerically.

The mean concentration, $\bar{c}$, of nanoparticles leaving the tube of length L, is the quantity of main interest. If $c_L(r)$ is the concentration at z = L, the dimensionless mean concentration $\bar{c}$ is defined by:

$$\frac{\bar{c}}{c_0}=\frac{\int_0^R c_L u2\pi r dr}{\int_0^R c_0 u2\pi r dr} \qquad \text{(E.9.2.7)}$$

The solution was found to be (Ingham 1975):

$$\frac{\bar{c}}{c_0}=0.819\exp(-14.63\Lambda)+0.0976\exp(-89.22\Lambda)$$
$$+0.0325\exp(-228\Lambda)+\ldots\ldots \qquad \text{(E.9.2.8)}$$

Graphs:

(a) Dimensionless nanopaticle concentrations in radial direction ($\Lambda = 0.005$)

(b) Dimensionless nanopaticle concentrations in radial direction ($\Lambda = 0.01$)

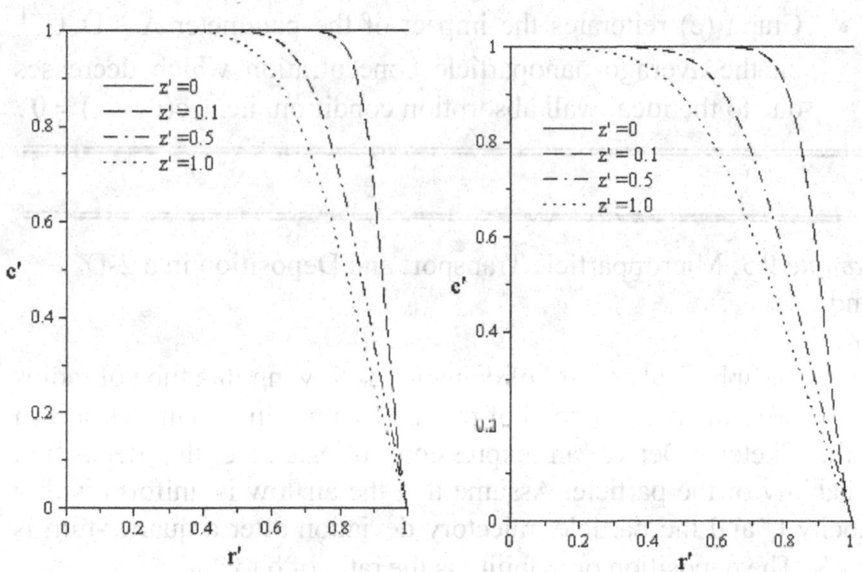

(c) Mean concentration as a function of Λ

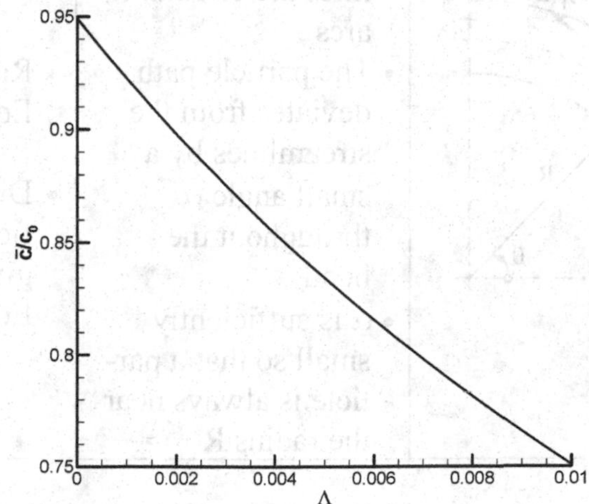

Comments

- Graphs (a) and (b) depict axial concentration profile developments with a measurable influence of the nanoparticle diffusivity which determines the Λ – parameters.
- Graph (c) reiterates the impact of the parameter $\Lambda \sim D, U^{-1}$ on the average nanoparticle concentration which decreases due to the ideal wall absorption condition, i.e., $c(r = r_0) = 0$.

Example 9.3: Micronparticle Transport and Deposition in a 2-D Bend

Consider a spherical aerosol of diameter d_p flowing in a tube of radius r_0 with two-dimensional bend of radius R with θ in radians, as shown in the Sketch. Derive an expression to calculate the deposition probability of the particle. Assume that the airflow is uniform with a velocity U and the particle trajectory deviation after a quarter-turn is $\delta \ll R$. The deposition probability is the ratio of δ to $2r_0$.

Sketch	*Assumptions*	*Concepts*
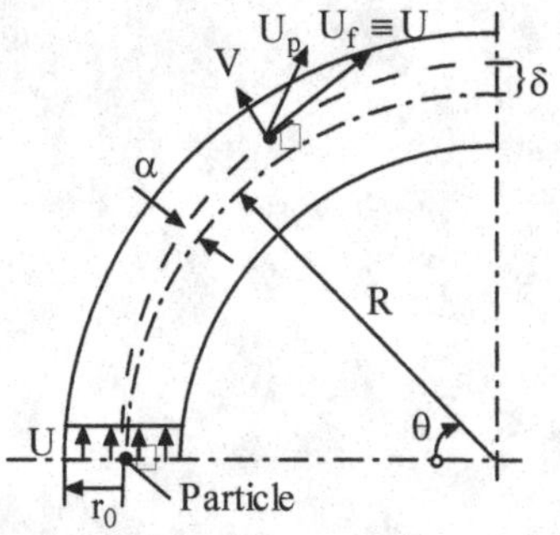 U_p, $U_f \equiv U$, V, δ, α, R, θ, U, r_0, Particle	• The fluid streamlines are circular arcs • The particle path deviates from the streamlines by a small angle α throughout the bend • α is sufficiently small so that a particle is always near the radius R	• Reduced Eq. (9.4) • Drag coefficient given in Eq. (9.6)

Solution:

With these assumptions, the equation of motion for a particle in the bend region reads

$$m_p a_r = m_p \frac{U^2}{R} = F_{drag} = \frac{1}{2}\rho V^2 C_D A \qquad \text{(E.9.3.1)}$$

where m_p is the particle mass, ρ is the fluid density, V is the "radial drift" velocity, C_D is the drag coefficient, and A is the projected particle area.

Assuming that the particle Reynolds number $Re_p = \rho V d_p / \mu << 1$ for micro-particles, we have

$$C_D = \frac{24}{Re_p} = \frac{24\mu}{\rho V d_p} \qquad \text{(E.9.3.2)}$$

For a sherical particle,

$$m_p = \frac{1}{6}\rho_p \pi d_p^3 \text{ and } A = \frac{1}{4}\pi d_p^2 \qquad \text{(E.9.3.3a, b)}$$

Substituting Eqs. (E. 9.3.2) and (E. 9.3.3) into Eq. (E. 9.3.1) yields for the radial particle velocity:

$$V = \frac{\rho_p d_p^2 U^2}{18\mu R} \qquad \text{(E.9.3.4)}$$

As shown in the Sketch, the particle is assumed to move with fluid velocity, i.e., $U_p \approx U$, in the near-tangential direction, but with a drift velocity V in the radial direction. This slip velocity V can be approximated as:

$$V = U \tan\alpha \approx U\alpha \quad (\alpha << 1) \qquad \text{(E.9.3.5)}$$

Combing Eq. (E. 9.3.5) with Eq. (E. 9.3.4), we have:

$$\alpha = \frac{\rho_p d_p^2 U}{18\mu R} \tag{E.9.3.6}$$

For a 2-D bend, the deposition probability of a particle can be expressed as:

$$\eta = \frac{\delta}{2r_0} \tag{E.9.3.7}$$

where δ is the radial displacement as the particle moves through the complete bend, which can be calculated as

$$\delta = R\theta\alpha \tag{E.9.3.8}$$

From Eqs. (E. 9.3.6)–(E9.3.8), we obtain:

$$\eta = \theta \frac{\rho_p d_p^2 U}{36\mu r_0} = \theta\, St \tag{E.9.3.9}$$

where the Stokes number $St = \frac{\rho_p d_p^2 U}{36\mu r_0}$.

Comment: In order to check the accuracy of Eq. (E. 9.3.8), an analytical solution to Dean's flow, i.e., $u(r,\theta)$ in a curved bend, should be used to find numerically the actual particle trajectory and hence δ.

Fluid–Structure Interaction (FSI) In conduits, a fluid flows because of pressure differentials while any elevated inside pressure exerts a force on the walls. In case of flexible walls, the structure and fluid interact (see Sect. 8.1 and Chap. 10). Coronary arteries, heart valves and alveoli in the deep lung region are prime examples of significant fluid–structure interactions. Circular pipes withstand high *liquid* flow pressure very well, while for relatively *low-pressure gas flow* thin-walled rectangular ducts are sufficient. Clearly, the solid mechanics of conduits, i.e., linearly elastic, hyperelastic, nonlinear anisotropic, etc., as well as the level of fluid pressure changes indicate the degree of fluid–structure interaction (FSI). Typically, wall distensibility has a relatively minor effect on the overall flow field; however, when highly elastic arteries, very soft tissue walls or flexible membranes are involved, FSI has to be considered.

For example, considering transient 1-D laminar axisymmetric incompressible flow in an elastic tube of radius R(t), the continuity equation $\nabla \cdot \vec{v} = 0$ reduces to:

$$\frac{\partial A}{\partial t} + \frac{\partial Q}{\partial x} = 0 \quad \text{or} \quad \frac{\partial R}{\partial t} + v\frac{\partial R}{\partial x} + \frac{R}{2}\frac{\partial v}{\partial x} = 0 \qquad (9.12a, b)$$

where $A = \pi R2$ is the cross-sectional tube area, Q = vA is the flow rate, and v is the cross-sectionally averaged velocity. The axial momentum equation for a Newtonian fluid, neglecting gravitational effects (see App. A), have been proposed in various forms (see for example, Steele et al. 2007 or Mabotuwana et al. 2007):

$$\frac{\partial Q}{\partial t} + \frac{\partial}{\partial x}\left(\frac{4}{3}\frac{Q^2}{A}\right) + \frac{A}{\rho}\frac{\partial p}{\partial x} = -8\,\pi\,\nu\frac{Q}{A} + \nu\frac{\partial^2 Q}{\partial x^2} \qquad (9.13a)$$

or

$$\frac{\partial v}{\partial t} + 2(1-\alpha)\frac{v}{R}\frac{\partial R}{\partial t} + \alpha v\frac{\partial v}{\partial x} + \frac{1}{\rho}\frac{\partial p}{\partial x} = \frac{2}{\rho R}\tau_w \qquad (9.13b)$$

where α is an axial profile shape factor and τ_W is the wall shear stress which can be approximated with a semi-analytical velocity distribution:

$$u(r) = \frac{\alpha}{2-\alpha}\,v\left[1-\left(\frac{r}{R}\right)^{\frac{2-\alpha}{\alpha-1}}\right] \qquad (9.14)$$

Using Eq. (9.14), Eq. (9.13b) can be rewritten as:

$$\frac{\partial v}{\partial t} + (2\alpha-1)v\frac{\partial v}{\partial x} + 2(\alpha-1)\frac{v^2}{R}\frac{\partial R}{\partial x} + \frac{1}{\rho}\frac{\partial p}{\partial x} = -2\frac{\alpha\nu}{\alpha-1}\frac{v}{R^2} \qquad (9.15)$$

The pressure-and-elastic-wall relation is assumed in the form:

$$p(R) = \beta\left[\left(\frac{R}{R_0}\right)^{\gamma} - 1\right] \tag{9.16}$$

where R_0 is the unstressed tube radius. For arterial blood flow with R_0 and $p_{in}(t)$ given, typical parameter values are $\alpha = 1.1$, $\beta = 21.2$ kPa, $\gamma = 2$, $\rho = 1.05$ g/cm^3 and $\nu = 3.2$ cm^2/s; hence, Eq. (9.15) can be solved numerically (see assignments Sect. 9.5).

Elastic Chamber or Windkessel Model A simpler example of FSI is the Windkessel model for the human arterial system or respiratory tract. Specifically, the blood vessels are lumped together as an elastic balloon (or chamber ≡ Windkessel) of time-dependent pressure and volume due to a transient inflow Q(t) that reflects the intermittent ventricular blood ejection. The chamber (or Windkessel) outflow encounters a hydraulic resistance due to the peripheral organs and/or vasculature. Clearly, the large elastic chamber smoothes the incoming pulsatile flow to more uniform flow.

Given a Windkessel distensibility κ which relates to the bulk modulus k, and assuming a hydraulic resistance R_h, a mass balance for incompressible flow reads (Fig. 9.4):

$$Q_{in} - Q_{out} = \frac{d\forall}{dt} = \kappa\frac{dp}{dt} \tag{9.17a, b}$$

where

$$\kappa = \frac{d\forall}{dp} = \frac{\forall}{k} \tag{9.18a, b}$$

With Young's module E and Poisson's ratio ν, the bulk modulus is (see Sect. 8.2):

$$k = \frac{E}{3(1-2\nu)} \tag{9.19}$$

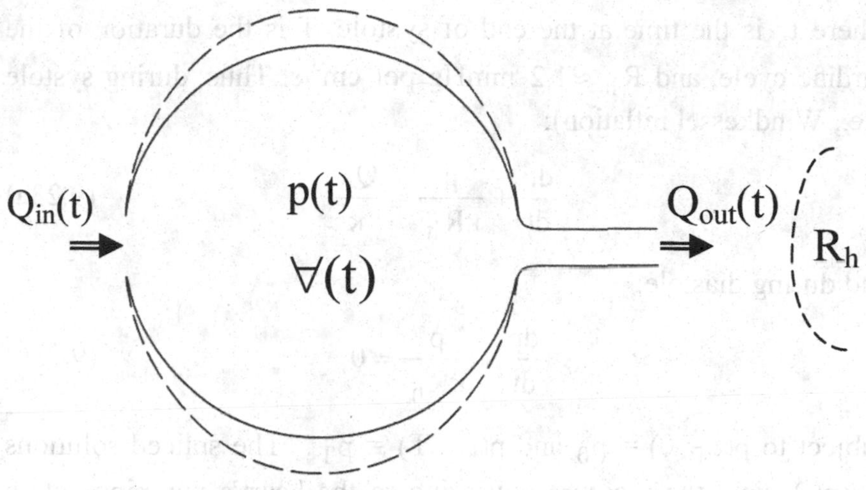

Fig. 9.4 Windkessel model

The blood outflow $Q_{out}(t)$ encounters a hydraulic resistance, R_h, due to all the arterioles and capillaries downstream, i.e.,

$$Q_{out} = \frac{\Delta p}{R_h}; \quad \Delta p = p(t)_{artery} - p_{venous} \tag{9.20a, b}$$

Assuming steady mean flow, $\Delta\overline{p} = \overline{p}_A - \overline{p}_v$ is the difference between the time-averaged arterial and venous pressures, i.e., $\overline{p}_A \approx \frac{1}{3}p_{systolic} - \frac{2}{3}p_{distolic}$, while $\overline{p}_v \approx 0$. For a person at rest, $\overline{p}_A \approx$ 100 mmHg and $\overline{Q}_{out} = 87\text{cm}^3/\text{s}$. Using Eqs. (9.20a) with $p_{venous} = 0$, Eq. (9.17b) now reads:

$$\underbrace{Q(t)}_{\text{Inflow}} - \underbrace{\frac{p(t)}{R_h}}_{\text{Outflow}} = \underbrace{\kappa\frac{dp}{dt}}_{\text{Rate of storage}} \tag{9.21}$$

The variable inflow can be approximated as:

$$Q_{in} \equiv Q(t) = \begin{cases} Q_0 & \text{for} \quad 0 \le t \le t_s \\ 0 & \text{for} \quad t_s < t \le T \end{cases} \tag{9.22a, b}$$

where t_s is the time at the end of systole, T is the duration of the cardiac cycle, and $R_h \approx 1.2$ mmHg per cm^3/s. Thus, during systole (i.e., Windkessel inflation):

$$\frac{dp}{dt} + \frac{p}{\kappa R_h} = \frac{Q_0}{\kappa} \tag{9.23a}$$

and during diastole:

$$\frac{dp}{dt} + \frac{p}{\kappa R_h} = 0 \tag{9.23b}$$

subject to $p(t = 0) = p_0$ and $p(t = T) = p_T$. The spliced solutions for p(t) yield the pressure pulse due to the heart's pumping action (see Fig. 9.5).

$$p(t) = \begin{cases} R_h Q_0 - (R_h Q_0 - p_0)\exp\left(-\dfrac{t}{\kappa R_h}\right) & 0 \le t \le t_s \\ p_T \exp\left(\dfrac{T-t}{\kappa R_h}\right) & t_s < t < T \end{cases} \tag{9.24a, b}$$

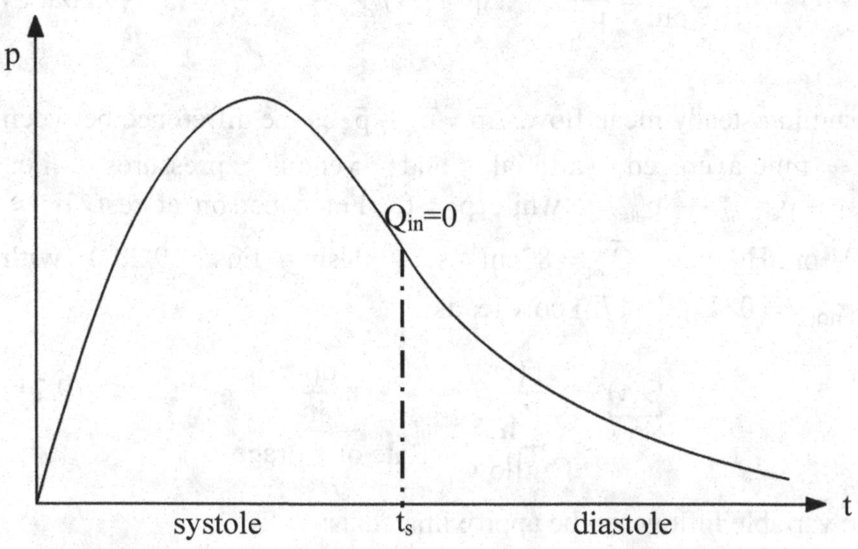

Fig. 9.5 Cardiac pressure pulse based on Windkessel theory

Clearly, the volume of blood pumped by the heart (i.e., the left ventricle) during the cardiac cycle is the stroke volume [mL/beat]:

$$\forall_s = \int_0^{t_s} Q(t)dt \approx Q_0 t_s \qquad (9.25a, b)$$

The cardiac output (CO [mL/min]) is then stroke volume times heart rate, i.e.,

$$CO = \forall_s * HR \qquad (9.26)$$

With typical values for $\forall_s = 70$ to $120\,\text{mL/beat}$ and HR = 60 to 80 beats/min during resting and exercise, respectively, the cardiac-output range can be $4{,}200 \leq CO \leq 21{,}600$ mL/min.

9.4 Lung-Aerosol Dynamics

The human respiratory system consists of the upper airways and tracheobronchial tree, called the conducting zone, as well as the bronchioli and alveolar region, i.e., the respiratory zone, where the $O_2 - CO_2$ gas exchange takes place (see Fig. 9.6). Approximate lung-airway geometries are given by Finlay (2001) and patient-specific image file conversion is discussed by van Ertbruggen et al. (2005).

Although geometrically more complex than the vascular system, the human respiratory tract has been approximated as a 1-D, trumped-like chamber for the purpose of estimating overall particle deposition (see Choi & Kim 2007), for which Kleinstreuer & Zhang (2009) provided a more detailed analysis. Inhalation rates may range from Q = 15 L/min under resting to Q = 60 L/min under moderate exercise conditions, with associated breathing frequencies of 14 and 25 cycles/min, respectively. Inhaled aerosols, i.e., droplets or (spherical) solids, are in the low nanometer range of $1 < d_p < 100$ nm with diffusivity $\mathcal{D} \sim d_p^{-1}$ (see Eq. (9.10)) or in the low micrometer range. While nanomaterial is deposited via convection and diffusion, micron particle deposition occurs mainly via impaction, including secondary airflow convecting particles to the airway walls. However, in regions of very low flow, e.g., in the alveolar region where the Reynolds number is of the order of 0.1, sedimentation becomes important as well. So, in addition to the inlet and local Reynolds numbers:

$$\mathrm{Re} = \frac{U\,D}{\nu} \tag{9.27}$$

the Stokes number is crucial in particle-deposition studies:

$$\mathrm{St} = \frac{\rho_p\, d_p^2\, U}{18\,\mu D} \tag{9.28}$$

where ρ_p and d_p are particle density and diameter, U is the characteristic air velocity, μ is the air viscosity, and D is the (local) tube diameter.

Typically, isothermal *dilute* spherical particle suspensions are assumed, i.e., the aerosols do not interact and influence the airflow. First, the momentum equation (Sect. 2.4) is solved and then the particle transport equation, i.e., Eq. (2.46) for nanomaterial convection-diffusion or Eqs. (6.35/9.29) for micron particle trajectories. Some experimental/computational results are depicted in Figs. 9.6 and 9.7 (see Kleinstreuer et al. 2008). Clearly, particles in the diameter range 20 nm $\le d_p \le 3\,\mu$m tend to travel past generation 3 into the deeper lung regions during normal breathing, while the oral airway (mouth to trachea) captures aerosols only at significant Stokes numbers, i.e., $\mathrm{St} > 10^{-2}$.

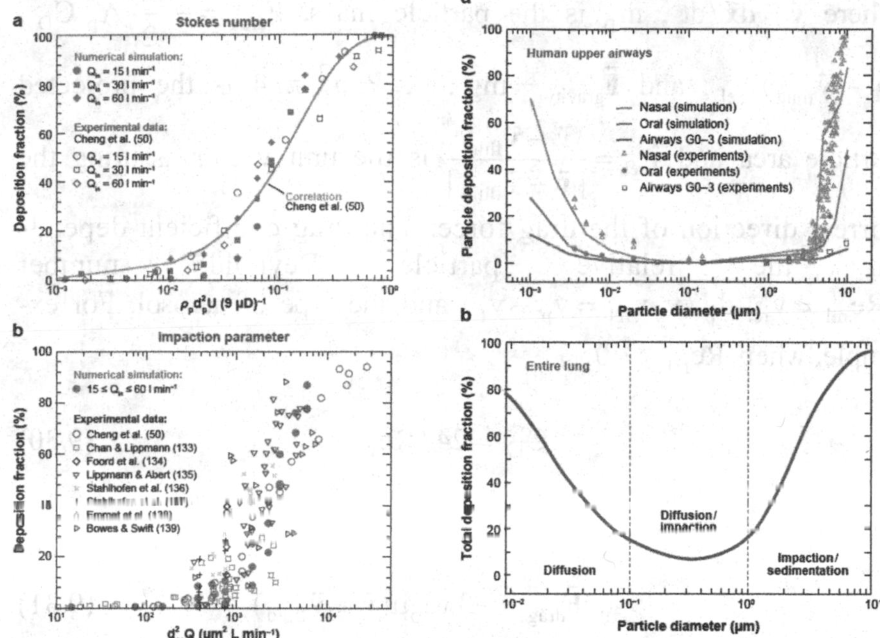

Fig. 9.6 Particle depositions in the oral airway model as a function of (a) Stokes number and (b) impaction parameter

Fig. 9.7 Examples of particle deposition fractions in (a) human upper airways and (b) the entire lung, as a function of particle size

Microparticle Dynamics In general, neglecting gravitational effects, particles with St << 1 follow fluid streamlines while particles with St ≥ 1 may cross streamlines, e.g., in curved airways.

When nondimensionalizing Newton's second law for particle drag and gravity, the Stokes number appears as the characteristic time ratio of particle relaxation to fluid-element residence as shown below. Specifically,

$$m_p \frac{d\vec{v}}{dt} = \vec{F}_{drag} + \vec{F}_{gravity} \tag{9.29}$$

where $\vec{\mathbf{v}} = d\vec{\mathbf{x}}/dt$, m_p is the particle mass, $\vec{\mathbf{F}}_{drag} = -\frac{\rho_f}{2} A_p C_D$, $(\vec{\mathbf{v}} - \vec{\mathbf{v}}_{fluid})\hat{v}_{rel}$, and $\vec{\mathbf{F}}_{gravity} = m\vec{\mathbf{g}}$. $A_p = d_p^2\,\pi/4$ is the projected particle area and $\hat{v}_{rel} = \frac{\vec{\mathbf{v}} - \vec{\mathbf{v}}_{fluid}}{|\vec{\mathbf{v}} - \vec{\mathbf{v}}_{fluid}|}$ is the unit vector assuring the correct direction of the drag force. The drag coefficient depends on the relative particle Reynolds number ($Re_{rel} = v_{rel}\, d_p / \nu$, $v_{rel} = v_p - v_f$) and the type of aerosol. For example, when $Re_{rel} < 1.0$,

$$C_D = 24 / Re_{rel} \tag{9.30}$$

so that

$$\vec{\mathbf{F}}_{drag} = -3\pi d_p \mu(\vec{\mathbf{v}} - \vec{\mathbf{v}}_{fluid}) \tag{9.31}$$

Hence, Eq. (9.29) can be rewritten as:

$$\frac{d\vec{\mathbf{v}}}{dt} = -\frac{\vec{\mathbf{v}} - \vec{\mathbf{v}}_{fluid}}{\tau} + \vec{\mathbf{g}} \tag{9.32}$$

where, deduced from the Stokes number St = $\tau_{particle} : \tau_{fluid}$, the particle characteristic time is:

$$\tau = \frac{\rho_p\, d_p^2}{18\mu} \tag{9.33}$$

Introducing the scales $\vec{\mathbf{v}} \sim U$, $t \sim \frac{D}{U}$ and $\vec{\mathbf{g}} = g$, leads to the dimensionless form of Eq. (9.32):

$$\frac{\tau\, U}{D}\frac{d\tilde{v}}{d\tilde{t}} = \frac{\tau\, g}{U}\tilde{g} - \tilde{v}_{rel} \tag{9.34a}$$

or

$$St\,\frac{d\tilde{v}}{d\tilde{t}} = \frac{v_{terminal}}{U}\,\tilde{g} - \tilde{v}_{rel} \tag{9.34b}$$

where $St = \tau_{particle} : \tau_{fluid} = \rho_p\, d_p^2\, U/(18\,\mu D)$ and v_{term} is the particle settling velocity in a stationary fluid. Thus, for a spherical non-rotating particle the gravity force is balanced by the vertical drag which is Eq. (9.31) with $\vec{v}_{fluid} = 0$:

$$mg = F_{drag} = 3\pi\,\mu\, d_p\, v_{term} \tag{9.35a}$$

so that

$$v_{term} = \frac{\rho_p\, g\, d_p^2}{18\,\mu} \tag{9.35b}$$

For submicron particles in air, the particle diameter may be somewhat near the mean-free-path of the air molecules, i.e., $\lambda_{air} = 0.067\ \mu m$ at STP conditions (see Sects. 7.1 and 7.2). Thus, particles with $d_p < 1.7\ \mu m$ may experience a relative motion around air molecules, which is accommodated by the Cunningham slip correction factor (Clift et al. 1978):

$$C_c \approx 1 + 2.52\,\frac{\lambda}{d_p} \quad \text{for} \quad 0.1 < dp < 1.7\mu m \tag{9.36}$$

Specifically, the drag coefficient Eq. (9.30) now reads:

$$C_D = \frac{24}{Re_{rel}}\, C_c \tag{9.37}$$

so that

$$v_{term} = \frac{\rho_p \, g \, d_p^2}{18\mu} C_c \tag{9.38}$$

Example 9.4: Variations in: (a) Stokes number and (b) Terminal Velocity of Micron-Particles in Air with and without the Cunningham Slip Correction Factor.

(A) For Stokes-number computations

$Q_{in} = 30\,\text{L/min}$, $d_p = 10\,\mu\text{m}$, $\rho_p = 10^3\,\text{kg/m}^3$, $\mu_{air} = 1.8 \times 10^{-5}\,\frac{\text{kg}}{\text{m}\cdot\text{s}}$, $D_{trachea} = 2\,\text{cm}$ and at, say, the 16th lung generation $D_{airway} = 0.06\text{cm}$.

Find: $St_{Trachea}$ and St_{Airway}

(B) For terminal velocity computation

$d_p = 0.5\,\mu\text{m}$, $\rho_p = 10^3\,\text{kg/m}^3$, $\mu_{air} = 1.8 \times 10^{-5}\,\frac{\text{kg}}{\text{m}\cdot\text{s}}$

Find: $v_{terminal}$

Part A	*Part B*	*Assumptions*	*Concepts*
		• 1-D steady motion • Spherical particle • Re < 1.0	• Stokes flow Regime • Newton's second law • Correlation for C_c

Solution Part A:

The Stokes number (see Eq. (9.28)) reads with the Cunningham correction factor included:

$$St = \frac{UC_c\rho_p d_p{}^2}{18\mu D}$$

where $U = (Q/A)_{local}$ and C_c given by Eq. (9.36), i.e., for air $C_c = 1 + 2.52\frac{0.067}{10} = 1.0169$. The mean velocity in the trachea is:

$$U_T = \frac{4Q_{in}}{\pi D_{trachea}{}^2} := 1.6\ m/s$$

However, Q_{in} bifurcates so that at Generation 16 the local flow rate is $Q_{in}/2^{16}$ and hence with $D_{airway} = 0.06$ cm,

$$U_A = \frac{4Q_{in}}{2^{16} D_A{}^2 \pi} := 0.0389\ m/s$$

Finally,

$$St = \begin{cases} 0.02511 & \text{in the trachea} \\ 0.02442 & \text{in the deep-lung region} \end{cases}$$

Comment: The relatively large St-values indicate that, when inhaled, inertial impaction is significant, mainly due to the micron-particle size because $St \sim d_p{}^2$

Note: For nanoparticles in the range $1nm < d_p < 1\mu m$ and with $St \ll 1$, inertial effects may be still important when $St > 5\times10^{-5}$.

Solution Part B:

The settling velocity is given by Eqs. (9.35b) or (9.38), so that with

$$C_c = 1 + 2.52\frac{\lambda}{d_p} := 1.34$$

$$v_{settling} = \begin{cases} 7.6\ \mu m/s & \text{without slip} \\ 10\ \mu m/s & \text{with air slip} \end{cases}$$

Comment: $C_c = 0$ implies that the air surrounding the falling particle is a continuum, i.e., mean free path $\lambda << d_p$. If $d_p = O(\lambda)$, the particle's settling trajectory is somewhat chaotic due to the bombardment of the surrounding air molecules, resulting in (submicron) particle Brownian motion.

Example 9.5: Horizontal Aerosol Acceleration in a Uniform Air Stream of Diameter D

Sketch	*Assumption*	*Concept*
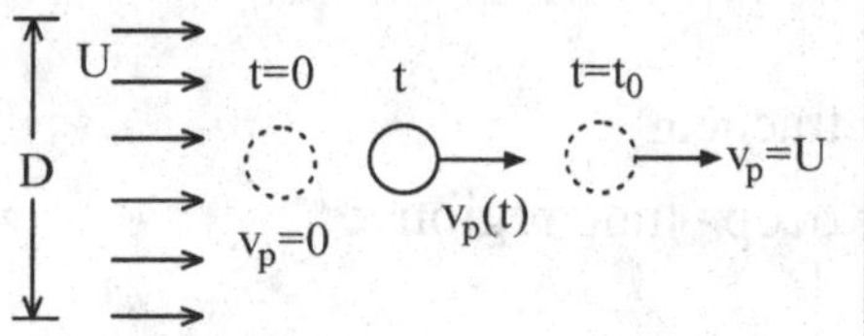 U; D; t=0; t; t=t₀ → $t=t_0$; $v_p=0$; $v_p(t)$; $v_p=U$	• Transient 1-D motion • Spherical particle • $Re_{rel} < 1.0$	• Solution to reduced Eq. (9.34b)

Solution:

Equation (9.34b) can be reduced to:

$$St\,\frac{d\tilde{v}_p}{d\tilde{t}} = -\,\tilde{v}_{rel}$$

where $St = \rho_p \, d_p^2 \, U/(18\mu D)$, $\tilde{v}_p = v_p / U$, $\tilde{t} = t\,U/D$, and $\tilde{v}_{rel} = (v_p - U)/U = \tilde{v}_p - 1$

Separation of variables and integration subject to $\tilde{v}_p(\tilde{t} = 0) = 0$, which implies $\tilde{v}_{rel}(\tilde{t} = 0) = -1$, yields:

$$|\,\tilde{v}_{rel}\,| = e^{-\tilde{t}/St}$$

Graph:

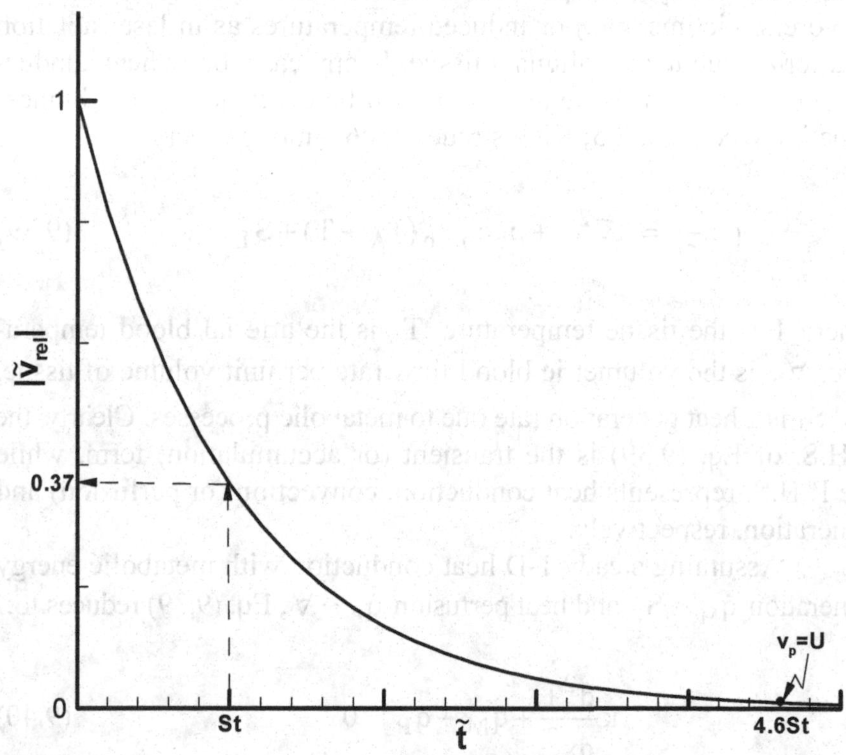

Comment:

The suddenly embedded sphere accelerates in the uniform air stream until $\tilde{\tau} = 4.6\text{St}$, which greatly depends on the St-value, i.e., the particle size and hence the inertia effect.

9.5 Bioheat Equation

So far, the arterial and pulmonary flow applications (see Sects. 9.3 and 9.4) assumed isothermal conditions. Generally, however, the human body may be exposed to extreme external temperatures (see explorers, victims, etc.) or induced temperatures as in laser ablation of arterial plaque or malignant tissue. In any case, body heat conduction-convection in tissue and blood can be described by the bioheat equation (see Sect. 2.5; Kleinstreuer 2006; among others):

$$\rho c \frac{\partial T}{\partial t} = k\nabla^2 T + \rho_b c_b \dot{\forall}_b (T_A - T) + S_T \tag{9.39}$$

where T is the tissue temperature, T_A is the arterial blood temperature, $\dot{\forall}_b$ is the volumetric blood flow rate per unit volume of tissue, and S_T is a heat generation rate due to metabolic processes. Clearly, the L.H.S. of Eq. (9.39) is the transient (or accumulation) term, while the R.H.S. represents heat conduction, convection (or perfusion) and generation, respectively.

Assuming steady 1-D heat conduction with metabolic energy generation $\dot{q}_M \sim S_T$ and heat perfusion $\dot{q}_P \sim \forall$, Eq. (9.39) reduces to:

$$k\frac{d^2T}{dx^2} + \dot{q}_M + \dot{q}_P = 0 \tag{9.40}$$

The perfusion term is due to the thermal energy exchange with the blood stream, i.e.,

$$\dot{q}_P = \dot{m}_b c_b (T_A - T) \tag{9.41}$$

where $\dot{m}_b$ is the blood-mass flow rate per unit volume, c_b is the blood specific heat, T_A is the blood (arterial) temperature, and T is the (variable) local (tissue) temperature.

Clearly, in reality the "heat sources" are not constant because the body adjusts $\dot{q}_M$ and $\dot{q}_P$ in order to keep $T_{body} \approx 37°C$. For example, if a person is too cold and shivers, $\dot{q}_M$ goes up. In contrast, if one is too hot the blood perfusion rate in the skin, and hence its local temperature, increases which in turn leads to a higher heat loss to the ambient.

Example 9.6: 1-D Heat Transfer in Composite Skin-Fat-Muscle Layer (after Incropera et al. 2007)

Given are the physiological and ambient data for the muscle layer L_m = 30 mm, $\dot{q}_m = 700 \text{ W/m}^3$ and $k_m = 0.5 \text{ W/m K}$; for the skin/fat layer $L_s = 3 \text{ mm}$ and $k_s = 0.3 \text{ W/m}\cdot\text{K}$; for the air convection and skin radiation $h_{air} = 2 \text{ W/m}^2\cdot\text{K}$ and $h_{rad} = 5.9 \text{ W/m}^2\cdot\text{K}$, also for the water convection $h_{water} = 30 \text{ W/m}^2\cdot\text{K}$ and the same $h_{rad} = 5.9 \text{ W/m}^2\cdot\text{K}$. The blood, equal to the body tissue temperature

Sketch	*Assumption*	*Concept*
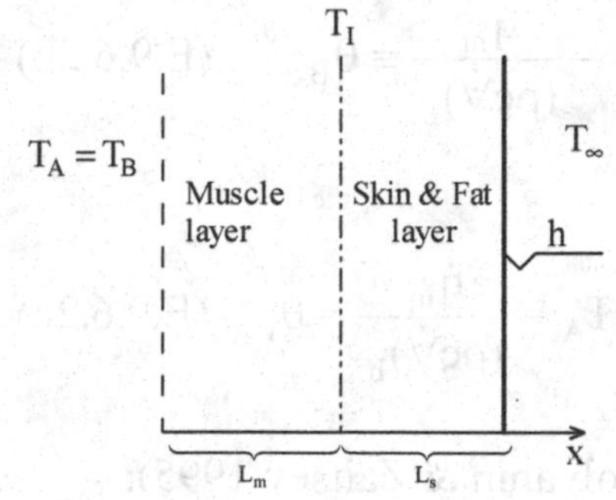 T_I; $T_A = T_B$; Muscle layer; Skin & Fat layer; T_∞; h; x; L_m; L_s	• Steady 1-D heat conduction • Constant heat generation and perfusion rates • Constant properties Constant T_A and T_∞	• Bioheat equation (9.40) for muscle layer • Heat transfer across skin/fat layer to ambient: $Q = \Delta T / R_{total}$

$T_A = T_B = 37°C$, the ambient $T_\infty = 24°C$, the perfusion rate $\dot{\forall}_b = 0.5 \times 10^{-3} \dfrac{m^3}{s \times m^3}$, $c_b = 3{,}600$ J/kg·K, and $\rho_b = 1{,}000$ kg/m^3. The skin surface is $A = 1.8$ m^2.

Solution:

- Muscle layer ($0 \le x \le L_m$): Equation (9.40) with Eq. (9.41) and $\dot{m}_b = (\rho\forall)_b$ can be rewritten as:

$$\frac{d^2T}{dx^2} + \frac{\dot{q}_m + \rho_b \dot{\forall}_b c_b (T_A - T)}{k} = 0 \qquad \text{(E.9.6.1)}$$

or

with $\theta \equiv T - T_A - \dfrac{\dot{q}_m}{(\rho c \dot{\forall})_b}$ and $\lambda^2 \equiv \dfrac{(\rho c \dot{\forall})_b}{k}$ we have

$$\frac{d^2\theta}{dx^2} - \lambda^2\theta = 0 \qquad \text{(E.9.6.2a)}$$

subject to

$$\theta(x = 0 \rightarrow T = T_A) = -\frac{\dot{q}_m}{(\rho c \dot{\forall})_b} \equiv \theta_B \qquad \text{(E.9.6.2b)}$$

and

$$\theta(x = L_m \rightarrow T = T_I) = T_I - T_A - \frac{\dot{q}_m}{(\rho c \dot{\forall})_b} \equiv \theta_I \qquad \text{(E.9.6.2c)}$$

The solution of Eq. (E. 9.6.2) is (see Polyanin & Zaitsev 1995):

$$\frac{\theta(x)}{\theta_B} = \frac{(\theta_I / \theta_B)\sinh(\lambda x) + \sinh[\lambda(L_m - x)]}{\sinh(\lambda L_m)} \qquad \text{(E.9.6.3)}$$

- Interface ($x = L_m$):

$$Q(x = L_m) = -k_m A \left.\frac{dT}{dx}\right|_{x=L_m}$$

$$= k_m \lambda A \theta_B \frac{1-(\theta_I/\theta_B)\cosh(\lambda L_m)}{\sinh(\lambda L_m)} \qquad \text{(E.9.6.4)}$$

Also, $Q(x = L_m)$ is equal to the heat flow rate through the skin/fat layer and into the ambient, lumped together as:

$$Q = \frac{T_I - T_\infty}{R_{thermal}} \qquad \text{(E.9.6.5a)}$$

where the total thermal resistance at $x = L_m + L_s$ is due to air stream convection (h_{conv}) and skin radiation (h_{rad}). Thus,

$$R_{thermal} = \frac{1}{A}\left(\frac{L_s}{k_s} + \frac{1}{h_{conv} + h_{rad}}\right) := 0.076\frac{K}{W} \qquad \text{(E.9.6.5b)}$$

For water,

$$R_{thermal} = \frac{1}{A}\left(\frac{L_s}{k_s} + \frac{1}{h_{conv} + h_{rad}}\right) := 0.021\frac{K}{W} \qquad \text{(E.9.6.5c)}$$

The interface temperature T_I can be obtained from setting Equation (E. 9.6.4) and (E. 9.6.5) equal, i.e., with $\lambda = \sqrt{(\rho c \dot{\forall})_b / k_m} := 60\ m^{-1}$, $\sinh(\lambda L_m) = 2.94$, $\cosh(\lambda L_m) = 3.11$, and $\theta_B = -0.389\ K$, we have

$T_I = 34.8°C$ ....for air stream

$T_I = 31.2°C$ ....for water stream

- Heat loss ($x = L_m + L_s$)

Equation (E. 9.6.5) provides the rate of heat transferred from the skin to the ambient, i.e.,

$$Q = \frac{T_I - T_\infty}{R_{thermal}} := 142.11 J/s \quad \text{for air stream}$$

$$Q = \frac{T_I - T_\infty}{R_{thermal}} := 342.86 J/s \quad \text{for water stream}$$

Graph:

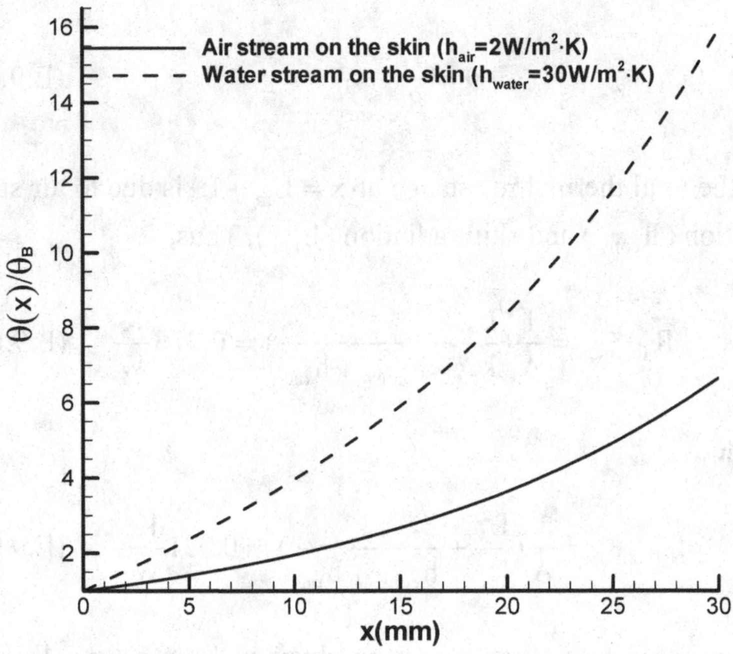

Comment:

The convection heat transfer coefficient on the skin surface, here $h_{water} = 15h_{air}$, greatly influences the temperature distribution inside the muscle layer $0 \leq x \leq L_m = 30$ mm.

9.6 Group Assignments and Course Projects

Solutions to homework problems done individually or in, say, three-person groups should help to further illustrate fluid dynamics concepts as well as approaches to problem solving, and in conjunction with App. A, sharpen the reader's math skills (see Fig. 2.1). Unfortunately, there is no substantial correlation between good HSA results and fine test performances, just *vice versa.* Table 1.1 summarizes three suggestions for students to achieve a good grade in fluid dynamics – for that matter in any engineering subject. The key word is "independence", i.e., equipped with an equation sheet (see App. A), the student should be able: (i) to satisfactory answer all concept questions and (ii) to solve correctly all basic fluid dynamics problems.

9.6.1 Guideline for Project Report Writing

A project is more than a homework problem. As a starter, it typically requires a brief literature review to document the state-of-the-art for the given task. The antagonistic interplay between a rather accurate mathematical description of an assigned problem and finding a tractable solution method puts a premium on making the right assumptions and selecting a suitable equation solver, often in terms of available software and hardware.

The resulting work has to be documented in an appropriate format, i.e., *project reports* should have the following features:

- **ToC**: Table of contents with page numbers (optional for smaller reports)
- **Abstract**: Summary of the fluid dynamics problem analyzed, solution method employed, and novel results with physical insight presented
- **Nomenclature**: Optional but recommended for complex problem solutions
- **Introduction**: Project objectives and task justification; System sketch with state-of-the-art description, basic approach or concept, and references of the *literature review* in terms of "author (year)," e.g., White (1998)
- **Theory**: Basic assumptions, governing equations, boundary conditions, property values, ranges of dimensionless groups, closure models, and references

- **Model Validations**: (a) Numerical accuracy checks, i.e., mesh-independence of the results as well as mass & momentum residuals less than, say, 10^{-4}; (b) Quantitative proof, e.g., comparison with an exact solution of (typically a much) simpler system, and/or experimental data comparisons, or (at least) qualitative justification that the solution is correct
- **Results and Discussion**: Graphs with interpretations, i.e., explain physical insight (note: number the figures and discuss them concisely); draw conclusions and provide applications
- **Future Work**
- **References**: List all literature cited in alphabetical order
- **Appendices**: Lengthy derivations, computer programs, etc.

9.6.2 Text-Embedded Insight Questions and Problems

9.6.1 Combining Fig. 9.1 and the essence of Sect. 9.2, develop BME modeling flow charts with supporting documents for the computational/experimental solutions to following problems:

(i) Brain aneurysm
(ii) Sleep apnea
(iii) Heat-valve replacement
(iv) Artificial knee joint
(v) Femoral-artery bypass
(vi) Your own problem choice

9.6.2 Discuss "blood rheology" in terms of advantages/disadvantages of property models and illustrations of blood flow dynamics.

9.6.3 Following Example 9.1, use the Quemada model (Eq. (9.3)) with different Ht-values and associated constants. Plote u(r, Ht) and Q(Ht), including $Q_{H\text{-}P}$.

9.6.4 Discuss modeling approaches for nanofluid flow when the nanoparticles are in the range $1 < d_p < 150\,\text{nm}$ and for $150\,\text{nm} \le d_p \le 1\,\mu\text{m}$.

9.6.5 Compare the drag coefficients for solid spheres and droplets in the range $1 < Re_p \le 400$ and demonstrate differences with a basic application of Eq. (9.5).

9.6.6 Provide a solution for Eq. (E. 9.2.4) and plot the results c(r,z) for different Γ – values.

9.6.7 Derive with insightful comments:

(i) Equations (9.12a, b) and (9.13a, b)
(ii) Equations (9.14) and (9.16)

9.6.8 Apply the Windkessel model (see Fig. 9.4 and Eq. (9.17)) to the human respiratory zone and plot the results for typical physiological parameter values from the open literature.

9.6.9 Solve Eq. (9.34) and plot the results using reasonable parameter values for (a) the trachea-bronchial zone and (b) the alveolar region.

9.6.10 Solve a 1-D version of Eq. (9.39) for convection heat transfer in:

(i) The liver
(ii) The kidney
(iii) A tumor

employing physiologically realistic parameter values and boundary conditions.

9.6.3 Projects

9.6.11 Analyze and simulate blood flow in the aortic arc, using Eq. (9.1).

9.6.12 Investigate real blood flow in capillaries and simulate a test case assuming steady flow in a rigid tube of inner diameter D = 5 μm.

9.6.13 Replicate Dean's flow (see Fig. 9.3a) for (i) plasma ($\mu_{eff} = \not\subset$) and (ii) blood when $Re_D = 1, 100, 1000$.

9.6.14 Retrace the 1954 Womersley solution of pulsatile flow and extend (numerical) analysis to (i) elastic tubes and (ii) transient blood flow in elastic tubes.

9.6.15 Provide a numerical analysis of the problem posed in Example 9.3 (see Comment).

9.6.16 Based on the 1-D FSI theory given with Eqs. (9.12)–(9.16), apply Eq. (9.15) to:

(i) Segment of the vasculature
(ii) Lung airways
(iii) Organ perfusion

9.6.17 Compare nanoparticle ($1\,\text{nm} \le d_p \le 500\,\text{nm}$) with micro particle ($1\,\mu\text{m} < d_p \le 10\,\mu\text{m}$) transport and deposition in a single bifurcation. Compare your results with experimental benchmark papers.

9.6.18 Research "cryogenics in BME" and provide a typical problem solution using Eq. (9.39).

Chapter 10

Computational Fluid Dynamics and System Design

10.1 Introduction

Computational fluid dynamics (CFD) methods are well documented (e.g., Versteeg & Malalasekera 1996; Tannehill et al. 1997; Ferziger & Peric 2002; Durbin & Medic 2007; Tu et al. 2008; among others). They are routinely applied to gain new physical insight and to improve engineering system design and hence performance. Nowadays in aircraft, automobile and machine-part design, CFD simulations have replaced wind-tunnel or other experimental tests, relying often on general-purpose (or problem-specific) commercial CFD software tools. Such software has to run efficiently on suitable computer platforms, which is especially important in light of parallel processing, computing speed and numerical accuracy. Needless to mention, experimental analysis will remain important for obtaining new discoveries, verifying theories, and validating computer models. Salient features of any CFD model are depicted in Fig. 10.1.

C. Kleinstreuer, *Modern Fluid Dynamics: Basic Theory and Selected Applications in Macro- and Micro-Fluidics*, Fluid Mechanics and Its Applications 87,
DOI 10.1007/978-1-4020-8670-0_10, © Springer Science+Business Media B.V. 2010

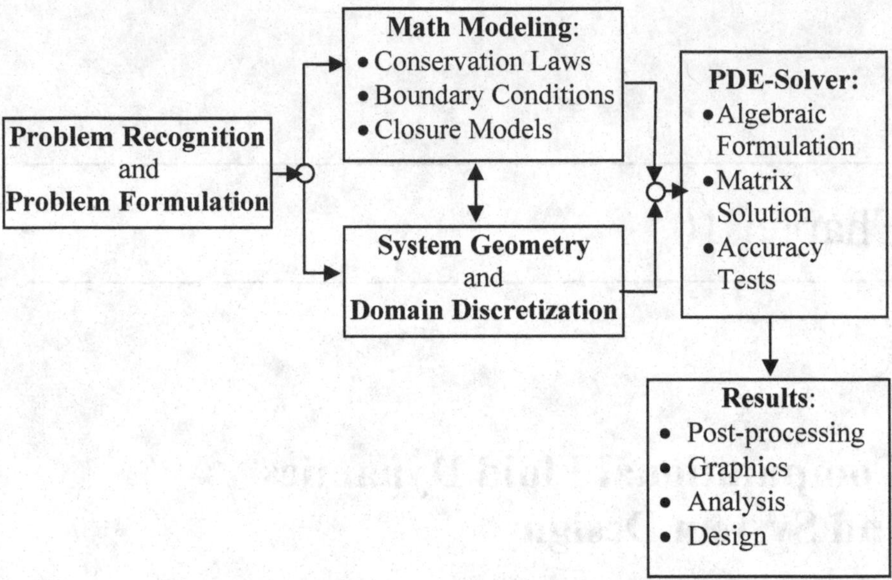

Fig. 10.1 Key elements of math modeling and computer simulation

Before undertaking any modeling/CFD work, two basic questions have to be answered for each new engineering problem:

- Are there sufficient information and resources available so that all important problem aspects are fully recognized and the modeling objectives can be achieved?
- Is it cost-effective and manageable to develop (or lease) an accurate and predictive computer simulation model which leads towards useful discoveries and practical applications?

The next sections deal with major aspects of mathematical modeling and computer simulations of engineering transport phenomena.

10.2 Modeling Obejctives and Numerical Tools

As already alluded to in Sect. 2.1, the *modeling objectives include an understanding of the system-specific transport phenomena and the ability to improve the design/function of a system or device.* In order to achieve these goals, powerful, flexible and accurate numerical tools, i.e., CFD software, in conjunction with fast, large-memory, multi-processor computer hardware are necessary. Complementary to Fig. 10.1, key steps in the development of a numerical model to execute CFD simulations are given in Fig. 10. 2.

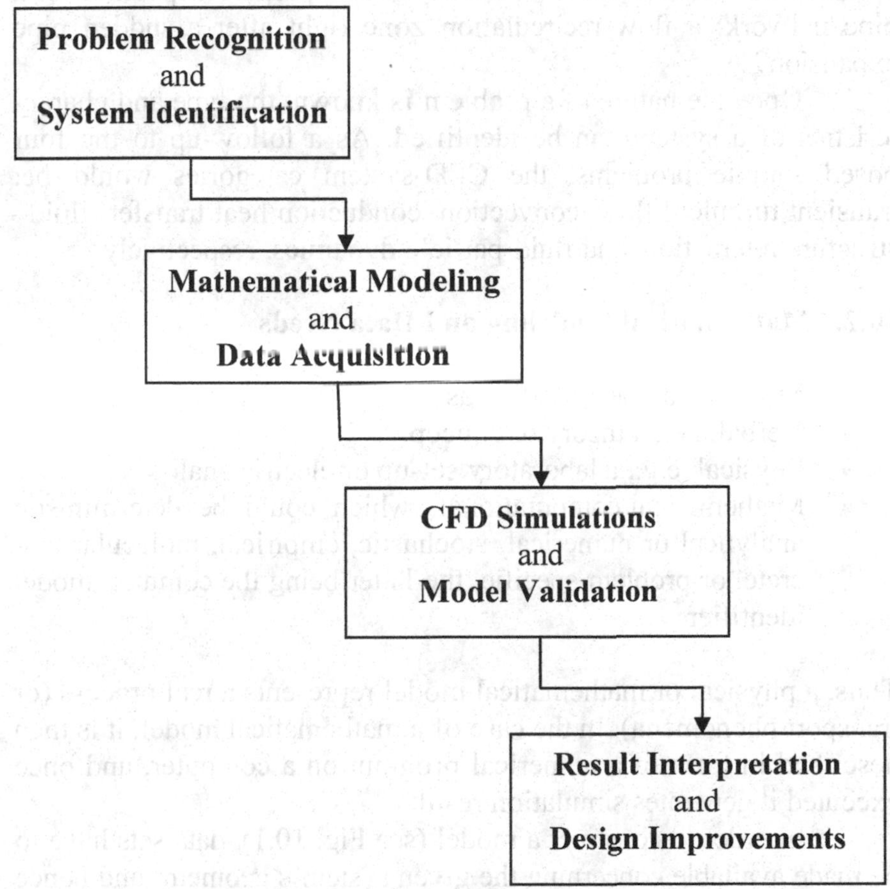

Fig. 10.2 Computational modeling steps

10.2.1 Problem Recognition and System Identification

To diagnose the cause of a problem correctly is especially challenging in the industrial environment when the system at hand is large and/or complex. For example, what is the cause of the periodic noise in a pipe network: component vibration due to sudden transient flow (also known as "water hammer") or turbulent flow? What is the cause of malfunction of an electronic device: a local "hot spot" temperature exceeding a thermal material tolerance? Why did a medical stent implant break: fatigue micro-cracks caused by pulsatile fluid pressure loads? What is the cause of excessive local particle deposition in a pipe network: a flow recirculation zone right after a sudden pipe expansion?

Once the nature of a problem is known, the type and characteristics of a system can be identified. As a follow-up to the four posed sample problems, the CFD-system categories would be: transient turbulent flow, convection–conduction heat transfer, fluid–structure interaction, and fluid-particle dynamics, respectively.

10.2.2 Mathematical Modeling and Data Needs

Models may be classified as

- Verbal, i.e., a theory or concept
- Physical, e.g., a laboratory set-up or electric analog
- Mathematical/computational, which could be deterministic analytical or numerical, stochastic, empirical, molecular discrete, or problem-specific, the latter being the common model identifier

Thus, a physical or mathematical model represents a real process (or transport phenomena). In the case of a mathematical model, it is then described in form of a numerical program on a computer, and once executed it generates simulation results.

In order to construct a model (see Fig. 10.1), data sets have to be made available concerning the given system's geometry and hence dimensionality, initial (if transient) and boundary (or operational) conditions, fluid properties, and necessary closure models in case of turbulent flow, heat and mass transfer, fluid–structure interaction, and/or fluid-particle dynamics.

10.2.3 Computational Fluid Dynamics

The first and very important step for geometrically complex systems is generating a grid (or mesh) that defines cells (or control volumes).

For such 2-D or 3-D cells all dependent variables (e.g., velocity, pressure, temperature, and/or species concentration) are computed throughout the system domain. Computational meshes are either *structured*, where the domain is discretized into large blocks made up of cells, or *unstructured* where more variable cells directly accommodate complex geometric domains (see Fig. 10.3). Mesh refinements are necessary until the results are independent of the cell density.

Boundary conditions are specified on each edge of the 2-D system or on each face of the 3-D domain. Values of material properties (e.g., fluid density and viscosity and/or Young's modulus and Poisson ratio) and sub-models in case of turbulent flow or particle suspension flows have to be given as well.

Starting with initial guesses for all dependent variables, the discretized forms of the governing equations are solved interactively, usually at each cell center. Various numerical methods (e.g., finite difference, finite volume, finite element) are available to discretize the system operations, typically coupled PDEs (see Chap. 2), into a set of algebraic equations. A suitable matrix solution algorithm and associated convergence parameters are then activated to obtain the solution. Clearly, a numerical solution is reached when, after hundreds (or thousands) of *iterations*, the mass and momentum residuals have decreased to, say, 10^{-4} or less.

(a) Structured (pseudo-2D) quadrilateral mesh

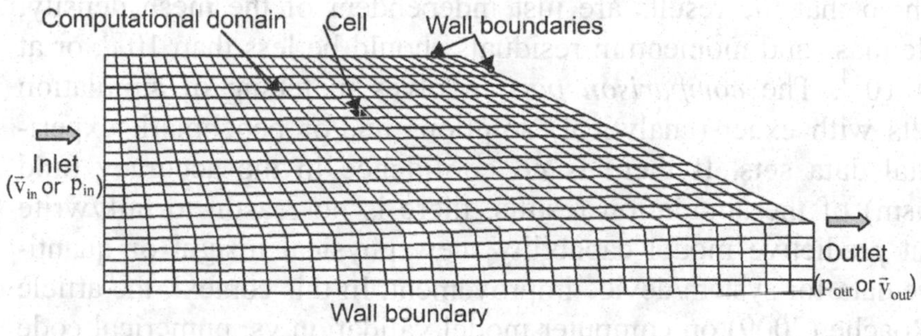

Notes:

- Identifiable grid lines discretize the computational domain
- Cell (in 3-D) has volume $\forall$ and surface S

(b) Unstructured triangular mesh

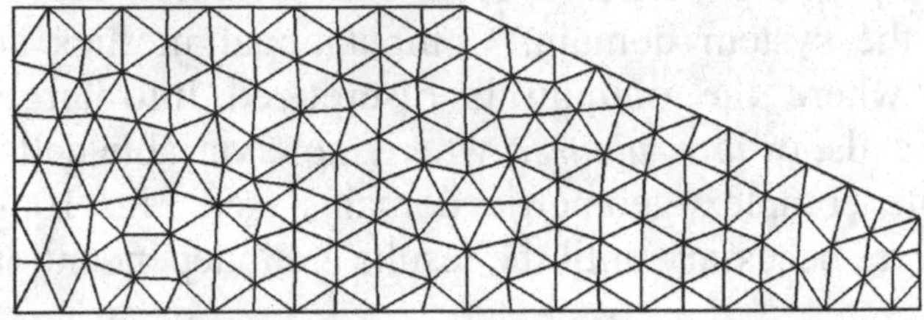

Note: Triangles of different sizes accommodate more easily complex domains

(c) Hybrid mesh

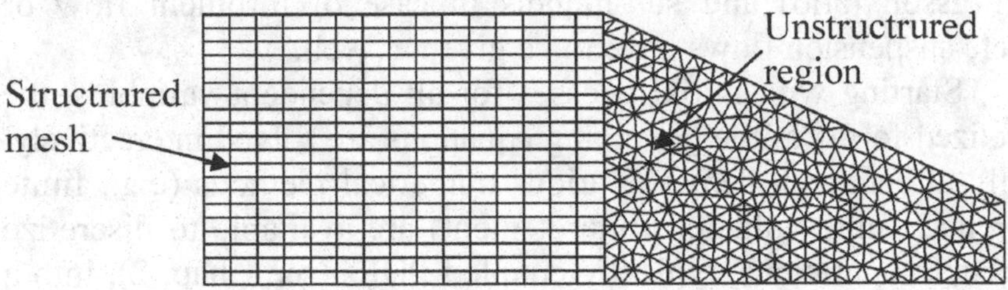

Note: Structured cell blocks are combined with a triangular grid in the more complex region or near walls

Fig. 10.3 Types of 2-D meshes

Simulation Accuracy Computer model validation has a numerical part and a comparison part. *Numerical validation* includes a refined mesh so that the results are just independent of the mesh density, while mass and momentum residuals should be less than 10^{-6}, or at least 10^{-4}. The *comparison part* involves matching of simulation results with exact (analytical) solutions and/or benchmark experimental data sets. If there is no confidence in the accuracy (and realism) of the simulation results, there is no reason to talk/write about predictive model capability, new physical insight or quantitative data for system/device improvement. In this context, the article by Roache (2009) on computer model validation vs. numerical code verification is of interest.

Computer simulation results are never exact for usually three reasons, i.e., incomplete mathematical description, simplified geometry/ flow data, and numerical errors. Potentially the most severe error sources are shortcomings in modeling when the physics of complex transport phenomena or property functions are mathematically not fully described. This may be very well the case when modeling turbulence, multiphase flows, bio-physical processes, and fluid-structure interactions. Input data as well as critical and/or boundary conditions can be erroneous when the system geometry inlet/outlet flows and fluid properties are only assumed and not measured for realistic scenarios. Intrinsic to all numerical methods is the discretization error which depends mainly on the degree of flow domain resolution with a fine mesh as well as the type and accuracy of the numerical method employed.

Obviously, the finer the mesh, the better are the system geometry and the local flow field represented. Simply put, a mesh which fills the flow region, bounded by walls or free surfaces, is the underlying grid for the discrete representation of the governing equations, typically PDEs (see Figs. 10.1 and 10.3). Indicative of a mesh (or grid) is the grid spacing h, i.e., the characteristic length scale of a 2-D or 3-D mesh element or cell. In fact, the order of numerical code accuracy is directly related to the power of h. Most numerical algorithms are second-order approximations of the governing equations.

As mentioned, there are basically structured and unstructured meshes (Thompson et al. 1999; Durbin and Medic 2007; among others). Structured meshes feature identifiable grid lines which can be numbered sequentially, say, in 3-D: i, j, k, forming grid points x (i, j, k), y (i, j, k), and z (i, j, k). Cells (or elements) of structured grids may be quite distorted when geometrically complex flow domains are meshed, i.e., discretized. However, straight wall geometries can be easily accommodated and boundary conditions can be readily enforced. In constrast, unstructured meshes, typically triangles for 2-D and tetrahedral elements for 3-D flow domains, map faithfully any physical space to the computational space (see Fig. 10.3).

Equation Discretization Along with flow domain discretization, i.e., meshing, it is required to discretize the governing equations, typically the Navier–Stokes equations, subject to given initial/ boundary conditions. Most codes use the finite volume method (FVM) as indicated in Sect. 10.2.3, employing either structured grids or unstructured meshes. The mesh decomposes the fluid-flow domain into connected cells. To each cell the flow variables are assigned; typically, velocity and pressure are stored at the center of each cell, i.e., the individual control volume. Specifically, integral forms of the conservation equations (see RTT in Sect. 2.2) are solved via an iteration method of successive approximations (see Durbin & Medic 2007; among others). For example, the linear momentum equation integrated over a cell (see Fig. 10.3a) of "control volume" $\forall$ with the "control surface" S and individual (inter) faces reads:

$$\frac{\partial}{\partial t}(\forall\rho\vec{v})_{cell} = -\sum_{f_i}(\rho\vec{v}\vec{v}\cdot\hat{n}S)_i - \sum_{f_i}(\rho\hat{n}S)_i + \sum_{f_i}(\mu\nabla\vec{v}\cdot\hat{n}S)_i \quad (10.1)$$

The LHS is the time-rate-of-change of momentum inside the cell (i.e., control volume) due to net momentum efflux across the cell (inter) faces f_{ij} as well as all the pressure and viscous forces acting on the cell faces f_i. Clearly, the equations of each cell (see (Eq. 10.1)) are coupled with all neighboring cells because the interface values needed are interpolated between the cell centers and so is the velocity gradient in the viscous force term generated with adjacent center velocity values. Furthermore, the local time derivative (see L.H.S. of Eq. (10.1)) is also replaced by a second-order (Euler) approximation. Discretization of the nonlinear convection term, $\rho\vec{v}\vec{v}$, is most crucial, requiring usually second-order (or higher) upwinding, such as the numerical QUICK scheme (see Patankar 1980; or Hoffman 2001).

Focusing on numerical solutions of fluid dynamics problems, standard *commercial software codes*, known as Navier–Stokes equation solvers, include CFD-ACE +, CFX, COMSOL, FLOW-3-D, FLUENT, and Star CD. Except Flow-3D, these codes are based on the finite volume method (Patankar 1980; Versteeg & Malaslaekera 1996) and feature their preferred mesh generators, i.e., CFD-GEOM,

ICEM-CFD, Flow-VU and Gambit, respectively. CFX and FLUENT belong to the ANSYS family of codes which, like COMSOL, can tackle a variety of multi-physics problems. Most software vendors offer mesh-size restricted versions for affordable student use. Linear PDEs and some PDE solvers are available free-of-charge via the Internet (see FlexPDE, Calculix, etc.). Solid mechanics problems can also be tackled with commercial software, typically based on the finite element method, such as Abaqus, Adina, ANSYS, I-Deas, and Nastran. Multi-Physics codes (e.g. ANSYS and COMSOL) allow for the solution of fluid–structure interaction (FSI) problems (see Chaps. 8 and 10).

Boundary Conditions The most common boundary conditions are prescribed velocity inlet, no-slip wall, and zero-gage pressure outlet conditions (see Fig. 10.3a). Alternatively, the inlet pressure is prescribed and a zero-velocity gradient, i.e., via a long exit conduit, is assumed at the extended outlet, implying fully-developed flow. Across a line/plane of symmetry, the gradients of all field variables are zero. For systems with repetitive geometries, such as turbine blades or heat exchanger tubes, periodic boundary conditions are imposed.

Mathematically, parabolic PDEs, e.g., Prandtl's boundary-layer equation, have solutions which march forward towards the open end, guided by the wall or edge boundary conditions. In contrast, for elliptic PDEs, e.g., the Laplace heat conduction equation, the domain-surrounding boundary conditions greatly determine the solution. If the magnitude of any field variable on a boundary (or inlet/outlet) is assigned, it is called a Dirichlet condition, while enforcing the gradient of a dependent variable, it is labeled a Neumann condition (or Neumann problem).

10.2.4 Result Interpretation

In a post-processing step, dependent variables or dimensionless groups are graphed for data interpretation or visualized in videos for physical insight and conference presentation. For proper result interpretation, i.e., new physical insight and useful applications, it is important that the flood of data sets is turned into smart 2-D or 3-D

- **Abstract**

  Summarize project objectives, approach taken, and new results.

- **Introduction**

  Problem statement with project goals, justified by a thorough literature review, system sketch, assumptions, concepts/approach and application of anticipated results.

- **Theory and (Numerical) Solution Method**

  List basic equations, postulates and reduced set of equations (equal the number of unknowns) with initial/boundary conditions; provide details of solution method, e.g., for numerical solutions the following is needed: information/discussion concerning the computer program used/developed, platform and mesh generation; numerical validation tests, i.e., mesh-independence of the results, as well as 10^{-6} residuals for both mass and momentum conservation.

- **Results and Discussion**

  For numerical solutions, *model validation* with experimental data sets, parametric sensitivity analyses, i.e., changes in boundary conditions, fluid properties, etc.; supply graphs with discussion of *basic research results*; graphs with explanations of *design applications*, and future work.

- **Conclusion**

  Similar to the abstract; but, add project result limitations and give future work goals.

- **References**

  Follow "author (year)" format; nomenclature with definitions (if necessary) and appendices (e.g., program listing, lengthy derivations, copies of key journal articles or relevant book pages, etc.)

Fig. 10.4 Elements of a research report

color figures. Graphing the results in terms of dimensionless groups and visualizing complex flow fields with streamlines or secondary velocity vector plots is most helpful. In any case, after successful computer simulation a comprehensive research report should be written, which is the point of departure for conference papers, journal articles, theses and dissertations (see Fig. 10.4).

10.2.5 Computational Design Aspects

After gaining sufficient physical insight into the mechanics of fluid flow, fluid-particle dynamics, and/or fluid–structure interactions, the next major task is new (or improved) engineering system design. That is accomplished via *virtual prototyping*, i.e., on the computer rather than in the laboratory (or wind tunnel) employing physical models.

In general, computational/experimental design analysis and simulation assist in evaluating the performance of a system described by a set of dimensions, material properties, (thermodynamic) loads, and operational requirements. Once established, validated computer simulations can also be used very efficiently in system parametric studies. A parametric analysis can be either deterministic or "probabilistic", the latter when taking measured uncertainties into account. In the deterministic case, all parameters vary continuously within the expected design/operational space, resulting in continuous responses, i.e., system performances. However, actual device dimensions and positioning, material properties and loads are not deterministic parameters, which poses the question of "design robustness." Again, computational analysis can swiftly assess how significant fluctuations in system parameter values may influence the design performance, including potential device failure.

Taking the methodology for a smart inhaler system, i.e., optimal drug-aerosol targeting as an example, the underlying idea was first tested via computer simulations of controlled air-particle streams (see Kleinstreuer & Zhang 2003; Kleinstreuer et al. 2008). Another example is uniform nanoparticle transport and mixing for micro-cooling devices or nanomedicine systems (see Kleinstreuer & Li 2008a, b). Still, for inventions to be acceptable, laboratory proof-of-concept has to follow any convincing computer simulation results, while for medical devices clinical outcome is the ultimate litmus test.

For the much more frequent case of device/system improvement, the transport phenomena have to be first fully understood via CFD analysis. Then, geometric and operational changes can be tested via computer experiments. Examples include automobile or aircraft design for drag reduction and/or (downward) lift modification, heat exchangers with improved mixing and reduced size/weight, fuel-injection systems generating higher combustion efficiencies, bio-MEMS for targeted drug delivery, medical implants to improve the quality of life, green-energy converters of high efficiency and reliability, etc.

10.3 Model Validation Examples

The previous sections highlighted the importance of validation studies prior to a detailed computational analysis of a system. Considering two-phase flows of dispersed particles for example (see Chap. 6), the particle trajectories and flow field solutions are often compared to experimental data of particle depositions in a specific geometry, or particle trajectories throughout a computational domain. Such validations establish confidence in the ability of computational modeling to analyze and optimize more complex systems.

10.3.1 Microsphere Deposition in a Double Bifurcation

Multiple experiments have been conducted on particle deposition in a symmetric double-bifurcation geometry. Thus, comparisons between the computationally predicted deposition efficiencies and experimental results are a common validation study. The computational procedure normally utilizes non-dimensional parameters such as the Reynolds number and the Stokes number to develop a simulation model that mimics the experimental conditions. For example, Fig. 10.5 compares predictions of particle depositions in a double bifurcation with the commercial finite volume solver CFX11 (ANSYS Inc., Canonsburg, PA) to the experimental analysis of Kim and Fisher (1999).

The computational simulations were conducted with varying particle diameters and at a constant Reynolds number (i.e., $3.1\,\mu m \leq d_p \leq 8.4\,\mu m$ and Re = 1200) to match the experimental Stokes numbers, $St = \frac{\rho_p d_p^2 U_0}{18 \mu D}$, where: D is the inlet tube diameter, d_p is the particle diameter, ρ_p is the particle density, μ is the fluid dynamics viscosity, and U_0 is the average velocity at the inlet defined by $Re_D = {U_0 D}/{\nu}$, where ν is the fluid kinematic viscosity. The deposition efficiency ($DE = \frac{\#\ of\ deposited\ particles}{\#\ of\ incoming\ particles}$) results illustrate the good matches with the experimental data, utilizing dimensionless parameters.

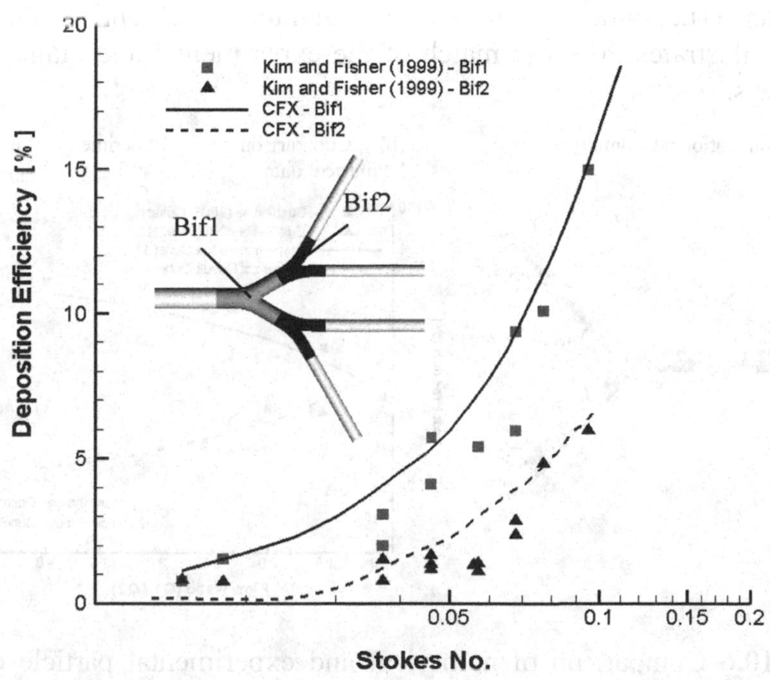

Fig. 10.5 Comparison between computational and experimental particle deposition data

10.3.2 Microsphere Transport Through an Asymmetric Bifurcation

An extension to the particle deposition validation is the evaluation of the amount of particles that exit the daughter branches of a bifurcation (see Fig. 10.6a). For the case of a symmetric bifurcation with different daughter branch diameters, the amount of particles exiting each branch under different flow ratios between the branches was measured by Bushi et al. (2005) in an attempt to model embolism trajectories through the human cardiovascular system. The exit percentages of neutrally buoyant solid particles of 0.6 mm diameter were obtained for a symmetric bifurcation where one daughter branch had the same diameter as the parent vessel and the other daughter branch had a diameter of 2/3 the parent vessel. Different ratios of flow (Q1/Q2) were established by varying the downstream resistance (i.e. outlet pressure) at the daughter branch outlets. Figure 10.6b illustrates the close match of the experimental and numerical data sets.

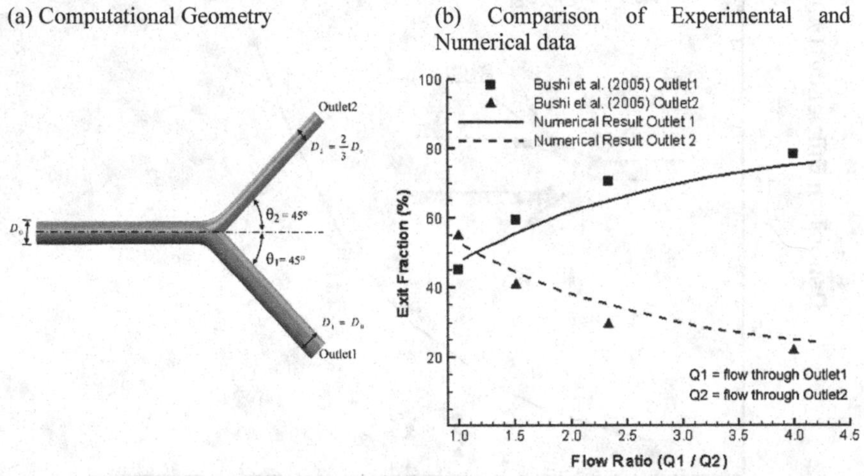

Fig. 10.6 Comparison of numerical and experimental particle exit percentage data

In addition to validating the computational modeling methodology, the experimental comparison also provide vital information and experience in determining appropriate meshes that result in a stable solution that is independent of further mesh refinements.

10.4 Example of Internal Flow

10.4.1 Introduction

Computational modeling of pathological cardiovascular conditions (constrictions, stenoses, aneurysms, atherosclerosis, etc.) and the resulting hemodynamics (flow of blood) continues to increase the understanding of the various conditions and in turn to improve healthcare (see Chap. 9 and Kleinstreuer 2006; among others). As computational simulations of hemodynamics become more widespread, standards of acceptable practices in computational biomechanical modeling are greatly needed. One fundamental example is the transport properties of human blood. Many highly cited hemodynamic investigations have modeled blood as a Newtonian fluid, although significant amounts of laminar, low shear-rate flow separation and recirculation occurred. This section documents a fundamental investigation of three different models of human blood (Newtonian, Quemada, and a modified model proposed by De Gruttola et al. (2005)) and their influence on the resulting hemodynamics of laminar flow separation and recirculation.

10.4.2 Methodology

As mentioned in Sect. 9.2, human blood contains four primary components: plasma, erythrocytes/red blood cells (RBCs), leukocytes/ white blood cells (WBCs), and platelets (Marieb 1998; Kleinstreuer 2006). Plasma is the liquid phase of blood and on average makes up approximately 55% of a given volume of blood, whereas RBCs make up approximately 44–45%, and WBCs and platelets make up approximately 0.7% (Marieb 1998). The concentration of RBCs within whole blood has been labeled the hematocrit (Ht) and is often used to discern the severity of chronic pathological conditions (e.g., diabetes). RBCs have a biconcave shape with a diameter of about 8 μm and are responsible for transporting nutrients to the body organs and tissue. WBCs are the "workhorse" cells of the immune system that attack foreign objects in the body and have a spherical shape with a diameter of 10–15 μm. Platelets are the smallest cells with a diameter of about 3 μm and are actively involved in the clotting abilities and thrombus formation of blood (Hochmuth 1986; Schmid-Schoenbein 1986). Figure 10.7 illustrates the three cells that are suspended in the blood plasma. The suspended cells within the blood plasma are responsible for the shear-thinning behavior exhibited by whole blood (Buchanan et al. 2001; Nichols and O'Rourke 2005). The shear-thinning behavior of whole blood has often been neglected in computational investigations of pathological conditions. A secondary goal of this section is to analyze the need of incorporating the non-Newtonian behavior of blood and at what situations is the shear-thinning behavior of whole blood vital to accurately model the behavior of whole blood.

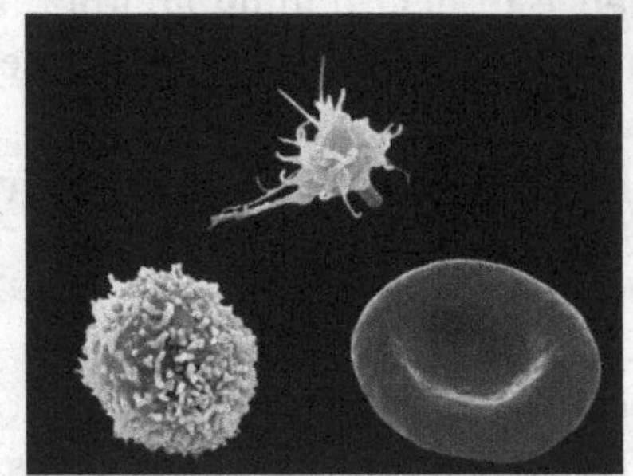

http://c.photoshelter.com/img-get/I0000bD6Ylr5NRqQ/s/600

Fig. 10.7 Primary cells in whole blood

A geometry containing a symmetric constriction of the vessel was used to determine the effects of blood viscosity models on flow separation and recirculation after the constriction. The constriction was modeled as a 75% area reduction cardiovascular stenosis and matched the dimensions used in Zhang and Kleinstreuer (2003). Figure 10.8 provides a visual representation of the geometry and its corresponding dimensions.

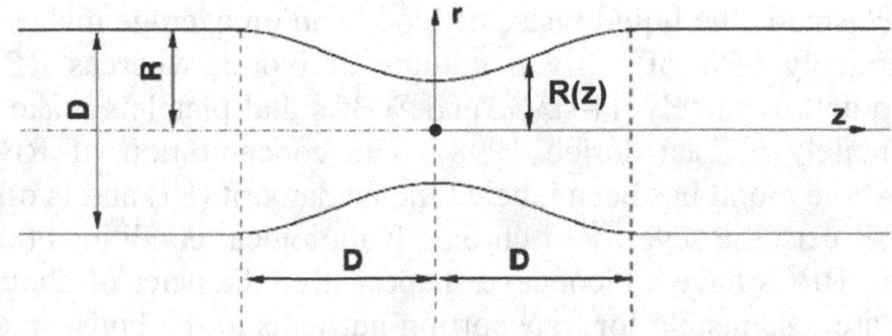

Fig. 10.8 Computational domain used in this study

The minimum point of the vessel radius is R/2 and occurs at the z-coordinate of zero. A structured grid was used for the 3D mesh, implementing an "O-Grid" orientation and clustered several layers of elements near the walls to capture high gradients of flow. Mesh independence was ensured by testing the velocity profiles of multiple locations throughout the domain for meshes of different element densities (cf. Zhang and Kleinstreuer 2003).

The governing equations for the analysis are the conservation of momentum and the constitutive viscosity law relating shear stress and apparent viscosity. Equations (10.2a, b) list the conservation of momentum for an incompressible fluid with a non-accelerating control volume and no body forces along with the constitutive relation between viscosity and shear stress. Specifically,

$$\rho\left[\frac{\partial v_i}{\partial t}+v_j\frac{\partial v_i}{\partial x_j}\right]=-\frac{\partial p}{\partial x_i}+\frac{\partial \tau_{ij}}{\partial x_j} \tag{10.2a}$$

with

$$\tau_{ij}=\eta\left[\frac{\partial v_i}{\partial x_j}+\frac{\partial v_j}{\partial x_i}\right] \tag{10.2b}$$

where x_i are the three orthogonal coordinate directions, t represents time, ρ is the fluid domain density, v_i is the fluid velocity vector, p is the fluid pressure, τ_{ij} is the fluid shear stress tensor, and η is the apparent dynamic viscosity of the fluid. The domain is governed by the function:

$$R(z)=\begin{cases} R-\dfrac{R}{2}\cos\left(\dfrac{\pi z}{2D}\right) & if \quad |z|\le D \\ R & if \quad |z|>D \end{cases} \tag{10.3}$$

where R is the non-constricted radius and D is the non-constricted diameter of the vessel. The computational domain has its origin at the center of the constricted portion of the vessel, a pre-constriction length of 7D and a post-constriction length of 17D. At the inlet a uniform velocity profile was applied, whereas a zero gauge pressure was assumed at the outlet. Three different models for the apparent viscosity of blood were used in this study. The first one is a Newtonian viscosity, which is approximated based on the experimental data of Merrill (1969), the second is the Quemada model with parameters given by Buchanan et al. (2001) that were fitted to the Merrill (1969) experimental data, and the third one is an exponential form presented by De Gruttola et al. (2005). The Newtonian model is shown in Eq. (10.4a), the full Quemada model can be simplified to a modified Casson model, whose final form is shown in Eq. (10.4b), and the De Gruttola et al. (2005) model (DGBP) is given as Eqs. (10.4c, d).

$$
\begin{aligned}
&\eta = \mu_{Newt} \\
&\eta = \eta_Q = \left(\sqrt{\eta_\infty} + \frac{\sqrt{\tau_0}}{\sqrt{\lambda} + \sqrt{\dot{\gamma}}} \right)^2 \\
&\eta \equiv \eta_{DGBP} = \mu_{plasma} \beta (1 - Ht) \exp\left(\frac{4.1 Ht}{1.64 - Ht} \right) \\
&\beta = 1 + \frac{b}{\dot{\gamma}^n}
\end{aligned}
\tag{10.4a–d}
$$

Here the term $\dot{\gamma}$ is the shear rate of the fluid and Ht is the blood hematocrit. Table 10.1 lists the parameters and their corresponding values for each viscosity model. An important note is that De Gruttola et al. (2005) listed the units of the curve-fitted parameter b as $[s^{-1}]$. However, that is inconsistent with exponential units of the shear rate and has been adjusted in Table 10.1.

Table 10.1 Parameters of the different viscosity models

Newtonian	$\mu_{Newt} = 0.0309 \left[\frac{g}{m \cdot s} \right]$
Quemada	$\eta_\infty = 0.02654 \left[\frac{g}{m \cdot s} \right], \tau_0 = 0.0436 \left[\frac{g}{m \cdot s^2} \right]$ $\lambda = 0.02181\ [s^{-1}]$
DGBP	$\mu_{plasma} = 0.014 \left[\frac{g}{m \cdot s} \right]$, Ht = 0.25,0.45,0.65 $b = 6.0[s^{-0.75}], n = 0.75$

The other transport property of blood is the density. Both the Newtonian and Quemada models used a constant density of 1060 kg/m^3, while the DGBP viscosity model included an equation for the density based on the Ht level and volume fractions of WBCs and platelets. The DGBP density model is shown in Eq. (10.5), with parameter values listed in Table 10.2:

$$\rho_{DGBP} = \rho_{plasma}(1 - Ht) + \rho_{RBC} Ht + Vf_{WBC}\rho_{WBC} + Vf_{platelet}\rho_{platelet} \tag{10.5}$$

where ρ_{plasma}, ρ_{RBC}, ρ_{WBC}, and $\rho_{platelet}$ are the density of plasma, RBCs, WBCs and platelets, respectively, while Vf_{WBC} and $Vf_{platelet}$ are the volume fraction of WBCs and platelets,

Table 10.2 Parameters of the DGBP density model

ρ_{plasma}	1,030 kg/m^3
ρ_{RBC}	1,096.5 kg/m^3
ρ_{WBC}	1,077.5 kg/m^3
$\rho_{platelet}$	1,040 kg/m^3
Ht	0.25, 0.45, 0.65
Vf_{WBC}	6.8293E-3
$Vf_{platelet}$	1.707E-4

Simulations of steady flow through the domain (see Fig. 10.8) at different Reynolds numbers were conducted to analyze laminar flow separation and recirculation in regions downstream of the vessel constriction. The fluid was introduced at the inlet with a parabolic velocity profile and inlet Reynolds numbers of 250, 500, and 750 were selected for the analysis. The range was based on physiological values of Reynolds numbers (Re) in the carotid artery (a common location of stenoses), which has an approximate mean Re of 250 and a maximum Re of about 1,100. Using the conservation of mass and volume flow rate, the Re value in the constriction can be calculated to be twice that of the upstream Re value for this geometry. Thus, the selected Re values ensured that the Re in the vessel constriction was well below 2,000 for each Re specified at the inlet.

The commercial finite-volume solver CFX V.11 (ANSYS Inc., Canonsburg, PA) was used to solve the governing and constitutive equations. A high resolution advection scheme was specified for the solver, where simulations concluded when root mean square error residuals were less than 1E-5. Simulations were run using a single Intel 3.59 GHz processor on a Dell Precision 670 workstation with 8 GB of RAM using Windows64 and using a single core on a quad-core Intel 3.0 GHz processor on a Dell T7400 workstation with 16 GB of RAM using Windows64. Simulation time varied between 13 min. (for Newtonian Re = 250 cases) to 1.5 h (for Re = 750 cases).

10.4.3 Results and Discussion

Apparent Viscosity The viscosity models were first plotted to illustrate their different behaviors at the same shear strain rates. Figure 10.9 depicts each viscosity model as a function of shear rate.

Figure 10.9 clearly illustrates significantly different responses of the different apparent viscosity models. Upon examining specific regions of the viscosity model behavior, when $\dot{\gamma}_{blood}$ is greater than 10^2, the different models all predict a nearly constant behavior and suggests a Newtonian fluid response. The two-phase nature of blood suggests that under high shear rates the particles move away from the high shear rate regions leaving primarily plasma, a Newtonian fluid with a lower viscosity than whole blood. Conversely, the models all predict an increase in viscosity as $\dot{\gamma}_{blood} \to 0$, which further illustrates shear-thinning behavior, but the Quemada model levels off to a maximum viscosity whereas the DGBP model continues to exponentially increase in value with no upper limit. Another important

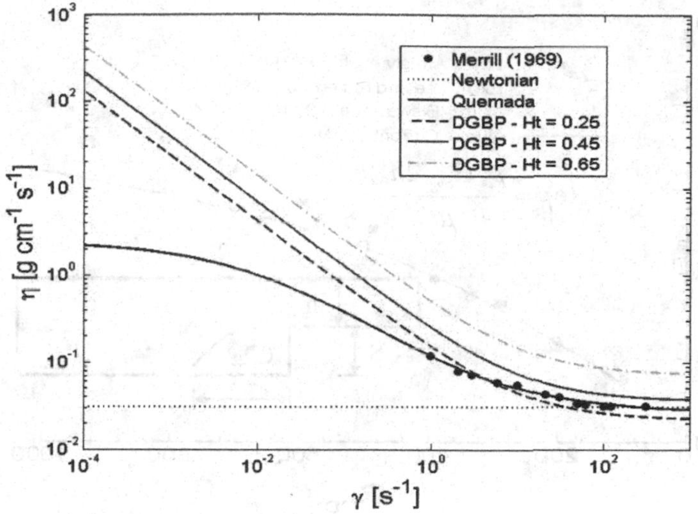

Fig. 10.9 The different viscosity models vs. shear strain rate

aspect is the influence of Ht on the different viscosity models. Figure 10.9 reveals similar overall shapes of the viscosity vs. $\dot{\gamma}$ curves, yet the multiple Ht-values offset the different curves to illustrate that an increase in RBC concentration is met with an increased baseline viscosity and a decrease in particle concentration is met with a decrease in baseline apparent viscosity.

Model Validation A model validation was conducted by comparing the primary reattachment length of laminar flow over a backstep with experimental and previously validated computational data (see Fig. 10.10).

The discrepancy between the experimental data and the computational simulations at the higher Reynolds numbers is due to the fact that the experiment encountered 3D physics whereas the simulation was only 2D. Such a trend is also seen with the 2D simulations of Biswas et al. (2004). The stenosis geometry is a full 3D geometry, and enables the range of Reynolds numbers to be modeled with confidence for accurate predictions of their reattachment lengths.

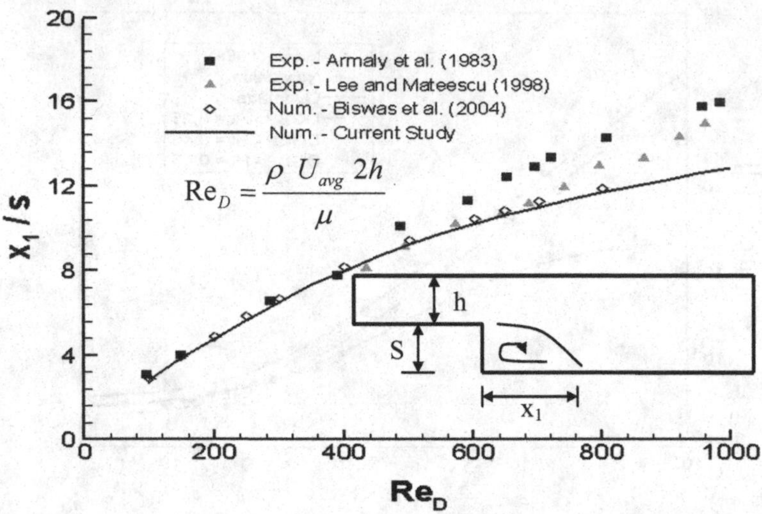

Fig. 10.10 Comparison of reattachment lengths for 2D flow over a backstep

Recirculation Regions The length of the recirculation after the vessel constriction is an important parameter for blood particle trajectories and future cardiovascular disease development. Different models of blood viscosity can have a pronounced effect on the recirculation zone and its length. Figure 10.11 illustrates the different lengths of flow separation calculated with the different viscosity models for different Reynolds numbers.

The reattachment-length trends of the different viscosity models are a direct result of the viscosity values of each model given in Fig. 10.9. Recirculation regions tend to have a very low shear rate, which increases the viscosity of a shear-thinning fluid (e.g. blood) in these regions of recirculation. An elevated viscosity provides an increased resistance to flow and limits the vortex development in regions of flow separation when compared to Newtonian fluids. Both the Quemada and DGBP viscosity models capture the shear-thinning behavior of blood, and hence produce higher viscosities in this region than the Newtonian fluid and consequently have lower reattachment lengths than the Newtonian model. Moreover, the DGBP model with

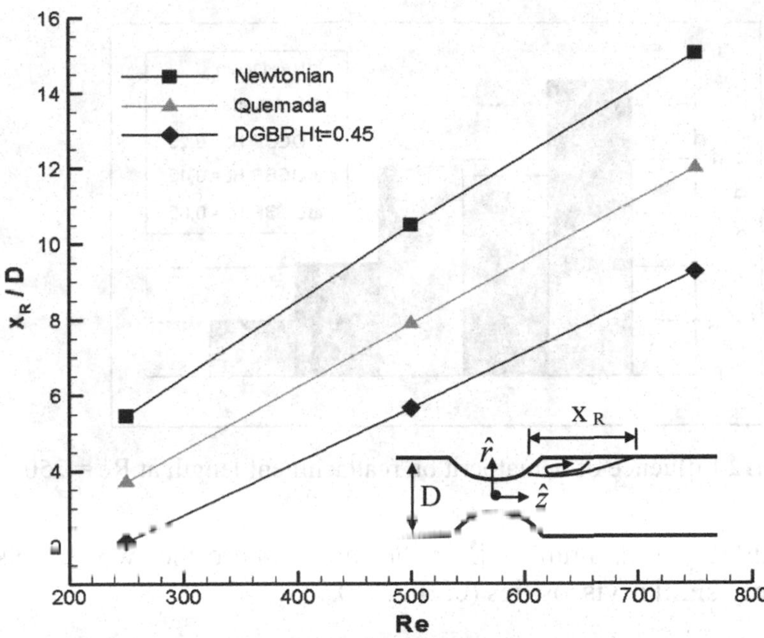

Fig. 10.11 Reattachment lengths of different viscosity models at different Reynolds numbers

Ht = 0.45 has a visibly higher viscosity than the Quemada model for almost all shear rates and consequently has a shorter reattachment length than the Quemada model.

The viscosity influence on the reattachment length is illustrated by manipulating the DGBP model parameter Ht at a constant Reynolds number. Figure 10.9 clearly shows that different values of Ht provide a significantly different viscosity for the same shear rate value. As a result, Fig. 10.12 shows the reattachment lengths for various Ht-values, i.e., 0.25, 0.45, and 0.65, along with the Quemada and Newtonian models at Re = 250.

The increased Ht level results in an increased viscosity which then decreases the reattachment length. An interesting note is that the Quemada model and the DGBP model with Ht = 0.25 are very close in reattachment length at Re of 250. This is because the shear

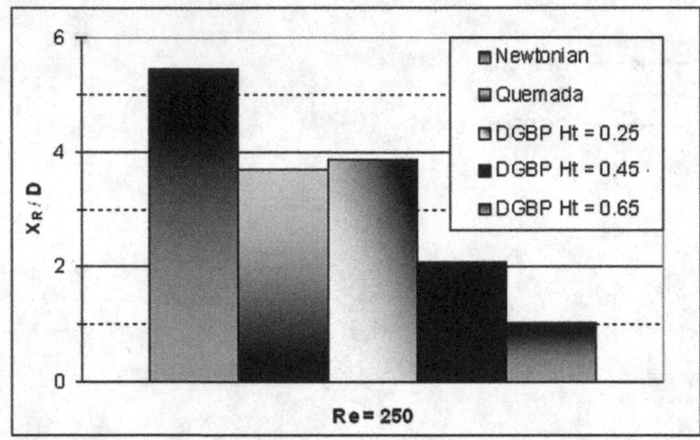

Fig. 10.12 Influence of hematocrit on reattachment length at Re = 250

rate at this Reynolds number is in the range where the two models predict very similar viscosities (cf. Fig. 10.9).

Velocity Profiles and Wall Shear Stress Velocity profiles (Fig. 10.13) at three locations downstream of the vessel constriction reveal a growing discrepancy between the viscosity models as the (axial) z-coordinate increases. Such a trend is again related to the viscosity effect, where the Newtonian viscosity having the lowest value and thus providing the least amount of flow resistance at the vessel centerline and near the wall.

It should be noted that the velocity profiles near the constriction are practically identical except for the centerline velocity. This is due to the high velocity of the fluid exiting the constriction, and is able to overcome the elevated viscosity of the non-Newtonian models.

An interesting result is observed when plotting the wall shear stress (WSS) through the vessel constriction up to the 5D location (z-coordinate = 10) downstream. The Quemada model actually predicts a peak WSS lower than the Newtonian model, while the DGBP Ht = 0.45 model has the highest peak WSS value. Figure 10.14 depicts the WSS trends over the specified range.

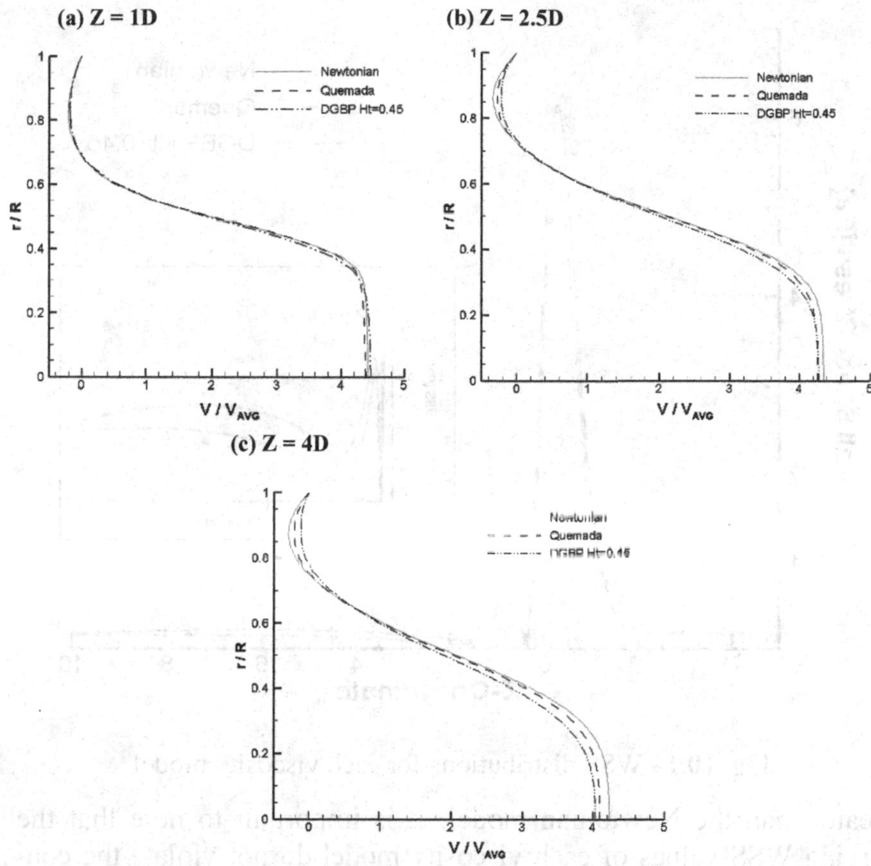

Fig. 10.13 Velocity profiles of the three viscosity models (a) z = D , (b) z = 2.5D , (c) z = 4D

The reason for these WSS distributions is that a lower limit was not placed on the Quemada model and at shear rates above 10^2 s^{-1} the Quemada model may actually predict a lower viscosity than the Newtonian model, due to its shear-thinning behavior. The DGBP Ht = 0.45 model will not predict a viscosity lower than the Newtonian value until higher shear rates occur. The magnified window of Fig. 10.14 reveals that in regions where the Quemada model would predict a higher viscosity than the Newtonian model, the WSS is

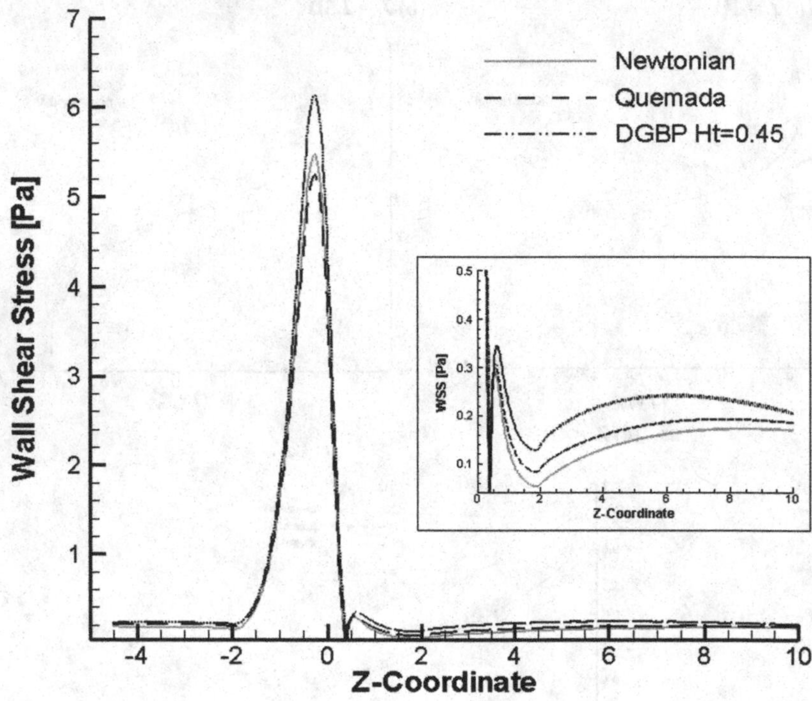

Fig. 10.14 WSS distributions for each viscosity model

greater than the Newtonian model. It is important to note that the varying WSS values of each viscosity model do not violate the conservation of momentum because the velocity is not fully developed at this point. The presence of the additional terms (specifically the dependence on the z-coordinate) creates discrepancies between the Newtonian and non-Newtonian WSS values.

10.4.4 Conclusions

The previous study analyzed the effects of non-Newtonian vs. Newtonian viscosity models on the local flow parameters in a constricted tube. The current study assumed constant hematocrit (Ht) values, where in actuality Ht is a function of time and space within the local domain. Such behavior can be incorporated into future analyses using a scalar transport equation for $Ht(\bar{x},t)$ that takes into account both diffusion and convection. Specifically, future simulations could be coupled with particles whose local concentrations would affect the value of the Ht parameter. Additionally, to analyze the time-dependent behavior of the non-Newtonian viscosity models, transient simulations would need to be conducted.

The results and discussions of the current study reveal that the local fluid dynamics is influenced by the type of non-Newtonian viscosity model. Specifically, the reattachment lengths, peak WSS values, and regions of the velocity field are significantly affected. The pressure gradient of the different viscosity models did not reveal significantly different distributions, which implies that the pressure drop across the constricted portion of the vessel is unchanged for each model. The recirculation zones proved to be highly dependent on the viscosity of the fluid and such knowledge should be considered when simulating pathological conditions where large recirculation regions occur (i.e. aneurysms, carotid artery, etc.).

De Gruttola et al. (2005) have shown that their model accurately predicts local concentrations of RBCs, WBCs, and platelets. However, when compared to the whole blood behavior as a function of shear rate, the model exhibits only similar results as the validated Quemada model in elevated shear rate regions, but is offset from the Quemada model by a factor related to the assumed Ht value. The low shear rate regions yield vast differences between all three models. In particular, the DGBP model predicts a much higher viscosity for the lower shear regions than the Quemada model. More experimental data of whole blood behavior in this low shear rate region is needed to determine which model best predicts actual blood behavior. An advantage of the DGBP model is the ability to incorporate a locally varying Ht value in future work, whereas the simplified Quemada model has the Ht term lumped within one of its parameters. For the utilized experimental data, the Quemada model still appears to be a more accurate predictor of whole blood behavior, but the DGBP model has the potential to come close to the Quemada model predictions given the right parameters, both curve-fitted and operational ones.

CFX V.11 (ANSYS Inc.) made implementation of both non-Newtonian models rather simple using the highly customizable CFX Expression Language (CEL) interface. The Newtonian model was the easiest to implement; however, it suffered the most from not matching the whole blood behavior under various shear rates.

In conclusion, the current study highlighted specific effects of non-Newtonian viscosity models on the local fluid dynamics. Physical insight gained from this study illustrates the need to incorporate non-Newtonian effects of blood where significant amounts of recirculation exist, regardless of the vessel size. Therefore, future analyses should incorporate and continue to investigate the non-Newtonian behavior of whole blood within low-shear regions and for different pathological conditions.

10.5 Example of External Flow

10.5.1 Background Information

An excellent example of external flow, relying on Chaps. 2, 5 and 8, is the multiphysics phenomenon of fluid-structure interaction (FSI). FSI is the coupling of solid mechanics and fluid dynamics as either two-way or one-way coupling. In one-way coupling, results are only passed on to one system (e.g., water flowing through a rigid pipe; where the flow of water results in pressure loading of the pipe, but the pipe does not deform and thus does not affect fluid flow). Conversely, in two-way coupling, the results are passed between both events. An example is flow through an artery, where the blood flow causes a pressure loading of the arterial wall, the arterial wall deforms under the pressure load which will in turn affect the artery's geometry and hence fluid flow. Figure 10.15 contains diagrams illustrating the different coupling of one and two-way fluid–structure interactions.

Important considerations of FSI simulations are the computational mesh and timestep iterations. Often times, FSI simulations are computationally demanding and require fine-tuned parameters to achieve desired accuracy and manageable simulation time.

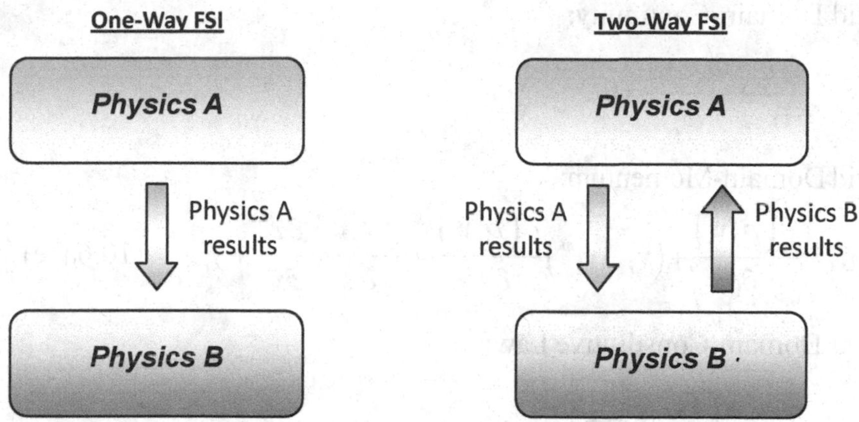

Fig. 10.15 Coupling diagrams of one and two-way FSI

10.5.2 Theory

The governing equations of most FSI systems are the conservation of momentum and mass for fluid flow coupled with the conservation of momentum (Newton's second law) for solid bodies and equations of mechanical equilibrium for solid bodies. The unknown quantities of the first set of equations include fluid pressure, orthogonal fluid velocity components, solid displacement/deformation, and solid forces/stresses. In order to solve for these unknowns, additional equations are needed, called *constitutive* equations. They link the fluid forces with fluid motion and solid forces with solid displacement/deformation. These equations often describe the material properties/parameters of the fluid or solid, and have a wide range of complexities (i.e. Newtonian vs. non-Newtonian viscosity models; isotropic Hooke's Law vs. anisotropic strain-energy functions). Specifically, Eqs. (10.6a–e) and (10.7a, b) provide the fluid, solid, and mesh displacement governing equations. They were solved with the commercial simulation products CFX and ANSYS (Ansys Inc. Canonsburg, PA):

Fluid Domain-Continuity:

$$\frac{\partial \rho_f}{\partial t}+\frac{\partial(\rho_f v_i)}{\partial x_i}=0$$

Fluid Domain-Momentum:

$$\frac{\partial(\rho_f v_i)}{\partial t}+(v_j - v_j{}^m)\frac{\partial(\rho_f v_i)}{\partial x_j}=-\frac{\partial p}{\partial x_i}+\frac{\partial \tau_{ij}}{\partial x_j}+f_i^V \qquad (10.6a\text{–}e)$$

Fluid Domain-Constitutive Law:

$$\tau_{ij}=\eta\left[\frac{\partial v_i}{\partial x_j}+\frac{\partial v_j}{\partial x_i}\right]$$

Fluid Domain-Mesh Displacement:

$$\nabla\cdot(\Gamma_{disp}\nabla\delta)=0$$

$$\Gamma_{disp}=f(\forall)$$

Solid Domain-Mechanical Equilibrium:

$$\frac{\partial \sigma_{ij}}{\partial x_j}+\rho_s f_i^{Vs}=\rho_s a_i \qquad (10.7a\text{–}d)$$

Solid Domain-Constitutive Laws:

- *Hooke's Law* ($\sigma_{ij}=C_{ijkl}E_{kl}$)
- *Strain-Energy Function* ($\sigma_{ij}=\dfrac{\partial\psi}{\partial E_{ij}}, \psi=\psi(E_{ij})$)
- *Viscoelastic Model* ($\sigma_{ij}=f(E_{ij},t)$)

Here, x_i is the three orthogonal coordinate directions, t represents time, ρ_f is the fluid domain density, v_i is the fluid velocity vector, v_i^m is the mesh velocity vector, p is the fluid pressure, τ_{ij} is the fluid shear stress tensor, η is the apparent dynamic viscosity of the fluid, δ is the displacement relative to the previous mesh position, $\forall$ is the cell/element control volume, f_i^v is a volumetric body force vector acting on the fluid domain, σ_{ij} is the solid stress tensor, f_i^{sV} is a volumetric body force vector acting on the solid domain, ρ_s is the solid domain density, a_i is the acceleration vector of the solid domain, C_{ijkl} is the general stiffness matrix, E_{ij} is the Lagrange strain tensor. It is worthy to note how the convective terms in the fluid conservation of momentum are the most altered as convective terms must take into account the deformable mesh and its velocity.

Further description is needed to understand the mesh displacement equations listed in Eqs. (10.6d, e). The overall methodology is to utilize a diffusion transport equation to define the mesh displacement at every control volume point throughout the domain. Mesh displacement relative to the previous mesh locations (δ) is the variable that is acted upon by a gradient and divergence operation and Γ_{disp} is a stiffness function that enables the option of a non-uniform diffusion of displacement throughout the computational fluid domain. Often, the stiffness parameter is a function of the element control volume, which enables the mesh to have different stiffness values at different points in time for transient analyses. Such a function is very important to prevent mesh foldings and negative volume elements from forming during FSI analyses with large displacements. It is important to note that Eqs. (10.6d, e) are utilized in CFX V.11 and should not be regarded as the only method to transfer the deformations calculated by the solid domain to the fluid mesh.

Two primary forms of coupling methodologies exist between the solid and fluid domains. The first is a direct coupling method, which solves both fluid and solid physics in a single computational step. The second is an iterative coupling methodology that solves the fluid and solid physics in a sequential process (i.e., fluid physics solved first, fluid loads are passed on as inputs to the solid physics simulation, resulting solid displacements are then mapped to the fluid domain mesh, fluid physics are then solved on the deformed mesh). Iterative coupling is more prevalently used in most situations in attempts to control the nonlinearities of both the fluid domains and ensure that physics are not eliminated through the coupling procedure. Figure 10.16 provides a flowchart illustrating the computational algorithm of the iterative coupling methodology per timestep, where the fluid domain is solved prior to the solid domain.

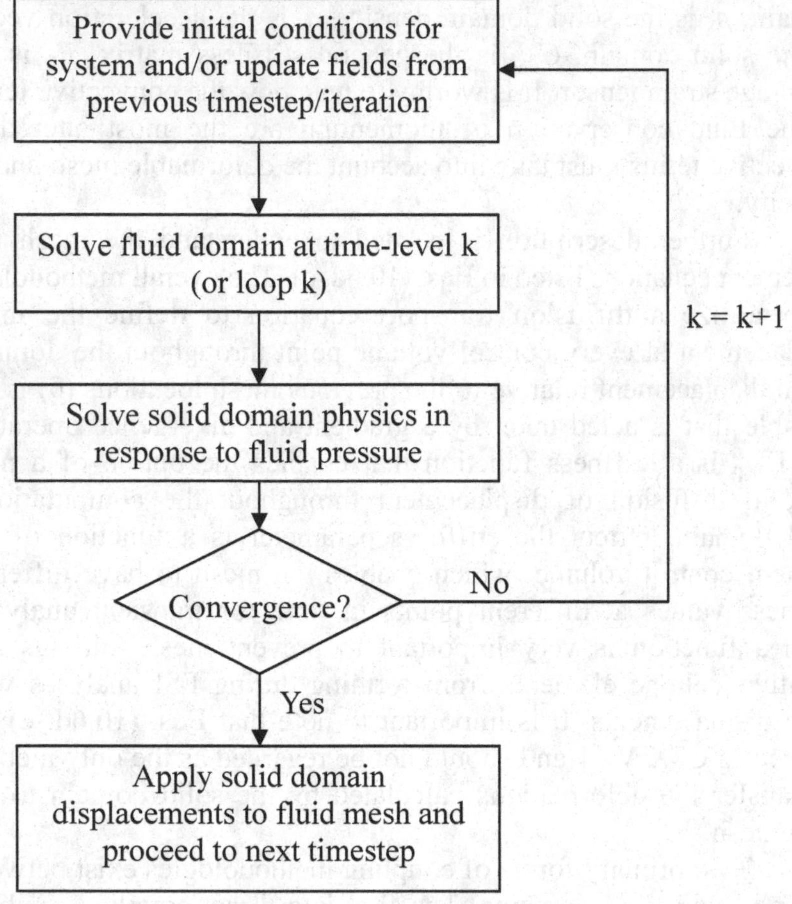

Fig. 10.16 Iterative coupling procedure per timestep with fluid physics being solved first

10.5.3 One-Way FSI Simulation of 2D-Flow over a Tall Building

System description An example of one-way FSI is the wind loading of a high rise, where building displacements do not significantly alter the surrounding airflow. A simplified 2D simulation of flow over a building is utilized for this illustration. Recent simulations of wind-loaded structures have shown the increased importance of 3D-simulations to capture all the physical phenomena; thus, the following analysis is used for illustration purposes and should not be used for more complex structural health design considerations. Figure 10.17

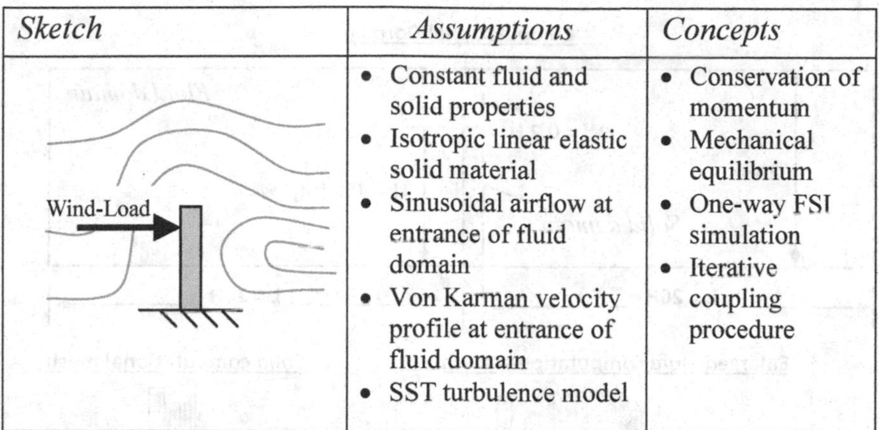

Sketch	*Assumptions*	*Concepts*
	• Constant fluid and solid properties • Isotropic linear elastic solid material • Sinusoidal airflow at entrance of fluid domain • Von Karman velocity profile at entrance of fluid domain • SST turbulence model	• Conservation of momentum • Mechanical equilibrium • One-way FSI simulation • Iterative coupling procedure

Fig. 10.17 Sketch, assumptions, and concepts of 2D-flow over an obstacle

summarizes the preliminary steps, i.e., system sketch, assumptions, and concepts.

Simulation Implementation ANSYS products along with Solid Works2008 were employed to develop the example. SolidWorks was used to create the fluid and solid domain geometry files, while ANSYS ICEM CFD generated the fluid mesh, ANSYS CFX carried out the fluid calculations, and ANSYS Workbench generated the solid mesh and carried out the solid calculations. Figure 10.18 illustrates the system dimensions and computational mesh. Notice regions of mesh refinement, which correspond to high gradients of the simulation variables (i.e., velocity, pressure, stress, and displacement). A caveat must be made when conducting FSI analyses, i.e., mesh refinement is a very sensitive task that can have a prominent effect on the final FSI solutions. In simulations where large displacements occur, the fluid mesh can be stretched and/or skewed in ways that cause mesh folding (negative volume elements) or can cause regions of originally sufficient mesh refinement to be stretched in ways that negate the mesh refinement and cause inaccurate solutions to regions of high gradients. Ways around these problems include knowledge of the system physics, using adaptive mesh refinement with element death throughout the solution procedure, or specifying a function of the mesh element size that diffuses the displacements to regions of the fluid mesh that can handle these displacements (cf. Eq. 10.6e).

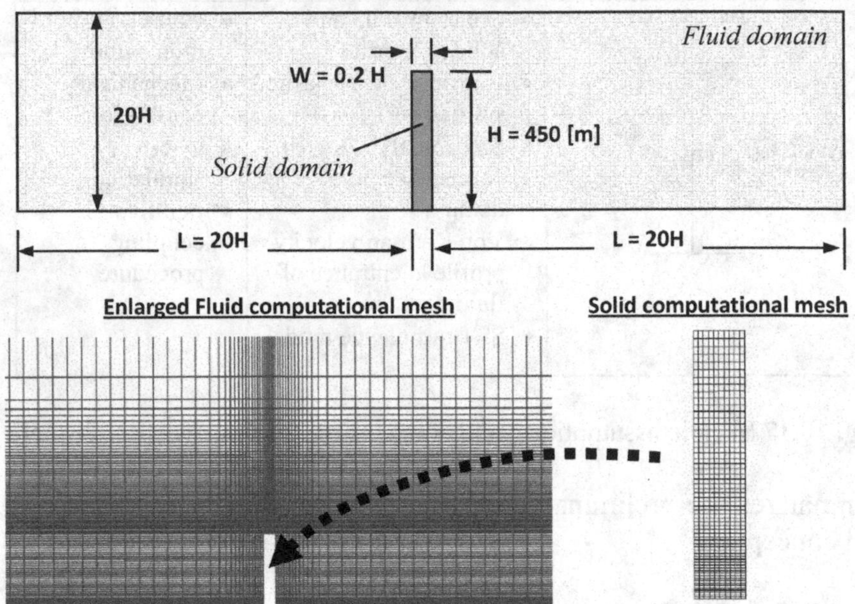

Fig. 10.18 Fluid and solid computational domains and meshes

Table 10.3 Computational settings of the coupled systems

Fluid domain	Solid domain
(i) Air at 25°C: • μ_{air} = 1.831E-5 Pa*s • ρ_{air} = 1.185 kg/m^3	(i) Mixture of Concrete and Steel • $E_{bldg} \approx$ 6.00E10 Pa • ν_{bldg} = 0.23
(ii) RMS Residual < 1E-4	(ii) L2 norm convergence for force and displacement
(iii) von Karman inlet velocity profile • $V_0\left(2\dfrac{y}{\delta}-\left(\dfrac{y}{\delta}\right)^2\right), \quad 0 \le y \le \delta, \quad \delta = 0.2H$ • $V_0 = 15[m/s]\cdot\sin(\omega\cdot t)$	

The computational settings applied to the coupled systems are summarized in Table 10.3 while the boundary conditions of each domain are illustrated in Fig. 10.19.

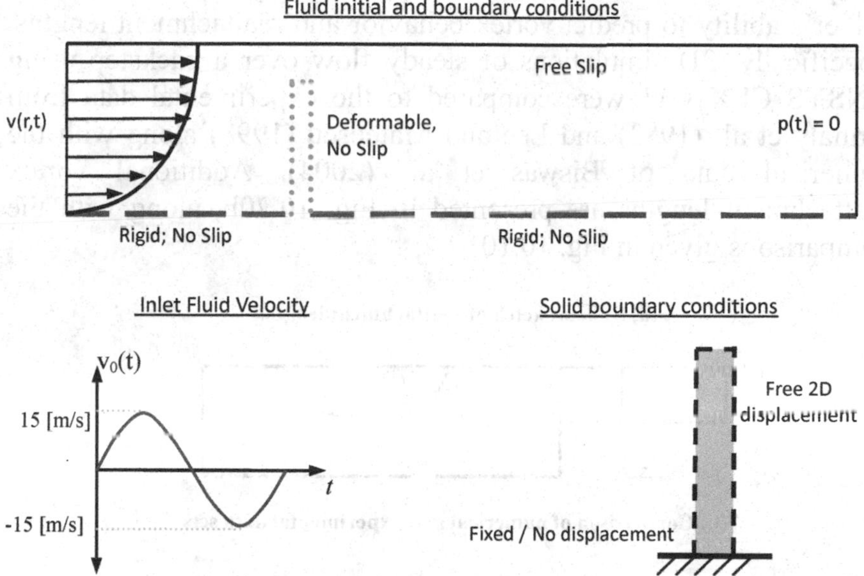

Fig. 10.19 Boundary conditions of fluid and solid systems

The iterative coupling methodology is implemented for this simulation, which requires additional parameters to be specified that are related to the number of internal loops ("k" in Fig. 10.16) and the convergence criteria of the variables passed between the two physical domains. The conditions are

- Max. number of iterations loops is 10.
- Min. number of iteration loops is 2.
- Iteration loops repeat until both domains' local physics converge and the coupling variables (fluid mesh displacement and applied loads) residuals < 1E-3.

Three types of simulations were conducted. The first one was a steady-state simulation with the inlet velocity at 15 m/s, second was a transient simulation that simulated developing flow until a near steady-state point was reached, and third was a transient simulation with a timestep of 5 s for a total time of 180 s. The amplitude and frequency of the inlet sinusoidal velocity for the transient case was set at 15 m/s and 2π rad/min, respectively.

Prior to the full solution of the FSI system, a validation comparison with flow over a backstep was completed to ensure the solver's ability to predict vortex behavior and reattachment lengths. Specifically, 2D simulations of steady flow over a backstep using ANSYS CFX v.11 were compared to the experimental data from Armaly et al. (1983) and Lee and Mateescu (1998) along with the numerical data of Biswas et al. (2004). Additional vortex reattachment lengths are presented in Fig. 10.20b, along with the comparisons given in Fig. 10.10.

(a) System sketch of reattachment lengths

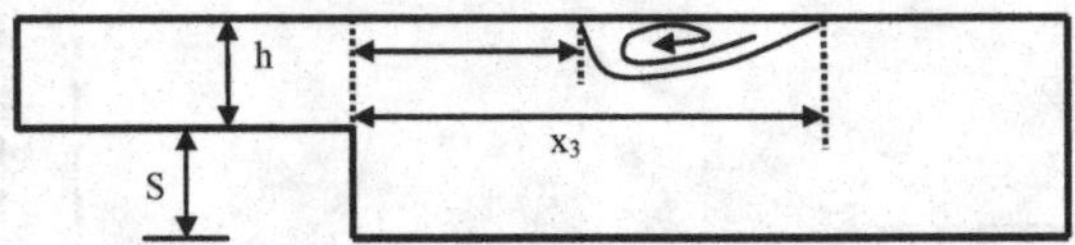

(b) Comparison of numerical and experimental data sets

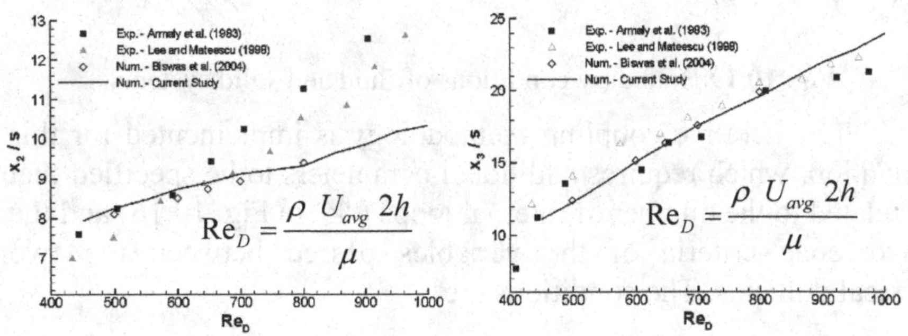

Fig. 10.20 (a) System sketch of reattachment lengths. (b) Comparison of experimental and numerical reattachment lengths of steady, laminar flow over a backstep

The additional vortices only develop at Re_D values larger than 400, which is where the 3D effects begin to have a prominent influence, and hence the 2D simulation accuracy diminishes. Despite the growing discrepancy illustrated in Fig. 10.20b for the x_2 reattachment length, the x_3 reattachment length data reveal a good match for all simulated Reynolds numbers. Thus, further validation of the 2D simulation environment for flow over an obstacle (i.e., a representative building) has been completed and suggests a specific region of Reynolds numbers where the 2D simulations are most valid.

Simulation Results The steady flow simulation can be compared against specific time points of the developing flow simulations to illustrate the characteristic time when the system's physics can be modeled as a time-independent process. Figure 10.21 illustrates the difference between the steady solution and the early start-up flow simulations.

The different plots in Fig. 10.21 are on different length scales to provide a clear view of the recirculation zone development with time for a constant approach stream. A measurable vortex develops immediately downstream of the building after 55 s and continues to increase in length and height (t =300 s). Additionally, at around 300 s, a small recirculation zone is noticeable at the base of the building structure. Even after t = 300 s, the steady-state recirculation zone has not been established. Therefore, the time needed to obtain the steady-state physics of fluid flow is greater than 5 min for a constant inflow velocity of 15 m/s.

Interestingly, the stress fields for the developing flow simulations also show some differences to the stress fields of the steady flow simulations. Figure 10.22 highlights the different levels of von Mises stress at different time levels for both start-up and steady flow simulations.

Figure 10.22 clearly illustrates that even if the flow over the building is constant, the resulting loads applied to the building vary with time and are the cause of the varying stress values with respect to passing time. Furthermore, an approximate 400% difference exists between the max von Mises stress level calculated by the steady flow simulation and the developing flow simulation. Thus, time effects should be a significant consideration when conducting analyses of flow over structures and questions regarding the validity of steady-state simulations to estimate structure loading should be investigated prior to utilizing steady-state simulations for detailed analyses of structures.

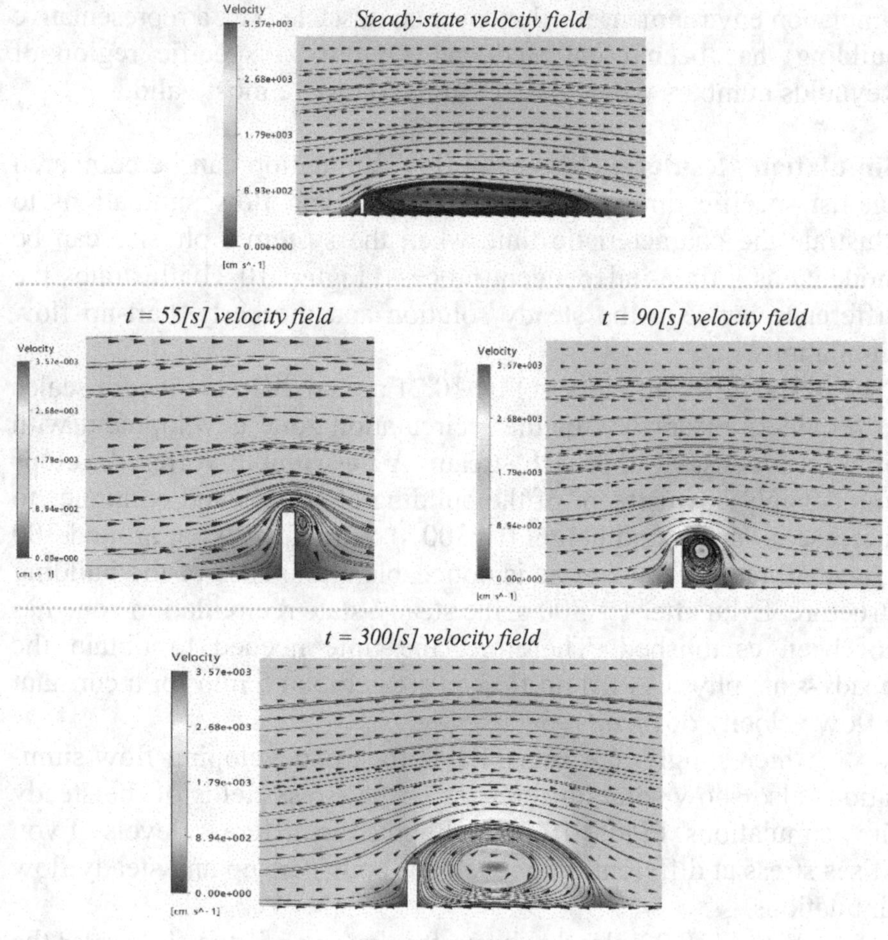

Fig. 10.21 Steady flow field compared to temporal flow field development at t = 55, 90, and 300 s

The timescales of all simulations varied significantly. Table 10.4 lists the different completion times of the various simulations completed for the external flow study. All simulations were completed using a single Intel 3.59 GHz processor on a Dell Precision 670 workstation with 8 GB of RAM using Windows64.

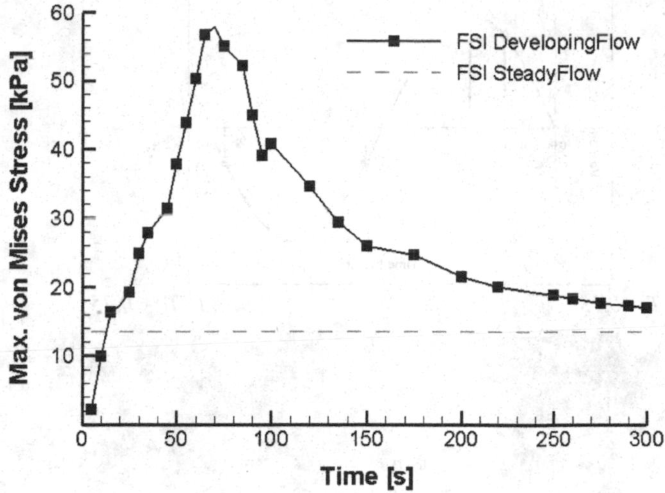

Fig. 10.22 Maximum von Mises stress vs. time

Table 10.4 Simulation run times

1-Way FSI simulation	Run time
2D steady flow	40 min
2D developing flow with constant inflow velocity	6 h 38min
2D oscillatory flow with sinusoidal inflow velocity	12 h 39 min

Results of the transient computational simulations exhibit measurable differences when compared to the constant-inlet-velocity simulation results. Specifically, vortices developed at different locations, primarily during the deceleration phases of the inlet velocity. Figure 10.23 illustrates the velocity fields around the building at different time levels during the third period. Only the third period is shown since the previous two periods do not exhibit visible dynamic equilibrium/cyclic repetition whereas a visible amount of dynamic equilibrium/cyclic repetition is observed in the third period.

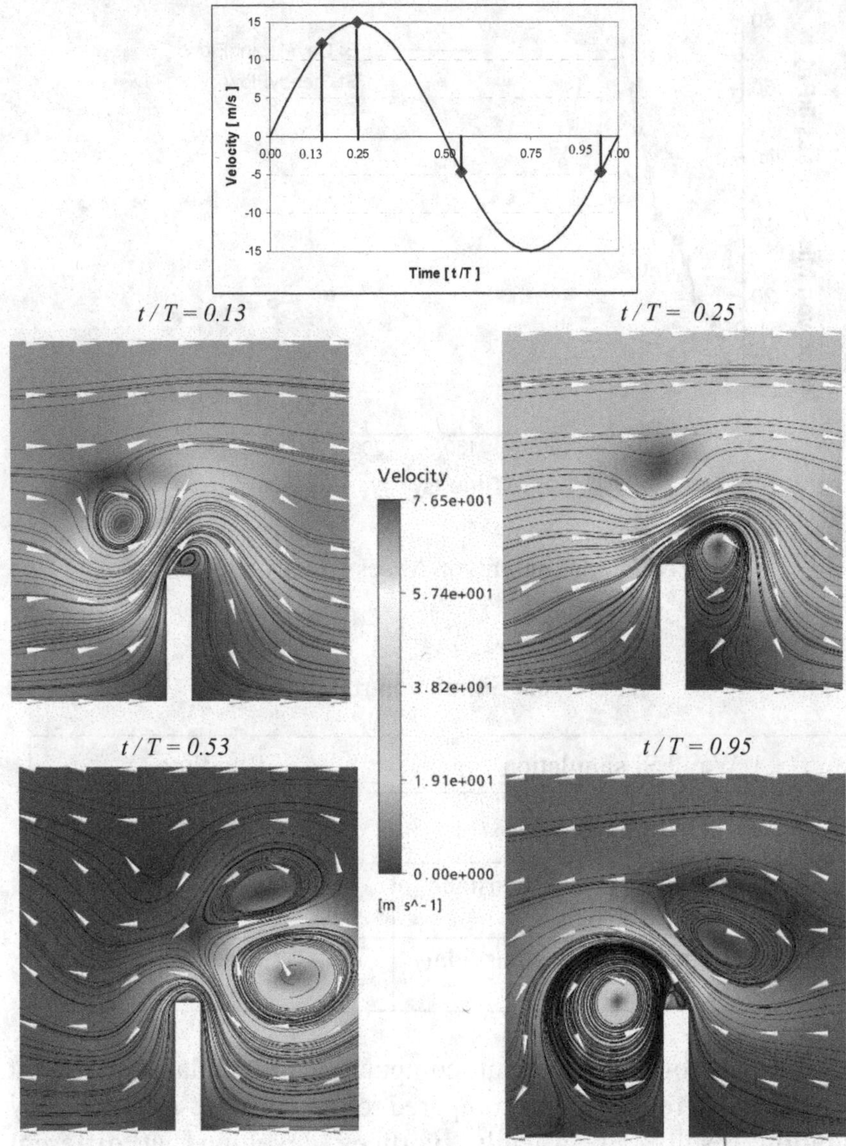

Fig. 10.23 Velocity fields around building at time points during the third period

An interesting result of the transient simulations is the development of multiple vortices at both upstream and downstream locations of the building. During the maximum levels of retrograde flow, the vortices significantly diminish and begin to return during the deceleration portion of negative inflow velocity, which continue to be present at the beginning of the acceleration of positive inflow. Throughout both the positive and negative inflows, a recirculation zone the size of the steady-state recirculation never develops during the three periods of oscillating flow.

Figure 10.24 illustrates the oscillatory behavior of the average force exerted by the wind onto the building's side surface.

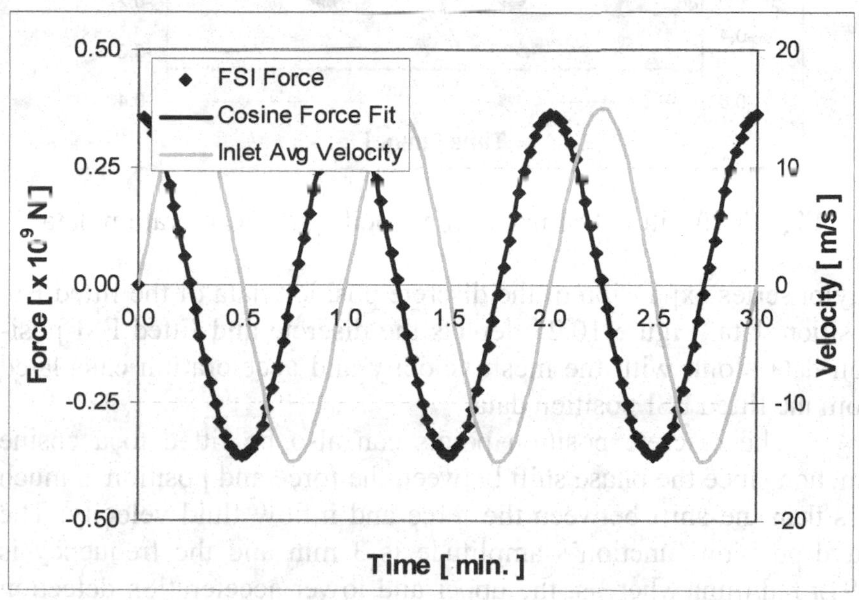

Fig. 10.24 Fitted FSI position, velocity, and acceleration data

The plot in Fig. 10.24 reveals that the average force applied to the building is approximately shifted 90° behind the average inlet fluid flow velocity. A cosine function can be used to fit the discrete force with an amplitude of 0.36×10^9 N and a frequency of 2π rad/min. Mesh position, velocity, and acceleration can also be calculated at the centerpoint of the building's upper surface. Mesh position is a direct output of the FSI simulation, whereas mesh velocity and acceleration are calculated using a first-order forward-difference.

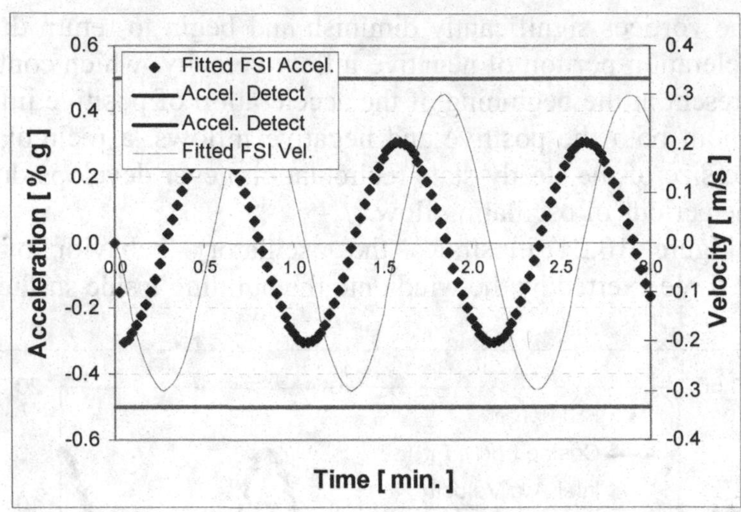

Fig. 10.25 Fitted FSI position, velocity, and acceleration data

Taylor series expansion of the discrete position data of the fitted FSI position data. Figure 10.25 depicts the discrete and fitted FSI position data along with the mesh velocity and acceleration calculated from the fitted FSI position data

The discrete position points can also be fitted to a cosine function since the phase shift between the force and position is much less than the shift between the force and inflow fluid velocity. The fitted position function's amplitude is 3 mm and the frequency is 1.91π rad/min; whereas, the upper and lower acceleration detection lines are based on the human levels of detectable acceleration (0.5%) from Simiu and Scanlan (1996).

The analytical single degree-of-freedom (SDOF) solution provided in Example 8.4 of Chap. 8 can be used to fit the analytic results to the oscillatory curves of the 2D FSI data shown in Figs. 10.24 and 10.25. It determines parameters of a simplified FSI model that can be used for extensive parametric analyses, which would require a large amount of simulation time if using the 2D FSI approach. Figure 10.26 depicts the fit between the 2D FSI position and the SDOF model, where the SDOF model has a stiffness coefficient of 1.18E11 kg/s^2 and a damping coefficient of 1.13E9.

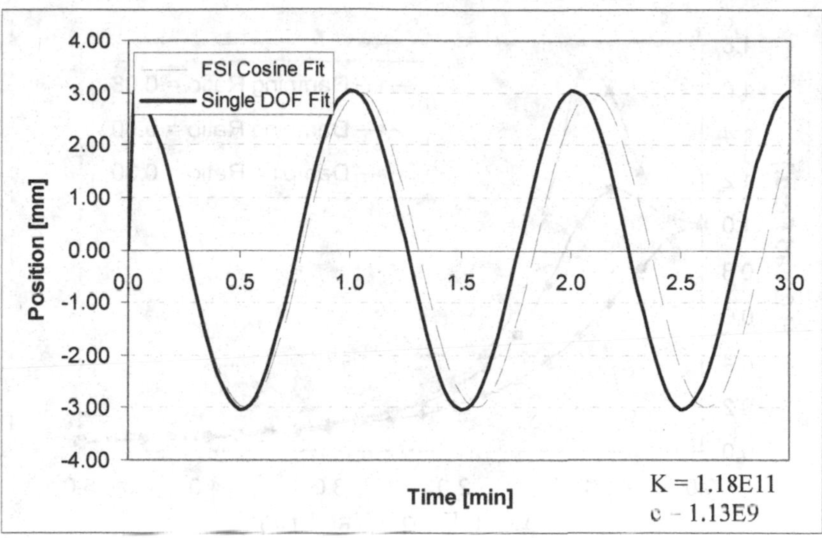

Fig. 10.26 Fitting the 2D FSI position data to the SDOF model in Example 8.4

Figure 10.26 reveals a growing discrepancy, in terms of a time shift, between the SDOF model and the 2D FSI data. However, it is important to remember that the SDOF model is a greater approximation of reality than the 2D FSI simulation. It should be recalled from Example 8.4 that the two pivotal system parameters defining the SDOF model are the natural frequency and the damping ratio. The stiffness coefficient, mass of the building model, and the damping coefficient yield a natural frequency of 1.57E2 rad/s and a damping ratio of 0.75. Utilizing these two parameters, an analysis of loading frequencies that would be most damaging to the structure can be obtained using the SDOF model. Upon further examination of the SDOF model, the primary term that will influence the resulting displacement to have a value much larger than the force is the H(ω)-function where ω is the loading frequency. Thus, Fig. 10.27 plots the H(ω)-function based on arbitrary natural frequencies, loading frequencies, and damping ratios.

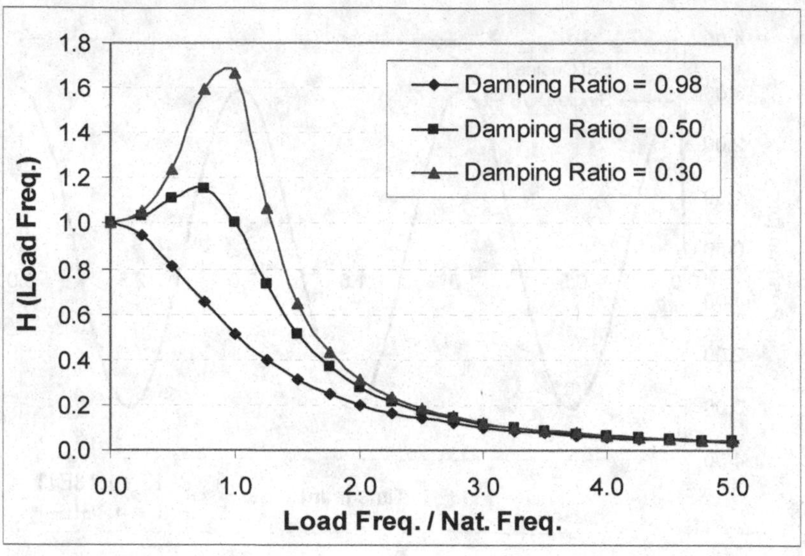

Fig. 10.27 H(ω)-functions for arbitrary system parameters

An interesting trend is observed with the H(ω)-function such that for damping ratios greater than or equal to 0.5, the maximum point does *not* occur when the loading frequency equals the natural frequency. However, when the damping ratio is closer to 0.3 (and below), the maximum H-value does occur when the loading frequency is equal to the natural frequency. Utilizing the understanding gained from this investigation, the loading conditions which would produce the maximum displacement under the derived SDOF natural frequency and damping ratio would be a high load (amplitude) under a low frequency. For instance, a cyclic loading with an amplitude of 36×10^9 N and a frequency of 0.5π rad/min would result in a maximum displacement of 0.304 m for the fitted SDOF parameters.

Summary and Conclusions Simulations of steady flow, developing flow and oscillatory flow over a simplified structure utilizing a one-way iterative FSI coupling has been completed and documented. Specific understanding regarding the characteristic time needed for steady-state physical phenomena have been investigated and shown to be greater than 5 min, and the potential to underestimate the maximum stress in structures by a factor of 400% when using a steady-state simulation. Oscillatory inlet flow has been shown to produce velocity fields and oscillatory loads not captured in any constant velocity simulation. Moreover, the SDOF model in Example 8.4 has been utilized to conduct a simplified parametric analysis of the system. The loading conditions that would yield the largest displacement (≈0.3 m) are high amplitude loads under low frequencies.

The previous example illustrates the utility of one-way FSI analyses and their ability to provide useful insight of complex physical phenomena that may not be possible with experimental analyses alone. It should be noted that to accurately determine the natural frequency and damping ratio of a system, additional experiments/simulations should be conducted to ensure that the fitted parameters enable matching of the system behavior in more than a single implementation. Additionally, the oscillatory loads and displacement of the vibrating structure should also include an analysis of fatigue to ensure that the strength of the structure under oscillatory loads does not decrease to a level that makes fatigue failure a high possibility.

10.6 Group Assignments and Project Suggestions

Again, the HWAs evolved directly out of Chap. 10, while the PSs are variations to the internal and external flow examples, i.e., Sects. 10.4 and 10.5. Research reports describing specific course projects should follow the format outlined in Fig. 10.4.

10.6.1 Group Assignments

10.6.1 Why is "problem recognition and system identification" so crucial in math modeling? Provide three additional examples (see Sect. 10.2.1) with sketches, assumptions and concepts.

10.6.2 In Sect. 10.2.2, three model categories are described. Give one detailed example each.

10.6.3 Provide illustrative examples and discuss

(i) Structured vs. instructured meshes
(ii) Methods of discretization of differential equations
(iii) Initial vs. inlet/outlet vs. wall conditions
(iv) Periodic, free-surface and permeable-wall boundary conditions
(v) Advantages of dealing with non-dimensionalized equations, system parameters, graphs, etc.

10.6.4 Develop a flow-chart for "prototyping in virtual reality" and give an insightful application

10.6.5 Interpret and discuss

(i) Figure 10.5 in Sect. 10.3
(ii) Figure 10.6, noting especially the cross-over at $Q_1 / Q_2 \approx 1.2$
(iii) Figure 10.7 in light of particle dynamics modeling
(iv) Figure 10.9 concerning $\eta(\dot{\gamma} \rightarrow 0)$ and the Ht influence

10.6.6 Update Fig. 10.9 with experimental data for whole blood in the shear-rate range $0 < \dot{\gamma} < 10^2\,s^{-1}$.

10.6.7 Discuss the pros and cons of Eqs. (10.4a–d) and suggest a PDE for Ht $(\vec{x}, t)$. How would you integrate the Ht-results into the DGBP model?

10.6.8 While two-way coupled fluid–structure interaction (FSI) is most realistic (see Sect. 10.5), it has been shown that in many industrial FSI applications one-way coupling is sufficient, i.e., the error is less than 1–3%. Provide two examples each for one-way vs. two-way FSIs.

10.6.9 Same as 10.6.8 but for fluid–particle interactions (FPIs).

10.6.10 In direct numerical simulations, moving structures (or particles) in a fluid flow field require constant mesh updates near the submerged body (see Eqs. (10.6d, e) for example). Provide a literature review with tutorial examples of alternative methods/ applications to obtain FSI or FPI results.

10.6.11 W.r.t. Fig. 10.20, why does a second "recirculation zone" of length $x_3 - x_2$ appear in the backward-facing step example?

10.6.12 W.r.t. Fig. 10.21, why does it take over 5 min after flow start-up to reach the steady-state velocity field?

10.6.13 Provide the underlying reasons and discuss:

(i) Figure 10.22, i.e., the maximum wall stress spike at $t \approx 70$ s.

(ii) Figure 10.23 w.r.t. the vortex-pair strength development and shift

(iii) Figure 10.24 in light of the $\pi/2$-shift of the resulting force function vs. the "input" function

(iv) Figures 10.26 and 10.27, concerning the pros and cons of using the SDOF model of Chap. 8

10.6.2 Project Suggestions

10.6.14 Similar to the validation examples of Sect. 10.3, select from the experimental literature steady 2-D (or axisymmetric) flow applications for computer simulations.

10.6.15 As 10.6.14 but add: (i) nanomaterial transport; or (ii) micro particle transport.

10.6.16 Study the new correlation for non-spherical particles (i.e., Eq. (10) in Hölzer & Sommerfeld (2008); Powder Technology; 84: 361–365) and apply it for, say, rigid fiber/ellipsoid dynamics in parallel-plate flow.

10.6.17 Considering flow in microchannels, select a suitable example from Chap. 7 to perform a numerical study, including variations in fluid (or mixture) properties and operational conditions.

10.6.18 Similar to 10.6.17, focus on the numerical solution of examples from Chap. 6.

10.6.19 Similar to 10.6.17, focus on the numerical solution of examples from Chap. 8.

References (Part C)

Armaly, B.F., Durst, F., Pereira, J.C.F., Schönung, B., 1983, Journal of Fluid Mechanics, Vol. 127, pp. 473–496.

Bejan, A., 1996, *Entropy Generation Minimization, the Method of Thermodynamic Optimization of Finite-Size System and Finite-time Processes*, CRC Press, Boca Raton, FL.

Bejan, A., 2002, International Journal of Energy Research, Vol. 26, pp. 545–565.

Bird, G.A., 1994, *Molecular Gas Dynamics and Direct Simulation of Gas Flow*, Oxford University Press, Oxford, UK.

Bird, R.B., Armstrong, R.C., Hassager, O., 1987, *Dynamics of Polymeric Liquids*, 2nd ed., Wiley, New York.

Bird, R.B., Stewart, W.E., Lightfoot, E.N., 2002, *Transport Phenomena*, 2nd ed., Wiley, New York.

Biswas, G., Breuer, M., Durst, F., 2004, Journal of Fluids Engineering (ASME), Vol. 126, pp. 362–374.

Brotherton, C.M., Sun, A.C., Davis, R.H., 2008, Microfluid Nanofluid, Vol. 5, pp. 43–53.

Brown, R., 1827, in: King, P.P. (ed), *Narrative of a Survey of the Intertropical and Westerncoasts of Australia, Performed Between Years 1818 and 1822*, John Murray, London, pp. 534–565.

Buchanan, J.R., Kleinstreuer, C., 1998, ASME Journal of Biomechanical Engineering, Vol.120 (3), pp. 446–454.

Buchanan, J.R., Kleinstreuer, C., Comer, J.K., 2000, Computers and Fluids, Vol. 29 (6), pp. 695–724.

Chandran, K.B., Rittgers, S.E., Yoganathan, A.P., 2007, *Biofluid Mechanics: the Human Circulation*, CRC Press, Boca Raton, FL.

Chang, C.-C., Yang, R.-J., 2007, Microfluid Nanofluid, Vol. 3, pp. 501–525.

Chung, C.K., Shih, T.R., 2008, Microfluid Nanofluid, Vol. 4, pp. 419–425.

Chein, R., Chuang, J., 2005, Applied Thermal Engineering, Vol. 25, pp. 3104–3114.

Choi, J.I., Kim, C.S., 2007, Inhalation Toxicology, Vol. 19, pp. 925–939.

Choi, S.U.S., 1995, Developments and Applications of Non-Newtonian Flows, in Sininer, D.A. and Wang, H.P. (eds), ASME, New York, FED-Vol. 231/MD-Vol. 66

Chopkar, M., Kumar, S., Bhandari, D.R., Das, P.K., Manna, I., 2007, Materials Science and Engineering B, Vol. 139, pp. 141–148.

Cimbala, J.M. and Cengel, Y.A., 2008, *Essentials of Fluid Mechanics: Fundamentals and Applications*, McGraw-Hill, New York, NY.

Clift, R., Grace, J.R., Weber, M.E., 1978, *Bubbles, Drops and Particles*, Academic, New York.

Crowe, C.T. (ed), 2006, *Multiphase Flow Handbook*, CRC Press, Boca Raton, FL.

Crowe, C.T., Sommerfeld, M., Tsuji, Y., 1998, *Multiphase Flows with Droplets and Particles*, CRC Press, Boca Raton, FL.

De Gruttola, S., Boomsma, K., Poulikakos, D., 2005, Artificial Organs, Vol. 29(12) pp. 949–959.

Durbin, P.A., Medic, G., 2007, *Fluid Dynamics with a Computational Perspective*, Cambridge University Press, New York, NY.

Faghri, A., Zhang, Y., 2006, *Transport Phenomena in Multiphase Systems with Phase Change*, Elsevier, Burlington, MA.

Ferziger, J.H., Peric, M., 2002, *Computational Methods for Fluid Dynamics*, 3rd ed., Springer-Verlag, New York, NY.

Finlay, W.H., 2001, *The Mechanics of Inhaled Pharmaceutical Aerosols: an Introduction*, Academic, London, UK.

Floyd-Smith, T.M., Golden, J.P., Howell, P.B., Ligler, F.S., 2006, Microfluid Nanofluid, Vol. 2, pp. 180–183.

Fung, Y., 1994, *A First Course in Continuum Mechanics: for Physical and Biological Engineers and Scientists*, Prentice Hall, Englewood Cliffs, NJ.

Hardt, S., Drese, K.S., Hessel, V., Schonfeld, F., 2005, Microfluid Nanofluid, Vol. 1, pp. 108–118.

Heris, S.Z., Etemad, S.Gh., Esfahany, Nasr, M., 2006, International Communications in Heat and Mass Transfer, Vol. 33, pp. 529–535.

Hölzer, A., Sommerfeld, M., 2008, Powder Technology, Vol. 184, pp. 361–365.

Hochmuth, R.M., 1986, Properties of red blood cells. In: Skalak, R., Chein, S., (eds), *Handbook of Bioengineering*, McGraw-Hill, New York, NY.

Hornyak, G.L., Tibbals, H.F., Dutta, J., Moore, J.J., 2009, *Introduction to Nanoscience and Nanotechnology*, CRC Press, Boca Raton, FL.

Humphrey, J.D., Delange, S.L., 2004, *An Introduction To Biomechanics: Solids and Fluids, Analysis and Design*, Springer Verlag, Berlin/Heidelberg/New York.

Ingham, D.B., 1975, Aerosal Science, Vol. 6, pp. 125–132.

Incropera, F.P., DeWitt, D.P., Bergman, T.L., Lavine, A.S., 2007, *Introduction to Heat Transfer*, Wiley, New York, NY.

Jang, S.P., Choi, S.U.S., 2006, Applied Thermal Engineering, Vol. 26, pp. 2457–2463.

Kang, T.G., Singh, M.K., Kwon, T.H., Anderson, P.D., 2008, Microfluid Nanofluid, Vol. 4, pp. 589–599.

Khan, W.A., Culham, J.R., and Yovanovich, M.M., 2007, Journal of Thermophysics and Heat Transfer, Vol. 21 (2), pp. 372–378.

Kim, D.S., Lee, S.W., Kwon, T.H., Lee S.S., 2004, Journal of Micromechanics and Microengineering, Vol. 14, pp. 798–805.

Kleinstreuer, C., 1997, *Engineering Fluid Dynamics*, Cambridge University Press, New York, NY.

Kleinstreuer, C., 2003, *Two-Phase Flow: Theory and Applications*, Taylor & Francis Books, New York.

Kleinstreuer, C., 2006, *Biofluid Dynamics – Principles and Selected Applications*, CRC Press/Taylor & Francis, Boca Raton/London/New York.

Kleinstreuer, C., Li, J., Koo, J., 2008, International Journal of Heat and Mass Transfer, Vol. 51, pp. 5590–5597.

Kleinstreuer, C., Li, J., 2008a, *Encyclopedia of Micro and Nanofluidics*, Edited by Li, D., pp. 1314–1325, Springer-Verlag, Heidelberg, DE.

Kleinstreuer, C., Li, J., 2008b, ASME Journal of Heat Transfer, Vol. 130, pp. 025501-1-3.

Kleinstreuer, C., Zhang, Z., Donohue, J.F., 2008, Annual Review of Biomedical Engineering, Vol. 10, pp. 195–220.

Kleinstreuer, C., Zhang, Z., Kim, C.S., 2007, Journal of Aerosol Science, Vol. 38 (10), pp. 1047–1061.

Kleinstreuer, C., Zhang, Z., 2009, Journal of Biomechanical Engineering, Vol. 131 (2), pp. 021007-1-10.

Ko, T.H., Wu, C.P., 2009, International Communications in Heat and Mass Transfer, Vol. 36 (1), pp. 25–31.

Koo, J., Kleinstreuer, C., 2003, Journal of Micromechanics and Microengineering, Vol. 13, pp. 568–579.

Koo, J., Kleinstreuer, C., 2004, Journal of Nanoparticle Research, Vol. 6, pp. 577–588.

Koo, J., Kleinstreuer, C., 2004, International Journal of Heat and Mass Transfer, Vol. 47, pp. 3159–3169.

Labhasetwar, V., Leslie-Pelecky, D.L., 2007, *Biomedical Applications of Nanotechnology*, Wiley-Interscience, A Wiley Publication.

Lee, T., Mateescu, D., 1998, Journal of Fluids and Structures, Vol. 12, pp. 703–716.

Li, J., Kleinstreuer, C., 2008, International Journal of Heat and Fluid Flow, Vol. 29(4), pp. 1221–1232.

Li, J., Kleinstreuer, C., 2009, Microfluid Nanofluid, Vol. 6 (5), pp. 661–668.

Li, J., Kleinstreuer, C., 2010, ASME Journal of Heat Transfer (In Press).

Mabotuwana, T.D.S., Cheng, L.K., Pullan, A.J., 2007, Biomedical Engineering online, Vol. 6, pp. 1–13.

Macosko, C.W., 1994, *Rheology: Principles Measurements and Applications*, VHC Publishers, New York.

Mahmud, S., Fraser, R.A., 2003, International Journal of Thermal Sciences, Vol. 42, pp. 177–186.

Mansour, R.B., Galanis, N., Nguyen, C.T., 2006, International Journal of Thermal Sciences, Vol. 45, pp. 998–1007.

Marieb, E.N., 1998, *Human Anatomy and Physiology*, 4th ed., Benjamin Cummings Science Publishing, Menlo Park, CA.

Merrill, E.W., 1969, Physiological Reviews, Vol. 26, pp. 863–888.

Michaelides, E.E., 1997, Journal of Fluids Engineering, Vol. 119(2), pp. 233–247.

Munson, M.S., Yager, P., 2004, Analytica Chimical Acta, Vol. 507, pp. 63–71.

Naterer, G.F., 2002, *Heat Transfer in Single and Multiphase Systems*, CRC Press, Boca Raton, FL.

Naterer, G.F., Camberos, J.A., 2008, *Entropy-Based Design and Analysis of Fluids Engineering Systems*, CRC Press, Taylor & Francis Group, Boca Raton, FL.

Nguyen, N-T., 2008, *Micromixers: Fundamentals, Design, and Fabrication*, William Andrew; Norwich, NY.

Nguyen, N-T., Wereley, S.T., 2006, *Fundamentals and Applications of Microfluidics*, Arten House, Boston, MA.

Nguyen, N-T., Wu, Z., 2005, Journal of Micromechanics and Microengineering, Vol. 15, pp. R1–R16.

Nichols, W.W., O'Rourke, M.F., 2005, *McDonalds Blood Flow in Arteries: Theoretical, Experimental, and Clinical Principles*, 5th ed., Hodder Arnold: Oxford University Press, New York, NY.

Ou, J., Perot, B., Rothstein, J.R., 2004, Physics of Fluids, Vol. 16 (12), pp. 4635–4643.

Païdoussis, M.P., 1998, *Fluid–Structure Interactions: Slender Structures and Axial Flow*, Vol. 1, Academic, London.

Palm, W.J., 2007, *Mechanical Vibration*, Wiley, Hoboken, NJ.

Probstein, R.F., 1994, *Physicochemical Hydrodynamics: An Introduction*, Wiley, New York.

Roark, R., Budynas, R., and Young, W., 2001, *Roark's Formulas for Stress and Strain*, 7th ed., McGraw-Hill, New York.

Sahin, A.Z., 1998, Heat Mass Transfer, Vol. 33, pp. 425–430.

Sahin, A.Z., 2000, International Journal of Heat Mass Transfer, Vol. 43, pp. 1469–1478.

Saliterman, S.S., 2006, *Fundamentals of BioMEMS and Medical Microdevices*, Wiley-Interscience, SPIE Press, Bellingham, Washington.

Schmid-Schoenbein, G.W., 1986, "Rheology of leukocytes." In: Skalak, R., Chien, S. (eds), *Handbook of Bioengineering*, McGraw-Hill, New York, NY.

Silverthorn, D.U., 2004, *Human Physiology, an Integrated Approach*, Prentice Hall, Upper Saddle River, NJ.

Simiu, E., Scanlan, R.H., 1996, *Wind Effects on Structures: Fundamentals and Applications to Design*, 3rd ed., Wiley, New York, NY.

Soo, S.L., 1990, *Multiphase Fluid Dynamics*, Science Press/Gower Technical, Brookfield.

Steele, B.N., Olufsen, M.S., Taylor, C.A., 2007, Computer Methods in Biomechanics and Biomedical Engineering, Vol. 10 (1), pp. 39–51.

Stone, H.A., Stroock, A.D., Ajdari, A., 2004, Annual Review of Fluid Mechanics, Vol. 36, pp. 381–411.

Stroock, A.D., Dertinger, S.K.W., Ajdari, A., et al., 2002, Science, Vol. 295, pp. 647–651.

Tabeling, P., 2005, *Introduction to Microfluidics*, Oxford Unversity Press, Oxford, U.K/New York.

Tanner, R.I., Walters, K., 1998, *Rheology: an Historical Perspective*, Elsevier, Amsterdam.

Tannehill, J.C., Anderson, D.A., Pletcher, R., 1997, *Computational Fluid Mechanics and Heat Transfer*, Second Edition (Series in Computional and Physical Processes in Mechanics and Thermal Sciences). Taylor & Francis, Davers, MA.

Tong, L.S., Tang, Y.S., 1997, *Boiling Heat Transfer and Two-Phase Flow*, Taylor & Francis, New York.

Tu, J., Yeoh, G.H., Liu, C., 2008, *Computational Fluid Dynamics: A Practical Approach*, Butterworth Heinemann, Burlington, MA.

Ugural, A.C., Fenster, S.K., 2003, *Advanced Strength and Applied Elasticity*, Prentice Hall PTR, Upper Saddle River, NJ.

van Ertbruggen, C., Hirsch, C., Paiva, M., 2005, Journal of Applied Physiology, Vol. 98, pp. 970–980.

Versteeg, H.K., Malalasekera, W., 1996, *an Introduction to Computational Fluid Dynamics: The Finite Volume Method*, Prentice Hall, Upper Saddle River, NJ.

White, F.M., 2006, *Viscous Fluid Flow*, McGraw-Hill, New York, NY.

Wu, H.Y., Cheng, P., 2003, International Journal of Heat and Mass Transfer, Vol. 46 (14), pp. 2519–2525.

Zapryanov, Z., Tabakova, S., 1999, *Dynamics of Bubbles, Drops and Rigid Particles*, 1st ed., Springer.

Zhang, Z., Kleinstreuer, C., 2003, AIAA Journal, Vol. 41 (5), pp. 831–840.

Zhang, Z.M., 2007, *Nano/microscale Heat Transfer*, McGraw-Hill, New York.

Appendices

A. Math Tools and Equations

B. Property Data and Charts

Appendix A

Review of Tensor Calculus, Differential Operations, Integral Transformations, and ODE Solutions Plus Removable Equation Sheets

A.1 Tensor Calculus

Here we restrict our review to tensor manipulations as needed in the text. Further information and solved examples can be found in Aris (1964), Schey (1973), and App. A of Bird et al. (2002).

A.1.1 Definitions

Recall: Tensors of rank n have 3^n components. For example:

- A tensor of rank "zero" is a *scalar* which has only one component, i.e., its magnitude (e.g., pressure).
- A tensor of rank "one" is a *vector* which has in general three components, i.e., three magnitudes and three directions (e.g., velocity).
- A tensor of rank "two" is usually labeled a *tensor* which has nine components, e.g., stress.

Coordinate Systems

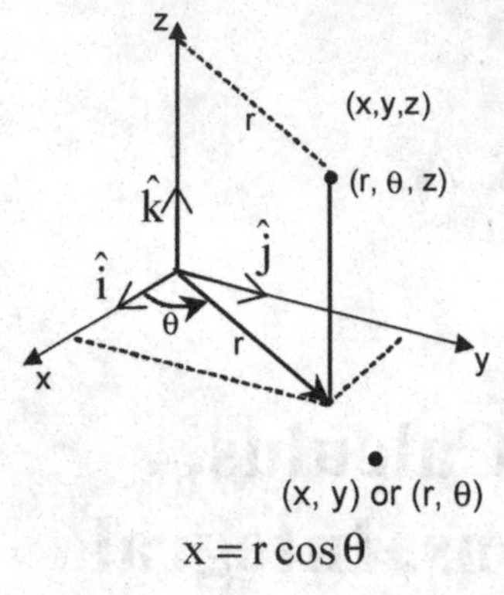

$x = r\cos\theta$

$y = r\sin\theta$

$z = z$

Vector Products

- Dot product

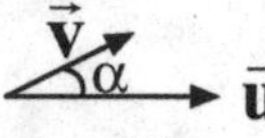

$$\vec{u}\cdot\vec{v} = \vec{v}\cdot\vec{u} = |\vec{u}||\vec{v}|\cos\alpha \ \rightarrow \text{scalar}$$

- Cross product

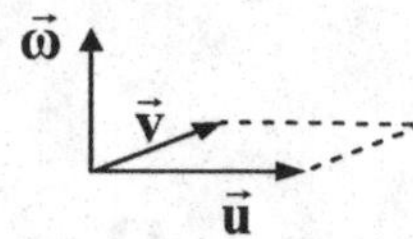

$$\vec{u}\times\vec{v} = -\vec{v}\times\vec{u} = \vec{\omega} \ \rightarrow \text{vector}$$

- Dyadic product
  $\vec{u}\vec{v} = \vec{\vec{a}} \rightarrow$tensor

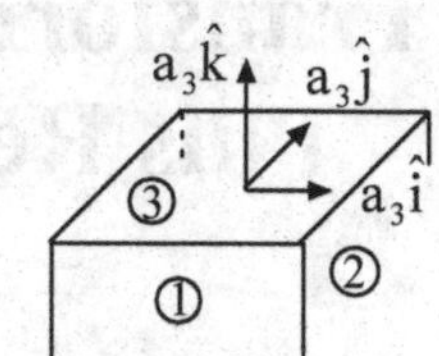

Clearly, the type of vector product may result in a scalar (see dot product) or a tensor of rank two with nine components (see dyadic product). This is further illustrated when using the *del operator* which has the characteristics of a vector (see Sect. A.1.2).

A.1.2 Operations with ∇

- By definition

(i) In rectangular coordinates: $$\nabla \equiv \hat{i}\frac{\partial}{\partial x} + \hat{j}\frac{\partial}{\partial y} + \hat{k}\frac{\partial}{\partial z} \tag{A.1.1a}$$

(ii) In cylindrical coordinates: $$\nabla \equiv \hat{e}_r\frac{\partial}{\partial r} + \hat{e}_\theta\frac{1}{r}\frac{\partial}{\partial \theta} + \hat{e}_z\frac{\partial}{\partial z} \tag{A.1.1b}$$

- When operating on a scalar, say, the pressure, ∇p generates a vector, i.e., the *pressure gradient*:

$$\nabla p = (\hat{i}\frac{\partial}{\partial x} + \hat{j}\frac{\partial}{\partial y} + \hat{k}\frac{\partial}{\partial z})p = \frac{\partial p}{\partial x}\hat{i} + \frac{\partial p}{\partial y}\hat{j} + \frac{\partial p}{\partial z}\hat{k} \tag{A.1.2}$$

- When operating on a vector, it can produce
  A scalar in case of a dot product, e.g., $\nabla \cdot \vec{\mathbf{v}}$
  A vector in case of a cross product, e.g., $\nabla \times \vec{\mathbf{v}}$
  A tensor in case of a dyadic product, e.g., $\nabla \vec{\mathbf{v}}$

Note: $\nabla \cdot \nabla = \nabla^2$ is the Laplacian operator (see Eqs. (A.1.8a, b))

Specifically,

(a) $\nabla \cdot \vec{\mathbf{v}} \equiv \text{div}\,\vec{\mathbf{v}}$ is the divergence of the velocity field:

$$\begin{aligned}\nabla \cdot \vec{\mathbf{v}} &= (\hat{i}\frac{\partial}{\partial x} + \hat{j}\frac{\partial}{\partial y} + \hat{k}\frac{\partial}{\partial z}) \cdot (u\hat{i} + v\hat{j} + w\hat{k}) \\ &= (\hat{i}\cdot\hat{i})\frac{\partial u}{\partial x} + (\hat{i}\cdot\hat{j})\frac{\partial v}{\partial x} + (\hat{i}\cdot\hat{k})\frac{\partial w}{\partial x} + \\ &\quad (\hat{j}\cdot\hat{i})\frac{\partial u}{\partial y} + (\hat{j}\cdot\hat{j})\frac{\partial v}{\partial y} + (\hat{j}\cdot\hat{k})\frac{\partial w}{\partial y} + \\ &\quad (\hat{k}\cdot\hat{i})\frac{\partial u}{\partial z} + (\hat{k}\cdot\hat{j})\frac{\partial v}{\partial z} + (\hat{k}\cdot\hat{k})\frac{\partial w}{\partial z} = \frac{\partial u}{\partial x} + \frac{\partial v}{\partial y} + \frac{\partial w}{\partial z}\end{aligned} \tag{A.1.3}$$

because $\hat{i}\cdot\hat{i} = |i||i|\cos\alpha$; $\alpha = 0 \Rightarrow \hat{i}\cdot\hat{i} = 1$, while $\hat{i}\cdot\hat{j} = 0$ $<\alpha = 90°>$

In general, $\hat{\delta}_i \cdot \hat{\delta}_j = \delta_{ij} := \begin{cases} 1 & \text{if } i = j \\ 0 & \text{if } i \neq j \end{cases}$, known as the *Kronecker delta*

(b) $\nabla \times \vec{\mathbf{v}} \equiv \text{curl}\,\vec{\mathbf{v}}$ is the curl (or rotation) of the velocity field:

$$\begin{aligned}\nabla \times \vec{\mathbf{v}} &= (\hat{i}\frac{\partial}{\partial x} + \hat{j}\frac{\partial}{\partial y} + \hat{k}\frac{\partial}{\partial z}) \times (u\hat{i} + v\hat{j} + w\hat{k}) \\ &= (\hat{i}\times\hat{i})\frac{\partial u}{\partial x} + (\hat{i}\times\hat{j})\frac{\partial v}{\partial x} + (\hat{i}\times\hat{k})\frac{\partial w}{\partial x} + \\ &\quad (\hat{j}\times\hat{i})\frac{\partial u}{\partial y} \ldots \text{etc.}\end{aligned} \tag{A.1.4}$$

Recalling the results for the cross products between unit (or base) vectors:

$$\hat{i}\times\hat{j}=\hat{k} \qquad \hat{j}\times\hat{i}=-\hat{k} \qquad \hat{i}\times\hat{i}=0$$
$$\hat{j}\times\hat{k}=\hat{i} \qquad \hat{k}\times\hat{j}=-\hat{i} \qquad \hat{j}\times\hat{j}=0$$
$$\hat{k}\times\hat{i}=\hat{j} \qquad \hat{i}\times\hat{k}=-\hat{j} \qquad \hat{k}\times\hat{k}=0$$

we obtain

$$\nabla\times\vec{\mathbf{v}}=\hat{i}\left(\frac{\partial w}{\partial y}-\frac{\partial v}{\partial z}\right)+\hat{j}\left(\frac{\partial u}{\partial z}-\frac{\partial w}{\partial x}\right)+\hat{k}\left(\frac{\partial v}{\partial x}-\frac{\partial u}{\partial y}\right)$$

or

$$\nabla\times\vec{\mathbf{v}}=\mathrm{curl}\,\vec{\mathbf{v}}=\begin{vmatrix}\hat{i} & \hat{j} & \hat{k}\\ \frac{\partial}{\partial x} & \frac{\partial}{\partial y} & \frac{\partial}{\partial z}\\ u & v & w\end{vmatrix}\equiv\vec{\zeta} \quad \text{(vorticity vector)} \qquad \text{(A.1.5)}$$

(c) $\nabla\vec{v}\equiv\mathrm{grad}\,\vec{v}$ is the dyadic product, or gradient, of the velocity field:

$$\nabla\vec{\mathbf{v}}=(\hat{i}\frac{\partial}{\partial x}+\hat{j}\frac{\partial}{\partial y}+\hat{k}\frac{\partial}{\partial z})(u\hat{i}+v\hat{j}+w\hat{k})$$
$$=(\hat{i}\hat{i})\frac{\partial u}{\partial x}+(\hat{i}\hat{j})\frac{\partial v}{\partial x}+(\hat{i}\hat{k})\frac{\partial w}{\partial x}+ \qquad \text{(A.1.6)}$$
$$(\hat{j}\hat{i})\frac{\partial u}{\partial y}+\ldots\text{ etc.}$$

Now, the unit vector dyadic product $\hat{\delta}_i\hat{\delta}_j$ indicates the *location* (i.e., $\hat{\delta}_i$ is the unit normal to the particular surface) and *direction* (i.e., $\hat{\delta}_j$ gives the direction) of a tensor. Thus,

$$\nabla\vec{\mathbf{v}}=\mathrm{grad}\,\vec{\mathbf{v}}=\begin{vmatrix}\frac{\partial u}{\partial x} & \frac{\partial v}{\partial x} & \frac{\partial w}{\partial x}\\ \frac{\partial u}{\partial y} & \frac{\partial v}{\partial y} & \frac{\partial w}{\partial y}\\ \frac{\partial u}{\partial z} & \frac{\partial v}{\partial z} & \frac{\partial w}{\partial z}\end{vmatrix}\sim\vec{\vec{\tau}} \quad \text{<stress tensor>} \qquad \text{(A.1.7)}$$

Notes:

- The use of $\nabla \cdot \vec{v}$, $\nabla \times \vec{v}$ and $\nabla \vec{v}$ plus illustrations were introduced in Chap. 1.
- The dot product *reduces* the rank of a tensor, e.g., $\nabla \cdot \vec{v} \rightarrow$ scalar and $\nabla \cdot \vec{\vec{\tau}} \rightarrow$ vector
- The dyadic (or gradient) product increase the rank, e.g., $\nabla p \rightarrow$ vector and $\nabla \vec{v} \rightarrow$ tensor.
- The divergence of a scalar gradient is $\nabla \cdot \nabla s = \nabla^2 s = \Sigma \frac{\partial^2 s}{\partial x^2}$,

  where ∇^2 *is the Laplacian operator* producing a scalar field:

  $$\text{Rectangular Coordinates: } \nabla^2 \equiv \frac{\partial^2}{\partial x^2} + \frac{\partial^2}{\partial y^2} + \frac{\partial^2}{\partial z^2} \tag{A.1.8a}$$

  $$\text{Cylindrical Coordinates: } \nabla^2 \equiv \frac{1}{r}\frac{\partial}{\partial r}\left(r\frac{\partial}{\partial r}\right) + \frac{1}{r^2}\frac{\partial^2}{\partial \theta^2} + \frac{\partial^2}{\partial z^2} \tag{A.1.8b}$$

- The transpose of a second-order tensor, $\vec{\vec{a}}$, with components a_{ij} is denoted by $\vec{\vec{a}}^{tr}$ and is defined by

$$\lfloor a^{tr} \rfloor_{ij} = a_{ji} \tag{A.1.9a}$$

For example,

$$\vec{\vec{a}} \equiv \vec{v}\vec{w} = \begin{pmatrix} v_1 w_1 & v_1 w_2 & v_1 w_3 \\ v_2 w_1 & v_2 w_2 & v_2 w_3 \\ v_3 w_1 & v_3 w_2 & v_3 w_3 \end{pmatrix} \tag{A.1.9b}$$

whereas

$$\vec{\vec{a}}^{tr} \equiv (\vec{v}\vec{w})^{tr} = \begin{pmatrix} v_1 w_1 & v_2 w_1 & v_3 w_1 \\ v_1 w_2 & v_2 w_2 & v_3 w_2 \\ v_1 w_3 & v_2 w_3 & v_3 w_3 \end{pmatrix} \tag{A.1.9c}$$

Sample Problem Solutions

To illustrate a few tensor manipulations, the following sample problems are solved. Given the components of a symmetric tensor $\vec{\vec{\tau}}$, i.e., $\tau_{ij} = \tau_{ji}$:

$$\tau_{xx} = 3, \qquad \tau_{xy} = 2, \qquad \tau_{xz} = -1,$$
$$\tau_{yy} = 2, \qquad \tau_{yz} = 1,$$
$$\tau_{zz} = 0$$

and the components of a vector $\vec{\mathbf{v}}$, e.g., $v_x = 5, v_y = 3, v_z = 7$; evaluate

(a) $\vec{\vec{\tau}} \cdot \vec{\mathbf{v}}$; (b) $\vec{\mathbf{v}} \cdot \vec{\vec{\tau}}$; (c) $\vec{\mathbf{v}}\vec{\mathbf{v}}$; and (d) $\vec{\vec{\tau}} \cdot \vec{\vec{\delta}}$

where $\vec{\vec{\delta}}$ is the unit tensor, i.e.,

$$\delta_{ij} = \begin{pmatrix} 1 & & \phi \\ & 1 & \\ \phi & & 1 \end{pmatrix}$$

Solution: A good preliminary exercise is to write down the vectors and tensors in component form.

(a) $$\vec{\vec{\tau}} \cdot \vec{\mathbf{v}} = \Sigma_i \vec{\boldsymbol{\delta}}_i \{\Sigma_j \tau_{ij} v_j\} = (\vec{\boldsymbol{\delta}}_1 \vec{\boldsymbol{\delta}}_2 \vec{\boldsymbol{\delta}}_3) \begin{pmatrix} 3 & 2 & -1 \\ 2 & 2 & 1 \\ -1 & 1 & 0 \end{pmatrix} \begin{pmatrix} 5 \\ 3 \\ 7 \end{pmatrix}$$

$$= 14\vec{\boldsymbol{\delta}}_1 + 23\vec{\boldsymbol{\delta}}_2 - 2\vec{\boldsymbol{\delta}}_3 \quad \text{<a vector>}$$

where $\vec{\delta}_i \mathrel{\hat{=}}$ unit vector in the i-direction, with i = 1, 2, 3.

(b) $\vec{\mathbf{v}} \cdot \vec{\vec{\tau}} = \vec{\vec{\tau}} \cdot \vec{\mathbf{v}}$ since $\vec{\vec{\tau}}$ is symmetric

(c) $$\vec{\mathbf{v}}\vec{\mathbf{v}} = \Sigma_i \Sigma_j \vec{\boldsymbol{\delta}}_i \vec{\boldsymbol{\delta}}_j v_i v_j = 25\vec{\boldsymbol{\delta}}_1\vec{\boldsymbol{\delta}}_1 + 15\vec{\boldsymbol{\delta}}_1\vec{\boldsymbol{\delta}}_2 + 35\vec{\boldsymbol{\delta}}_1\vec{\boldsymbol{\delta}}_3 + 15\vec{\boldsymbol{\delta}}_2\vec{\boldsymbol{\delta}}_1 + 9\vec{\boldsymbol{\delta}}_2\vec{\boldsymbol{\delta}}_2 + 21\vec{\boldsymbol{\delta}}_2\vec{\boldsymbol{\delta}}_3 + 35\vec{\boldsymbol{\delta}}_3\vec{\boldsymbol{\delta}}_1 + 21\vec{\boldsymbol{\delta}}_3\vec{\boldsymbol{\delta}}_2 + 49\vec{\boldsymbol{\delta}}_3\vec{\boldsymbol{\delta}}_3$$

(d) $$\vec{\vec{\tau}} \cdot \vec{\vec{\delta}} = \Sigma_i \Sigma_l \vec{\boldsymbol{\delta}}_i \vec{\boldsymbol{\delta}}_l (\Sigma_j \tau_{ij} \delta_{jl}) = \vec{\vec{\tau}} = 3\vec{\boldsymbol{\delta}}_1\vec{\boldsymbol{\delta}}_1 + 2\vec{\boldsymbol{\delta}}_1\vec{\boldsymbol{\delta}}_2 - 1\vec{\boldsymbol{\delta}}_1\vec{\boldsymbol{\delta}}_3 + 2\vec{\boldsymbol{\delta}}_2\vec{\boldsymbol{\delta}}_1 + 2\vec{\boldsymbol{\delta}}_2\vec{\boldsymbol{\delta}}_2 + 1\vec{\boldsymbol{\delta}}_2\vec{\boldsymbol{\delta}}_3 - 1\vec{\boldsymbol{\delta}}_3\vec{\boldsymbol{\delta}}_1 + 1\vec{\boldsymbol{\delta}}_3\vec{\boldsymbol{\delta}}_2 + 0\vec{\boldsymbol{\delta}}_3\vec{\boldsymbol{\delta}}_3$$

A.1.3 Some Tensor Identities

$$\nabla rs = r\nabla s + s\nabla r \tag{A.1.10}$$

$$(\nabla \cdot s\vec{v}) = (\nabla s \cdot \vec{v}) + s(\nabla \cdot \vec{v}) \tag{A.1.11}$$

$$(\nabla\cdot[\vec{v}\times\vec{w}])=(\vec{w}\cdot[\nabla\times\vec{v}])-(\vec{v}\cdot[\nabla\times\vec{w}]) \tag{A.1.12}$$

$$[\nabla\times s\vec{v}]=[\nabla s\times\vec{v}]+s[\nabla\times\vec{v}] \tag{A.1.13}$$

$$[\nabla\cdot\nabla\vec{v}]=\nabla(\nabla\cdot\vec{v})-[\nabla\times[\nabla\times\vec{v}]] \tag{A.1.14}$$

$$[\vec{v}\cdot\nabla\vec{v}]=\tfrac{1}{2}\nabla(\vec{v}\cdot\vec{v})-[\vec{v}\times[\nabla\times\vec{v}]] \tag{A.1.15}$$

$$[\nabla\cdot\vec{v}\vec{w}]=[\vec{v}\cdot\nabla\vec{w}]+\vec{w}(\nabla\cdot\vec{v}) \tag{A.1.16}$$

$$[\nabla\cdot s\vec{\vec{\delta}}]=\nabla s \tag{A.1.17}$$

$$[\nabla\cdot s\vec{\vec{\tau}}]=[\nabla s\cdot\vec{\vec{\tau}}]+s[\nabla\cdot\vec{\vec{\tau}}] \tag{A.1.18}$$

$$\nabla(\vec{v}\cdot\vec{w})=[(\nabla\vec{v})\cdot\vec{w}]+[(\nabla\vec{w})\cdot\vec{v}] \tag{A.1.19}$$

A.2 Differentiation

A.2.1 Differential Time Operators

In order to understand and solve fluid mechanics problems, the basic skills in linear algebra, as well as differentiating and integrating functions, graphing and analyzing functions, as well as curve fitting are definitely *prerequisites*. If a review is necessary, the reader may want to consult M. Spiegel (1971), Schaum's Outline series, or M. D. Greenberg (1998), Prentice-Hall; among many other texts.

The different notations and the physical meanings of various *time* derivatives (i.e., differential operators) are presented as follows:

- Partial time derivative: $\partial\#/\partial t \mathrel{\hat{=}}$ Changes in variable "#" with time observed from a *fixed* position in space, i.e., stationary observer.
- Substantial or material time derivate: $D\#/Dt \mathrel{\hat{=}}$ Changes of variable "#" with time *following* the fluid/material element in motion. The Stokes, or material derivative is defined as:

$$\frac{D\#}{Dt}\equiv\frac{\partial\#}{\partial t}+(\vec{v}\cdot\nabla)\# \tag{A.2.1}$$

In Eq. (A.2.1) the Lagrangian time rate of change is expressed in terms of Eulerian derivatives. For example, c being a species concentration $[M/L^3]$, the material time derivative is

$$\frac{Dc}{Dt}\equiv\frac{\partial c}{\partial t}+(\vec{v}\cdot\nabla)c \tag{A.2.2a}$$

In rectangular coordinates

$$\frac{Dc}{Dt}=\frac{\partial c}{\partial t}+u\frac{\partial c}{\partial x}+v\frac{\partial c}{\partial y}+w\frac{\partial c}{\partial z} \tag{A2.2b}$$

while in tensor notation

$$\frac{Dc}{Dt}=\frac{\partial c}{\partial t}+v_k\frac{\partial c}{\partial x_k};\ k=1,2,3 \tag{A.2.2c}$$

where $\frac{\partial c}{\partial t} \hat{=}$ local time derivative (i.e., accumulation of species *c*) and

$v_k\frac{\partial c}{\partial x_k} \hat{=}$ convective derivatives (i.e., mass transfer by convection).

Note: Repeated indices imply summation of these terms, here $k = 1, 2, 3$ <Einstein convention>.

- Total time derivative: $d\#/dt \hat{=}$ Changes of # with respect to time observed from a point *moving differently* than the flow field. For example:

$$\frac{dc}{dt}=\frac{\partial c}{\partial t}+\frac{dx}{dt}\frac{\partial c}{\partial x}+\frac{dy}{dt}\frac{\partial c}{\partial y}+\frac{dz}{dt}\frac{\partial c}{\partial z} \tag{A.2.3}$$

where $dx/dt=u$, $dy/dt=v$, and $dz/dt=w$ are the velocity components of the moving observer.

A.2.2 The Total Differential

Dependent variables describing transport phenomena, such as fluid velocity, pressure and species concentration, are often a function of more than one independent variable. For example, the fluid velocity is a function of three spatial coordinates, say, x, y, and z and, if the flow is unsteady, time t. Thus, $\vec{v}=\vec{v}(x,y,z,t)$. The total differential is defined as

$$d\vec{v}=\frac{\partial\vec{v}}{\partial x}dx+\frac{\partial\vec{v}}{\partial y}dy+\frac{\partial\vec{v}}{\partial z}dz+\frac{\partial\vec{v}}{\partial t}dt \tag{A.2.4}$$

If the spatial coordinates x, y, and z are also functions of time, then the total (particle) time derivative is:

$$\frac{d\vec{v}}{dt}=\frac{\partial\vec{v}}{\partial x}\frac{dx}{dt}+\frac{\partial\vec{v}}{\partial y}\frac{dy}{dt}+\frac{\partial\vec{v}}{\partial z}\frac{dz}{dt}+\frac{\partial\vec{v}}{\partial t} \tag{A.2.5}$$

Such differentiation can be extended to the calculation of fluid acceleration and mass transport where the local quantities change with time. For a scalar, say, the pressure, we have

$$dp = \frac{\partial p}{\partial x}dx + \frac{\partial p}{\partial y}dy + \frac{\partial p}{\partial z}dz \tag{A.2.6}$$

A.2.3 Truncated Taylor Series Expansions and Binomial Theorem

In order to approximate a function, say, y(x), around some point $x = x_0$, we employ two or three terms of the Taylor series. For one independent variable,

$$y(x) = y\Big|_{x=x_0} + \frac{dy}{dx}\Big|_{x=x_0}(x - x_0) + \frac{1}{2}\frac{d^2y}{dx^2}\Big|_{x=x_0}(x - x_0)^2 + \frac{1}{6}\frac{d^3y}{dx^3}\Big|_{x=x_0}(x - x_0)^3 + \ldots \tag{A.2.7}$$

Clearly, the first two terms provide a straight-line fit and the first three a parabolic fit of y(x) at $x = x_0$.

If we want to estimate the value of y a very small distance away from the known y(x), i.e., what is $y(x+\varepsilon)$ where $\varepsilon << 1$, we can write with Eq. (A.2.7):

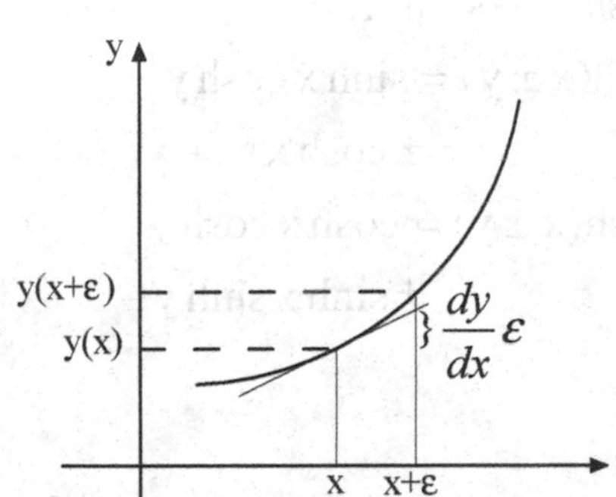

$$y(x+\varepsilon) \approx y(x) + \frac{dy}{dx}\Big|_{x=\varepsilon}\varepsilon \tag{A.2.8}$$

As the graph indicates, the step size ε and local curvature of y(x) determine the accuracy of Eq. (A.2.8). Equation (A.2.8) is employed in Sect. 1.3 to derive equations in differential form.

For functions of two variables, e.g., $f(x, y)$, we write

$$f(x+\varepsilon, y+\delta) \approx f(x, y) + \frac{\partial f}{\partial x}\Big|_{x=\varepsilon}\varepsilon + \frac{\partial f}{\partial y}\Big|_{y=\delta}\delta \tag{A.2.9}$$

When dealing with rational *fractional* functions, it is often advantageous to express them in terms of partial fractions and then expand them, using the *binomial theorem*. For example, the expansion

$$(c+\varepsilon)^n = c^n + nc^{n-1}\varepsilon + \frac{n(n-1)}{2!}c^{n-2}\varepsilon^2$$

$$+\frac{n(n-1)(n-2)}{3!}c^{n-3}\varepsilon^3+\cdots \qquad \text{(A.2.10)}$$

is valid for all values of n if $|\varepsilon|<|c|$. If $|c|<|\varepsilon|$, the expansion is valid only if n is a non-negative integer.

A.2.4 Hyperbolic Functions

Next to the circular trigonometric functions, the hyperbolic functions sinhx, coshx and tanhx appear frequently in science and engineering and hence in the present text.

Graphs:

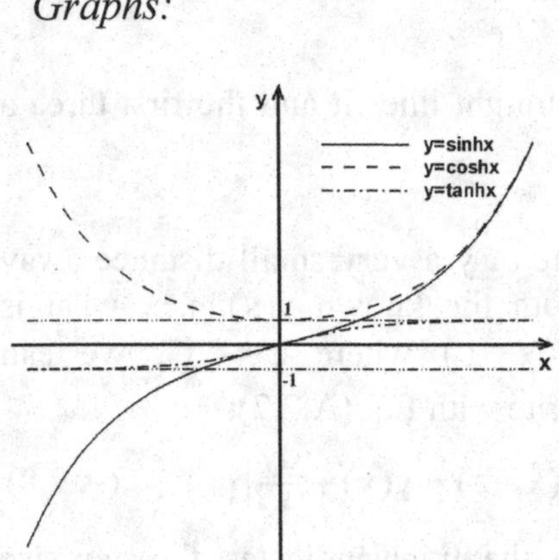

Relations:

$$\sinh x=\frac{1}{2}(e^x-e^{-x})$$

$$\cosh x=\frac{1}{2}(e^x+e^{-x})$$

$$\cosh^2 x-\sinh^2 x=1$$

$$\sinh(x\pm y)=\sinh x\cosh y \pm \cosh x\sinh y$$

$$\cosh(x\pm y)=\cosh x\cosh y \pm \sinh x\sinh y$$

A.3 Integral Transformations

A.3.1 The Divergence Theorem

As established by Gauss, the divergence theorem states that the integration over the dot product of a vector field, $\vec{\mathbf{v}}$, with a closed regular area element, $d\vec{\mathbf{A}}$, is equal to the integration of the divergence of $\vec{\mathbf{v}}$, i.e., $\nabla\cdot\vec{\mathbf{v}}$, over the interior volume, $\forall$:

$$\iint_A \vec{\mathbf{v}}\cdot d\vec{\mathbf{A}}=\iiint_\forall (\nabla\cdot\vec{\mathbf{v}})d\forall \qquad \text{(A.3.1)}$$

Equation (A.3.1) is being used in Sect. 1.3 when converting all surface integrals in the Reynolds Transport Theorem into volume integrals in order to express the conservation laws of mass, momentum, and energy in *differential form*.

Sample Problem: Given $\vec{v} = 4xz\hat{i} - y^2\hat{j} + yz\hat{k}$, i.e., $u = 4xz$, $v = -y^2$, and $w = yz$, in a unit cube, i.e., $0 \le x \le 1$, $0 \le y \le 1$, and $0 \le z \le 1$, show that Eq. (A.3.1) holds

Solution:

(a) Six-sided surface integral: $\iint_A \vec{v} \cdot d\vec{A} = \int_0^1\int_0^1 \vec{v} \cdot \vec{n} dA$

where

$\hat{n} = \hat{i}$ at $x = 1$ and $\hat{n} = -\hat{i}$ at $x = 0$ with $dA = dydz$

$\hat{n} = \hat{j}$ at $y = 1$ and $\hat{n} = -\hat{j}$ at $y = 0$ with $dA = dxdz$

and

$\hat{n} = \hat{k}$ at $z = 1$ and $\hat{n} = -\hat{k}$ at $z = 0$ with $dA = dxdy$.

Thus,

$$\int_0^1\int_0^1 \vec{v} \cdot \hat{i} dA = \int_0^1\int_0^1 4xz\big|_{x=1} dydz = 4y\big|_0^1 \left(\frac{1}{2} z^2\big|_0^1\right) = 2$$

Similarly,

$$\int_0^1\int_0^1 \vec{v} \cdot \hat{j} dA = -1; \qquad \int_0^1\int_0^1 \vec{v} \cdot \hat{k} dA = \frac{1}{2}; \qquad \text{and}$$

the other three negative surface integrals are zero. Hence,

$$\iint_A \vec{v} \cdot \hat{n} dA = 2 - 1 + \frac{1}{2} + 0 + 0 + 0 = \frac{3}{2}$$

(b) Volume integral:

$$\iiint_{\forall} \nabla \cdot \vec{v} d\forall = \int_0^1\int_0^1\int_0^1 \left(\frac{\partial u}{\partial x} + \frac{\partial v}{\partial y} + \frac{\partial w}{\partial z}\right) dxdydz$$

$$\int_0^1\int_0^1\int_0^1 (4z - 2y + y) dxdydz = \left|\frac{4z^2}{2} - \frac{z}{2}\right|_0^1 = \frac{3}{2}$$

Comment: It is evident that such integral operations over *variable* vector fields result in scalars (i.e., numbers); this implies that the Reynolds Transport Theorem generates scalar quantities, i.e., numerical values for flow rates, averaged velocities, forces, wall stresses, pressure drops, etc.

A.3.2 Leibniz Rule

A switch in the order of operation is justified with Leibniz' Rule: If $F(t) = \int_a^b f(x,t)dx$ and a, b are constants, then

$$\frac{dF}{dt} \equiv \frac{d}{dt}\left[\int_a^b f(x,t)dx\right] = \int_a^b \frac{\partial f}{\partial t}dx \qquad \text{(A.3.2)}$$

Equation (A.3.2) is occasionally applied when dealing with the transient term <volume integral> in the RTT. In general, Leibniz Rule reads

$$\frac{d}{dt}\int_{a(t)}^{b(t)} f(x,t)dx = \int_a^b \frac{\partial f}{\partial t}dx + b'f[b(t),t] - a'f[a(t),t]$$

A.3.3 Error Function

Numerous natural phenomena follow exponential functions. For example,

$$\text{erf}(x) = \frac{2}{\sqrt{\pi}}\int_0^x e^{-\xi^2}d\xi \qquad \text{(A.3.3a)}$$

where

$$\text{erf}(0) = 0 \text{ and } \text{erf}(\infty) = 1 \qquad \text{(A.3.3b, c)}$$

The integrand $\exp(-\xi^2)$ is a normal probability distribution. Thus, Eq. (A.3.3) is a solution part of processes governed by Gaussian-type distributions. The "complementary error function" is defined as

$$1 - \text{erf}(x) \equiv \text{erfc}(x) \qquad \text{(A.3.4)}$$

A.3.4 Integral Methods

Two solution techniques dealing with integral equations are briefly discussed. The first method starts with the integration of a given set of partial differential equations that describe a given flow system, known as the integral method. The second approach starts with balance equations in integral form, i.e., the Reynolds Transport Theorem, which assures the conservation laws for a control volume.

Von Karman Integral Method In contrast to separation of variables and similarity theory, the integral method is an *approximation* method. The Von Karman integral method is the most famous member of the family of integral relations, which in turn is a special case of the method of weighted residuals (MWR). Specifically, a transport equation in normal form can be written as [cf. Eq. (1.7)]

$$L(\phi) \equiv \frac{\partial \phi}{\partial t} + \nabla \cdot (\vec{v}\phi) - \nu \nabla^2 \phi - S = 0 \tag{A.3.5}$$

where $L(\bullet)$ is a (nonlinear) operator, ϕ is a dependent variable, and S represents sink/source terms. Now, the unknown ϕ-function is replaced by an *approximate* expression, i.e., a "profile" or functional $\tilde{\phi}$ that satisfies the boundary conditions, but contains a number of unknown coefficients or parameters. As can be expected,

$$L(\tilde{\phi}) \neq 0 \text{, i.e., } L(\tilde{\psi}) = R \tag{A.3.6a, b}$$

where R is the residual. In requiring that

$$\int_{\Omega} W R \, d\Omega = 0 \tag{A.3.6c}$$

we force the weighted residual over the computational domain Ω to be zero and thereby determine the unknown coefficients or parameters in the assumed $\tilde{\phi}$ -function. The type of weighing function W determines the special case of the MWR, e.g., integral method, collocation method, Galerkin finite element method, control volume method, etc. (cf. Finlayson 1978).

The Von Karman method is best applicable to laminar/turbulent similar or nonsimilar *boundary-layer type* flows for which appropriate velocity, concentration, and temperature profiles are known, i.e., thin and thick wall shear layers as well as plumes, jets, and wakes. Solutions of such problems yield global or integral system parameters, such as flow rates, fluxes, forces, boundary-layer thicknesses, shape factors, drag coefficients, etc.

In general, a two-dimensional partial differential equation is integrated in one direction, typically normal to the main flow, and thereby transformed into an ordinary differential equation, which is then solved analytically or numerically. Implementation of the integral method rests on two general characteristics of boundary-layer type problems: (i) the boundary conditions for a particular system simplify the integration process significantly so that a simpler differential equation is obtained, and (ii) all extra unknown functions, or parameters, remaining in the governing differential equation are approximated on physical grounds or by empirical relationships. Thus, closure is gained using, for example, the entrainment concept for plumes, jets, and wakes or by expressing velocity and temperature profiles with power expansions for high Reynolds number flows past submerged bodies.

Sample Problem: Integral method applied to the Blasius problem ($U_0 = u_\infty = ¢$, i.e., $\partial p/\partial x = 0$)

Sketch:

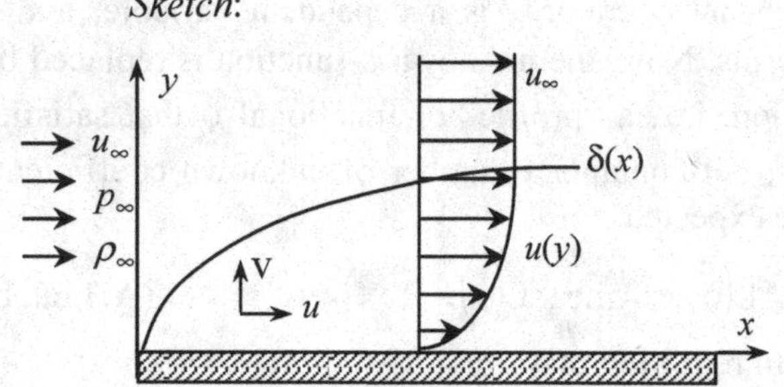

Assumptions:

- Steady laminar 2-D flow with constant fluid properties

Recall: For the Blasius problem, the boundary-layer equations reduce to

$$\frac{\partial u}{\partial x} + \frac{\partial v}{\partial y} = 0 \quad \text{and} \quad \frac{\partial u}{\partial t} + u\frac{\partial u}{\partial x} + v\frac{\partial u}{\partial y} = \frac{1}{\rho}\frac{d\tau_{yx}}{dx}$$

- Solving for v in the continuity equation yields

$$v = -\int \frac{\partial u}{\partial y} dy + f(x) := -\int_0^y \frac{\partial u}{\partial x} dy$$

- Integration across the *x*-momentum equation yields

$$\int_{y=0}^{\delta(t)} \left(u\frac{\partial u}{\partial x} + v\frac{\partial u}{\partial y} \right) dy = \frac{1}{\rho}\int_0^\delta \frac{d\tau_{yx}}{dx} dy := -\frac{\tau_w}{\rho}$$

- Inserting v(*x*,*y*) and integrating the term $\int_0^\delta v\frac{\partial u}{\partial y} dy$ by parts, i.e., $\int u dv = uv - \int v du$, yields

- $$\int_0^\delta \frac{\partial}{\partial x}[u(U - u)] dy = \frac{\tau_w}{\rho}$$

  where $U = u_\infty$. Now, a suitable *u*(*x*,*y*) profile has to be postulated, which matches the boundary conditions at $y = 0$ and $y = \delta(x)$.

Solution: For laminar flow over a flat plate,

$$\frac{u}{u_\infty} = 2\frac{y}{\delta} - \left(\frac{y}{\delta}\right)^2$$

is a suitable profile where $\delta(x)$ is the key unknown.
With the previously derived *momentum integral relation*,

$$\tau_w(x) = \rho \frac{d}{dx}\int_0^\delta u(U-u)\,dy := 0.133\,\rho\,U^2 \frac{d\delta}{dx}$$

We also know that

$$\tau_w(x) = \mu \left.\frac{\partial u}{\partial y}\right|_{y=0} := \mu U \frac{2}{\delta}$$

Combining both results leads to an ODE for $\delta(x)$:

$$\frac{\mu dx}{0.0665\rho U} = \delta d\delta \text{ subject to } \delta(x=0)=0.$$

Integration yields

- $$\delta(x) = 5.48\sqrt{\frac{\mu x}{\rho U}}$$

and hence

- $$\tau_w(x) = 0.356\rho U^2\sqrt{\frac{\mu}{\rho U x}}$$

Graph:

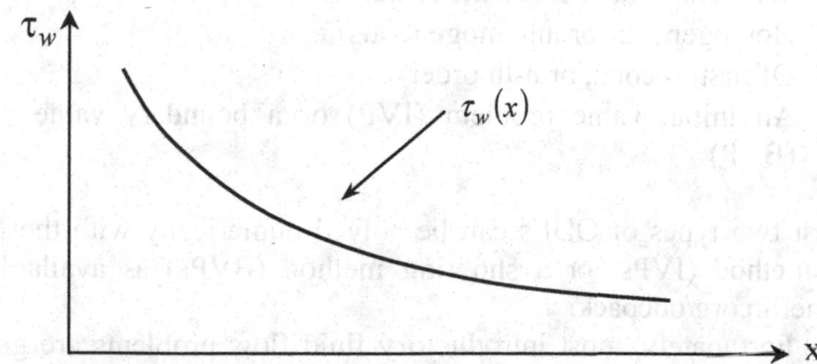

A.4 Ordinary Differential Equations

For most applications in biofluids, the key differential equations are the equation of continuity, i.e., conservation of fluid mass, and the Navier–Stokes equations, i.e., conservation of linear momentum for constant fluid properties, as well as the scalar transport equations for species mass and heat transfer. They reflect the conservation laws in terms of differential balances for fluid mass, momentum, species concentration, and energy.

If the dependent variable, say, the velocity, is a function of more than one independent variable (e.g., x, y, z, t or r, θ, z, t) then the describing equation is a partial differential equation (PDE); otherwise, it is an ordinary differential equation (ODE). Clearly, solving PDEs requires usually elaborate transformations or numerical algorithms (see Ozisik 1993; Hoffman 2001). For that reason and to gain direct physical insight, *simplified base-case problems* are discussed (see Chap. 1) where the continuity equation is fully satisfied and the Navier–Stokes equations are reduced to ODEs. Exact solutions for ODEs are listed in Polyamin & Zaitsev (1995). Numerical ODE solutions may be obtained with commercial software such as Matlab and Mathcad, which for their underlying finite difference approximations rely on selected terms of the Taylor Series (see Sect. A.2.3).

After developing a mathematical model describing approximately the fluid flow problem at hand, the resulting ODE (or system of coupled ODEs) has to be classified. One has to determine if the ODE, say, for y(x) is:

- Linear or nonlinear (e.g., y^2, yy', $\sqrt{y''}$, etc.)
- With constant coefficients or not
- Homogeneous or inhomogeneous
- Of first, second, or n-th order
- An initial value problem (IVP) or a boundary value problem (BVP)

The last two types of ODEs can be solved numerically with the Runge–Kutta method (IVPs) or a shooting method (BVPs) as available from www.netlib.org/odepack.

Fortunately, most introductory fluid flow problems are governed by ODEs of the form

$$\frac{d^n}{dy^n}[f(y)] = g(y) \tag{A.4.1}$$

which can be solved by direct integration, subject to n boundary conditions (BCs).

Typically, n = 2 and $g(y) \equiv K$, a constant, so that Eq. (A.4.1) can be rewritten as

$$\frac{d}{dy}\left[\frac{df}{dy}\right] = K \tag{A.4.2a}$$

or

$$d\left[\frac{df}{dy}\right] = Kdy \tag{A.4.2b}$$

Hence,

$$\frac{df}{dy} = Ky + C_1 \tag{A.4.2c}$$

and integrating again yields

$$f(y) = \frac{K}{2}y^2 + C_1 y + C_2 \tag{A.4.3}$$

where the integration constants are determined with two given boundary conditions for f(y).

In cylindrical coordinates, we may encounter an ODE somewhat similar to Eq. (A.4.1) of the form

$$\frac{1}{r}\frac{df(r)}{dr} + \frac{d^2 f(r)}{dr^2} = g(r) \tag{A.4.4a}$$

which can be rewritten for direct integration as:

$$\frac{1}{r}\frac{d}{dr}\left(r\frac{df(r)}{dr}\right) = g(r) \tag{A.4.4b}$$

For example, with $g(r) \equiv K$ =constant and $f \equiv v$, say the fully-developed axial velocity in a tube of radius R, we have

$$d\left(r\frac{dv(r)}{dr}\right) = Krdr \tag{A.4.5a}$$

so that after integration:

$$\frac{dv(r)}{dr} = \frac{K}{2}r + \frac{C_1}{r} \tag{A.4.5b}$$

Integrating again yields

$$v(r) = \frac{K}{4}r^2 + C_1 \ln r + C_2 \tag{A.4.6}$$

The differences between the solutions of Eqs. (A.4.3) and (A.4.6) as well as the impact of different BCs on Eq. (A.4.6) were discussed in Sect. 1.3.

Another interesting case where term contraction allows for direct integration is as an ODE of the form

$$\frac{df(x)}{dx}+\frac{n}{x}f(x)\equiv\frac{1}{x^n}\frac{d}{dx}\left[x^n f(x)\right]=g(x); \qquad n=1,2 \tag{A.4.7a}$$

which yields after the first integration:

$$x^n f(x)=\int x^n g(x)dx+C_1 \tag{A.4.7b}$$

The term $\int x^n g(x)dx$ could be solved via integration by parts, i.e.,

$$\int_a^b udv=(uv)\Big|_a^b-\int_a^b vdu \tag{A.4.8}$$

Numerous natural processes can be described by a linear, inhomogeneous second-order ODE with constant coefficients, i.e.,

$$f''+Af'+Bf=F(x) \tag{A.4.9}$$

where F(x) is a prescribed (forcing) function. Typically,

$$f(x)=f_{\text{hom og.}}+f_{\text{in hom og.}} \tag{A.4.10a}$$

In general, $f_{\text{hom}\,og.}$ admits exponential solutions, e.g.,

$$f(x)\sim e^{\lambda x} \tag{A.4.10b}$$

where λ can be obtained from the quadratic equation

$$\lambda^2+A\lambda+B=0 \tag{A.4.10c}$$

so that

$$f(x)=C_1e^{\lambda_1 x}+C_2e^{\lambda_2 x} \tag{A.4.10d}$$

Table A.4.1 summarizes typical ODEs describing transport phenomena, where f and g are functions of x and the quantities a, b, and c are real constants.

Table A.4.1 Typical ODEs and their general solutions

Equation	General solution
$\frac{dy}{dx}=\frac{f(x)}{g(y)}$	$\int g dy=\int f dx+C_1$
$\frac{dy}{dx}+f(x)y=g(x)$	$y=e^{-\int f dx}(\int e^{\int f dx} g dx+C_1)$
$\frac{d^2y}{dx^2}+a^2y=0$	$y=C_1\cos ax+C_2\sin ax$
$\frac{d^2y}{dx^2}-a^2y=0$	$y=C_1\cosh ax+C_2\sinh ax$ or $y=C_3e^{+ax}+C_4e^{-ax}$
$\frac{1}{x}\frac{d}{dx}\left(x^2\frac{dy}{dx}\right)+a^2y=0$	$y=\frac{C_1}{x}\cos ax+\frac{C_2}{x}\sin ax$

A.5 Transport Equations (Continuity, Momentum and Heat Transfer)

A.5.1 Continuity Equation

$$\frac{\partial \rho}{\partial t}+\nabla\cdot(\rho\vec{v})=0$$

Note: For $\rho=¢ \Rightarrow \nabla\cdot\vec{v}=0$

- Rectangular Coordinates: $\frac{\partial \rho}{\partial t}+\frac{\partial}{\partial x}(\rho u)+\frac{\partial}{\partial y}(\rho v)+\frac{\partial}{\partial z}(\rho w)=0$
- Cylindrical Coordinates:

$$\frac{\partial \rho}{\partial t}+\frac{1}{r}\frac{\partial}{\partial r}(\rho r v_r)+\frac{1}{r}\frac{\partial}{\partial \theta}(\rho v_\theta)+\frac{\partial}{\partial z}(\rho v_z)=0$$

A.5.2 Equation of Motion (or Linear Momentum Equation)

Cauchy Equation

$$\rho\frac{D\vec{\mathbf{v}}}{Dt}=-\nabla p+\nabla\cdot\vec{\vec{\tau}}+\rho\vec{\mathbf{g}}$$

where

$$\frac{D\#}{Dt}\equiv\frac{\partial\#}{\partial t}+(\vec{\mathbf{v}}\cdot\nabla)\#$$

- Rectangular Coordinates

$$\rho\left(\frac{\partial u}{\partial t}+u\frac{\partial u}{\partial x}+v\frac{\partial u}{\partial y}+w\frac{\partial u}{\partial z}\right)=-\frac{\partial p}{\partial x}+\left[\frac{\partial}{\partial x}\tau_{xx}+\frac{\partial}{\partial y}\tau_{yx}+\frac{\partial}{\partial z}\tau_{zx}\right]+\rho g_x$$

$$\rho\left(\frac{\partial v}{\partial t}+u\frac{\partial v}{\partial x}+v\frac{\partial v}{\partial y}+w\frac{\partial v}{\partial z}\right)=-\frac{\partial p}{\partial y}+\left[\frac{\partial}{\partial x}\tau_{xy}+\frac{\partial}{\partial y}\tau_{yy}+\frac{\partial}{\partial z}\tau_{zy}\right]+\rho g_y$$

$$\rho\left(\frac{\partial w}{\partial t}+u\frac{\partial w}{\partial x}+v\frac{\partial w}{\partial y}+w\frac{\partial w}{\partial z}\right)=-\frac{\partial p}{\partial z}+\left[\frac{\partial}{\partial x}\tau_{xz}+\frac{\partial}{\partial y}\tau_{yz}+\frac{\partial}{\partial z}\tau_{zz}\right]+\rho g_z$$

- Cylindrical Coordinates

$$\rho\left(\frac{\partial v_r}{\partial t}+v_r\frac{\partial v_r}{\partial r}+\frac{v_\theta}{r}\frac{\partial v_r}{\partial \theta}+v_z\frac{\partial v_r}{\partial z}-\frac{v_\theta^2}{r}\right)=-\frac{\partial p}{\partial r}$$

$$+\left[\frac{1}{r}\frac{\partial}{\partial r}(r\tau_{rr})+\frac{1}{r}\frac{\partial}{\partial \theta}\tau_{\theta r}+\frac{\partial}{\partial z}\tau_{zr}-\frac{\tau_{\theta\theta}}{r}\right]+\rho g_r$$

$$\rho\left(\frac{\partial v_\theta}{\partial t}+v_r\frac{\partial v_\theta}{\partial r}+\frac{v_\theta}{r}\frac{\partial v_\theta}{\partial \theta}+v_z\frac{\partial v_\theta}{\partial z}+\frac{v_r v_\theta}{r}\right)=-\frac{1}{r}\frac{\partial p}{\partial \theta}$$

$$+\left[\frac{1}{r^2}\frac{\partial}{\partial r}\left(r^2\tau_{r\theta}\right)+\frac{1}{r}\frac{\partial}{\partial \theta}\tau_{\theta\theta}+\frac{\partial}{\partial z}\tau_{z\theta}+\frac{\tau_{\theta r}-\tau_{r\theta}}{r}\right]+\rho g_\theta$$

$$\rho\left(\frac{\partial v_z}{\partial t}+v_r\frac{\partial v_z}{\partial r}+\frac{v_\theta}{r}\frac{\partial v_z}{\partial \theta}+v_z\frac{\partial v_z}{\partial z}\right)=-\frac{\partial p}{\partial z}$$

$$+\left[\frac{1}{r}\frac{\partial}{\partial r}(r\tau_{rz})+\frac{1}{r}\frac{\partial}{\partial \theta}\tau_{\theta z}+\frac{\partial}{\partial z}\tau_{zz}\right]+\rho g_z$$

A.5.3 Momentum Equation for Constant-Property Fluids

Navier–Stokes Equation

$$\rho\frac{D\vec{v}}{Dt}=-\nabla p+\mu\nabla^2\vec{v}+\rho\vec{g}$$

- Rectangular Coordinates

$$\rho\left(\frac{\partial u}{\partial t}+u\frac{\partial u}{\partial x}+v\frac{\partial u}{\partial y}+w\frac{\partial u}{\partial z}\right)=-\frac{\partial p}{\partial x}+\mu\left[\frac{\partial^2 u}{\partial x^2}+\frac{\partial^2 u}{\partial y^2}+\frac{\partial^2 u}{\partial z^2}\right]+\rho g_x$$

$$\rho\left(\frac{\partial v}{\partial t}+u\frac{\partial v}{\partial x}+v\frac{\partial v}{\partial y}+w\frac{\partial v}{\partial z}\right)=-\frac{\partial p}{\partial y}+\mu\left[\frac{\partial^2 v}{\partial x^2}+\frac{\partial^2 v}{\partial y^2}+\frac{\partial^2 v}{\partial z^2}\right]+\rho g_y$$

$$\rho\left(\frac{\partial w}{\partial t}+u\frac{\partial w}{\partial x}+v\frac{\partial w}{\partial y}+w\frac{\partial w}{\partial z}\right)=-\frac{\partial p}{\partial z}+\mu\left[\frac{\partial^2 w}{\partial x^2}+\frac{\partial^2 w}{\partial y^2}+\frac{\partial^2 w}{\partial z^2}\right]+\rho g_z$$

- Cylindrical Coordinates

$$\rho\left(\frac{\partial v_r}{\partial t}+v_r\frac{\partial v_r}{\partial r}+\frac{v_\theta}{r}\frac{\partial v_r}{\partial \theta}+v_z\frac{\partial v_r}{\partial z}-\frac{v_\theta^2}{r}\right)=-\frac{\partial p}{\partial r}$$

$$+\mu\left[\frac{\partial}{\partial r}\left(\frac{1}{r}\frac{\partial}{\partial r}(rv_r)\right)+\frac{1}{r^2}\frac{\partial^2 v_r}{\partial \theta^2}+\frac{\partial^2 v_r}{\partial z^2}-\frac{2}{r^2}\frac{\partial v_\theta}{\partial \theta}\right]+\rho g_r$$

$$\rho\left(\frac{\partial v_\theta}{\partial t}+v_r\frac{\partial v_\theta}{\partial r}+\frac{v_\theta}{r}\frac{\partial v_\theta}{\partial \theta}+v_z\frac{\partial v_\theta}{\partial z}+\frac{v_r v_\theta}{r}\right)=-\frac{1}{r}\frac{\partial p}{\partial \theta}$$

$$+\mu\left[\frac{\partial}{\partial r}\left(\frac{1}{r}\frac{\partial}{\partial r}(rv_\theta)\right)+\frac{1}{r^2}\frac{\partial^2 v_\theta}{\partial \theta^2}+\frac{\partial^2 v_\theta}{\partial z^2}+\frac{2}{r^2}\frac{\partial v_r}{\partial \theta}\right]+\rho g_\theta$$

$$\rho\left(\frac{\partial v_z}{\partial t}+v_r\frac{\partial v_z}{\partial r}+\frac{v_\theta}{r}\frac{\partial v_z}{\partial \theta}+v_z\frac{\partial v_z}{\partial z}\right)=-\frac{\partial p}{\partial z}$$

$$+\mu\left[\frac{1}{r}\frac{\partial}{\partial r}\left(r\frac{\partial v_z}{\partial r}\right)+\frac{1}{r^2}\frac{\partial^2 v_z}{\partial \theta^2}+\frac{\partial^2 v_z}{\partial z^2}\right]+\rho g_z$$

A.5.4 Heat Transfer Equation for Constant-Property Fluids

$$\rho c_p\frac{DT}{Dt}=k\nabla^2 T+\mu\Phi;\quad \frac{k}{\rho c_p}\equiv\alpha \text{ < thermal diffusivity >}$$

Note: For species-mass transport of concentration c $[M/L^3]$:

$$\frac{Dc}{Dt}=D_{AB}\nabla^2 c+S_c;\ D_{AB}\triangleq \text{binary mass diffusivity}$$

Note: The equations of Sect. A.5.3 in conjunction with Sects. A.5.1 plus A.5.4 are nowadays summarized as the Navier–Stokes equations.

- Rectangular Coordinates (see Sect. A.5.6 for viscous dissipation function Φ)

$$\frac{\partial T}{\partial t}+u\frac{\partial T}{\partial x}+v\frac{\partial T}{\partial y}+w\frac{\partial T}{\partial z}=\alpha\left[\frac{\partial^2 T}{\partial x^2}+\frac{\partial^2 T}{\partial y^2}+\frac{\partial^2 T}{\partial z^2}\right]+\frac{\mu}{\rho c_p}\Phi$$

- Cylindrical Coordinates

$$\frac{\partial T}{\partial t}+v_r\frac{\partial T}{\partial r}+\frac{v_\theta}{r}\frac{\partial T}{\partial \theta}+v_z\frac{\partial T}{\partial z}=\alpha\left[\frac{1}{r}\frac{\partial}{\partial r}\left(r\frac{\partial T}{\partial r}\right)+\frac{1}{r^2}\frac{\partial^2 T}{\partial \theta^2}+\frac{\partial^2 T}{\partial z^2}\right]+\frac{\mu}{\rho c_p}\Phi$$

A.5.5 Stresses: $\vec{\vec{\tau}} = \mu\left[\nabla\vec{v} + (\nabla\vec{v})^{tr}\right]$ and Fluxes: $\vec{q}_{cond} = -k\nabla T$

Note: Incompressible fluids

- Rectangular Coordinates

$$\tau_{xx} = 2\mu\frac{\partial u}{\partial x};\ \tau_{yy} = 2\mu\frac{\partial v}{\partial y};\ \tau_{zz} = 2\mu\frac{\partial w}{\partial z}$$

$$\tau_{xy} = \tau_{yx} = \mu\left[\frac{\partial v}{\partial x} + \frac{\partial u}{\partial y}\right]$$

$$\tau_{yz} = \tau_{zy} = \mu\left[\frac{\partial w}{\partial y} + \frac{\partial v}{\partial z}\right]$$

$$\tau_{zx} = \tau_{xz} = \mu\left[\frac{\partial u}{\partial z} + \frac{\partial w}{\partial x}\right]$$

$$q_x = -k\frac{\partial T}{\partial x};\ q_y = -k\frac{\partial T}{\partial y};\ q_z = -k\frac{\partial T}{\partial z}$$

- Cylindrical Coordinates

$$\tau_{rr} = 2\mu\frac{\partial v_r}{\partial r};\ \tau_{\theta\theta} = 2\mu\left(\frac{1}{r}\frac{\partial v_\theta}{\partial \theta} + \frac{v_r}{r}\right);\ \tau_{zz} = 2\mu\frac{\partial v_z}{\partial z}$$

$$\tau_{r\theta} = \tau_{\theta r} = \mu\left[r\frac{\partial}{\partial r}\left(\frac{v_\theta}{r}\right) + \frac{1}{r}\frac{\partial v_r}{\partial \theta}\right]$$

$$\tau_{\theta z} = \tau_{z\theta} = \mu\left[\frac{1}{r}\frac{\partial v_z}{\partial \theta} + \frac{\partial v_\theta}{\partial z}\right]$$

$$\tau_{zr} = \tau_{rz} = \mu\left[\frac{\partial v_r}{\partial z} + \frac{\partial v_z}{\partial r}\right]$$

$$q_r = -k\frac{\partial T}{\partial r};\ q_\theta = -k\frac{1}{r}\frac{\partial T}{\partial \theta};\ q_z = -k\frac{\partial T}{\partial z}$$

A.5.6 Dissipation Function for Newtonian Fluids

- Rectangular Coordinates

$$\Phi = 2\left[\left(\frac{\partial u}{\partial x}\right)^2 + \left(\frac{\partial v}{\partial y}\right)^2 + \left(\frac{\partial w}{\partial z}\right)^2\right] + \left[\frac{\partial v}{\partial x} + \frac{\partial u}{\partial y}\right]^2 + \left[\frac{\partial w}{\partial y} + \frac{\partial v}{\partial z}\right]^2 + \left[\frac{\partial u}{\partial z} + \frac{\partial w}{\partial x}\right]^2$$

- Cylindrical Coordinates

$$\Phi = 2\left[\left(\frac{\partial v_r}{\partial r}\right)^2 + \left(\frac{1}{r}\frac{\partial v_\theta}{\partial \theta} + \frac{v_r}{r}\right)^2 + \left(\frac{\partial v_z}{\partial z}\right)^2\right] + \left[r\frac{\partial}{\partial r}\left(\frac{v_\theta}{r}\right) + \frac{1}{r}\frac{\partial v_r}{\partial \theta}\right]^2 + \left[\frac{1}{r}\frac{\partial v_z}{\partial \theta} + \frac{\partial v_\theta}{\partial z}\right]^2 + \left[\frac{\partial v_r}{\partial z} + \frac{\partial v_z}{\partial r}\right]^2$$

A.6 Equation Sheets

- **Format**

Sketch, Assumptions, Approach/Concepts, Solution, Graphs and Comments

- **Derivatives and Vectors**

$\nabla \equiv (\partial/\partial x)\hat{i} + (\partial/\partial y)\hat{j} + (\partial/\partial z)\hat{k}$ (Cartesian del operator)
$\nabla \equiv (\partial/\partial r)\hat{e}_r + 1/r(\partial/\partial\vartheta)\hat{e}_\theta + (\partial/\partial z)\hat{e}_z$ (cylindrical del operator)
$\nabla \cdot \vec{v} = \partial u/\partial x + \partial v/\partial y + \partial w/\partial z$ (dot product)
$\nabla^2 \equiv \partial^2/\partial x^2 + \partial^2/\partial y^2 + \partial^2/\partial z^2$ (Cartesian Laplace operator)
$\nabla^2 \equiv (1/r)\partial/\partial r(r\partial/\partial r) + (1/r^2)\partial^2/\partial\theta^2 + \partial^2/\partial z^2$ (cyl. Laplace op.)
$\partial/\partial x(\rho u) = \rho(\partial u/\partial x) + u(\partial\rho/\partial x)$ (chain rule)
$(\vec{v}\cdot\nabla)\vec{v}$ $(2-\text{D; u}-\text{comp.}) \rightarrow u\,\partial u/\partial x + v\,\partial u/\partial y$
$\nabla^2\vec{v}$ $(2-\text{D; scalar}) \rightarrow \partial^2 u/\partial x^2 + \partial^2 v/\partial y^2$
$\nabla\cdot\vec{v} \Rightarrow$ scalar $\nabla\vec{v} \Rightarrow$ tensor $\nabla\times\vec{v}$;
$\nabla p;\ \nabla\cdot\vec{\vec{\tau}} \Rightarrow$ vectors

- **Some Differential and Integral Relations**

Force: $dF = \tau_w dA \Rightarrow F_{drag} = \int \tau_{wall} dA_{surf}$;

Torque: $T = \int rdF = \int_A r\tau_w dA;\ dA = 2\pi rdr$

$\tau_{yx} = \mu\, du/dy$ (1-D laminar flow)

Power: $P = F\cdot v$

Flow Rate:
$Q = v_{avg}A = \int_A udA;\ v_{avg} \equiv \bar{u} = (1/A)\int_A udA$

Streamlines:
$dy/dx = v/u$ (planar flow) $\Rightarrow y(x)$

- **Fluid Statics**

$-\nabla p - \rho g\hat{k} = \rho\vec{a}$
$p_2 - p_1 = \rho g(z_1 - z_2)$ (const. fluid properties); or
$p = \gamma h;\quad \gamma \equiv \rho g,\quad h = -z$;
$F_p = \iint pdA$

Plane surfaces $F_{normal} = p_{CG}A$ (CG ≡ center of gravity)
$F\cdot y_{cp} = \int y\cdot pdA$ and $F\cdot x_{cp} = \int x\cdot pdA$

Curved surfaces
$F_V = \rho g\forall$ (fluid weight acts through centroid of fluid column)
$F_H = \int_{A_H} p\,dA_H$ (vert. plane projection to find horizontal force)

- **Kinematics**

$\vec{v} = u\hat{i} + v\hat{j} + w\hat{k}$ (Cartesian velocity vector)
$\vec{v} = v_r\hat{e}_r + v_\theta\hat{e}_\theta + v_z\hat{e}_z$ (cylindrical velocity vector)

Fluid element acceleration
$$\vec{a} \equiv \frac{D\vec{v}}{Dt} = \frac{\partial\vec{v}}{\partial t} + u\frac{\partial\vec{v}}{\partial x} + v\frac{\partial\vec{v}}{\partial y} + w\frac{\partial\vec{v}}{\partial z}$$
$$= \partial\vec{v}/\partial t + (\vec{v}\cdot\nabla)\vec{v}$$
$\vec{a}_{total} = \vec{a}_{local} + \vec{a}_{convective}$
$\vec{a} = a_x\hat{i} + a_y\hat{j} + a_z\hat{k}$

Rotation rate vector
$\vec{\omega} = \omega_x\hat{i} + \omega_y\hat{j} + \omega_z\hat{k}$
$$\omega_x = 0.5(\partial w/\partial y - \partial v/\partial z)$$
$$\omega_y = 0.5(\partial u/\partial z - \partial w/\partial x)$$
$$\omega_z = 0.5(\partial v/\partial x - \partial u/\partial y)$$
Vorticity Vector: $\zeta = \nabla\times\vec{v} = 2\vec{\omega}$
Deformation tensor: $\vec{\vec{\varepsilon}} = 1/2(\nabla\vec{v} + \nabla\vec{v}^T)$;
Shear rate tensor: $\vec{\vec{\gamma}} = 2\vec{\vec{\varepsilon}}$
Shear stress tensor: $\vec{\vec{\tau}} = 2\mu\vec{\vec{\varepsilon}}$ (Newtonian fluids)
$\mu \rightarrow \eta = \eta(\dot{\gamma})$ for non-Newtonian fluids;
e.g., $\eta = K\dot{\gamma}^{n-1}$

- **Reynolds Transport Theorem (RTT)**

$$\left.\frac{DB}{Dt}\right|_{sys} = \frac{d}{dt}\iiint_{CV}(\rho\beta)dV + \iint_{CS}\rho\beta\vec{v}_{rel}\cdot d\vec{A}\ ;$$

$$B_{sys} = \begin{cases} m \\ m\vec{v} \end{cases} \Rightarrow \beta = \begin{cases} 1 \\ \vec{v} \end{cases}$$

Relative velocity: $\vec{v}_{rel} = \vec{v}_{fluid} - \vec{v}_{CV}$

Green's theorem:

$$\iint_{CS}\rho\vec{v}\cdot d\vec{A} = \iiint_{CV}\nabla\cdot(\rho\vec{v})dV$$

Conservation of mass (RTT for fixed C.V.)

$0 = \frac{\partial}{\partial t}\iiint_{CV}\rho dV + \iint_{CS}\rho\vec{v} = \cdot d\vec{A}$, can give:

$\sum_i \dot{m}_i = 0$ or $Q_1 = Q_2$

Conservation of momentum (RTT for an accelerating CV)

$$\vec{F}_B + \vec{F}_S - \iiint_{CV}\vec{a}\rho dV = \frac{\partial}{\partial t}\iiint_{CV}\vec{v}\rho\, dV + \iint_{CS}\vec{v}\rho\vec{v}_{rel}\cdot d\vec{A}$$

Governing equations as they appear after derivation from RTT

$\nabla\cdot\vec{v} = 0$ (incompressible fluid)

$(\partial/\partial t)(\rho\vec{v}) + \nabla\cdot(\rho\vec{v}\vec{v}) = -\nabla p + \nabla\cdot\vec{\tau} + \rho\vec{g}$

(cf. pg. A-14 for expansion)

- **Stream Functions**

$u \equiv \partial\psi/\partial y$ and $v \equiv -\partial\psi/\partial x$ such that

Continuity Eqn. $\partial^2\psi/\partial x\partial y - \partial^2\psi/\partial y\partial x = 0$

$\psi = \int u\partial y + f(x)$, $\psi = -\int v\partial x + g(y)$

ψ = const. Note: y(x) streamline equation

$$Q_{1\to2} = \int_1^2(\vec{v}\cdot\vec{n})dA = \int_1^2 d\psi = \psi_2 - \psi_1$$

- **Bernoulli Equation**

$p/\rho + (1/2)v^2 + gz = \text{const}$. (along a streamline)

Kinetic energy correction factor

$\alpha = 2.0$ (laminar); ≈ 1 (turbulent)

- **Pi Theorem**

1. Physical Insight – select relevant variables, ex. A=fct(B,C,D,E)
2. Count variables = n, ex. n = 5
3. List dimensions of each variable
4. Select primary dimensions, k = 3 (FLT or MLT)
5. Det. # of pi-terms required = n – k
6. Select repeating variables (=k) used to generate pi terms (never choose the dependent variable; choose one from every category, i.e., kinematic, dynamic, geometric, fluid properties)
7. Derive pi-terms (or proceed by Inspection)

Dimensions

ρ: $FT^2/L^4 = M/L^3$	μ: M/LT
F: ML/T^2	Δp: $M/LT^2 = F/L^2$
γ: M/L^2T^2	σ: M/T^2
Q: L^3/T	$\nu \equiv \mu/\rho : L^2/T$

- **Non-dimensionalization**

1. Select reference quantity for each *variable*, ex. U_o, l, P_o
2. Non-dim., all variables, ex. $\hat{u} = u/U_o, \hat{x} = x/l$
3. Insert $u = U_o\hat{u}, etc.$ into governing equations (watch second derivatives)
4. Divide entire eq. by one coefficient to non-dim. entire equation
5. Check resulting dim.-less groups

Dim-less Groups

Reynolds: $Re_L = \frac{uL}{\nu} = \frac{\text{InertiaForce}}{\text{ViscousForce}}$; L := D (for a pipe); ℓ (for a plate)

Euler: $Eu = 2\Delta p/\rho v^2$ Mach: $M = v/c = \sqrt{\rho v^2/E}$

Froude: $Fr^2 = v^2/gl$ Weber: $We = \rho v^2 l/\sigma$

Prototypes (real) and models: Set $\pi_{i,p} = \pi_{i,m}$ for similarity.

- **Internal Flow**

Extended Bernoulli

$$p_1/\rho + 0.5\alpha_1 V_1^2 + gz_1 = p_2/\rho + 0.5\alpha_2 V_2^2 + gz_2 + gh$$

h_{lT} is total losses due to friction (major) and form changes (minor) as expressed via K-values.

$$h_{lT} = h_f + \sum h_m = \frac{v^2}{2g}\left(f(\frac{L}{D}) + \sum K \right)$$

Laminar pipe flow $f = \dfrac{64}{Re_D}$; $Re_D = \dfrac{v_{avg} D}{\nu}$

Turbulent: use the Moody chart: $f = fct(Re_D, e/D)$

Hydraulic diameter $D_H = 4A/P$ ($P \hat{=}$ perimeter)

Laminar (Poiseuille) pipe flow

$$u(r) = u_{max}\left[1 - (r/R)^2\right];$$

$$u_{max} = 2v_{avg} = \left(-R^2/4\mu\right)(dp/dx)$$

$$Q = v_{avg} A = \int_0^R u(r) dA = \pi \Delta p D^4/(128\mu L)$$

$$\tau_{rx} = \mu\, du/dr = \frac{r}{2}\frac{\partial p}{\partial x}$$

Steady turbulent flow

$$u = \bar{u} + u';$$

where, $\bar{u} = (1/T)\int_0^T u dt$, and $\overline{u'} = 0$

$$\tau_{total} = \tau_{lam} + \tau_{turb} = \mu_{lam}(d\bar{u}/dy) + \mu_{turb}(d\bar{u}/dy)$$

$$\tau_{turb} = -\rho\overline{u'v'};$$

$$\mu_{turb} = fct(k, \varepsilon, \nabla\vec{v}, geo., etc.)$$

$u^+ \equiv u/u_\tau$; $u_\tau \equiv \sqrt{\tau_w/\rho}$ <friction velocity>; $y^+ \equiv y u_\tau/(\mu/\rho)$

Laminar wall layer: $u^+ = y^+$ $(0 \le y^+ \le 5)$

Overlap log-law layer $\dfrac{u}{u_\tau} = \dfrac{1}{\kappa}\ln\left(\dfrac{y u_\tau}{(\mu/\rho)}\right) + B$

$\kappa \approx 0.41$; $B \approx 5.0$

Empirical smooth pipe flow approximation $\bar{u}/u_{max} = (y/R)^{1/n} = (1 - r/R)^{1/n}$,

$n \approx 7$ for $Re_D \approx 10^5$ when

$$v_{av}/u_{max} = 0.817$$

$\dfrac{v_{av}}{u_{max}} \approx (1 + 1.3\sqrt{f})^{-1}$; $f = f(Re_D, e/D)$

Appendix B

B.1 Conversion Factors

Dimension	Metric	Metric/English
Acceleration	1 m/s^2 = 100 cm/s^2	1 m/s^2 = 3.2808 ft/s^2 1 ft/s^2 = 0.3048[a] m/s^2
Area	1 m^2 = 10^4 cm^2 = 10^6 mm^2 = 10^{-6} km^2	1 m^2 = 1550 in.^{2} = 10.764 ft^2 1 ft^2 = 144 in.^{2} = 0.09290304[a] m^2
Density	1 g/cm^3 = 1 kg/L = 1,000 kg/m^3	1 g/cm^3 = 62.428 lbm/ft^3 = 0.036127 lbm/in.^{3} 1 lbm/in.^{3} = 1728 lbm/ft^3 1 kg/m^3 = 0.062428 lbm/ft^3
Energy, heat, work, internal energy, enthalpy	1 kJ = 1,000 J = 1,000 N · m = 1 kPa · m^3 1 kJ/kg = 1,000 m^2/s^2 1 kWh = 3,600 kJ 1 cal[b] = 4.184 J 1 IT cal[b] = 4.1868 J 1 cal[b] = 4.1868 kJ	1 kJ = 0.94782 Btu 1 Btu = 1.055056 kJ = 5.40395 psia · ft^3 = 778.169 lbf · ft 1 Btu/lbm = 25,037 ft^2/s^2 = 2.326[a] kJ/kg 1 kJ/kg = 0.430 Btu/lbm 1 kWh = 3412.14 Btu 1 therm = 10^5 Btu = 1.055 × 10^5 kJ (natural gas)
Force	1 N = 1 kg · m/s^2 = 10^5 dyn 1 kgf = 9.80665 N	1 N = 0.22481 lbf 1 lbf = 32.174 lbm · ft/s^2 = 4.44822 N
Heat flux	1 W/cm^2 = 10^4 W/m^2	1 W/m^2 = 0.3171 Btu/h · ft^2
Heat transfer coefficient	1 W/m^2 · °C = 1 W/m^2 · K	1 W/m^2 · °C = 0.17612 Btu/h · ft^2 · °F
Length	1 m = 100 cm = 1,000 mm = 10^{6}µm 1 km = 1,000 m	1 m = 39.370 in. = 3.2808 ft = 1.0926 yd 1 ft = 12 in. = 0.3048[a] m 1 mi. = 5280 ft = 1.6093 km 1 in. = 2.54[a] cm
Mass	1 kg = 1,000 g 1 metric ton = 1,000 kg	1 kg = 2.2046226 lbm 1 lbm = 0.45359237[a] kg 1 oz = 28.3495 g 1 slug = 32.174 lbm = 14.5939 kg 1 short ton = 2000 lbm = 907.1847 kg
Power, heat transfer rate	1 W = 1 J/s 1 kW = 1,000 W = 1.341 hp 1 hp[c] = 745.7 W	1 kW = 3412.14 Btu/h = 737.56 lbf · ft/s 1 hp = 550 lbf · ft/s = 0.7068 Btu/s = 42.41 Btu/min = 2544.5 Btu/h = 0.74570 kW 1 boiler hp = 33,475 Btu/h 1 Btu/h = 1.055056 kJ/h 1 ton of refrigeration = 200 Btu/min
Pressure	1 Pa = 1 N/m^2 1 kPa = 10^3 Pa = 10^{-3} MPa 1 atm = 101.325 kPa = 1.01325 bar = 760 mm Hg at 0°C = 1.03323 kgf/cm^2 1 mm Hg = 0.1333 kPa	1 Pa = 1.4504 × 10^{-4} psia = 0.020886 lbf/ft^2 1 psi = 144 lbf/ft^2 = 6.894757 kPa 1 atm = 14.696 psia = 29.92 in. Hg at 30°F 1 in. Hg = 3.387 kPa
Specific heat	1 kJ/kg · °C = 1 kJ/kg · K = 1 J/g · °C	1 Btu/lbm · °F = 4.1868 kJ/kg · °C 1 Btu/lbmol · R = 4.1868 kJ/kmol · K 1 kJ/kg · °C = 0.23885 Btu/lbm · °F = 0.23885 Btu/lbm · R

[a] Exact conversion factor between metric and English units.

[b] Calorie is originally defined as the amount of heat needed to raise the temperature of 1 g of water by 1°C, but it varies with temperature. The international steam table (IT) calorie (generally preferred by engineers) is exactly 4.1868 J by definition and corresponds to the specific heat of water at 15°C. The thermochemical calorie (generally preferred by physicists) is exactly 4.184 J by definition and corresponds to the specific heat of water at room temperature. The difference between the two is about 0.06%, which is negligible. The capitalized calorie used by nutritionists is actually a kilocalorie (1,000 IT cal).

Dimension	Metric	Metric/English
Specific volume	1 m^3/kg = 1,000 L/kg = 1,000 cm^3/g	1 m^3/kg = 16.02 ft^3/lbm 1 ft^3/lbm = 0.062428 m^3/kg
Temperature	T(K) = T(°C) + 273.15 ΔT(K) = Δ T(°C)	T(R) = T(°F) + 459.67 = 1.8T(K) T(°F) = 1.8 T(°C) + 32 ΔT(°F) = ΔT(R) = 1.8 ΔT(K)
Thermal conductivity	1 W/m · °C = 1 W/m · K	1 W/m · °C = 0.57782 Btu/h · ft · °F
Velocity	1 m/s = 3.60 km/h	1 m/s = 3.2808 ft/s = 2.237 mi/h 1 mi/h = 1.4667 ft/s 1 mi/h = 1.6093 km/h
Volume	1 m^3 = 1,000 L = 10^6 cm^3 (cc)	1 m^3 = 6.1024 × 10^4 $in.^3$ = 35.315 ft^3 = 264.17 gal (U.S.) 1 U.S. gal = 231 $in.^3$ = 3.7854 L 1 fl oz = 29.5735 cm^3 = 0.0295735 L 1 U.S. gal = 128 fl oz
Volume flow rate	1 m^3/s = 60,000 L/min = 10^6 cm^3/s	1 m^3/s = 15,850 gal/min (gpm) = 35.315 ft^3/s = 2118.9 ft^3/min (cfm)

[c] Mechanical horsepower. The electrical horsepower is taken to be exactly 746 W.

Some Physical Constants

Universal gas constant	R_u = 8.31447 kJ/kmol · K = 8.31447 kPa · m^3/kmol · K = 0.0831447 bar · m^3/kmol · K = 82.05 L · atm/kmol · K = 1.9858 Btu/lbmol · R = 1545.37 ft · lbf/lbmol · R = 10.73 psia · ft^3/lbmol · R
Standard acceleration of gravity	g = 9.80665 m/s^2 = 32.174 ft/s^2
Standard atmospheric pressure	1 atm = 101.325 kPa = 1.01325 bar = 14.696 psia = 760 mm Hg (0°C) = 29.9213 in Hg (32°F) = 10.3323 m H_2O (4°C)
Stefan–Boltzmann constant	σ = 5.6704 × 10^{-8} W/m^2 · K^4 = 0.1714 × 10^{-8} Btu/h · ft^2 · R^4
Boltzmann's constant	k = 1.380650 × 10^{-23} J/K
Speed of light in vacuum	c_o = 2.9979 × 10^8 m/s = 9.836 × 10^8 ft/s
Speed of sound in dry air at 0°C and 1 atm	c = 331.36 m/s = 1089 ft/s
Heat of fusion of water at 1 atm	h_{if} = 333.7 kJ/kg = 143.5 Btu/lbm
Enthalpy of vaporization of water at 1 atm	h_{fg} = 2256.5 kJ/kg = 970.12 Btu/lbm

B.2 Properties

Table B.2-1 Molar mass, gas constant, and critical-point properties

Substance	Formula	Molar mass, M kg/kmol	Gas constant, R kJ/kg · K[a]	Critical-point properties: Temperature, K	Pressure, MPa	Volume, m^3/kmol
Air	–	28.97	0.2870	132.5	3.77	0.0883
Ammonia	NH_3	17.03	0.4882	405.5	11.28	0.0724
Argon	Ar	39.948	0.2081	151	4.86	0.0749
Benzene	C_6H_6	78.115	0.1064	562	4.92	0.2603
Bromine	Br_2	159.808	0.0520	584	10.34	0.1355
n-Butane	C_4H_{10}	58.124	0.1430	425.2	3.80	0.2547
Carbon dioxide	CO_2	44.01	0.1889	304.2	7.39	0.0943
Carbon monoxide	CO	28.011	0.2968	133	3.50	0.0930
Carbon tetrachloride	CCl_4	153.82	0.05405	556.4	4.56	0.2759
Chlorine	Cl_2	70.906	0.1173	417	7.71	0.1242
Chloroform	$CHCl_3$	119.38	0.06964	536.6	5.47	0.2403
Dichlorodifluoromethane (R-12)	CCl_2F_2	120.91	0.06876	384.7	4.01	0.2179
Dichlorofluoromethane (R-21)	$CHCl_2F$	102.92	0.08078	451.7	5.17	0.1973
Ethane	C_2H_6	30.070	0.2765	305.5	4.48	0.1480
Ethyl alcohol	C_2H_5OH	46.07	0.1805	516	6.38	0.1673
Ethylene	C_2H_4	28.054	0.2964	282.4	5.12	0.1242
Helium	He	4.003	2.0769	5.3	0.23	0.0578
n-Hexane	C_6H_{14}	86.179	0.09647	507.9	3.03	0.3677
Hydrogen (normal)	H_2	2.016	4.1240	33.3	1.30	0.0649
Krypton	[illegible]	[illegible]	[illegible]	209.4	5.50	0.0924
Methane	CH_4	16.043	0.5182	191.1	4.64	0.0993
Methyl alcohol	CH_3OH	32.042	0.2595	513.2	7.95	0.1180
Methyl chloride	CH_3Cl	50.488	0.1647	416.3	6.68	0.1430
Neon	Ne	20.183	0.4119	44.5	2.73	0.0417
Nitrogen	N_2	28.013	0.2968	126.2	3.39	0.0899
Nitrous oxide	N_2O	44.013	0.1889	309.7	7.27	0.0961
Oxygen	O_2	31.999	0.2598	154.8	5.08	0.0780
Propane	C_3H_8	44.097	0.1885	370	4.26	0.1998
Propylene	C_3H_6	42.081	0.1976	365	4.62	0.1810
Sulfur dioxide	SO_2	64.063	0.1298	430.7	7.88	0.1217
Tetrafluoroethane (R-134a)	CF_3CH_2F	102.03	0.08149	374.2	4.059	0.1993
Trichlorofluoromethane (R-11)	CCl_3F	137.37	0.06052	471.2	4.38	0.2478
Water	H_2O	18.015	0.4615	647.1	22.06	0.0560
Xenon	Xe	131.30	0.06332	289.8	5.88	0.1186

[a] The unit kJ/kg · K is equivalent to kPa · m^3/kg · K. The gas constant is calculated from $R = R_u/M$, where $R_u = 8.31447$ kJ/kmol · K and M is the molar mass.

Source: K. A. Kobe and R. E. Lynn, Jr., *Chemical Review* 52 (1953), pp. 117–236; and ASHRAE, *Handbook of Fundamentals* (Atlanta, GA: American Society of Heating, Refrigerating and Air-Conditioning Engineers, Inc., 1993), pp. 16.4 and 36.1.

Table B.2-2 Ideal-gas specific heats of various common gases

(a) At 300 K

Gas	Formula	Gas constant, R kJ/kg · K	C_p kJ/kg · K	C_v kJ/kg · K	k
Air	–	0.2870	1.005	0.718	1.400
Argon	Ar	0.2081	0.5203	0.3122	1.667
Butane	C_4H_{10}	0.1433	1.7164	1.5734	1.091
Carbon dioxide	CO_2	0.1889	0.846	0.657	1.289
Carbon monoxide	CO	0.2968	1.040	0.744	1.400
Ethane	C_2H_6	0.2765	1.7662	1.4897	1.186
Ethylene	C_2H_4	0.2964	1.5482	1.2518	1.237
Helium	He	2.0769	5.1926	3.1156	1.667
Hydrogen	H_2	4.1240	14.307	10.183	1.405
Methane	CH_4	0.5182	2.2537	1.7354	1.299
Neon	Ne	0.4119	1.0299	0.6179	1.667
Nitrogen	N_2	0.2968	1.039	0.743	1.400
Octane	C_8H_{18}	0.0729	1.7113	1.6385	1.044
Oxygen	O_2	0.2598	0.918	0.658	1.395
Propane	C_3H_8	0.1885	1.6794	1.4909	1.126
Steam	H_2O	0.4615	1.8723	1.4108	1.327

Note: The unit kJ/kg · K is equivalent to kJ/kg · °C.

Source: *Chemical and Process Thermodynamics* 3/E by Kyte, B. G., © 2000 (Adapted by permission of Pearson Education, Inc., Upper Saddle River, NJ)

(b) At various temperatures

Temperature, K	c_p kJ/kg · K	c_v kJ/kg · K	k	c_p kJ/kg · K	c_v kJ/kg · K	k	c_p kJ/kg · K	c_v kJ/kg · K	k
	Air			**Carbon dioxide, CO_2**			**Carbon monoxide, CO**		
250	1.003	0.716	1.401	0.791	0.602	1.314	1.039	0.743	1.400
300	1.005	0.718	1.400	0.846	0.657	1.288	1.040	0.744	1.399
350	1.008	0.721	1.398	0.895	0.706	1.268	1.043	0.746	1.398
400	1.013	0.726	1.395	0.939	0.750	1.252	1.047	0.751	1.395
450	1.020	0.733	1.391	0.978	0.790	1.239	1.054	0.757	1.392
500	1.029	0.742	1.387	1.014	0.825	1.229	1.063	0.767	1.387
550	1.040	0.753	1.381	1.046	0.857	1.220	1.075	0.778	1.382
600	1.051	0.764	1.376	1.075	0.886	1.213	1.087	0.790	1.376
650	1.063	0.776	1.370	1.102	0.913	1.207	1.100	0.803	1.370
700	1.075	0.788	1.364	1.126	0.937	1.202	1.113	0.816	1.364
750	1.087	0.800	1.359	1.148	0.959	1.197	1.126	0.829	1.358
800	1.099	0.812	1.354	1.169	0.980	1.193	1.139	0.842	1.353
900	1.121	0.834	1.344	1.204	1.015	1.186	1.163	0.866	1.343
1,000	1.142	0.855	1.336	1.234	1.045	1.181	1.185	0.888	1.335
	Hydrogen, H_2			**Nitrogen, N_2**			**Oxygen, O_2**		
250	14.051	9.927	1.416	1.039	0.742	1.400	0.913	0.653	1.398
300	14.307	10.183	1.405	1.039	0.743	1.400	0.918	0.658	1.395
350	14.427	10.302	1.400	1.041	0.744	1.399	0.928	0.668	1.389
400	14.476	10.352	1.398	1.044	0.747	1.397	0.941	0.681	1.382
450	14.501	10.377	1.398	1.049	0.752	1.395	0.956	0.696	1.373
500	14.513	10.389	1.397	1.056	0.759	1.391	0.972	0.712	1.365
550	14.530	10.405	1.396	1.065	0.768	1.387	0.988	0.728	1.358
600	14.546	10.422	1.396	1.075	0.778	1.382	1.003	0.743	1.350
650	14.571	10.447	1.395	1.086	0.789	1.376	1.017	0.758	1.343
700	14.604	10.480	1.394	1.098	0.801	1.371	1.031	0.771	1.337
750	14.645	10.521	1.392	1.110	0.813	1.365	1.043	0.783	1.332
800	14.695	10.570	1.390	1.121	0.825	1.360	1.054	0.794	1.327
900	14.822	10.698	1.385	1.145	0.849	1.349	1.074	0.814	1.319
1,000	14.983	10.859	1.380	1.167	0.870	1.341	1.090	0.830	1.313

Source: Kenneth Wark, *Thermodynamics*, 4th ed. (New York: McGraw-Hill, 1983), p. 783, Table A-4M. Originally published in *Tables of Thermal Properties of Gases*, NBS Circular 564, 1955.

Table B.2.3 Properties of common liquids, solids, and foods

(a) Liquids

	Boiling data at 1 atm		*Freezing data*		*Liquid properties*		
Substance	Normal boiling point, °C	Latent heat of vaporization h_{fg}, kJ/kg	Freezing point, °C	Latent heat of fusion h_{if}, kJ/kg	Temperature, °C	Density, ρ, kg/m^3	Specific heat c_p, kJ/kg · K
Ammonia	−33.3	1357	−77.7	322.4	−33.3	682	4.43
					−20	665	4.52
					0	639	4.60
					25	602	4.80
Argon	−185.9	161.6	−189.3	28	−185.6	1394	1.14
Benzene	80.2	394	5.5	126	20	879	1.72
Brine (20% sodium chloride by mass)	103.9	–	−17.4	–	20	1150	3.11
n-Butane	−0.5	385.2	−138.5	80.3	−0.5	601	2.31
Carbon dioxide	−78.4[a]	230.5 (at 0°C)	−56.6		0	298	0.59
Ethanol	78.2	838.3	−114.2	109	25	783	2.46
Ethyl alcohol	78.6	855	−156	108	20	789	2.84
Ethylene glycol	198.1	800.1	−10.8	181.1	20	1109	2.84
Glycerine	179.9	974	18.9	200.6	20	1261	2.32
Helium	−268.9	22.8	–	–	−268.9	146.2	22.8
Hydrogen	−252.8	445.7	−259.2	59.5	−252.8	70.7	10.0
Isobutane	−11.7	367.1	−160	105.7	−11.7	593.8	2.28
Kerosene	204–293	251	−24.9	–	20	820	2.00
Mercury	356.7	294.7	−38.9	11.4	25	13,560	0.139
Methane	−161.5	510.4	−182.2	58.4	−161.5	423	3.49
					−100	301	5.79
Methanol	64.5	1100	−97.7	99.2	25	787	2.55
Nitrogen	−195.8	198.6	−210	25.3	−195.8	809	2.06
					−160	596	2.97
Octane	124.8	306.3	−57.5	180.7	20	703	2.10
Oil (light)					25	910	1.80
Oxygen	−183	212.7	−218.8	13.7	−183	1141	1.71
Petroleum	–	230–384			20	640	2.0
Propane	−42.1	427.8	−187.7	80.0	−42.1	581	2.25
					0	529	2.53
					50	449	3.13
Refrigerant-134a	−26.1	217.0	−96.6	–	−50	1443	1.23
					−26.1	1374	1.27
					0	1295	1.34
					25	1207	1.43

Water	100	2257	0.0	333.7	0	1000	4.22
					25	997	4.18
					50	988	4.18
					75	975	4.19
					100	958	4.22

[a] Sublimation temperature, (At pressures below the triple-point pressure of 518 kPa, carbon dioxide exists as a solid or gas. Also, the freezing-point temperature of carbon dioxide is the triple-point temperature of −56.5°C.)

Table B.2-3 Properties of common liquids, solids, and foods (concluded)

(b) Solids (values are for room temperature unless indicated otherwise)

Substance	Density, ρ kg/m^3	Specific heat, c_p kJ/kg · K	Substance	Density, ρ kg/m^3	Specific heat, c_p kJ/kg . K
Metals			**Nonmetals**		
Aluminum			Asphalt	2110	0.920
200 K		0.797	Brick, common	1922	0.79
250 K		0.859	Brick, fireclay (500°C)	2300	0.960
300 K	2,700	0.902	Concrete	2300	0.653
350 K		0.929	Clay	1000	0.920
400 K		0.949	Diamond	2420	0.616
450 K		0.973	Glass, window	2700	0.800
500 K		0.997	Glass, pyrex	2230	0.840
Bronze (76% Cu, 2% Zn, 2% Al)	8,280	0.400	Graphite	2500	0.711
			Granite	2700	1.017
Brass, yellow (65% Cu, 35% Zn)	8,310	0.400	Gypsum or plaster board	800	1.09
			Ice		
Copper			200 K		1.56
−173°C		0.254	220 K		1.71
−100°C		0.342	240 K		1.86
−50°C		0.367	260 K		2.01
0°C		0.381	273 K	921	2.11
27°C	8,900	0.386	Limestone	1650	0.909
100°C		0.393	Marble	2600	0.880
200°C		0.403	Plywood (Douglas Fir)	545	1.21
Iron	7,840	0.45	Rubber (soft)	1100	1.840
Lead	11,310	0.128	Rubber (hard)	1150	2.009
Magnesium	1,730	1.000	Sand	1520	0.800
Nickel	8,890	0.440	Stone	1500	0.800
Silver	10,470	0.235	Woods, hard (maple, oak, etc.)	721	1.26
Steel, mild	7,830	0.500	Woods, soft (fir, pine, etc.)	513	1.38
Tungsten	19,400	0.130			

(c) Foods

Food	Water content, % (mass)	Freezing point, °C	Specific heat, kJ/kg . K Above freezing	Below freezing	Latent heat of fusion, kJ/kg	Food	Water content, % (mass)	Freezing point, °C	Specific heat, kJ/kg . K Above freezing	Below freezing	Latent heat of fusion, kJ/kg
Apples	84	-1.1	3.65	1.90	281	Lettuce	95	−0.2	4.02	2.04	317
Bananas	75	−0.8	3.35	1.78	251	Milk, whole	88	−0.6	3.79	1.95	294
Beet round	67	–	3.08	1.68	224	Oranges	87	−0.8	3.75	1.94	291
Broccoli	90	−0.6	3.86	1.97	301	Potatoes	78	−0.6	3.45	1.82	261
Butter	16	–	–	1.04	53	Salmon fish	64	−2.2	2.98	1.65	214
Cheese, swiss	39	−10.0	2.15	1.33	130	Shrimp	83	−2.2	3.62	1.89	277
Cherries	80	−1.8	3.52	1.85	267	Spinach	93	−0.3	3.96	2.01	311
Chicken	74	−2.8	3.32	1.77	247	Strawberries	90	−0.8	3.86	1.97	301
Corn, sweet	74	−0.6	3.32	1.77	247	Tomatoes, ripe	94	−0.5	3.99	2.02	314
Eggs, whole	74	−0.6	3.32	1.77	247	Turkey	64	–	2.98	1.65	214
Ice cream	63	−5.6	2.95	1.63	210	Watermelon	93	−0.4	3.96	2.01	311

Source: Values are obtained from various handbooks and other sources or are calculated. Water content and freezing-point data of foods are from *ASHRAE, Handbook of Fundamentals*, SI version (Atlanta, GA: American Society of Heating, Refrigerating and Air-Conditioning Engineers, Inc., 1993), Chapter 30, Table 1. Freezing point is the temperature at which freezing starts for fruits and vegetables, and the average freezing temperature for other foods.

B.3 Drag Coefficient: (A) smooth sphere and (B) an infinite cylinder as a function of Reynolds number

(A)

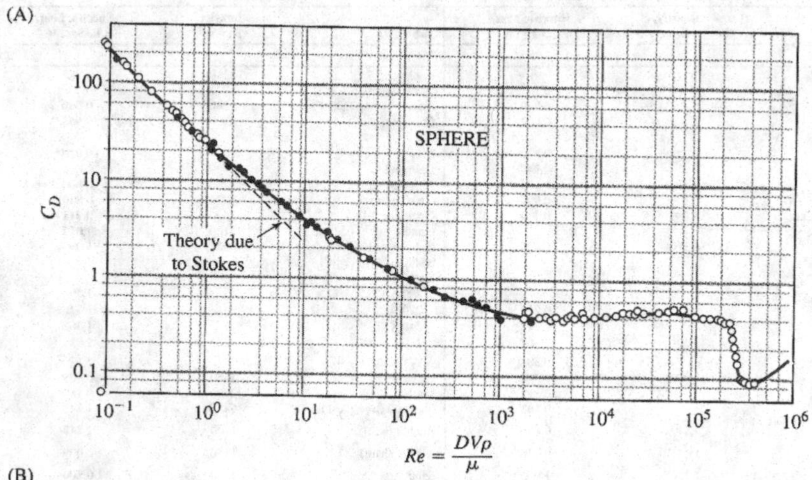

(B)

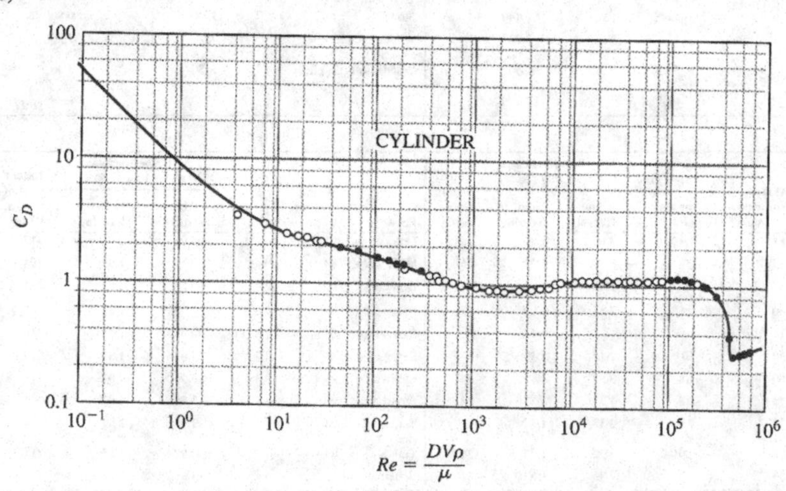

B.4 Moody Chart

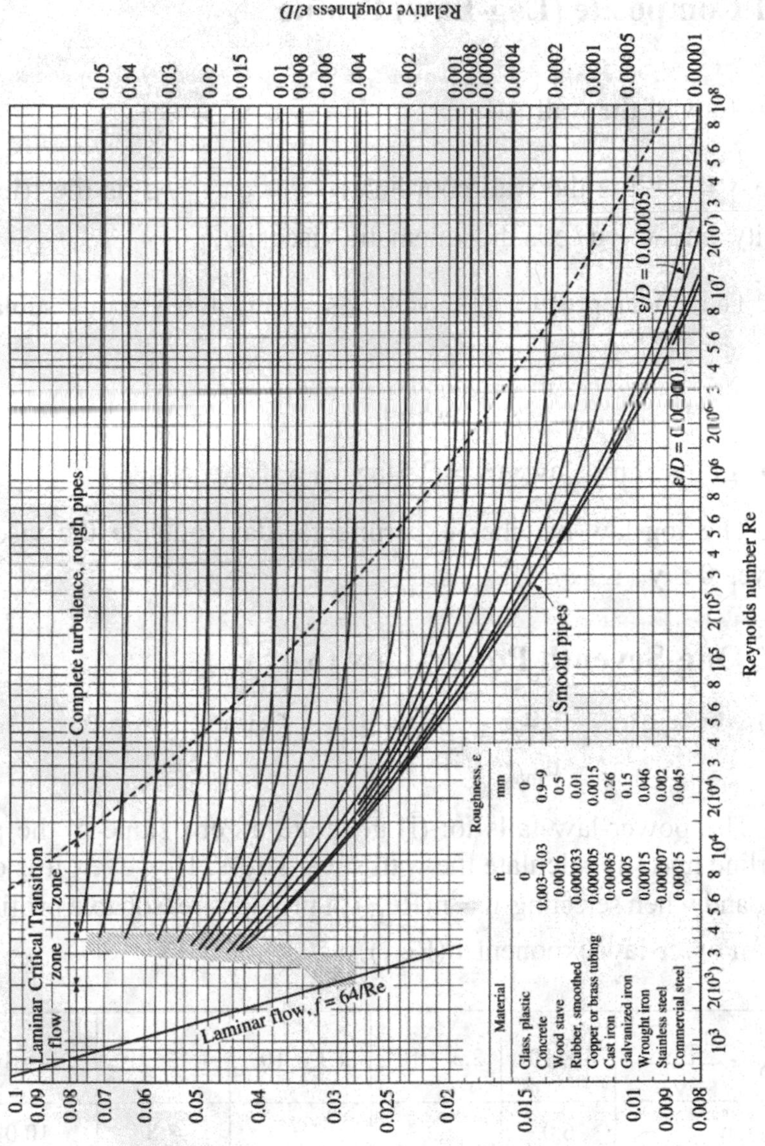

Material	Roughness, ε ft	Roughness, ε mm
Glass, plastic	0	0
Concrete	0.003–0.03	0.9–9
Wood stave	0.0016	0.5
Rubber, smoothed	0.000033	0.01
Copper or brass tubing	0.000005	0.0015
Cast iron	0.00085	0.26
Galvanized iron	0.0005	0.15
Wrought iron	0.00015	0.046
Stainless steel	0.000007	0.002
Commercial steel	0.00015	0.045

B.5 Turbulent Velocity Profiles in Pipes

B.5.1 Composite (Log-Law) Profile

- Law of the wall: $u^+ = y^+$, $u^+ \equiv \frac{u}{u_\tau}$, $y^+ \equiv \frac{y u_\tau}{\nu}$

where $y = R - r$ is the wall coordinate, $u_\tau \equiv \sqrt{\tau_{wall}/\rho}$ is the friction velocity, and $\nu = \mu/\rho$ is the kinematic viscosity.

Note: The linear profile $u^+ = y^+$ holds within the viscous sublayer $0 \le y^+ \le 5$, i.e.,

$$y = \delta_{sublayer} = 5\nu / u_\tau$$

- Logarithmic law: $u^+ = 2.5 \ln y^+ + 5.0$; $y^+ > 5.0$

Note: The log-law matches experimental data outside the viscous sublayer $0 \le y^+ \le 5$.

B.5.2 One-Seventh Power-Law ($n \approx 7$):

$$\frac{u}{u_{max}} = \left(\frac{y}{R}\right)^{1/n} = \left(1 - \frac{r}{R}\right)^{1/n}$$

Note: The power-law fails to: (i) generate a zero slope at the pipe centerline and (ii) calculate the wall shear stress. However, it is easy to use, and when selecting $n = n(Re_D)$ it provides reasonable profiles.

- Power-law exponent $n(Re_D)$

$Re_D = \frac{u_{avg} D}{\nu}$	4×10^3	2.3×10^4	1.1×10^5	1.1×10^6	3.2×10^5
n	6.0	6.6	7.0	8.8	10.0
$\frac{u_{average}}{u_{max}}$	0.791	0.807	0.817	0.850	0.865

Index

U

V

W

Y

Printed in the United States
By Bookmasters